On Einstein's Path

Springer Science+Business Media, LLC

Alex Harvey
Editor

On Einstein's Path

Essays in Honor of Engelbert Schucking

With 11 Illustrations

Springer

Alex Harvey
Department of Physics
Queens College
Flushing, NY 11367
USA

Library of Congress Cataloging-in-Publication Data
Harvey, Alex, Professor Emeritus.
On Einstein's path: essays in honor of Engelbert Schucking / Alex Harvey.
p. cm.
"Friends and colleagues of Engelbert Schucking came together in a symposium on the 12th and 13th of December 1996 at New York University . . . "—Introduction.
Includes bibliographical references
ISBN 978-0-387-98564-0 ISBN 978-1-4612-1422-9 (eBook)
DOI 10.1007/978-1-4612-1422-9
1. General relativity (Physics)—Congresses. 2. Quantum theory—Congresses. 3. Cosmology—Congresses. 4. Space and time—Congresses. 5. Astrophysics—Congresses. 6. Schucking, E.L. (Engelbert L.) I. Schucking, E.L. (Engelbert L.) II. Title.
QC173.5.H37 1998
530.11—dc21 98-20292

Printed on acid-free paper.

Production managed by A. Orrantia; manufacturing supervised by Joe Quatela.
Typeset by The Bartlett Press, Inc., Marietta, GA.

9 8 7 6 5 4 3 2 1

ISBN 978-0-387-98564-0

Preface

Friends and colleagues of Engelbert Schucking came together in a symposium on the 12th and 13th of December 1996 at New York University to celebrate and express to him their respect, admiration, and affection. They came to celebrate his scientific and scholarly achievements, the inspirational quality of his teaching, his graciousness as a colleague, his thoughtful guidance of graduate students, his service to the department, the university and the physics community at large—and, not least, his open, courteous, easy accessibility to anyone needing his counsel or expertise. The announcement was

A SYMPOSIUM

In Honor of

PROF. ENGELBERT SCHUCKING

Physics Department—New York University

On December 12th and 13th there will be a Symposium to honor Professor Engelbert Schucking for his service to the University, the Department, and the Physics Community. The December 12th session will run from 1 to 6 PM followed by a reception. The following morning the session will run from 9 AM to 1 PM. Attendance (including the reception) is open to all friends and colleagues of Professor Schucking and anyone interested in General Relativity. The talks will be presented in Room 121, 4 Washington Place; the reception will be in the office of Dean Furmankis, 5 Washington Square North from 6:15 to 8:00 PM

Thursday Afternoon:

Greetings
Alice S. Huang
Dean for Science - New York University

Greetings
Professor Peter Levy

Chairman - Physics Department - New York University

Introductions
Alex Harvey
Professor Emeritus - Queens College CUNY

Professor Ivor Robinson
University of Texas, Dallas
"Bel-Robinson Revisited"

Professor Istvan Ozsvath
University of Texas, Dallas
"Working with Engelbert"

Coffee Recess

Professor Andrzej Trautman
University of Warsaw
"Complex Numbers in Physics"

Prof. Edward Spiegel
Columbia University
"$\mathcal{S}$ and $\mathcal{S}^3$ — A Felicitous Duality"

Reception

Friday Morning:

Professor Jürgen Ehlers
Max Planck Institute, Potsdam
"Recent Developments in Newtonian Cosmology"

Dr. Anne Kinney
Space Telescope Science Institute, Baltimore
"The Dynamic Universe"

Coffee Recess

Professor Roger Penrose
Mathematical Institute, University of Oxford
"Where is Twistor Theory Going?"

Professor Wolfgang Rindler
University of Texas, Dallas
"Some Thoughts on Elementary Relativity"

This symposium, devoted to honoring Engelbert Schucking, had its origin almost four decades earlier. In July of 1955 the "Jubilee of Relativity Theory" was celebrated at an international conference in Bern hosted by Switzerland. The organizing committee was composed exclusively of members of the Swiss academic establishment. Although "Jubilee" referred to Albert Einstein's 1905 paper on "The Electrodynamics of Moving Bodies," the papers presented were largely devoted to the theory of general relativity. Persons attending came mostly from Europe, with a few from the United states. It was noteworthy as the first international conference devoted exclusively to relativity theory.

The Bern conference was such a success that it was followed soon after by a "Conference on the Role of Gravitation in Physics" at the University of North Carolina, Chapel Hill, in January 1957. Though smaller in number of attendees, it too was deemed a success. The next such conference was in June 1959. It was sponsored by the French Centre National de la Recherche Scientifique and was held at Royaumont, some miles north of Paris. The participants were from all across Europe and the United States.

It was a watershed for general relativists; it confirmed a pattern for open, broadly-based international conferences on gravitational physics. A "steering committee" with international representation was organized to coordinate subsequent meetings. There was general agreement to hold such conferences on a triennial basis and they have been held in this fashion ever since. The "steering committee" was succeeded by the International Society on General Relativity and Gravitation in the early '70s and has since, with a formal host country, sponsored the meetings.

In addition to being the start of the series of triennial meetings, the meeting at Royaumont has a special significance for our symposium. At the Royaumont meeting were not only our honoree and members of the audience, but four of our speakers as well: Professors Robinson, Trautman, Ehlers, and Penrose. Of these four, the latter three presented talks at Royaumont. There were less than 120 participants in the meeting; it was small enough that many friendships and collegial relationships were generated. These have formed an enduring, strengthening, mutually nourishing web. This is attested to in part by our symposium, which has almost the atmosphere of a family gathering.

Engelbert was born May 23, 1926, in Hoerde, Germany. Some people are born with silver spoons in their mouths. Engelbert's blessing was a small silver telescope. Very early this was manifested by an active interest in astronomy. By the age of fourteen he was an observational astronomer engaged in counting sunspots. Later, as an undergraduate at the University of Muenster and then Goettingen he studied mathematics and physics: He learned nuclear physics from Werner Heisenberg, electricity and magnetism from Richard Becker, and advanced experimental physics from Kopfermann and Pohl. He did not get the degree in physics because he refused to do analytic chemistry.

He advanced to graduate studies at the University in Hamburg, where in 1955 he obtained his PhD in mathematics under the guidance of Pascual Jordan. During this period he assisted Pascual Jordan in the preparation of the second edition of *Schwerkraft und Weltall.* Hamburg was at that time and has since continued to be an important center for research in the theory of general relativity. It was here and in this discipline that Engelbert found the locus of his life's work—broadly, the geometric aspects of the Einstein field equations. Engelbert spent the next six years as Assistant at the University of Hamburg. Simultaneously, he worked at the Hamburg Observatory with Otto Heckmann, the leading cosmologist of the day, where they produced among other things the article on "Newtonian and Einsteinian Cosmology" for the *Encyclopedia of Physics*.

In 1961, Engelbert became an intellectual bird of passage. He traveled across the Atlantic, becoming in turn a Research Associate first at Syracuse and then at Cornell. Perhaps in search of a more salubrious climate, he left the Finger Lakes for the University of Texas at Austin in 1962. His appointment was as Associate Professor of Mathematics, and a year later, was promoted to Professor of Physics. Together with Alfred Schild he bolted out of the Mathematics Department and started a group that worked on general relativity. This remarkable group included Roger Penrose, Ray Sachs, Roy Kerr, and Jürgen Ehlers among others. Together with Alfred Schild and Ivor Robinson, he organized the first of the highly successful Texas Symposia on Relativistic Astrophysics, now known as the "Texas Conferences."

After half a decade Engelbert returned to the Northeast and during a short stint as Visiting Research Professor of Physics at Yeshiva University in uptown Manhattan, took thought, took the No. 6 IRT downtown to Astor Place, and wandered to Washington Square, where he found a permanent home in the Physics Department at New York University in 1967. After acclimating he instituted in the 1970s a series of outreach courses to expose non-science students to physics with topics such as the Origins of Astronomy [alias Astronomy and Astrology], The Universe, and Intelligent Life in the Universe. His audience broadened to the general public through his astronomy course in the television series "Sunrise Semester." Over the years he has taught approximately 5,000 students and supervised about 20 successful PhD candidates.

During these years Engelbert has acquired a substantial and well-deserved reputation as a cosmologist, astrophysicist, and after-dinner speaker at scientific meetings. His penetrating critical ability is a continuing source of inspiration to all fortunate enough to know him.

New York ALEX HARVEY

Contents

Preface v

Contributors xix

1 Jordan, Pauli, Politics, Brecht . . . and a Variable Gravitational Constant 1
Engelbert L. Schucking

2 Thomson Scattering in an Expanding Universe 15
James L. Anderson
2.1 Introduction 15
2.2 EIH Surface Integrals 16
2.3 Approximation Procedures 17
2.4 Thomson Scattering 19

3 Geometrical Formulation of Quantum Mechanics 23
Abhay Ashtekar, Troy A. Schilling
3.1 Introduction 23
3.2 Geometric Formulation of Quantum Mechanics 28
3.2.1 The Hilbert Space as a Kähler Space 28
3.2.2 The Quantum Phase Space 32
3.2.3 Riemannian Geometry and Measurement Theory . . . 35
3.2.4 The Postulates of Quantum Mechanics 42
3.3 A Unified Framework for Generalizations of Quantum Mechanics 44
3.3.1 Generalized Dynamics 45
3.3.2 Characterization of the Standard Quantum Kinematics . 49
3.4 Semiclassical Considerations 53
3.4.1 Kinematics 53
3.4.2 Dynamics: Oscillators 57

3.4.3 Dynamics: WKB Approximation 58
3.5 Discussion 61

4 General Covariance is Bose-Einstein Statistics **67**
James Baugh, David Ritz Finkelstein, Heinrich Saller, Zhong Tang
4.1 Quantum Relativity 67
4.2 General Covariance and Bosonic Statistics 68
4.3 The Two-Point Paradox 69
4.4 The Quantum Causal Relation 70
4.5 Quantum is Simpler 71
4.6 Limits to Spacetime 72
4.7 The Elephantine Chronon 73
4.8 Topological Nature of Gauge 73
4.9 Paradox Lost 75
4.10 Summation 79

5 The Split and Propagation of Light Rays in Relativity **81**
Stanisław L. Bażański
5.1 Introduction 81
5.2 Traditional Approach 81
5.3 Modified Approach 85
5.4 Sagnac-Like Effects 87
5.5 Examples 89

6 How to Define a Unique Vacuum in Cosmology **95**
Lluís Bel
6.1 Klein–Gordon Equation 95
6.2 Quantization of a Scalar Field 96
6.3 Robertson–Walker Models 97
6.4 Modes 97
6.5 Reduction of the Evolution Equation 98
6.6 Approximations to the Regular Solutions 99
6.7 Critical Points at $t = \infty$ 101
6.8 Special Cases 102
6.9 Positive and Negative Energy Modes 102
6.10 Concluding Remarks 104

7 EIH Theory and Noether's Theorem **107**
Peter G. Bergmann
7.1 Introduction 107
7.2 Invariance Group and Noether's Theorem 108
7.3 An Example 108
7.4 The Generalized EIH Theory 110
7.5 Concluding Remarks 111

8 The Static Cylinder in General Relativity **113**
W.B. Bonnor
8.1 Introduction . . . 113
8.2 The HD Solution . . . 115
8.3 Matching to LC Spacetime . . . 115
8.4 Physical Interpretation . . . 117
8.5 The Whittaker Mass per Unit Length . . . 117
8.6 Conclusion . . . 119

9 Gravity and the Tenacious Scalar Field **121**
Carl H. Brans
9.1 Scalar Gravity? . . . 121
9.2 Kaluza–Klein Theories . . . 126
9.3 Dirac's Numbers . . . 128
9.4 Scalar–Tensor Theories . . . 129
9.5 Dilatons . . . 134
9.6 Inflatons . . . 136
9.7 Conclusion . . . 137

10 The Cavendish Experiment in General Relativity **139**
Dieter Brill
10.1 Introduction . . . 139
10.2 Planar Symmetry . . . 140
10.3 Exact, Static Solutions . . . 141
10.4 Test Particle Motion . . . 144
10.5 Conclusion . . . 145

11 Wave Maps in General Relativity **147**
Yvonne Choquet-Bruhat
11.1 Introduction . . . 147
11.2 Definitions . . . 148
11.3 Wave Maps—the Cauchy Problem . . . 149
11.4 Harmonic Gauges in General Relativity . . . 153
11.4.1 Existence of a Solution of the Vacuum Einstein Equations . . . 154
11.4.2 Uniqueness Theorem . . . 154
11.5 Global Problem—the First Energy Estimate . . . 155
11.6 Second Energy Inequality . . . 158
11.7 Wave Map from the Outside of a Black Hole . . . 163

12 General Relativity and Experiment **171**
Thibault Damour
12.1 Introduction . . . 171
12.2 Experimental Tests of the Coupling Between Matter and Gravity . . . 172

12.3 Tests of the Dynamics of the Gravitational Field in the Weak-Field Regime 174
12.4 Tests of the Dynamics of the Gravitational Field in the Strong-Field Regime 176
12.5 Cosmological Tests 180
12.6 Was Einstein 100% Right ? 181

13 Some Developments in Newtonian Cosmology **189**
Jürgen Ehlers
13.1 Introduction: The Curious History of Newtonian Cosmology . 189
13.2 Two Formulations of Newtonian Cosmology and its Relationship to Relativistic Cosmology 190
13.2.1 The Heckmann–Schucking Formulation 190
13.2.2 The Cartan–Friedrichs Formulation 192
13.2.3 Electrodynamics and Optics in Newtonian Cosmology . 193
13.2.4 Relations Between Relativistic and Newtonian Cosmological Models 194
13.3 Observer-Homogenous, Bianchi-Type Models 195
13.4 Averaging in Cosmology 196
13.5 Lagrangian Perturbation Theory 199

14 Deviation of Geodesics in FLRW Spacetime Geometries **203**
George F.R. Ellis, Henk van Elst
14.1 Introduction 203
14.1.1 The Cosmological Context 205
14.2 The Riemann Curvature Tensor 205
14.3 The Geodesics 207
14.3.1 Timelike 209
14.3.2 Spacelike 209
14.3.3 Null 210
14.4 The Geodesic Deviation Equation 210
14.4.1 The Deviation Vectors 210
14.4.2 Geodesic Deviation for a Fundamental Observer 212
14.4.3 Past Directed Null Vector Fields 217
14.4.4 Generic Geodesic Vector Fields 220
14.5 Conclusion 223

15 Poincaré Pseudosymmetries in Asymptotically Flat Spacetimes **227**
Simonetta Frittelli, Ezra T. Newman
15.1 Introduction 227
15.2 Minkowski Space 229
15.3 Asymptotically Flat Spacetimes 232
15.4 Discussion 236

16 Taub Numbers and Asymptotic Invariants **241**
Edward N. Glass
16.1 Introduction 241
16.2 Taub Numbers and Superpotential 243
16.3 Null Infinity 245
16.4 Kerr–Schild Solutions 246
16.5 Bondi–Sachs Solutions 247
16.6 Summary 249

17 Second-Class Constraints **251**
Joshua N. Goldberg
17.1 Introduction 251
17.2 The Mechanical System 252
17.3 The Self-Dual Maxwell Field 254
17.4 Conclusion 255

18 On the Structure of the Energy-Momentum and the Spin Currents in Dirac's Electron Theory **257**
Friedrich W. Hehl, Alfredo Macías, Eckehard W. Mielke, Yuri N. Obukhov
18.1 Introduction 257
18.2 Dirac–Yang-Mills Theory 259
18.3 Gordon Decomposition of Energy-Momentum and Spin Currents 261
18.4 Relocalization of Energy-Momentum and Spin 264
18.5 Trivial Lagrangians and Relocalization 266
18.6 Belinfante Symmetrization of the Energy-Momentum Current . 268
18.7 Properties of the Gravitational Moments and Nonrelativistic Limit 270
18.8 Discussion 271

19 The Physical Reality of the Quantum Wave Function **275**
Arthur Komar
19.1 Introduction 275
19.2 The Thought Experiment 277
19.3 Schrödinger's Cats 280

20 The Ultimate Extension of the Bianchi Classification for Rotating Dust Models **283**
Andrzej Krasiński
20.1 Introduction and Summary 283
20.2 The Classification of Differential Forms of First Order and the Darboux Theorem 284
20.3 Geodesically Moving Fluids 285
20.4 The Killing Vector Fields Compatible with Rotation 288

20.5 The Case of Two Generators Spanned on u^α and w^α 289
20.6 The Case of One Generator Spanned on u^α and w^α 293
20.7 The Case of All Three Generators Being Linearly Independent of u^α and w^α 296
20.8 Conclusion 297

21 On the Classification of the Real Four-Dimensional Lie Algebras 299
M.A.H. MacCallum
21.1 Introduction 300
21.2 An Enumeration of the 4-Dimensional Algebras 302
21.3 Comparison with Other Enumerations 307
21.4 Extensions, Applications and Other Work 310

22 Spinning Universes in Newtonian Cosmology 319
Jayant V. Narlikar
22.1 Introduction 319
22.2 Homogeneous and Anisotropic Cosmologies 320
22.2.1 The Potential Function Approach 321
22.2.2 The Gravitational Force Approach 322
22.3 The Work of Davidson and Evans 324
22.4 Concluding Remarks 327

23 Relativistic Gravitational Fields with Close Newtonian Analogs 329
Pawel Nurowski, Engelbert Schucking, Andrzej Trautman
23.1 Introduction 329
23.2 Notation 330
23.3 The Metric, the Curvature, and the Ricci Tensors 330
23.4 The Comoving Coordinate System 331
23.5 Flat Space-Times 331
23.6 Nontrivial Solutions 332
23.6.1 Equations for a Perfect Fluid 332
23.6.2 Spherically Symmetric Spaces 332
23.6.3 The Kasner Solution 333
23.6.4 Perfect Fluid Generalizations of the Kasner Solution . . 334
23.7 Congruences of Null Geodesics 336

24 Working with Engelbert 339
István Ozsváth
24.1 Introduction 339
24.2 Exact Solutions 339
24.2.1 Remarks by Kurt Gödel 339
24.2.2 The Schucking Equations 340
24.2.3 The Schucking Solution 341
24.2.4 The Finite Rotating Universe 341
24.2.5 The Anti-Mach Metric 342

24.3 More on Exact Solutions . . . 342
24.3.1 All Homogeneous Vacuum Solutions with Λ term . . . 342
24.3.2 All Type N Vacuum Solutions with Λ term Vanishing Shear and Expansion . . . 343
24.3.3 Finite Rotating Universe Revisited . . . 343
24.4 Embedding Problems . . . 345
24.4.1 Embedding of Dantes into S^4 . . . 346
24.4.2 A Special Case for $\lambda = 0$ . . . 346
24.4.3 Embedding of Dantes into S^5 . . . 347
24.4.4 Embedding of Dantes into S^8 . . . 347
24.4.5 Embedding the General Dantes into S^8 . . . 348
24.5 The SU_3 Group . . . 349
24.6 In Closing . . . 350

25 Some Remarks on Twistor Theory **353**
Roger Penrose
25.1 Historical Comments . . . 353
25.2 Twistors and the Einstein Equations . . . 358

26 Critique of the Wheeler-DeWitt Equation **367**
Asher Peres
26.1 Introduction . . . 367
26.2 A Simple Example of Constrained Dynamics . . . 369
26.3 Definition of a Dynamical Time . . . 371
26.4 Quantization of a Minisuperspace . . . 373

27 A New Version of the Heavenly Equation **381**
Jerzy F. Plebański, Maciej Przanowski
27.1 Introduction . . . 381
27.2 An Expanding Congruence of Null Strings . . . 382
27.3 Hermitian and Kählerian Structures on $\mathcal{H}$ Space in Euclidean Relativity . . . 393
27.4 The Heavenly Equation . . . 398

28 A Plain Man's Guide to Bivectors, Biquaternions, and the Algebra and Geometry of Lorentz Transformations **407**
Wolfgang Rindler, Ivor Robinson
28.1 Introduction . . . 407
28.2 Basic Algebra of Real Bivectors and Complex Scalars . . . 408
28.3 Basic Geometry of Bivectors . . . 411
28.4 Biquaternions and the Bivector Transformations They Generate . . . 416
28.5 Lorentz Matrices . . . 419
28.6 The Geometry of Lorentz Transformations . . . 423
28.7 t-Real Biquaternions, t-Rotations, and t-Boosts . . . 428

28.8 The Thomas Precession 431

29 Leon Lichtenstein's Work on Rotating Fluids **435**
Bernd Schmidt
29.1 Introduction 435
29.2 Rigidly Rotating Fluids in Newtonian Theory 435
29.3 Rigidly Rotating Fluids in Einstein's Theory of Gravity 439
29.4 Speculations 440

30 Decaying Neutrinos and the Flattening of the Galactic Halo **443**
Dennis W. Sciama
30.1 Introduction 443
30.2 The Neutrino Density Near the Sun 444
30.3 τ_{23} and the Hα Data 446
30.4 τ_{23} and the Extragalactic Background at 1500 Å 446
30.5 Conclusions 447

31 The Kasner Condition and Inhomogeneous Perfect Fluid Cosmologies **449**
Jim E.F. Skea
31.1 Introduction 449
31.2 Mathematical Background 452
31.3 The Basic Quantities 454
31.4 The Cosmological Models 456
31.5 Application to Space-Times Admitting a G_1 459
31.6 Aspects of Invariant Classification 461
31.7 Discussion 462

32 Gravitational Screening **465**
E.A. Spiegel
32.1 Gravitational Stopping Power 465
32.2 A Drag Crisis 466
32.3 Saved by Self-Gravity 469
32.4 The Message Is The Medium 473

33 On the Interpretation of the Einstein–Cartan Formalism **475**
John Stachel
33.1 Introduction 475
33.2 The Analogy: Electromagnetism 476
33.3 The Analogy: Gravitation 478
33.4 Discussion 481

34 On Complex Structures in Physics **487**
Andrzej Trautman
34.1 Introduction 487
34.2 Definitions and Notation 488

34.3 A Complex Structure Defined by Differentiation 490
34.4 Complex Structures Associated with Pseudo-Euclidean Vector Spaces 491
34.5 Charge Conjugation 493
34.6 CR Structures Associated with Integrable Optical Geometries . 495

35 The Engelbert Experience: Pathways from the Past **503**
C. V. Vishveshwara

36 Curriculum Vita **515**
Engelbert Schucking

Contributors

James L. Anderson
Department of Physics
Stevens Institute of Technology
Hoboken, NJ 07030
USA

Abhay Ashtekar
Center for Gravitational Physics and Geometry
Department of Physics
Pennsylvania State University
University Park, PA 16802–6300
USA

and

Erwin Schrödinger International Institute for Mathematical Sciences
Boltzmanngasse 9
A-1090 Vienna
Austria

James Baugh
School of Physics
Georgia Institute of Technology
Atlanta, GA 30332–0430
USA

Stanisław L. Bazanski
Institute of Theoretical Physics
University of Warsaw
ul. Hoza 69
00–681 Warszawa
Poland

Lluís Bel
Laboratorie de Gravitation et Cosmologie Relativistes
CNRS/URA 769, Université Pierre et Marie Curie
4, place Jussieu, Tour 22–12
75252 Paris Cedex 05
France

Peter G. Bergmann
Physics Department
New York University
4 Washington Place
New York, NY 10003
USA

and

Department of Physics
Syracuse University
Syracuse, NY 13244–1130
USA

William B. Bonnor
School of Mathematical Sciences
Queen Mary and Westfield College, University of London
Mile End Road
London E1 4NS
UK

Carl H. Brans
Physics Department
Loyola University
New Orleans, LA 70118
USA

Dieter Brill
Department of Physics
University of Maryland
College Park, MD 20742
USA

Yvonne Choquet-Bruhat
Department of Mechanics
University Paris VI
5 Place Jussieu
Paris 75232
France

Thibault Damour
Institut des Hautes Etudes Scientifiques
91440 Bures-sur-Yvette
France

and

DARC
CNRS—Observatorie de Paris
92195 Meudon
France

Jürgen Ehlers
Max-Planck-Institut für
Gravitiationsphysik
Albert-Einstein-Institut
Schlaatzweg 1
14473 Potsdam
Germany

George F.R. Ellis
Department of Mathematics and
Applied Mathematics
University of Cape Town
Rondebosch 7700
Cape Town
South Africa

David Ritz Finkelstein
School of Physics
Georgia Institute of Technology
Atlanta, GA 30332–0430
USA

Simonetta Frittelli
Department of Physics and
Astronomy,
University of Pittsburgh
3941 O'Hara Street
Pittsburgh, PA 15260
USA

Edward N. Glass
Physics Department
University of Windsor
Windsor, Ontario N9B 3P4
Canada

Joshua N. Goldberg
Department of Physics
Syracuse University
Syracuse, NY 13244–1130
USA

Friedrich W. Hehl
Institute for Theoretical Physics
University of Cologne
D-50923 Köln
Germany

Arthur Komar
Physics Department
Yeshiva University
New York, NY 10033
USA

Andrzej Krasiński
N. Copernicus Astronomical Center
and College of Science
Polish Academy of Sciences
Bartycka 18
00 716 Warszawa
Poland

Malcolm A.H. MacCallum
School of Mathematical Sciences
Queen Mary and Westfield College
London E1 4NS
UK

Alfredo Macías
Departamento de Física
Universidad Autónoma
Metropolitana—Iztapalapa
P.O. Box 55–534
09340 Mexico D.F.
Mexico

Eckehard W. Mielke
Departamento de Física
Universidad Autónoma
Metropolitana—Iztapalapa
P.O. Box 55-534
09340 Mexico D.F.
Mexico

Jayant V. Narlikar
Inter-University Centre for
Astronomy
Ganseshkhind, Post Bag 4
Pune 411 007
India

Ezra T. Newman
Department of Physics and
Astronomy,
University of Pittsburgh
3941 O'Hara Street
Pittsburgh, PA 15260
USA

Pawel Nurowski
Dipartimento di Scienze
Matematiche
Universita Studi di Trieste
Piazzale Europa 1
Trieste
Italy

Yuri N. Obukhov
Department of Theoretical Physics
Moscow State University
117234 Moscow
Russia

Istvan Ozsvath
Mathematical Sciences EC35
The University of Texas at Dallas
P.O. Box 830688
Richardson, TX 75083–0688
USA

Roger Penrose
Mathematical Institute
24–29 St. Giles
Oxford OX1 3LB
UK

Asher Peres
Department of Physics
Technion - Israel Institute of
Technology
32 000 Haifa
Astrophysics Israel

Jerzy F. Plebanski
Department of Physics
Centro de Investigación y de
Estudios Avanzados del IPN
Apdo. Postal 14-740
07000 Mexico D.F.

Maciej Przanowski
Institute of Physics
Technical University of Lódź
Wólczanska 219
93-005 Lódź
Poland

Wolfgang Rindler
Physics Program
University of Texas, Dallas
Richardson, TX 75083
USA

Ivor Robinson
Physics Program
University of Texas, Dallas
Richardson, TX 75083
USA

Heinrich Saller
Heisenberg Institute of Theoretical Physics
Max-Planck-Institut für Physik
D-80805 Munich
Germany

Troy A. Schilling
Center for Gravitational Physics and Geometry
Department of Physics
Pennsylvania State University
University Park, PA 16802–6300
USA

and

Institute for Defense Analysis
1801 North Beauregard Street
Alexandria, VA 22311-1772
USA

Engelbert Schucking
Department of Physics
New York University
4 Washington Place
New York, NY 10003
USA

Bernd G. Schmidt
Max-Planck-Institut für Gravitiationsphysik
Albert-Einstein-Institut
Schlaatzweg 1
14473 Potsdam
Germany

Dennis Sciama
International School of Advanced Studies SISSA
Strada Costiera 11
I-34014 Trieste
Italy

Jim E.F. Skea
Symbolic Computation Group
Departamento de Física Teórica
Instituto de Física
Universidade do Estado do Rio de Janeiro
Rua São Francisco Xavier, 524
Maracanã
22-013 Rio de Janeiro, RJ
Brazil

Edward A. Spiegel
Department of Astronomy
Columbia University
New York, NY 10027
USA

John Stachel
Department of Physics
Center for Einstein Studies
Boston University
Boston, MA 02215
USA

Zhong Tang
School of Physics
Georgia Institute of Technology
Atlanta, GA 30332–0430
USA

Andrzej Trautman
Instytut Fizyki Teoretyczne
Uniwersytet Warszawski
ul. Hoza 69
Warszawa
Poland

Henk van Elst
Department of Mathematics and
Applied Mathematics
University of Cape Town
Rondebosch 7700
Cape Town
South Africa

C.V. Vishveshwara
Indian Institute of Astrophysics
Bangalore 560 034
India

1

Jordan, Pauli, Politics, Brecht . . . and a Variable Gravitational Constant

Engelbert L. Schucking

November 1952. *Die Welt*'s headline read "EISENHOWER ELECTED PRESIDENT." Why did they do that, those crazy Americans? This meant the Dulles brothers in power and Adenauer's rearmament of Germany. I put politics out of my mind and pushed the doorbell of an apartment on Hamburg's Bundesstrasse. A maid with a little white cap opened the door.

"My name is Schücking."

"I shall announce you to Herr Professor," she said and vanished. The stout Herr Professor Dr. Jordan appeared after a while, tried to introduce himself with some difficulty because of his stammer—but in a very friendly way—and motioned me into his study.

I told him I wanted to work in relativity and had come to him because he was now active in this field. His book *Schwerkraft und Weltall* [1] had just come out. He gave me some proofs from his book and suggested as a problem to integrate a differential equation in his theory, which had a variable gravitational scalar. I learned later there was a problem with the first copies of the book [2]: the publisher, Vieweg, had produced copies with all pages blank. When Pauli received such a copy from the publisher he had remarked, "Jordan knows that I could think of what should be in it myself."

My interview was finished when a dolled-up woman with butterfly-shaped glasses appeared, who informed me that I should rise because a lady had entered the room. I did and was introduced to Frau Professor Jordan. She held out her hand to be kissed, but I shook it and said good-bye.

My solution to the problem found Jordan's approval, and, with a tiny stipend that he got for me from Hamburg's Rotarians, I became his student.

Pascual Jordan—he had a Spanish great-grandfather who served in Napoleon's army and got stuck in Germany on the retreat from Moscow—was one of the greatest physicists of this century. He was the originator of the theory which we now believe is the basis for all of physics: the quantum theory of fields. He was the first to realize that all things in the universe—photons, electrons, protons, atoms, and elephants—are field quanta. Out of the triumvirate (with Born and Heisenberg) that formulated quantum mechanics in 1925, Jordan was the principal architect of the theory. But in spite of his revolutionary contributions he never achieved the fame of his colleagues Heisenberg and Pauli, who patronized him. "Herr Jordan was always a formalist," Pauli told me once, meaning he was not a true physicist but only a mathematician, thus a lower form of life. But it was the formalism that contained the true physics.

A persistent stutter and bad luck seriously hampered his career. When I visited Max Born once in Bad Pyrmont to help him in his attempt to debunk the photon rocketeer Eugen Sänger, I mentioned that I was working on Jordan's theories. Born told me, "I hate Jordan's politics, but I can never undo what I did to him. And that came about like this: In December of 1925 I went to America to give lectures at MIT. I was editor of the *Zeitschrift für Physik* and Jordan gave me a paper to be published in the journal. I didn't find time to read it and put it in my suitcase. I forgot about it, and when I returned half a year later and unpacked, I found the paper at the bottom of the suitcase. It contained the Fermi-Dirac statistics. Meanwhile both Fermi and Dirac had discovered it. But Jordan was first."

When I returned to Hamburg I asked Jordan about it. "Was that so?" He just laughed and stuttered agreement. When I came home I looked up Jordan's book *Statistische Mechanik auf Quantentheoretischer Grundlage* [3]. The names Fermi and Dirac did not appear in it. Nor did he mention himself in this connection. He had called his brainchild "the Pauli statistic."

Born was right in hating Jordan's politics. In May 1933 Jordan had joined the NSDAP together with a million German opportunists [4]. One time when Pauli came to Hamburg to consult on the succession of his former boss Wilhelm Lenz (of the Lenz-Ising model), I remember that during a reception at the university Pauli exclaimed, "Richard Kuhn (Nobelist in chemistry and Pauli's schoolmate in Vienna) had no excuse for having been a Nazi, but Herr Jordan had: he was a professor in Rostock!" (Rostock was considered the Paraguay of German academe.) Jordan winced. He had been a full professor of theoretical physics in Rostock from 1935 to 1944.

Before the Nazis came to power, Jordan had been a conservative nationalist who published his elitist views in the right-wing journal *Deutsches Volkstum* (German Heritage) under the pseudonym "Domeier" [5]. My Göttingen teacher Hans Kopfermann may have had Jordan in mind when he wrote to Niels Bohr in May 1933, "There is a tendency among the non-Jewish younger scientists to try to join the movement and to act as much as possible as a moderating element, instead of standing disapprovingly on the sidelines" [6]. But Jordan became a strange moderator. In November 1933 he joined an SA unit and now became a stormtrooper who, in his brown uniform with jackboots and swastika badge, defended Einstein, Freud, the positivists, and other arch-enemies of Nazi ideology, against their official detractors. He tried to sell the Nazi regime on relativity and quantum acausality as idealistic theories that would be veritable weapons in the battle against Bolshevist materialism. He ridiculed Bieberbach's racist "German" mathematics by writing, "The differences between German and French mathematics are not any more essential than the differences between German and French machine guns" [7].

Although he advertised science in militaristic language as the great weapon of the Nazi realm, his own contribution to the war was modest. He volunteered to join the air force in 1939 and worked mostly as a meteorologist at air fields, including the German rocket center in Peenemünde. Apparently unwittingly he helped the Allied cause by spilling certain vital secrets of the Peenemünde operation to the Springer editor Paul Rosbaud, who was a British spy [8]. In spite of his bizarre political activities, he was personally a shy and kind man. With Pauli's support Jordan was rehabilitated after the war. In 1953, thanks to Pauli's intercession, he advanced from visiting to full professor at Hamburg University. "It would be incorrect," Pauli commented at the time, "for West Germany to ignore a person like P. Jordan" [9].

The problem that Jordan had in mind for me was to solve the differential equation [10]

$$\left[3\dot{R}+(\zeta-3)\frac{R}{t}\right]\left[\dot{R}+(\zeta-1)\frac{R}{t}\right](2c^2+\ddot{R}R+2\dot{R})$$
$$-\zeta(2c^2+\ddot{R}R+2\dot{R}^2)^2+3\ddot{R}R\left[\dot{R}+(\zeta-1)\frac{R}{t}\right]^2=0 \qquad (1)$$
$$\dot{R}=\frac{dR}{dt}, \qquad \zeta=\text{const.}, \qquad c=\text{const.}$$

This equation was his generalization of Friedmann's equation for the radius $R(t)$ of an expanding or contracting universe. The universe is considered to be a hypersphere filled with incoherent matter and a time-dependent scalar field—his variable gravitational "constant."

Jordan had apparently tried quite hard to solve this monstrous equation but had not succeeded; and therefore, in keeping with a hallowed tradition, he posed it as a problem for this graduate student. I was not keen on the idea since this was clearly an insoluble problem. All one could hope to do for such a horribly nonlinear equation was to construct a theory for the qualitative behavior of its solutions and find various approximate solutions to be patched together with appropriate numerical bounds. In Göttingen I had listened to Franz Rellich spend a whole semester doing nothing else but developing the theory of van der Pol's equation [11]

$$\ddot{R}-\lambda(1-R^2)\dot{R}+R=0, \quad R=R(t), \quad \dot{R}=\frac{dR}{dt}, \quad \lambda=\text{const.}, \qquad (2)$$

which was a kitten compared with Jordan's monster.

I had ideas of my own for a thesis—for instance to develop quaternionic quantum mechanics. I had read the thesis of Otto Teichmüller, an ingenious SS Obersturmführer and M.I.A. of the war, who had done the spectral theory of self-adjoint operators in quaternionic Hilbert space [12]. Or I wanted to work on spinors in the Einstein theory. But Jordan would have nothing to do with it. "This cannot give anything new," he declared. "Nobody works on *my* theory. If you want to do a thesis you have to work on *my* theory." I thought his theory was crazy. It was an attempt to turn Dirac's numerology into a field theory.

Gamow recalled [13] that in 1937 he was sitting in his room at Bohr's institute in Copenhagen, where he was a guest, when Bohr came in waving the latest issue of *Nature*. Bohr exclaimed, "Look what happens to people when they get married!" He was referring to the fact that Dirac had just married Wigner's sister Margit and concurrently had proposed in *Nature* [14] a radical hypothesis: that since the immense number 10^{39}, the ratio of electric to gravitational attraction in the hydrogen atom, was roughly equal to the age of the universe in elementary times (10^{-23} sec.), it could thus be explained, without the slightest observational evidence, by a time dependence of Newton's G, with $G \sim t^{-1}$. While Dirac had soon thereafter shelved this weird idea for some thirty years, Jordan apparently had been the only one to take it seriously by trying to turn G into a scalar field.

During the war at the rocket center at Peenemünde, Jordan [15] developed a generalization of Pauli's projective relativity that gave rise to a scalar field which he identified with G. His book *Schwerkraft und Weltall* was largely an elaboration of this theory, which turned out to be mathematically equivalent to a generalization of Kaluza's theory, with a variable g_{55}, that had been considered by Peter Bergmann and Albert Einstein [16].

I thought this theory was pretty ugly, since one had no obvious choice for a Lagrange density. Moreover, a second-order partial differential equation for G would hardly lead to the special solution $G \sim t^{-1}$. Quite apart from other constants, the first derivative of G with respect to time would now be a new cosmological parameter undetermined by theory.

But Jordan could not be stopped. "My husband is like a tank," said Mrs. Jordan. He thought diminishing gravity would make the earth expand and thus account for continental drift. Moreover, since his cosmos had some 10^{78} particles, its mass should go as t^2 (10^{78} being the square of 10^{39}). Since Pauli had known Jordan as a lean youngster, he remarked that the now stout professor had projected his increase of mass onto the cosmos.

Jordan conjectured that the mass increase would occur explosively through white hole formation, the creation of baby universes that would turn into Ambarzumian's stellar associations. Later my task was to look for time-dependent solutions of his very unpleasant field equations that might support such wild surmises.

When Willibald Jentschke's accelerator center DESY was started in a suburb of Hamburg, a reporter for the *Hamburger Abendblatt* familiar with Jordan's ideas asked him whether the collision of high energy particles in the machine might not start the birth of a star in Bahrenfeld, with all its dire consequences. Jordan said—to the relief of Jentschke—that the energy was not high enough.

Whenever I visited Jordan at Ise Strasse 132, his new quarters near the Aussenalster in Hamburg's posh neighborhood, to report about progress in my calculations, he supplied me with stacks of paper consisting of letters he had received whose backs were blank. From these I gathered that a major part of his income apparently derived from honoraria for talks to all sorts of societies, church groups, and adult education centers—about physics and God, society, the universe, ESP, psychology, biology and whatnot. His estate contains manuscripts or notes for over 135 such talks, not counting lecture notes. His notations on an outline for a talk on cosmology indicate that he gave this one particular lecture over 100 times. From 1945 to his death in 1980 he also published 17 popular books, apart from new editions of his older ones [17]. Through his lectures, publications, and radio talks, Jordan not only interested many people in physics but became the best known popularizer of physics in Germany, whose advice was widely sought. He certified that radioactive watch dials did no harm and came out for German rearmament because he felt the same Einstein who wrote the letter to Roosevelt would have recommended just that to the Germans. From the front of my calculation sheets I got the impression that there was no crackpot in Germany who did not write to Pascual Jordan. While answering these letters he had often made doodles of fascinating complexity. I should have saved the stuff.

It was only recently, when I saw von Meyenn's publication of Pauli's letters from the early fifties, that I learned he had encouraged Jordan to pursue his theory. In October 1952 he wrote to Jordan [18], "In itself, Dirac's idea of a variable κ [$\kappa = 8\pi G$] appears to me as a natural one, and I am convinced at the moment that the action principles of the type used in your book on p.132 are the only reasonable formulation of the "Dirac idea." I do not dare yet judge whether this corresponds to physical reality or not." Once when Pauli visited Hamburg Wilhelm Lenz introduced him by saying that Pauli had often judged something that was right as wrong but never something that was wrong as right. But Pauli's early judgment in Jordan's case came as a surprise to me. Still, when Jordan gave a talk about his theory with Pauli sitting in the first row, swaying rhythmically, Pauli got up after the lecture, pointed his finger at Jordan, and said into the expectant silence, "Herr Jordan, a theory does not become right just by talking about it." Jordan laughed. It was his defense that he couldn't get a word in quickly.

But Jordan could think fast. Pauli once quoted to him an awful sentence, dripping with Nazi ideology, from Jordan's popular writings in the thirties, and said, "Herr Jordan, how could you write such a thing?" To which Jordan retorted, "Herr Pauli, how could you read such a thing?"

In the evening after Jordan's talk we had dinner with Pauli in a posh restaurant on the Rothenbaum Chaussee, the avenue where Pauli had had his epiphany in 1925: the discovery of the exclusion principle. I remember how Pauli relished his *Karpfen Blau* with horseradish and told his Zermelo anecdotes from his Göttingen days.

Ernst Zermelo, who created a system of axioms for set theory, was a *privatdozent* at Göttingen when the mathematics department of the university was under the sway of Herr Geheimrat Felix Klein. Pauli said, "Zermelo taught a course on mathematical logic and stunned his students by posing the following question: All mathematicians in Göttingen belong to one of two classes. In the first class belong those mathematicians who do what Felix Klein likes but what they dislike. In the second class are those mathematicians who do what they like but what Felix Klein dislikes. To what class does Felix Klein belong?"

Jordan, having listened intently, broke into roaring laughter. Pauli paused, taking a sip of wine, and said disapprovingly, "Herr Jordan, you have laughed too early."

He continued, "None of the awed students could solve this iconoclastic question. Zermelo then crowed in his high-pitched voice, 'But meine Herren, this is very simple. Felix Klein isn't a mathematician.' " Jordan laughed again and Pauli drained his wine glass approvingly.

His second Zermelo anecdote went like this: Klein taught a course in Göttingen with the title "Precision and Approximation Mathematics," a subject very dear to him. He wanted to show that for every theorem in exact mathematics there was a corollary in which the conclusion was approximately correct. On one day, when Zermelo had come to the lecture and was sitting in the back of the auditorium, Klein had chosen as an example for his thesis Pascal's famous theorem on conics. It says that given five points on a conic, the sixth point constructed by means of Pascal's line will also lie on the same conic. Klein wanted to prove that when the five points lie approximately on a conic, the sixth point constructed à la Pascal will also lie approximately on the same conic. After having explained Pascal's precise construction with a beautifully drawn figure, Klein set out to formulate his theorem in approximation mathematics. He began, "Let us take . . . " and here Zermelo's falsetto from the back-benches finished Klein's sentence with "approximately five points."

"Zermelo was not offered a professorship in Göttingen," concluded Pauli.

By studying the books on differential equations by Ince [19] and Kamke [20] I succeeded in reducing equation (1) to a more manageable first-order differential equation and worked out the theory of its solutions. Jordan asked Professor König in Clausthal to plot numerically the solutions of this equation for a special value of a parameter [21]. He invited König and me for lunch in the gourmet restaurant of the Hamburg main railway station to compare results. It was pleasing to see that my sketch and König's graph (Fig. 1.1) were identical. Jordan was exuberant about the pretty picture. He wrote, "The fact that the extended gravitational theory could give rise to such beautiful mathematical investigations strengthens my hope that it has succeeded in eavesdropping on nature's secret harmonies." I wasn't so optimistic. After lunch König and Jordan seemed to have some secret project at the Iron Curtain, which started a few miles from Clausthal, and I left them to their plots.

I read Jordan's book *Schwerkraft und Weltall* carefully. In its first part he had tried to give a new introduction to relativity from an algebraic point of view. Since the subject was apparently new to him, some of it was original, like an axiomatic characterization of the covariant derivative that was also discovered later by Koszul. But there were also many misprints and some mistakes. I had studied Einstein's papers and the books by Weyl, Eddington, and Levi-Civita, while his source of information had been mainly Pauli's *Enzyklopädie* article.

Pauli's report in the *Enzyklopädie der Mathematischen Wissenschaften* was one of the best treatments of the subject, but it was not perfect, as Pauli was the first to point out. I said to him, "Your article is unique. One can think of examples of great inventions by the young, such as Galois, but I know of no case where a young man

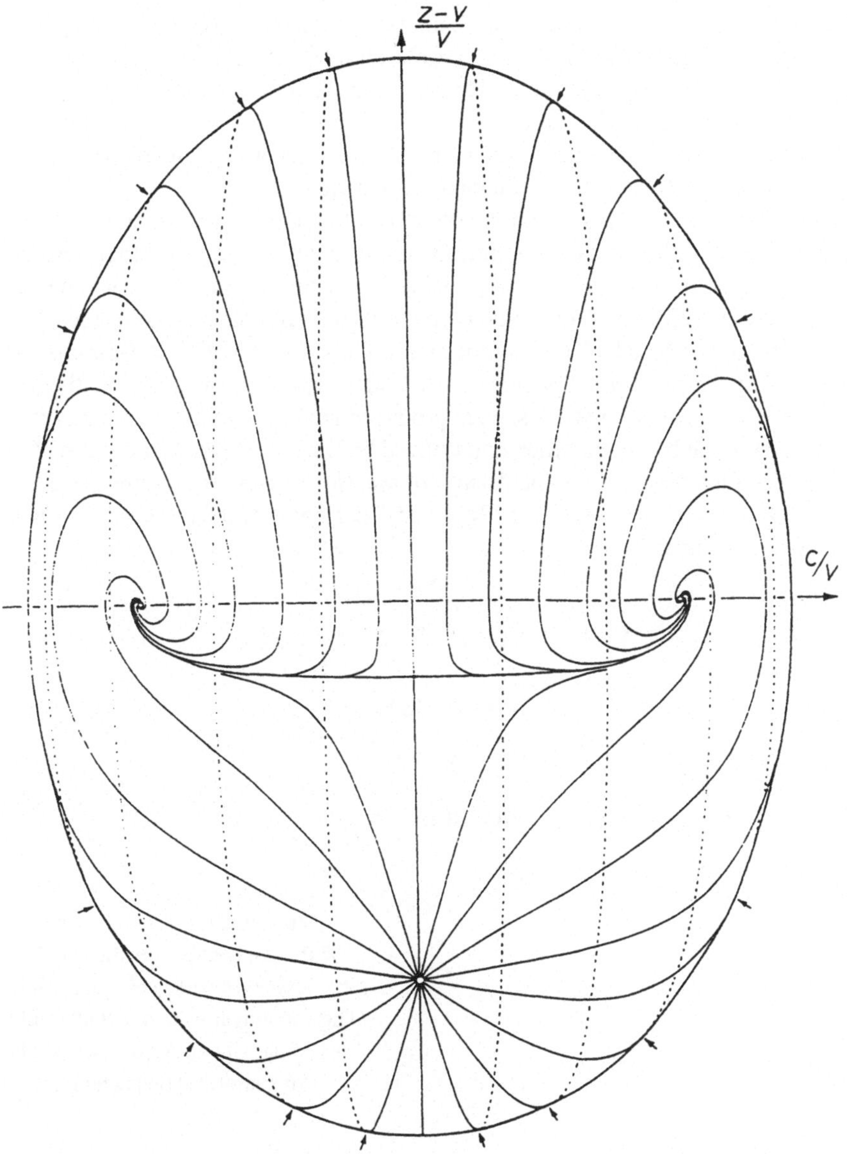

Figure 1.1

reviewed a theory in such a magisterial and comprehensive way." Pauli swayed back and forth like a Hasid in prayer. "There is an example . . . " (pause and more swaying) ". . . Gauss." He added, "It isn't perfect. I missed the Bianchi identities." I hadn't noticed that; but if one saw the Riemann tensor as the field strength of the gravitational field, he had missed the analog of half the Maxwell equations. He rectified this lapse in an appendix to the English translation of his article.

I learned about his *modus operandi* in mending such mishaps when he prepared a new edition of his 1933 *Handbuch* article on quantum mechanics, known by cognoscenti as the "New Testament." He took time off from watching the elephants, his favorite animals, at Hagenbeck's Zoo in Hamburg and went to the physics department's library, where he looked up his article and scanned the margins, page by page, for comments and corrections by its readers.

While preparing Jordan's book for a second edition I learned from him about an aspect of Einstein's theory that was entirely new to me: its Cauchy problem. Jordan, a former student of Courant, was now praising the second volume of "the wonderful book by Courant-Hilbert." This had led him to the remarkable papers on causality and gravitational waves by Karl Stellmacher in the 1937 and 1938 *Mathematische Annalen* [22]. These seminal papers on the Cauchy problem in Einstein's theory, still widely unknown to relativists, were produced in Göttingen's Mathematisches Institut, terrorized by Teichmüller, and utilized by Bertolt Brecht as the setting for Segment 7 of his play, *Fright and Misery of the Third Reich*, which premiered in Paris in 1938. Segment 7 is entitled "Physicists" [23]; Segment 8 is the well-known "The Jewish Woman."

‡ ‡ ‡ ‡ ‡ ‡ ‡ ‡ ‡ ‡

PHYSICISTS

by

Bertolt Brecht

Here come the Teutonic scholars
With fake beards over their molars
And eyes wide and scared sick.
They no longer want one that's right
Nay, one with an Aryan face bright
An authorized deutsche Physik.

In an institute of physics. Two scientists, X and Y. Y has just come in. He acts like a conspirator.

Y. I've got it!

X. What?

Y. The answer to the questions to Mikowsky in Paris.

X. About the gravitational waves?

Y. Yes.

X. And?

Y. Do you know who wrote us about it, giving us exactly what we need?

X. Well?

Y writes a name on a piece of paper and gives it to X. After X has read it, Y takes the paper back and tears it into small pieces and throws them into the oven.

Y. Mikowsky passed our question on to him. Here is the answer.

X. (*grabs for it*) Give it to me! (*Suddenly he stops.*) But if they catch us corresponding with him . . .

Y. They must not, on any account!

X. But without it we can't go on. Come on, give it to me.

Y. You cannot read it. I have transcribed it into my stenographic system that is safer. I'll read it to you.

X. You must be careful!

Y. Is Rollkopf in the lab? (*He points to the right.*)

X. (*points to the left*) No, but Reinhardt is. Sit over here.

Y. (*reads*) We are dealing with two arbitrary contravariant vectors phi and nu and a contravariant vector t. With their help the components of a mixed tensor of the second rank are formed whose structure is given by

$$\Sigma_l^r = C_{lhi}\phi^h \nu^i t^r$$

X. (*X, who has taken notes, suddenly motions him to be quiet.*) Just a moment!

He gets up and tiptoes to the wall on the left. Apparently he does not hear anything suspicious and returns. Y continues reading but is sometimes interrupted in a similar manner. Then they investigate the telephone, open the door suddenly, and so on.

Y. For incoherent pressureless matter at rest, $T = \mu$ is the only component of the energy tensor different from zero. Therefore, a static gravitational field is generated whose equation gives, by taking the constant of proportionality factor to be $8\pi\kappa$,

$$\nabla^2 f = 4\pi\kappa\mu.$$

With a suitable choice of space coordinates the deviation from $c^2\,dt^2$ is very small...

Since somewhere a door is banged they want to hide their notes. But then it doesn't seem to be necessary. From now on, however, they become absorbed in the matter and seem to forget how dangerous their activity is.

Y. (*reads on*) ...On the other hand the masses in question are very small compared with the field generating mass at rest; thus the motion of the bodies embedded in the gravitational field is given by a geodesic world line in this static gravitational field. As such it obeys the variational principle

$$\delta\int ds = 0,$$

where the ends of the world line segment in question remain fixed.

X. But what does Einstein say to...
From Y's horror X realizes his lapse and becomes rigid with fright. Y rips the notes from his hand and pockets all papers.
Y. (*very loudly toward the left wall*) Yes, a typical Jewish sophistry!
What does that have to do with physics?
Relieved, they go back to their notes and go on working silently, with the greatest precaution.

‡ ‡ ‡ ‡ ‡ ‡ ‡ ‡ ‡ ‡

It was much to Jordan's credit that he didn't shun the dangerous E-word when writing about relativity during the Nazi era, while a circumspect Heisenberg managed to avoid it. The Hamburg theoretician Wilhelm Lenz tried a different tack. He had found a (spurious) derivation of the Schwarzschild line element from special relativity that he wanted to publish in *Naturwissenschaften*. To avoid the E-word, he tried to enlist the help of Max von Laue in Aryanizing relativity by crediting the theory to Poincaré. Laue was disgusted and in 1939 wrote to Einstein that he found that "as reprehensible as [it was] foolish" [24]. In 1944 Lenz turned his notes over to Sommerfeld, who published them in 1949 in his lectures on electrodynamics [25]. They were then taken up by Leonard Schiff [26], refuted by Alfred Schild [27], and, hopefully, thus laid to rest.

I contributed a bit to Jordan's theory, which he incorporated into the second edition of *Schwerkraft und Weltall* that Vieweg published in 1955. I earned my doctorate by producing some exact nonstatic solutions of his theory with spherical symmetry, whose physical interpretation has remained obscure [28].

To find a solution for the fusion of two separate universes into one—I called such a model a "four-dimensional trouser world"—I concocted metrics whose space sections were three-dimensional generalizations of lemniscates, with a time-dependent parameter given by the equations

$$((x-a)^2+y^2+z^2+u^2)((x+a)^2+y^2+z^2+u^2)=R^4(t), \qquad a=\text{const.} \tag{3}$$

$$dl^2 = dx^2+dy^2+dz^2+du^2 \tag{4}$$

For the unwieldy calculations of these double-bang models Jordan's discarded correspondence was no longer adequate, so I used the back of wallpaper. I ran into contradictions that I interpreted not as mistakes but as the nonexistence of exact solutions for this *ansatz* in Jordan's theory. The scrolls are lost and have, perhaps, reverted to their original destination.

In the summer of 1953 Pauli gave a course on relativity at the E.T.H. in Zürich—*on revient toujours à son premier amour*. His student and later assistant Charles Enz turned his lecture notes into a 111-page manuscript. As Pauli wrote to Oskar Klein in July, 1953, he had given the course "to learn the Jordan theory" [29]. However, this learning process took a distinctly Paulinic form. The French mathematician Yves Thiry, who had independently developed the mathematical framework of the Jordan theory, had sent his two *Thèses* [30] to Pauli. Pauli wrote to Jordan that he had resolved not to open them. "The Thèses of Thiry lie on my table; they are so terribly bulky (and also have no reasonable summary) that it is much simpler not to open the book and just figure out what must be in it." Jordan agreed. He replied, "Your method for treating the Thèse of Thiry appears to me quite sound" [31].

While studying Jordan's theory Pauli discovered the important role of conformal transformations for its metric and pointed out that Jordan had misinterpreted his scalar field [32]. According to Pauli, Jordan's κ was not Newton's gravitational constant G times 8π but its inverse. Adding to this confusion was a paper by Markus Fierz [33], who found that the scalar field had nothing to do with gravitation but was the dielectric constant of the vacuum.

At the relativity conference in Bern that celebrated the 50th anniversary of Einstein's 1905 paper, Jordan reported on his theory without talking about its physical interpretation. In his masterful summary of the whole conference Pauli buried Jordan's theory as follows [34]:

"Now we come to the *extensions of the theory*. There is the extension with five-dimensional methods and the extension with nonsymmetrical quantities. Herr Prof. Fock told me that he did not believe that these extensions have any physical sense, and I too have very strong doubts about the physical correctness of these extensions if one does not simultaneously introduce quantum theory in a deeper way. It could well be that those who are not entirely serious about accepting these extensions physically have a very good chance of being right in the end. Herr Jordan has unfortunately prevented us by the magic of his mathematical theorems from hearing something about what his physical reasons actually are for assuming a variability of the gravitational constant. This certainly would have interested everybody, and he would have been certain of no competition from our distinguished speaker Lichnerowicz. Of course he has written something about it; but to me, after what we have heard from Robertson, there seem no longer to remain many arguments for a variability of the gravitational constant in the empirical results on the expansion of the universe. I do not want to claim that it is impossible to assume a variability of the gravitational constant. It is possible, perhaps, to interpret certain phenomena on earth more easily; but it so happens that I do not quite see that this theory has enough foundation. I, like others, have strong doubts that here one is physically on the right track."

Meanwhile I was no longer alone in working for Jordan. Jürgen Ehlers and Wolfgang Kundt had become Jordanians and had also begun contributing to Jordan's theory. Founding the Jordan Seminar, we met every week to instruct ourselves in relativity. Jordan, who had hoped we would all work on his theory, gave in to our demands to study the old Einstein theory first and bore our lack of enthusiasm for his pet project with grace.

The impression that Einstein's theory of gravitation had not been properly developed and understood occurred in the fifties to many theoretical physicists and mathematicians, who created a renaissance of the theory. In many countries, independently of each other, young researchers tried to develop the theory with modern mathematical tools, and Hamburg's Jordan Seminar became one of less than a dozen centers where relativity began to flourish anew. Jordan supplied the resources and his collaborators produced, with his help, what became known as the "Hamburg Bible" as their contribution to the relativistic renaissance. I hope that Jürgen Ehlers, who became the leading spirit in this development, will some day write the history of the Jordan Seminar.

Pauli's monumental unpublished contributions to physics and his dreams of a universal science encompassing mind and matter are now slowly emerging through von Meyenn's erudite edition of his correspondence and show him as one of this century's greatest minds. But the depth of Jordan's contributions to physics has not yet found its proper appreciation. While his work on quantum biology, inspired by Bohr's mysticism, has not stood the test of time, his work on the foundations of physics is regrettably still widely unknown.

Max Jammer, one of the foremost historians of quantum mechanics, did point out that the bulk of the monumental Born-Jordan paper, "Zur Quantenmechanik," was written by Jordan [35]. In it he had used the mathematical tools of the Courant-Hilbert book that had just come out. He knew these techniques intimately because he had just written this book together with Courant. It has also been argued that Jordan's *habilitation* lecture was crucial for Heisenberg's discovery of the uncertainty relations [36]. And clearly, even Jordan's pioneering work in quantum field theory was not immediately appreciated. His formalism of creation and annihilation operators, now the basic language of physics, was in 1933 still viewed with suspicion by Pauli. He wrote [37]. "It is doubtful whether a really deep-seated physical connection lies at the root of this approach, and it can be shown that all the results of quantum mechanics can be obtained without applying these methods."

In a seminal paper in 1935 Jordan [38] showed how his formalism could treat the physics of multiparticle systems—now the standard treatment of condensed-matter physics—and create the representations of the unitary and permutation groups that are now used by the particle theoreticians. In 1979 Wigner proposed Jordan for the Nobel prize in physics, but the Swedish Academy awarded the prize to three practitioners of the art that Jordan had invented.

After his development of quantum field theory Jordan began, in 1930, a quest for a new mathematics that would make it possible to overcome the contradictions of quantum field theory. He discovered a new version of quantum mechanics based on commutative nonassociative algebras, the Jordan algebras, that created a new universe of mathematics [39]. This led to the discovery of one of the most beautiful symmetric structures, the Albert algebra of 3x3 Hermitian matrices of octonions. After the war Jordan found that its idempotents are the points and lines of the Moufang plane [40]. In his search for the ultimate formalism of physics he became engaged in creating the theory of the *schrägverbände* (skew lattices), till death caught up with him at his home while he was filling in formulae in his manuscript at the kitchen table.

References

[1] Jordan, P., *Schwerkraft und Weltall*, Vieweg (Braunschweig), 1952.

[2] von Meyenn, Karl, ed., W. Pauli, *Scientific Correspondence*, Vol. 4, Part 1, Springer (New York), 1996, p. 737.

[3] Jordan, P., *Statistische Mechanik*, Vieweg (Braunschweig), 1944, p. 100.

[4] Henschel, K., ed., *Physics and National Socialism*, Birkhäuser (Boston), 1996, Appendix F, p. xxxv.

[5] Beyler, Richard H., "From Positivism to Organicism: Pascual Jordan's Interpretations of Modern Physics in Cultural Context," Ph.D. thesis, Harvard Univ., 1994, p. 207.

[6] Henschel, K., *op.cit.*, p. 57.

[7] Wise, M. Norton, "Quantum Mechanics, Psychology, National Socialism," in Renneberg, Monika, and Walker, Mark, eds., *Science, Technology and National Socialism*, Cambridge Univ. Press, 1994, p. 226.

[8] Kramish, A., *The Griffin*, Houghton Mifflin (Boston), 1986.

[9] von Meyenn, Karl, "Jordan," in *Dictionary of Scientific Biography*, Suppl. Vol. II, Scribner's (New York), 1990, p. 452.

[10] Jordan, ref. 1., p. 163.

[11] van der Pol, B., *Philos. Mag.* **6**, No.43, 700–719 (1922); *ibid.* 7, No. 2, 978–992 (1926).

[12] Teichmüller, Otto, *J. Reine Angew. Mathematik* **174**, 73 (1935).

[13] Sullivan, Walter, *Continents in Motion*, McGraw-Hill (New York), 1974, p. 50.

[14] Dirac, P. A. M., "The Cosmological Constants," *Nature* **139**, 323 (1937).

[15] Jordan, P., "Erweiterung der projektiven Relativitätstheorie," *Annalen der Physik*, 6 Folge, Band 1, p. 219 (1947).

[16] Bergmann, P. G., *Annals of Mathematics* **49**, 255 (1948).

[17] Beyler, Richard H., *op. cit.*, pp. 486–489.

[18] von Meyenn, ref. 2, p. 737.

[19] Ince, E. L., *Ordinary Differential Equations*, Longmans (New York), 1927.

[20] Kamke, E., *Differentialgleichungen*, Akademie Verlag (Leipzig), 1944.

[21] Jordan, P., *Schwerkraft und Weltall*, zweite, erweiterte Auflage, bearbeitet unter Mitwirkung von E. Schücking, Vieweg (Braunschweig), 1955, p. 202.

[22] Stellmacher, K., *Mathematische Annalen* **115**, 136 (1937); **115**, 740 (1938).

[23] Brecht, B., *Gesammelte Werke*, Stücke 4, Suhrkamp (Berlin), 1988, p. 197.

[24] Beyerchen, Alan D., *Scientists under Hitler*, Yale Univ. Press (New Haven), 1977, p. 170.

[25] Sommerfeld, A., *Elektrodynamik*, Akademie Verlag (Leipzig), 1949, p. 323.

[26] Schiff, L., *Amer. J. Phys.* **28**, 340 (1960).

[27] Schild, A., *Amer. J. Phys.* **28**, 778 (1960).

[28] Schücking, E., *Zeitschrift für Physik* **148**, 72–92 (1957).

[29] von Meyenn, ref. 2, p. 800.

[30] Thiry, Y., "Étude mathématique des equations d'une théorie unitaire à quinze variables de champ," Thèse de l'Université de Paris, 1951; *Journ. math. p. et appl.* **30**, 275–396 (1951).

[31] von Meyenn, Karl, ed., W. Pauli, *Scientific Correspondence*, Vol. 4, Part 2, Springer (New York), to be published, pp. 163, 167.

[32] *Ibid.*, pp. 191, 202.

[33] Fierz, M., *Helvetica Physica Acta* **XXIX**, 128 (1956).

[34] Pauli, W., in Mercier, A., and Kervaire, M., eds., *Fünfzig Jahre Relativitätstheorie*, [*Helvetica Physica Acta*, Supplementum IV], Birkhäuser Verlag (Basel), 1956, p. 265.

[35] Jammer, M., *Naturwissenschaftliche Rundschau*, 1984, p. 1.

[36] Beller, Mara, *Archive for the History of Exact Sciences* **33**, 337 (1985).

[37] Pauli, W., "Die allgemeinen Prinzipien der Wellenmechanik," in *Handbuch der Physik*, zweite Auflage, Band XXIV, Teil 1, Springer (Berlin), 1933, p. 198.

[38] Jordan, P., *Zeitschrift für Physik* **94**, 531 (1935).

[39] Jordan, P., von Neumann, J., and Wigner, E., *Annals of Mathematics* **35**, 29 (1934).
[40] Jordan, P., *Abhandlungen aus dem Mathematischen Seminar der Universität Hamburg*, Vol. 16, Part 1, (Göttingen), 1949, p. 74.

Acknowledgments

I am grateful to Peter Bergmann, Jürgen Ehlers, and especially to Bill Wallace for reading the manuscript and suggesting improvements and corrections in fact and style. However, remaining errors or misjudgments, and what little I remember, belong to me.

2

Thomson Scattering in an Expanding Universe

James L. Anderson

ABSTRACT The Thomson cross section for scattering of electromagnetic waves by a free electron in an expanding universe is derived here. The equations of motion of the electron are obtained from the Einstein-Maxwell field equations of general relativity using the Einstein-Infeld-Hoffmann surface integral method. These integrals are evaluated approximately by perturbing off an Einstein–de Sitter cosmological field. It is found that the Thomson cross section does not vary with cosmic time.

2.1 Introduction

The question of whether or not the cosmic expansion of the universe effects local systems is usually answered in the negative. Thus Misner, Thorne, and Wheeler [1] have likened the situation to that of blowing up a balloon with pennies affixed to its surface. The pennies recede from each other but do not themselves grow in size. One might however equally well imagine that the pennies are painted on the balloon so that they also expand as the balloon expands. If one were to place pennies side by side between us and the Coma cluster of galaxies, would gaps appear between them as the universe expanded or not? In order to decide this question it is necessary to know the dynamics of the system with which one is dealing. One cannot of course simply postulate this dynamics since one could obtain either answer in this way. One needs to be able to derive this dynamics from some more fundamental theory. In a recent work [2] (referred to hereafter as A1), I showed how such a dynamics could in fact be derived for gravitationally and electromagnetically bound systems directly from the Einstein-Maxwell field equations of general relativity without additional assumptions by using the Einstein-Infeld-Hoffmann (EIH) [3] surface integral method. In deriving these dynamical laws, I considered systems whose periods T_S were large compared to the light travel time T_L across them but small compared to the Hubble time T_H, so that $T_L \ll T_S \ll T_H$. What I was able to show was that the "size" of all such systems remain constant as the universe expands, so that indeed gaps would appear between the pennies. Furthermore, the characteristic time scales of such systems do not change with time, so that model gravitational and electromagnetic clocks will measure cosmic time and will, consequently, observe red shifts in the light from the Coma cluster.

In this paper I address a different problem associated with physical processes in an expanding universe, namely Thomson scattering. At first glance it might appear to the reader strange to imagine that gravitational effects associated with the expanding universe could have any effect whatsoever on what is seemingly a purely electrodynamical process. But in fact Thomson scattering is not a purely electromagnetic process. The inertial force term in the equations of motion for the scattering electron is of gravitational origin as EIH first showed and, in the case of an expanding universe, involves the cosmic scale factor $R(t)$. As in A1 the exact dependence of the inertial force on $R(t)$ as well as the form of the electromagnetic force due to the incident wave can be determined from the EIH surface integrals by perturbing the gravitational and electromagnetic fields off the Einstein–de Sitter field.

2.2 EIH Surface Integrals

The EIH surface integrals can most easily be derived from the Landau and Lifshitz [4] form of the field equations and have the form (in what follows I will use units in which $G = c = 1$, Latin indicies run from 1 to 3, Greek indicies run from 0 to 3, and I employ both the Einstein summation convention and the comma notation to denote partial derivatives, that is $_{,\mu} = \partial_{x^\mu}$)

$$U^{\mu\nu\rho}{}_{,\rho} = \Theta^{\mu\nu}, \tag{1}$$

where

$$U^{\mu\nu\rho} = -U^{\mu\rho\nu} = \frac{1}{16\pi}\{\mathfrak{g}^{\mu\nu}\mathfrak{g}^{\rho\sigma} - \mathfrak{g}^{\mu\rho}\mathfrak{g}^{\nu\sigma}\}_{,\sigma}, \tag{2}$$

$$\Theta^{\mu\nu} = (-g)(T^{\mu\nu} + \mathfrak{t}^{\mu\nu}_{LL}), \tag{3}$$

and

$$\mathfrak{F}^{\mu\nu}{}_{,\nu} = 0 \quad \text{and} \quad F_{|\mu\nu,\rho|} = 0. \tag{4}$$

In these equations $g = \det(g_{\mu\nu})$, $\mathfrak{g}^{\mu\nu} = \sqrt{-g}g^{\mu\nu}$, $\mathfrak{F}^{\mu\nu} = \sqrt{-g}F^{\mu\nu}$, $\mathfrak{t}^{\mu\nu}_{LL}$ is the Landau-Lifshitz energy-stress pseudotensor and $T^{\mu\nu}$ is the electromagnetic energy-stress tensor given by

$$T^{\mu\nu} = \frac{1}{16\pi}\left(g^{\mu\nu}F_{\rho\sigma}F^{\rho\sigma} - 4g_{\rho\sigma}F^{\mu\rho}F^{\nu\sigma}\right). \tag{5}$$

Note that in our expression for $\Theta^{\mu\nu}$ above, there is no matter contribution from the sources, since they are assumed to be compact and to vanish on and outside the EIH surfaces. This feature of the EIH method thereby avoids having to make specific assumptions about the form of the matter energy-stress tensor or the need to introduce singular source terms.

Because of the antisymmetry of $F^{\mu\nu}$ and $U^{\mu\nu\rho}$ in their indicies, it follows that $F^{rs}{}_{,s}$ and $U^{\mu rs}{}_{,s}$ are 3-dimensional curls whose integrals over a closed surface vanish. As a consequence, integration of Eq. (1) over a closed 2-surface in a $t = $ const. hypersurface gives

$$\oint \left(U^{\mu r0}{}_{,0} - \Theta^{\mu r}\right) n_r \, dS = 0 \tag{6}$$

where n_r is a unit surface normal. In a similar way one gets from Eq. (4) the result

$$\oint \mathfrak{F}^{r0}{}_{,0} \, n_r \, dS = 0. \tag{7}$$

It is these last two equations that yield equations of motion for the sources of the gravitational and electromagnetic fields. When the surfaces over which the integrals are taken surround a source, the requirement that the surface-independent contributions vanish yield these equations. (The surface-dependent terms will in all cases vanish either identically or as a consequence of the field equations [5].) These equations will be used to derive equations of motion for a compact charge in the presence of an incident electromagnetic field whose characteristic length scale is large compared to the size of the EIH surfaces needed to enclose the charge. The near electromagnetic field produced by this charge is then matched to a far radiation field. The total flux of this scattered field is finally divided by the incident flux to yield an expression for the Thomson scattering cross section.

2.3 Approximation Procedures

To evaluate the fields appearing in Eqs. (6) and (7) I follow the methods developed in A1. However, since one is dealing with radiation fields here, it turns out to be more convenient to employ conformal coordinates (τ, x, y, z) in which the background gravitational field has the form

$$g_{\mu\nu} = R^2(\tau)\,\mathrm{diag}(1, -1, -1, -1) \tag{8}$$

and where τ is related to the cosmic time t by $dt = R(\tau)\, d\tau$. For convenience the total field is written as

$$\mathfrak{g}^{\mu\nu} = R^2 \tilde{\mathfrak{g}}^{\mu\nu}, \tag{9}$$

and $\tilde{\mathfrak{g}}^{\mu\nu}$ as well as the electromagnetic 4-potential A^μ is expanded in a double series in two small dimensionless parameters $\varepsilon = T_L/T_S$ and $\delta = T_L/T_H$. These fields in addition are assumed to depend on the conformal time τ through their dependence on $\varepsilon\tau$ and $\delta\tau$ while R is a function of δt. (In higher orders of approximation they will of course depend on higher order multiple times.) In addition, charges and masses are scaled so as to be $O(\varepsilon^2)$. In what follows we will need an accuracy of ε^2 and $\varepsilon\delta$, since we are only concerned here with the modifications in the Newtonian dynamics in the background cosmological field.

The lowest order correction to the gravitational field is taken to be [6]

$$\widetilde{\mathfrak{g}}^{00} = 1 + \varepsilon^2 h \tag{10}$$

and satisfies

$$\nabla^2 h = 0. \tag{11}$$

For compact spherical sources (and Schwarzschild black holes) the solution in the weak-field zone on and outside the EIH surfaces has the form

$$h = 4 \sum \frac{\widetilde{m}_A}{r_A}, \tag{12}$$

where the index A labels the sources in the system and the sum is over all A. The $\widetilde{m}_A$ are as yet to be determined functions of $\varepsilon\tau$ and $\delta\tau$ and $\mathbf{r}_A = \mathbf{x} - \mathbf{x}_A$, where the $\mathbf{x}_A$ are the coordinates of the Ath particle and are also functions of $\varepsilon\tau$ and $\delta\tau$. When the surface integrals in Eq. (6) with $\mu = 0$ are evaluated using this field, one finds that

$$\partial_{\varepsilon\tau} \widetilde{m}_A = 0 \quad \text{and} \quad \partial_{\delta\tau} \widetilde{m}_A = -\partial_{\delta\tau} R \widetilde{m}_A, \tag{13}$$

so that, to this order of accuracy,

$$\widetilde{m}_A = \frac{m_A}{R}, \tag{14}$$

where the m_A are constants. In a like manner one constructs the lowest order contribution to the scalar potential

$$A^0 = \varepsilon^2 \phi, \tag{15}$$

with ϕ satisfying

$$\nabla^2 \phi = 0. \tag{16}$$

When the spherically symmetric solution

$$\phi = \frac{\widetilde{q}_A}{r_A} \tag{17}$$

is substituted into the surface integral (7), one obtains the result that

$$\partial_{\varepsilon\tau} \widetilde{q}_A = \partial_{\delta\tau} \widetilde{q}_A = 0, \tag{18}$$

so that

$$\widetilde{q}_A = q_A, \tag{19}$$

where the q_A are constants.

To derive equations of motion from the surface integrals (6) we need the lowest order corrections to $\widetilde{\mathfrak{g}}^{0r}$, which are $O(\varepsilon^3)$ and $O(\varepsilon^2\delta)$, so we set

$$\widetilde{\mathfrak{g}}^{0r} = \varepsilon^3 h^r_\varepsilon + \varepsilon^2 \delta\, h^r_\delta. \tag{20}$$

There will of course be additional corrections in higher orders of approximation that are small compared to the first correction but large compared to the second, e.g., post-Newtonian corrections of order ε^5, but here we are only interested in the first-order effects of the expanding universe on Newtownian physics. In order to determine h^r_ε and h^r_δ, it is necessary to impose some kind of coordinate conditions, which we take to be

$$h^r_{\varepsilon},_r + \partial_{\varepsilon\tau}\phi = 0 \quad \text{and} \quad h^r_{\delta},_r + \sum \widetilde{m}_A \partial_{\delta\tau}\left(\frac{1}{r_A}\right) = 0. \tag{21}$$

As a consequence, the field equations determining h^r_ε and h^r_δ are

$$\nabla^2 h^r_\varepsilon = 0 \quad \text{and} \quad \nabla^2 h^r_\delta = 0 \tag{22}$$

and have solutions

$$h^r_\varepsilon = 4\sum \frac{\widetilde{m}_A}{r_A} x^r_A,_{\varepsilon\tau} \quad \text{and} \quad h^r_\delta = 4\sum \frac{\widetilde{m}_A}{r_A} x^r_A,_{\delta\tau}\ . \tag{23}$$

When the above fields are used to evaluate the surface integrals in Eq. (6), one obtains the equations

$$m_A x^r_A,_{\varepsilon\tau\varepsilon\tau} = \frac{1}{R}\sum{}' (-q_A q_B + m_A m_B)\frac{r^r_{AB}}{{r_{AB}}^3} \tag{24}$$

and

$$m_A x^r_A,_{\varepsilon\tau\delta\tau} = 0, \tag{25}$$

where the prime on the sum indicates the sum is over all B not equal to A and $r^r_{AB} = x^r_A - x^r_B$. For two particles, charged or uncharged, moving in circular orbits about each other, these equations yield the results that

$$r^3\omega^2 = \frac{a}{R} \quad \text{and} \quad r\omega = b, \tag{26}$$

where r is the relative coordinate radius of the circle, ω is the angular frequency of the motion, and a and b are constants. As a consequence one has that

$$r = \frac{r_0}{R} \quad \text{and} \quad \omega = \omega_0 R, \tag{27}$$

where r_0 and ω_0 are constants. Thus we see that at this order of approximation, although electric and gravitational clocks remain synchronous, they do not measure the conformal time τ but rather, as I showed in A1, they measure the cosmic time t.

2.4 Thomson Scattering

Since my purpose here is to examine Thomson scattering in an expanding universe rather than the behavior of clocks, I will confine my attention to a single charged source and take the electromagnetic field used to evaluate $\Theta^{\mu\nu}$ in Eq. (6) to be a plane wave. Maxwell's equations (4) have the solution

$$\mathfrak{F}^{0s} = \delta^s_3 E_0 e^{i(\omega\varepsilon\tau - kx)} \tag{28}$$

for a slowly varying wave propogating in the x-direction with $(\omega\varepsilon)^2 - k^2 = 0$. For a wave whose spatial variation is large compared to the size of the EIH surface surrounding the charge, the surface integral equation (6) yields the equation of motion

$$m_1 x^r_{1,\varepsilon\tau\varepsilon\tau} = -\frac{1}{R} q E_0 e^{i\omega\varepsilon\tau}, \tag{29}$$

which has the solution

$$x^3_1 = \frac{1}{R}\frac{q_1}{m_1}\frac{1}{\omega^2} E_0 e^{i\omega\varepsilon\tau}. \tag{30}$$

It remains finally to compute the scattered field produced by the motion of our charge. One possible way to do this would be to introduce a model source term into the right-hand side of the first of Eqs. (4) whose motion is characterized by x^3_1 above. However, in keeping with the EIH philosophy of not specifically introducing sources into the field equations and because the EIH procedure makes it unnecessary, we will proceed by constructing a radiation solution whose inner expansion matches on to the outer expansion of the near field ϕ in Eq. (17) with r_A computed using x^3_1 above. This outer expansion in inverse powers of r is given by

$$\begin{aligned}\phi &= \frac{q_1}{\left(r^2 + r_1^2 - 2rr_1\cos\theta\right)^{1/2}} \\ &= \frac{q_1}{r} + \frac{q_1}{r^2} r_1 \cos\theta + O(1/r^3).\end{aligned} \tag{31}$$

The corresponding outer dipole field is given by

$$\phi = \frac{a}{\varepsilon r} + \left\{\frac{W'(\varepsilon(\tau - r))}{\varepsilon r} + \frac{W(\varepsilon(\tau - r))}{(\varepsilon r)^2}\right\}\cos\theta, \tag{32}$$

whose inner expansion is

$$\phi = \frac{a}{\varepsilon r} + \frac{W(\varepsilon\tau)}{(\varepsilon r)^2} + O(1), \tag{33}$$

where a and $W(\varepsilon\tau)$ are to be determined by matching. Comparing Eqs. (31) and (33), we see that

$$a = \varepsilon q_1 \quad \text{and} \quad W(\varepsilon\tau) = \varepsilon^2 q_1 r_1. \tag{34}$$

The gauge condition $A^\mu{}_{,\mu} = 0$ used to derive the wave equations for A^μ allows us to derive an expression for the vector potential part of A^μ corresponding to ϕ given by

$$A_3 = -\frac{\varepsilon q_1 r_{1,\varepsilon\tau}}{r}. \tag{35}$$

With the above expressions for ϕ and A_3 we can now calculate $F_{\mu\nu}$, which in turn allows us to determine the Poynting vector $S^r = T^{0r}$ to be

$$S^r = \frac{1}{4\pi}\frac{q_1^2}{r^2}\left\{(1+n_3)n_1, (1+n_3)n_2, -(n_1^2+n_2^2)\right\}(r_{1,\tau\tau})^2 + O\left(\frac{1}{r^3}\right). \tag{36}$$

where n_r is a unit vector. To calculate the scattering crosssection, we need to calculate the total integrated flux f by integrating S^r over a sphere surrounding the charge. In order that observers at different times integrate over spheres of the same physical radius as measured by local systems, we see from Eq. (27) that the element of surface area must be taken to be $dS = (Rr)^2\, d\Omega$. As a consequence the total flux is calculated to be

$$f = \frac{2}{3} q_1^2 (r_{1,\tau\tau})^2 = \frac{2}{3}\left(\frac{q_1^2}{m_1} E_0\right)^2 . \tag{37}$$

The Thomson scattering cross section σ_τ, equal to f divided by the incoming flux $E_0^2/4\pi$, is thus given by the usual expression

$$\sigma_T = \frac{8\pi}{3}\left(\frac{q_1^2}{m_1}\right)^2 , \tag{38}$$

and so does not change with cosmic time.

REFERENCES

[1] C. W. Misner, K. S. Thorne, and J. A. Wheeler, *Gravitation*, (W. H. Freeman & Co., San Francisco, 1973).

[2] J. L. Anderson, *Phys. Rev. Lett.* **75**, 3602 (1995).

[3] A. Einstein, L. Infeld, and B. Hoffmann, *Ann. Math.*, Ser. 2 **39**, 65 (1939); A. Einstein and L. Infeld, *Ann. Math.*, Ser 2 **40**, 455 (1940); A. Einstein and L. Infeld, *Can. J. Math.* **3**, 209 (1949).

[4] L. D. Landau and E. M. Lifshitz, *The Classical Theory of Fields* (Pergamon Press, Oxford, 1985), 4th ed., p. 282.

[5] For a more detailed anaysis, see J. N. Goldberg, *Phys. Rev.* **89**, 263 (1953).

[6] For more details, see J. L. Anderson, *Phys. Rev.* **D36**, 2301 (1987) and J. L. Anderson, *Phys. Rev.* **D56**, 4675.

3

Geometrical Formulation of Quantum Mechanics

Abhay Ashtekar
Troy A. Schilling

ABSTRACT States of a quantum mechanical system are represented by rays in a complex Hilbert space. The space of rays has, naturally, the structure of a Kähler manifold. This leads to a geometrical formulation of the postulates of quantum mechanics which, although equivalent to the standard algebraic formulation, has a very different appearance. In particular, states are now represented by points of a symplectic manifold (which happens to have in addition a compatible Riemannian metric), observables are represented by certain real-valued functions on this space, and the Schrödinger evolution is captured by the symplectic flow generated by a Hamiltonian function. There is thus a remarkable similarity with the standard symplectic formulation of classical mechanics. Features—such as uncertainties and state vector reductions—which are specific to quantum mechanics can also be formulated geometrically but now refer to the Riemannian metric—a structure which is absent in classical mechanics. The geometrical formulation sheds considerable light on a number of issues such as the second quantization procedure, the role of coherent states in semiclassical considerations, and the WKB approximation. More importantly, it suggests generalizations of quantum mechanics. The simplest among these are equivalent to the dynamical generalizations that have appeared in the literature. The geometrical reformulation provides a unified framework to discuss these and to correct a misconception. Finally, it also suggests directions in which more radical generalizations may be found.

3.1 Introduction

Quantum mechanics is probably the most successful scientific theory ever invented. It has an astonishing range of applications—from quarks and leptons to neutron stars and white dwarfs—and the accuracy with which its underlying ideas have been tested is equally impressive. Yet, from its very inception, prominent physicists have expressed deep reservations about its conceptual foundations, and leading figures continue to argue that it is incomplete in its core. Time and again, attempts have been made to extend it in a nontrivial fashion. Some of these proposals have been phenomenological (see, for example, [1, 2, 3]), aimed at providing a "mechanism" for the state reduction process. Some have been more radical, e.g., invoking hidden variables (see, for example, [4]). Yet others involve nonlinear generalizations of the Schrödinger equation [5, 6, 7]. Deep discomfort has been expressed at the tension between objective descriptions of happenings provided by the space time geometry of special relativity and the quantum measurement theory [8, 9]. Further conceptual issues arise when one brings general relativity into picture, issues that go under the heading of "problem of time" in quantum gravity [10, 11, 12]. Thus, while there is universal agreement that quantum mechanics is an astonishingly powerful working tool, in the "foundation of physics circles" there has also been a strong sentiment that sooner or later one would be forced to generalize it in a profound fashion [13, 14, 15].

It is often the case that while an existing theory admits a number of equivalent descriptions, one of them suggests generalizations more readily than others. Furthermore, typically, this description is not the most familiar one, i.e., not the one that seems simplest from the limited perspective of the existing theory. An example is provided by Cartan's formulation of Newtonian gravity. While it played no role in the invention of the theory (it came some two and a half centuries later!) at a conceptual level, Cartan's framework provides a deeper understanding of Newtonian gravity and its relation to general relativity. A much more striking example is Minkowski's geometric reformulation of special relativity. His emphasis on hyperbolic geometry seemed abstract and abstruse at first; at the time, Einstein himself is said to have remarked that it made the subject incomprehensible to physicists. Yet, it proved to be an essential stepping stone to general relativity.

The purpose of this article is to present, in this spirit, a reformulation of the mathematical framework underlying standard quantum mechanics (and quantum field theory). The strength of the framework is that it is extremely natural from a geometric perspective and succinctly illuminates the essential difference between classical and quantum mechanics. It has already clarified certain issues related to the second quantization procedure and semiclassical approximations [16]. It also serves to unify in a coherent fashion a number of proposed generalizations of quantum mechanics; in particular, we will see that generalizations that were believed to be distinct (and even incompatible) are in fact closely related. More importantly, this reformulation may well lead to viable generalizations of quantum mechanics which are more profound than the ones considered so far. Finally, our experience from seminars and discussions has shown that ideas underlying this reformulation lie close to the heart of geometrically oriented physicists. It is therefore surprising that the framework is not widely known among relativists. We are particularly happy to be able to rectify this situation in this volume honoring Engelbert and hope that the role played by the Kähler geometry, in particular, will delight him.

Let us begin by comparing the standard frameworks underlying classical and quantum mechanics. The classical description is *geometrical:* States are represented by points of a symplectic manifold Γ, the phase space. The space of observables consists of the (smooth) real-valued functions on this manifold. The (ideal) measurement of an observable f in a state $p \in \Gamma$ yields simply the value

$f(p)$ at the point p; the state is left undisturbed. These outcomes occur with complete certainty. The space of observables is naturally endowed with the structure of a commutative, associative algebra, the product being given simply by pointwise multiplication. Thanks to the symplectic structure, it also inherits a Lie bracket—the Poisson bracket. Finally to each observable f is associated a vector field X_f called the Hamiltonian vector field of f. Thus, each observable generates a flow on Γ. Dynamics is determined by a preferred observable, the Hamiltonian H; the flow generated by X_H describes the time evolution of the system.

The arena for quantum mechanics, on the other hand, is a Hilbert space $\mathcal{H}$. States of the system now correspond to rays in $\mathcal{H}$, and the observables are represented by selfadjoint linear operators on $\mathcal{H}$. As in the classical description, the space of observables is a real vector space equipped with with two algebraic structures. First, we have the the Jordan product—i.e., the anticommutator—which is commutative but now fails to be associative. Second, we have $(1/2i$ times) the commutator bracket, which endows the space of observables with the structure of a Lie algebra. Measurement theory, on the other hand, is strikingly different. In the textbook description based on the Copenhagen interpretation, the (ideal) measurement of an observable $\hat{A}$ in a state $\Psi \in \mathcal{H}$ yields an *eigenvalue* of $\hat{A}$ and, immediately after the measurement, the state is thrown into the corresponding eigenstate. However, the specific outcome can only be predicted probabilistically. As in the classical theory, each observable $\hat{A}$ gives rise to a flow on the state space. But now, the flow is generated by the 1-parameter group $\exp(i\hat{A}t)$ and respects the linearity of $\mathcal{H}$. Dynamics is again dictated by a preferred observable, the Hamiltonian operator $\hat{H}$.

Clearly, the two descriptions have several points in common. However, there is also a striking difference: While the classical framework is *geometric and nonlinear*, the quantum description is intrinsically *algebraic and linear*. Indeed, the emphasis on the underlying linearity is so strong that none of the standard textbook postulates of quantum mechanics can be stated without reference to the linear structure of $\mathcal{H}$.

From a general perspective, this difference seems quite surprising. For linear structures in physics generally arise as approximations to the more accurate nonlinear ones. Thus, for example, we often encounter nonlinear equations which correctly capture a physical situation. But, typically, they are technically difficult to work with and we probe properties of their solutions through linearization. In the present context, on the other hand, it is the deeper, more correct theory that is linear and the nonlinear, geometric, classical framework is to arise as a suitable limiting case.

However, deeper reflection shows that quantum mechanics is in fact not as linear as it is advertised to be. For the space of physical states is *not* the Hilbert space $\mathcal{H}$ but the space of rays in it, i.e., the *projective* Hilbert space $\mathcal{P}$. And $\mathcal{P}$ is a genuine, nonlinear manifold. Furthermore, it turns out that the Hermitian inner product of $\mathcal{H}$ naturally endows $\mathcal{P}$ with the structure of a Kähler manifold. Thus, in particular, like the classical state space Γ, the correct space of quantum states, $\mathcal{P}$, is a symplectic manifold! We will therefore refer to $\mathcal{P}$ as the *quantum phase space*. Given any selfadjoint operator $\hat{H}$, we can take its expectation value to obtain a real function on $\mathcal{H}$. It is easy to verify that this function admits an unambiguous projection h to the projective Hilbert space $\mathcal{P}$. Recall, now, that every phase space function gives rise to a flow through its Hamiltonian vector field. What then is the interpretation of the flow X_h? It turns out [17] to be exactly the (projection to $\mathcal{P}$ of the) flow defined by the Schrödinger equation (on $\mathcal{H}$) of the quantum theory. Thus, Schrödinger evolution is precisely the Hamiltonian flow on the quantum phase space!

As we will see, the interplay between the classical and quantum ideas stretches much further. The overall picture can be summarized as follows. Classical phase spaces Γ are, in general, equipped only with a symplectic structure. Quantum phase spaces, $\mathcal{P}$, on the other hand, come with an additional structure, the Riemannian metric provided by the Kähler structure. Roughly speaking, features of quantum mechanics which have direct classical analogues refer only to the symplectic structure. On the other hand, features—such as quantum uncertainties and state vector reduction in a measurement process—refer also the Riemannian metric. This neat division lies at the heart of the structural similarities and differences between the (mathematical frameworks underlying the) two theories.

Section 3.2 summarizes this geometrical reformulation of standard quantum mechanics. We begin in 3.2.1 by showing that the quantum Hilbert space can be regarded as a (linear) Kähler space and discuss the roles played by the symplectic structure and the Kähler metric. In 3.2.2, we show that one can naturally arrive at the quantum state space $\mathcal{P}$ by using the Bergmann-Dirac theory of constrained systems. (This method of constructing the quantum phase space will turn out to be especially convenient in Section 3.3 while analyzing the relation between various generalizations of quantum mechanics.) Section 3.2.3 provides a self-contained treatment of the various issues related to observables—associated algebraic structures, quantum uncertainty relations and measurement theory—in an intrinsically geometric fashion. These results are collected in Section 3.2.4 to obtain a geometric formulation of the postulates of quantum mechanics, *a formulation that makes no reference to the Hilbert space $\mathcal{H}$ or the associated linear structures*. In practical applications, except while dealing with simple cases such as spin systems, the underlying quantum phase space $\mathcal{P}$ is *infinite*-dimensional (since it comes from an infinite-dimensional Hilbert space $\mathcal{H}$). *Our mathematical discussion encompasses this case.* Also, in the discussion of measurement theory, we allow for the possibility that observables may have continuous spectra.

In Section 3.3, we consider possible generalizations of quantum mechanics. These generalizations can occur in two distinct ways. First, we can retain the original kinematic structure but allow more general dynamics, for example, by replacing the Schrödinger equation by a suitable nonlinear one (see [5]). In Section 3.3.1 we show that the geometrical reformulation of Section 3.2 naturally suggests a class of such extensions which encompasses those proposed by Birula and Mycielski [5] and by Weinberg [7]. In Section 3.3.2 we consider the possibility of more radical extensions in which the kinematical setup itself is changed. Although (to our knowledge) there do not exist interesting proposals of this type, such generalizations would be much more interesting. In particular, it is sometimes argued that the linear structure underlying quantum mechanics would have to be sacrificed in a subtle but essential way to obtain a satisfactory quantum theory of gravity and/or to cope satisfactorily with the "measurement problem" [13]. To implement such ideas, the underlying kinematic structure will have to be altered. A first step in this direction is to obtain a useful *characterization* of the kinematical framework of standard quantum mechanics. Section 3.3.2 provides a "reconstruction theorem" which singles out quantum mechanics from its plausible generalizations. The theorem provides powerful guidelines: it spells out directions along which one can proceed to obtain a genuine extension.

Section 3.4 is devoted to semiclassical issues. Consider a simple mechanical system, such as a particle in $\mathbb{R}^3$. In this case, the classical phase space Γ is six-dimensional while the quantum phase space $\mathcal{P}$ is infinite-dimensional. Is there a relation between the two? In Section 3.4.1 we show that the answer is in the affirmative: $\mathcal{P}$ is a bundle over Γ. Furthermore, the bundle is trivial. Thus, through each quantum state $p \in \mathcal{P}$, there is a cross section, i.e., a copy of Γ. It turns out that the quantum states that lie on any one cross section are precisely the *generalized coherent states* [18, 20, 21]. In the remainder of Section 3.4, we use this interplay between $\mathcal{P}$ and Γ to discuss the relation between classical and quantum dynamics. Section 3.4.2 is devoted to the correspondence in terms of Ehrenfest's theorem, while 3.4.3 discusses the problem along the lines of the WKB approximation. Somewhat interestingly, it turns out that WKB dynamics is an example of generalized dynamics of the Weinberg type [7].

Our conventions are as follows. If the manifold under consideration is infinite-dimensional, we will assume that it is a Hilbert manifold. (Projective Hilbert spaces are naturally endowed with this structure; see, for example, [16].) Riemannian metrics and symplectic structures on these manifolds will be assumed to be everywhere defined, smooth, strongly nondegenerate fields. (Thus, they define *isomorphisms* between the tangent and cotangent spaces at each point). In detailed calculations we will often use the abstract index notation of Penrose's [22, 23]. Note that, in spite of the appearance of indices, this notation is well defined also on infinite-dimensional manifolds. (For example, if V^a denotes a contravariant vector field; the subscript a does not refer to its components but is only a label telling us that V is a specific type of tensor field, namely a contravariant vector field. Similarly, $V^a\omega_a$ is the function obtained by the action of the 1-form ω on the contravariant vector field V.) Finally, due to space limitation, we have not included detailed proofs of several technical assertions; they can be found in [16]. Our aim here is only to provide a thorough overview of the subject.

This work was intended to be an extension of a paper by Kibble [17] which pointed out that the Schrödinger evolution can be regarded as an Hamiltonian flow on $\mathcal{H}$. However, after completing this work, we learned that many of the results contained in Sections 3.2 and 3.3.2 were obtained independently by others (although the viewpoints and technical proofs are often distinct.) In 1985, Heslot [24] observed that quantum mechanics admits a symplectic formulation in which the phase space is the projective Hilbert space. That discussion was, however, restricted to the finite-dimensional case and did not include a discussion of the role of the metric, probabilistic interpretation and quantum uncertainties. Anandan and Aharonov [25] rediscovered some of these results and also discussed some of the probabilistic aspects. This work was also restricted to finite-dimensional systems and focused on the issue of evolution. Similar observations were made by Gibbons [26], who also discussed density matrices (which are not considered here) and raised the issue of characterization of quantum mechanics (which is resolved in Section 3.4.2). An essentially complete treatment of the finite-dimensional case was given by Hughston [27]. (This work was done in parallel to ours. However, it also contains some proposals for mechanisms for state reduction [28] which are not discussed here.) The only references (to our knowledge) which treat the infinite-dimensional case are [29, 30], which also discuss the issue of characterization of standard quantum mechanics. Finally, since the geometric structures that arise here are so natural, it is very possible that they were independently discovered by other authors that we are not aware of.

3.2 Geometric Formulation of Quantum Mechanics

The goal of this section is to show that quantum mechanics can be formulated in an intrinsically geometric fashion, without any reference to a Hilbert space or the associated linear structure. We will assume that the reader is familiar with basic symplectic geometry.

3.2.1 The Hilbert Space as a Kähler Space

Let us begin with the standard Hilbert space formulation of quantum mechanics. In this subsection we will view the Hilbert space $\mathcal{H}$ as a Kähler space and examine the role played by the associated symplectic structure and the Riemannian metric. This discussion will serve as a stepping stone to the analysis of the quantum phase space $\mathcal{P}$ in Section 3.2.2.

The similarities between classical and quantum mechanics can be put in a much more suggestive form with an alternative, but equivalent, description of the Hilbert space. We view $\mathcal{H}$ as a *real* vector space equipped with a *complex structure* J. The complex structure is a preferred linear operator which represents multiplication by i; hence $J^2 = -I$. Initially, this change of notation seems rather trivial; the element which is typically written $(a + ib)\Psi$ is now denoted $a\Psi + bJ\Psi$ and (external) multiplication of vectors by complex numbers is not permitted. However, this slight change of viewpoint will come with a reward—a symplectic formulation of quantum mechanics.

Since $\mathcal{H}$ is now viewed as a real vector space, the Hermitian inner product is slightly unnatural. We therefore decompose it into real and imaginary parts,

$$\langle \Phi, \Psi \rangle := \frac{1}{2\hbar} G(\Phi, \Psi) + \frac{i}{2\hbar} \Omega(\Phi, \Psi). \tag{1}$$

(The reason for the factors of $1/2\hbar$ will become clear shortly.) Properties of the Hermitian inner product imply that G is a positive definite, real inner product and that Ω is a symplectic form, both of which are strongly nondegenerate. Moreover, since $\langle \Phi, J\Psi \rangle = i\langle \Phi, \Psi \rangle$, one immediately observes that the metric, symplectic structure, and complex structure are related as

$$G(\Phi, \Psi) = \Omega(\Phi, J\Psi). \tag{2}$$

That is, the triple (J, G, Ω) equips $\mathcal{H}$ with the structure of a *Kähler space*. Therefore, every Hilbert space may be naturally viewed as a Kähler space.

Next, by use of the canonical identification of the tangent space (at any point of $\mathcal{H}$) with $\mathcal{H}$ itself, Ω is naturally extended to a strongly nondegenerate, closed, differential 2-form $\mathcal{H}$, which we will denote also by Ω. Any Hilbert space is therefore naturally viewed as the simplest sort of symplectic manifold, i.e., a *phase space*. The inverse of Ω may be used to define Poisson brackets and Hamiltonian vector fields. As we are about to see, these notions are just as relevant in quantum mechanics as in classical mechanics.

The symplectic form.

In classical mechanics, observables are real-valued functions, and to each such function is associated a corresponding Hamiltonian vector field. In quantum mechanics, on the other hand, the observables themselves may be viewed as vector fields, since linear operators associate a vector to each element of the Hilbert space. However, the Schrödinger equation, which in our language is written as $\dot{\Psi} = -\frac{1}{\hbar} J\hat{H}\Psi$, motivates us to associate to each quantum observable $\hat{F}$ the vector field

$$Y_{\hat{F}}(\Psi) := -\frac{1}{\hbar} J\hat{F}\Psi. \tag{3}$$

This *Schrödinger vector field* is defined so that the time evolution of the system corresponds to the flow along the Schrödinger vector field associated to the Hamiltonian operator. (Note that if the Hamiltonian is unbounded, it is only densely defined and so is the vector field. The (unitary) flow, however, is defined on all of $\mathcal{H}$. See, for example, [31].)

Natural questions immediately arise. Let $\hat{F}$ be any bounded, selfadjoint operator on $\mathcal{H}$. Is the corresponding vector field, $Y_{\hat{F}}$, Hamiltonian on the symplectic space $(\mathcal{H}, \Omega)$? If so, what is the real-valued function which generates this vector field? What is the physical meaning of the Poisson bracket? In particular, how is it related to the commutator Lie algebra?

The answers to these questions are remarkably simple. As we know from standard quantum mechanics, $\hat{F}$ generates a 1-parameter family of unitary mappings on $\mathcal{H}$. By definition, $Y_{\hat{F}}$ is the generator of this 1-parameter family and therefore preserves both the metric G and symplectic form Ω. It is therefore locally Hamiltonian, and, since $\mathcal{H}$ is a linear space, also globally Hamiltonian! In fact, the function which generates this Hamiltonian vector field is of physical interest; it is simply the expectation value of $\hat{F}$.

Let us see this explicitly. Denote by $F : \mathcal{H} \longrightarrow \mathbb{R}$ the expectation value function

$$F(\Psi) := \langle \Psi, \hat{F}\Psi \rangle = \frac{1}{2\hbar} G(\Psi, \hat{F}\Psi). \tag{4}$$

We will continue to use this notation; expectation value functions will be denoted by simply "un-hatting" the corresponding operators. Now, if η is any tangent vector at Ψ, then[1]

$$\begin{aligned}(\mathrm{d}F)(\eta) &= \frac{\mathrm{d}}{\mathrm{d}t}\langle \Psi + t\eta \hat{F}(\Psi + t\eta)\rangle\big|_{t=0} = \langle \Psi, \hat{F}\eta\rangle + \langle \eta, \hat{F}\Psi\rangle \\ &= \frac{1}{\hbar} G(\hat{F}\Psi, \eta) = \Omega(Y_{\hat{F}}, \eta) = \left(i_{Y_{\hat{F}}}\Omega\right)(\eta),\end{aligned} \tag{5}$$

where we have used the selfadjointness of $\hat{F}$, Eq. (2) and the definition of $Y_{\hat{F}}$. Therefore, the Hamiltonian vector field X_F generated by the expectation value function F coincides with the Schrödinger vector field $Y_{\hat{F}}$ associated to $\hat{F}$. As a particular consequence, the time evolution of any quantum mechanical system may be written in terms of Hamilton's equation of classical mechanics; the Hamiltonian *function* is simply the expectation value of the Hamiltonian *operator*. *Schrödinger's equation is Hamilton's equation in disguise!*

Next, let $\hat{F}$ and $\hat{K}$ be two quantum observables, and denote by F and K the respective expectation value functions. It is natural to ask whether the Poisson bracket of F and K is related in a simple manner to an algebraic operation involving the original operators. Performing a calculation as simple as Eq. (5), one finds that

$$\{F, K\}_\Omega = \Omega(X_F, X_K) = \left\langle \frac{1}{\mathrm{i}\hbar}[\hat{F}, \hat{K}] \right\rangle. \tag{6}$$

Notice that the quantity inside the brackets on the right side of Eq. (6) is precisely the quantum Lie bracket of $\hat{F}$ and $\hat{K}$. The algebraic operation on the expectation value functions, which is induced by the commutator bracket is *exactly* a Poisson bracket! Note that this is *not* Dirac's correspondence principle; the Poisson bracket here is the quantum one, determined by the imaginary part of the Hermitian inner product.

The basic features of the classical formalism appear also in quantum mechanics. The Hilbert space, as a real vector space, is equipped with a symplectic form. To each quantum observable is associated a real-valued function on $\mathcal{H}$, and the time evolution is determined by the Hamiltonian vector field associated to a preferred function. Moreover, the Lie bracket of two quantum observables corresponds precisely to the Poisson bracket of the corresponding functions.

---

[1] With our conventions for symplectic geometry, the Hamiltonian vector field X_f generated by the function f satisfies the equation $i_{X_f}\Omega = \mathrm{d}f$, and the Poisson bracket is defined by $\{f, g\} = \Omega(X_f, X_g)$.

Uncertainty and the real inner product.

Let us now examine the role played by the metric G. Clearly, G enables us to define a real inner product, $G(X_F, X_K)$, between any two Hamiltonian vector fields X_F and X_K. One may expect that this inner product is related to the Jordan product in much the same way that the symplectic form corresponds to the commutator Lie bracket. It is easy to verify that this expectation is correct. Operating just as in Eq. (6), we obtain

$$\{F, K\}_+ := \frac{\hbar}{2}G(X_F, X_K) = \left\langle \frac{1}{2}[\hat{F}, \hat{K}]_+ \right\rangle. \tag{7}$$

The operation defined by the first equation above will be called the *Riemann bracket* of F and K. Up to the factor of $\hbar/2$, the Riemann bracket of F and K is simply given by the (real) inner product of their Hamiltonian vector fields, and corresponds precisely to the Jordan product of the respective operators.

Since the classical phase space is, in general, not equipped with a Riemannian metric, the Riemann product does *not* have an analogue in the classical formalism; it does, however, admit a physical interpretation. In order to see this, note that the uncertainty of the observable $\hat{F}$ at a state with unit norm is given by

$$(\Delta\hat{F})^2 = \langle \hat{F}^2 \rangle - \langle \hat{F} \rangle^2 = \{F, F\}_+ - F^2. \tag{8}$$

Thus, the uncertainty of an operator $\hat{F}$, when written in terms of the expectation value function F, involves the Riemann bracket. Moreover, this expression for the uncertainty is quite simple.

In fact Heisenberg's famous uncertainty relation also assumes a nice form when expressed in terms of the expectation value functions. It is very well known that the familiar uncertainty relation between two quantum observables may be written in a slightly stronger form (see, for example, [32]):

$$(\Delta\hat{F})^2(\Delta\hat{K})^2 \geq \left\langle \frac{1}{2i}[\hat{F}, \hat{K}] \right\rangle^2 + \left\langle \frac{1}{2}[\hat{F}_\perp, \hat{K}_\perp]_+ \right\rangle^2, \tag{9}$$

where $\hat{F}_\perp$ is the *nonlinear* operator defined by

$$\hat{F}_\perp(\Psi) := \hat{F}(\Psi) - F(\Psi),$$

so that $\hat{F}_\perp(\Psi)$ is orthogonal to Ψ if $\|\Psi\| = 1$.

Using the above results, we may immediately rewrite Eq. (9) in the form

$$(\Delta\hat{F})^2(\Delta\hat{K})^2 \geq \left(\frac{\hbar}{2}\{F, K\}_\Omega\right)^2 + \left(\{F, K\}_+ - FK\right)^2. \tag{10}$$

Incidentally, the last expression in Eq. (10) may be interpreted as the "quantum covariance" of $\hat{F}$ and $\hat{K}$; see [16] for an explanation.

3.2.2 *The Quantum Phase Space*

We have seen that to each quantum observable is associated a smooth real-valued (expectation value) function on the Hilbert space. Further, the familiar operations involving quantum operators correspond to simple "classical-looking" operations on the corresponding functions. These observations suggest a formulation of standard quantum mechanics in the language of classical mechanics. However, there are two difficulties. First, although the Hilbert space is a symplectic space, because two state vectors related by multiplication by any complex number define the same state, it is *not* the space of physical states, i.e., the quantum analog of the classical phase space. Second, the description of the measurement process in a manner intrinsic to the Kähler structure on $\mathcal{H}$ turns out not to be natural.

The true state space of the quantum system is the space of *rays* in the Hilbert space, i.e., the *projective* Hilbert space, which we shall denote $\mathcal{P}$. It should not be surprising that $\mathcal{P}$ is a Kähler manifold, and hence, in particular, a symplectic manifold. After all, for the special case in which $\mathcal{H}$ is $\mathbb{C}^{n+1}$, $\mathcal{P}$ is the complex projective space CP^n—the archetypical Kähler manifold. In this section, we present a particularly useful description of the projective Hilbert space which is valid for the *infinite-dimensional case* and which illuminates the role of its symplectic structure. These developments will enable us to handle the above complications and allow an elegant geometric formulation of quantum mechanics.

Gauge reduction.

The standard strategy to handle the ambiguity of the state vector is to consider only those elements of the Hilbert space which are normalized to unity. We will adopt this approach by insisting that the only physically relevant portion of the Hilbert space is that on which the *constraint function*,

$$C(\Psi) := \langle \Psi, \Psi \rangle - 1 = \frac{1}{2\hbar} G(\Psi, \Psi) - 1, \tag{11}$$

vanishes. The attitude adopted here is one in which the the *constraint surface*—the unit sphere, with respect to the Hermitian inner product—is the only portion of the Hilbert space which is accessible to the system. The rest of the Hilbert space is often quite convenient, but is viewed as an artificial element of the formalism.

Let us consider the above restriction from the point of view of the Bergmann-Dirac theory of constrained systems (see, e.g., [33]). In other words, we will pretend, for a moment, that we are dealing with a *classical* theory with the constraint $C = 0$. First notice that since the time evolution preserves the constraint surface, no further (secondary) constraints arise. Since we have only a single constraint, it is trivially of *first-class* in Dirac's terminology; i.e., the constraint generates a motion which preserves the constraint surface ($\pounds_{X_C} C = \{C, C\}_\Omega = 0$).

Recall that to every first-class constraint on a Hamiltonian system is associated a gauge degree of freedom; the associated gauge transformations are defined by the flow along the Hamiltonian vector field generated by the constraint function.

In our case, the gauge directions are simply given by

$$X^a_C = \Omega^{ab} D_b C = \frac{1}{\hbar}\Omega^{ab}\Psi_b = -\frac{1}{\hbar}J^a{}_b\Psi^b, \tag{12}$$

where D_a denotes the Levi-Cività derivative operator. For later convenience, let us define

$$\mathcal{J}^a := \hbar X_C{}^a\big|_S = -J^a{}_b\Psi^b\big|_S. \tag{13}$$

Notice that $\mathcal{J}^a$ is the generator of phase rotations on S. Therefore, the gauge transformations generated by the constraint are exactly what they ought to be; they represent the arbitrariness in our choice of phase!

Thus, we now see the relevance of the description in terms of constrained Hamiltonian systems. By taking the quotient of the constraint surface of any constrained system by the action of the gauge transformations, one obtains the true phase space of the system—often called the *reduced phase space*. The projective Hilbert space may therefore be interpreted as the "reduced phase space" of our constrained Hamiltonian system. In order to emphasize both its physical role and geometric structure, we will refer to the projective Hilbert space $\mathcal{P}$ as the *quantum phase space*. One can explicitly show that if $\mathcal{H}$ is infinite-dimensional, $\mathcal{P}$ is an infinite-dimensional Hilbert manifold [16].

As the terminology suggests, any reduced phase space is equipped with a natural symplectic structure. This fact may be seen as follows. Denote by $i : S \to \mathcal{H}$ and $\pi : S \to \mathcal{P}$ the inclusion mapping and projection to the quantum phase space, respectively. By restricting the symplectic structure Ω to the constraint surface, one obtains a closed 2-form $i^*\Omega$ on S. This 2-form is degenerate, but only in the gauge direction. Fortunately, since gauge transformations are defined by the Hamiltonian vector field generated by the constraint function, $i^*\Omega$ is constant along its directions of degeneracy. As a result, there exists a symplectic form ω on $\mathcal{P}$ whose pull-back via π agrees precisely with $i^*\Omega$. This is a standard construction in the theory of systems with first-class constraints; the only novelty here lies in its application to ordinary quantum mechanics. Finally, we note that this result applies to the typical case of interest, in which the original Hilbert space is infinite-dimensional [16]. The symplectic structure ω is then a smooth, strongly nondegenerate field (i.e., defines an *isomorphism* from the tangent space to the cotangent space at each point).

Before discussing the geometry of the quantum phase space, we should point out that the viewpoint adopted in this section in fact generalizes quantum mechanics, but in a very trivial way. In Section 3.2.1 we observed that each quantum observable defines a real-valued function on the entire Hilbert space, and that the quantum evolution is given by the Hamiltonian flow defined by a preferred function. In viewing a quantum system as a constrained Hamiltonian system, we must concede that it is only the constraint surface S that is of physical relevance. In particular, the restriction $F|_S$ of the expectation value function F contains all gauge-invariant information about the observable; one may extend F off the constraint surface in any desirable manner without affecting the corresponding flow on the projective space. The particular extensions defined by expectation values of (bounded) self-adjoint operators may be viewed as mere convention. We will make use of this point in Section 3.3.1.

Symplectic geometry.

Our method of arriving at the quantum phase space $\mathcal{P}$ by the reduced phase space construction of constrained systems immediately suggests further definitions and constructions.

Recall that to each bounded, selfadjoint operator $\hat{F}$ on $\mathcal{H}$, we have associated the function $F(\Psi) := \langle \Psi, \hat{F}\Psi \rangle$ on the Hilbert space. In fact, we may go a short step further. First, let us restrict the expectation value function F to the constraint surface, thereby obtaining the function $i^*F : S \to \mathbb{R}$. i^*F is clearly gauge invariant (i.e., independent of phase), and therefore defines the function $f : \mathcal{P} \to \mathbb{R}$ for which $\pi^* f = i^* F$. Therefore, to each quantum observable is associated a smooth, real-valued function on the quantum phase space. The functions obtained in this manner will represent the observables in the geometric formulation of quantum mechanics. Let us therefore make

Definition 3.2.1 *Let $f : \mathcal{P} \to \mathbb{R}$ be a smooth function on $\mathcal{P}$. If there exists a bounded, selfadjoint operator $\hat{F}$ on $\mathcal{H}$ for which $\pi^* f = \langle \hat{F} \rangle \big|_S$, then f is said to be an* observable function.

Note that we consider the set of quantum observables to consist of the *bounded* selfadjoint operators on $\mathcal{H}$. At first sight this appears to be a severe restriction. However, further reflection shows that it is not. In any actual experiment, one deals only with a finite range of relevant parameters and hence in practice one only measures observables of the type considered here. Thus, there is by definition a one-to-one correspondence between quantum observables and the observable functions on $\mathcal{P}$. As we will see below, the set of observable functions is a very small subset of the entire function space.

A natural question arises: What is the relationship between the Hamiltonian vector fields X_F (on $\mathcal{H}$) and X_f (on $\mathcal{P}$)? Given any point $\Psi \in S$, we may push-forward the vector $X_F\big|_\Psi$ to obtain a tangent vector at $\pi(\Psi)$. Since F is gauge invariant, it commutes with C; therefore

$$\pounds_{X_C} X_F = X_{\{F,C\}_\Omega} \equiv 0.$$

As a consequence, X_F is "constant along the integral curves of $\mathcal{J}$." Thus, by pushing-forward X_F at each point of S, one obtains a well-defined (smooth) vector field on all of $\mathcal{P}$. As is known to those familiar with the analysis of constrained systems, this vector field is also Hamiltonian; in fact, it agrees precisely with X_f. The flow on $\mathcal{P}$, which is induced by the Schrödinger vector field of $\hat{F}$, corresponds exactly to the Hamiltonian flow determined by the observable function f.

Next, consider the Poisson bracket $\{,\}_\omega$ defined by the reduced symplectic structure ω. Let $F, K : \mathcal{H} \to \mathbb{R}$ be expectation value functions of two quantum observables and denote by $f, k : \mathcal{P} \to \mathbb{R}$ the corresponding observable functions on $\mathcal{P}$. As a consequence of the above result,

$$\begin{aligned}\pi^*\{f,k\}_\omega &= \pi^*(\omega(X_f, X_k)) = \omega(\pi_* X_F, \pi_* X_K) \\ &= \Omega(X_F, X_K)\big|_S = \{F, K\}_\Omega\big|_S.\end{aligned} \tag{14}$$

Therefore, the Poisson bracket defined by ω *exactly* reflects the commutator bracket on the space of quantum observables.

In summary, to each quantum observable $\hat{F}$ is associated a real-valued function $f : \mathcal{P} \to \mathbb{R}$ on the quantum phase space. The Schrödinger vector field determined by $\hat{F}$ determines a flow on $\mathcal{P}$; this flow is generated by the Hamiltonian vector field associated to the observable function f. Further, the mapping $\hat{F} \mapsto f$ is one-to-one and respects the Lie algebraic structures provided by the commutator and Poisson bracket on $\mathcal{P}$, respectively.

3.2.3 Riemannian Geometry and Measurement Theory

Any quantum mechanical system may be described as an infinite-dimensional Hamiltonian system. However, the structure of the quantum phase space is much richer than that of classical mechanics. $\mathcal{P}$ is also equipped with a natural Riemannian metric. As we will see, the probabilistic features of quantum mechanics are conveniently described by the Riemannian structure.

The quantum metric may be described in much the same way as the symplectic structure. The restriction i^*G of G to the unit sphere is a strongly nondegenerate Riemannian metric on S. Recall that the gauge generator $\mathcal{J}$ is (up to the constant factor of $\hbar$) the Schrödinger vector field associated to the identity operator. Since any Schrödinger vector field preserves the Hermitian inner product, it preserves both the symplectic structure Ω and the metric G. As a consequence, $\mathcal{J}$ is a Killing vector field on S;

$$£_{\mathcal{J}}(i^*G) \equiv 0. \tag{15}$$

Therefore, $\mathcal{P}$ may also be described as the *Killing reduction* [34] of S with respect to the Killing field $\mathcal{J}$.

A manifold which arises in this way is always equipped with a Riemannian metric of its own.[2] Although i^*G is "constant" on the integral curves of $\mathcal{J}$, it is not degenerate in that direction. However, by subtracting off the component in the direction of $\mathcal{J}$,

$$\tilde{g} := \left[G - \frac{1}{2\hbar}(\Psi \otimes \Psi + \mathcal{J} \otimes \mathcal{J})\right]\bigg|_S, \tag{16}$$

we obtain a symmetric tensor field which agrees with i^*G when acting on vectors orthogonal to $\mathcal{J}$, is constant along $\mathcal{J}$, and is degenerate only in the direction of $\mathcal{J}$. Therefore $\tilde{g}$ defines a *strongly nondegenerate Riemannian metric* g on $\mathcal{P}$. It is

---

[2]One must require that the integral curves of the Killing vector field do not come arbitrarily close to one another. This condition is satisfied in our case.

a simple matter to verify that g, when combined with the symplectic structure ω, equips the quantum phase space with the structure of a *Kähler manifold.*

Quantum Observables.

According to Definition 3.2.1, quantum mechanical observables may be represented by real-valued functions on the quantum phase space. Unfortunately, these functions have still been defined in terms of selfadjoint operators on the Hilbert space. Our goal, however, is a formulation of quantum mechanics which is intrinsic to the projective space; we wish to avoid any explicit reference to the underlying Hilbert space. We now explain how this deficiency may be overcome.

Since the Schrödinger vector field $Y_{\hat{F}} = X_F$ generates a one-parameter family of unitary transformations on $\mathcal{H}$, X_F preserves not only the symplectic structure Ω, but the metric G as well; X_F is also a *Killing vector field.* This fact also holds for the corresponding observable function f; the Hamiltonian vector field X_f associated to any observable function f is also a Killing vector field on $(\mathcal{P}, g)$. We will see that it is this property which characterizes the set of observable functions on the quantum phase space.

Let us begin by recalling a general property of Killing vector fields. Since the calculations are somewhat involved, we will now use Penrose's abstract index notation (which, as already pointed out, is meaningful also on infinite-dimensional Hilbert manifolds). Let X^α be any Killing vector field on $\mathcal{P}$. Then, by definition, $\nabla_\alpha X_\beta + \nabla_\beta X_\alpha = 0$, where ∇ denotes the (Levi-Cività) derivative operator associated to the metric g. Therefore, $K_{\alpha\beta} := \nabla_\alpha X_\beta$ is necessarily skew-symmetric. As one can easily verify, [35], K satisfies the identity

$$\nabla_\alpha K_{\beta\gamma} = R_{\gamma\beta\alpha}{}^{\delta} X_\delta. \tag{17}$$

(Our conventions are such that for any 1-form k_γ, $R_{\alpha\beta\gamma}{}^{\delta} k_\delta = (\nabla_\alpha \nabla_\beta - \nabla_\beta \nabla_\alpha) k_\gamma$.) As a consequence, the Killing vector field X is *completely determined* by its value and first covariant derivative at a single point. (See [36] or Appendix C of [35] for a discussion of this useful fact.)

Now suppose that the above Killing vector field is generated by the observable function f. (Below, it will be understood that $X = X_f$.) Since

$$\omega_\alpha{}^\gamma K_{\gamma\beta} = -\omega_\alpha{}^\gamma \nabla_\beta X_\gamma = \nabla_\alpha \nabla_\beta f, \tag{18}$$

K satisfies the additional property that $\omega_\alpha{}^\gamma K_{\gamma\beta}$ is symmetric. (It then defines a bounded, skew-selfadjoint operator on each tangent space.) By considering the coupled differential equations

$$\begin{aligned} c\nabla_\alpha f &= \omega_{\gamma\alpha} X^\gamma, \\ \nabla_\alpha X_\beta &= K_{\alpha\beta}, \\ \nabla_\alpha K_{\beta\gamma} &= R_{\gamma\beta\alpha}{}^{\delta} X_\delta, \end{aligned} \tag{19}$$

we see [16] that *any observable function is completely determined by its value and first two derivatives at a single point!* This fact motivates

Definition 3.2.2 *For each point $p \in \mathcal{P}$, let $\mathcal{S}_p$ consist of all triples, $(\lambda, X_\alpha, K_{\alpha\beta})$, where λ is a real number, X_α is a covector at p, and $K_{\alpha\beta}$ is a 2-form at p for which $\omega_\alpha{}^\gamma K_{\gamma\beta} = \omega_\beta{}^\gamma K_{\gamma\alpha}$. We call $\mathcal{S}_p$ the* algebra of symmetry data at *p.*

Thus any observable function f determines an element $(\lambda, X, K) \in \mathcal{S}_p$ (λ provides the value of f at p), and f is completely determined by this symmetry data. The algebra of quantum observables is then isomorphic to a subset of $\mathcal{S}_p$. The converse to this statement is provided by [16]

Theorem 3.2.1 *For any element (λ, X, K) of $\mathcal{S}_p$, there exists an observable function $f : \mathcal{P} \to \mathbb{R}$ such that $f(p) = \lambda$, $(X_f)_\alpha = X_\alpha$, and $\nabla_\alpha(X_f)_\beta = K_{\alpha\beta}$.*

Therefore, the space of observable functions is isomorphic to the algebra of symmetry data $\mathcal{S}_p$ for any $p \in \mathcal{P}$. We will utilize this result in Section 3.3.2. For the purposes of this section, however, the main use of the above theorem is the following. Recall that the Hamiltonian vector field of any observable function is a Killing vector field. Conversely, let f be any smooth, real-valued function on $\mathcal{P}$ for which X_f is also a Killing vector field. Of course, the value and first two covariant derivatives of f at $p \in \mathcal{P}$ determine an element of $\mathcal{S}_p$. Therefore, by Theorem 3.2.1, f is an observable function. This important result is expressed in

Corollary 3.2.1 *A smooth function $f : \mathcal{P} \to \mathbb{R}$ is observable if and only if its Hamiltonian vector field is also Killing.*

Therefore, as in classical mechanics, the space of quantum observables is isomorphic to the space of smooth functions on the phase space whose Hamiltonian vector fields are infinitesimal symmetries of the available structure. However, now the available structure includes not just the symplectic structure but also the metric. Hence, unlike in classical mechanics, this function space is an extremely small subset of the set of all smooth functions on $\mathcal{P}$. This should not be terribly surprising; as an example consider the finite-dimensional case, for which the function space is infinite-dimensional while the algebra of Hermitian operators is finite-dimensional.

Quantum uncertainty.

In analogy with our considerations of the symplectic geometry of $\mathcal{P}$, let us consider the Riemann bracket defined by the quantum metric g. We denote the g-Riemann bracket of $f, k : \mathcal{P} \to \mathbb{R}$ by

$$(f, k) := \frac{\hbar}{2} g(X_f, X_k) = \frac{\hbar}{2} (\nabla_\alpha f) g^{\alpha\beta} (\nabla_\beta k). \tag{20}$$

If f and k correspond to the expectation value functions $F, K : \mathcal{H} \to \mathbb{R}$, then, by Eq. (16),

$$\begin{aligned} \{F, K\}_+ &= \frac{\hbar}{2} G(X_F, X_K) \\ &= \frac{\hbar}{2} \tilde{g}(X_F, X_K) + \frac{1}{4} G(\mathcal{J}, X_F) G(\mathcal{J}, X_K) \\ &= \pi^*\big[(f, k) + fk\big], \end{aligned} \tag{21}$$

where we have used the fact that for any observable $\hat{F}$,

$$G(\mathcal{J}, X_F)|_\Psi = -G(J\Psi, X_F) = \Omega(X_F, \Psi) = \Psi \circ (\mathrm{d}F) = 2F. \qquad (22)$$

Therefore, the observable function which corresponds to the Jordan product of $\hat{F}$ and $\hat{K}$ is given not by the g-Riemann bracket of f and k, but by the quantity

$$\{f, k\}_+ := (f, k) + fk, \qquad (23)$$

where we have utilized a minor notational abuse to emphasize the relationship with Eq. (7). The above quantity will be called the *symmetric bracket* of f and k.

Note that the g-Riemann bracket of two observables, while not necessarily an observable itself, is a physically meaningful quantity; it is simply the quantum covariance function (see the comment following Eq. (10)). In particular, $(f, f)(p)$ is exactly the squared uncertainty of f at the quantum state labeled by p;

$$(\Delta f)^2(p) := (\Delta \hat{F})^2(\pi^{-1} p) = (f, f)(p). \qquad (24)$$

In order to obtain a feeling for the physical meaning of the quantum covariance, notice that the uncertainty relation of Eq. (10) assumes the form

$$(\Delta f)(\Delta k) \geq \left(\frac{\hbar}{2}\{f, k\}_\omega\right)^2 + (f, k)^2. \qquad (25)$$

As one can easily show, the standard uncertainty relation

$$(\Delta f)(\Delta k) \geq \left(\frac{\hbar}{2}\{f, k\}_\omega\right)^2 \qquad (26)$$

is saturated at the state $p \in \mathcal{P}$ if and only if $(f, k)(p) = 0$ *and* $X_f \propto j(X_k)$, where j is the complex structure on $\mathcal{P}$ (compatible with ω and g). The quantum covariance $(f, k)(p)$ therefore measures the "coherence" of the state p with respect to the observables f and k.

Let us conclude this subsection by reproducing a result of Anandan and Aharanov [25]. Suppose that the Hamiltonian operator of a quantum system is bounded and let h be the corresponding observable function. By definition of the Riemann bracket, the uncertainty of h is given by $(\Delta h)^2 = \frac{\hbar}{2} g(X_h, X_h)$. Therefore, apart from the constant coefficient, the uncertainty in the energy is exactly the length of the Hamiltonian vector field which generates the time evolution. Thus, the energy uncertainty can be thought of as "the speed with which the system moves through the quantum phase space"; during its evolution, the system passes quickly through regions where the energy uncertainty is large and spends more time in states where it is small.

The measurement process.

Some of the most significant aspects of quantum mechanics involve the measurement process and, without a complete geometric description of these issues, our program would be incomplete. In this subsection we will sketch the desired geometric description, including the case when the spectrum of the operator is continuous. A more complete discussion may be found in [16].

Let $\Psi_0 \in S$ be an arbitrary normalized element of the Hilbert space, and let $p_0 = \pi(\Psi_0)$ be its projection to $\mathcal{P}$. Of obvious interest, in the context of measurement, is the function $\tilde{\delta}_{\Psi_0} : \Psi \in S \mapsto |\langle\Psi_0, \Psi\rangle|^2$. Since $\tilde{\delta}_{\Psi_0}$ is independent of the phase of Ψ, it defines a function δ_{p_0} on $\mathcal{P}$ via

$$\delta_{p_0}(p) := \tilde{\delta}_{\Psi_0}(\pi^{-1} p) = \left|\langle\pi^{-1}(p_0), \pi^{-1}(p)\rangle\right|^2. \tag{27}$$

If the quantum system is in the state labeled by p_0 when a measurement is performed, the relevant quantum mechanical probability distribution is determined by the function δ_{p_0}. We therefore desire a description of this function which does not rely explicitly on the underlying Hilbert space. This is provided by

Theorem 3.2.2 *Given arbitrary points $p_0, p \in \mathcal{P}$ there exists a (closed) geodesic which passes through p_0 and p. Further, $d_{p_0}(p) = \cos^2\left(\sigma(p_0, p)/\sqrt{2\hbar}\right)$, where $\sigma(p_0, p)$ denotes the geodesic separation of p_0 and p.*

A few comments regarding Theorem 3.2.2 are in order. First, if $\pi^{-1}(p_0)$ and $\pi^{-1}(p)$ are nonorthogonal, then the above-mentioned geodesic is unique, up to reparameterization. Next, since all geodesics on $\mathcal{P}$ are *closed*, the geodesic distance $\sigma(p_0, p)$ is, strictly speaking, ill-defined. Due to the periodicity of the cosine function, however, the "transition amplitude function," d_{p_0}, is insensitive to this ambiguity. For the sake of precision, by $\sigma(p_0, p)$, we will mean the *minimal* geodesic distance separating p_0 and p.

Suppose one is dealing with the measurement of an operator $\hat{F}$ with discrete, nondegenerate spectrum, and let f be the corresponding observable function. Each eigenspace of $\hat{F}$ is one complex-dimensional, and therefore determines a single point of $\mathcal{P}$. Denote these eigenstates by p_i. Suppose the system is in the state labeled by the point p_0 when an ideal measurement of f is performed. We know that the system will "collapse" to one of the states p_i. Theorem 3.2.2 provides the corresponding probabilities. It is interesting to notice that the probability of collapse to an eigenstate p_i is a monotonically decreasing function of the geodesic separation of p_0 and p_i; the system is more likely to collapse to a nearby state than a distant one.

We now have a description of the probabilities associated with the measurement process, but there are two deficiencies to be remedied. The eigenstates p_i above have been defined in terms of the algebraic properties of the operator $\hat{F}$. For our program to be complete, we require a definition of these eigenstates which does not refer to the Hilbert space explicitly. Next, the above discussion was limited to the measurement of an observable with discrete, nondegenerate spectrum. We will describe the generic situation in two steps. First, we consider the measurement of observables with discrete, but possibly degenerate, spectra. This will require the aforementioned description of the eigenstates. We will then be prepared to consider measurement of observables with continuous spectra.

Let us first examine the notions of eigenstates and eigenvalues of an observable operator $\hat{F}$. Let $F : \mathcal{H} \to \mathbb{R}$ and $f : \mathcal{P} \to \mathbb{R}$ be the expectation value and corresponding observable function, respectively. A vector $\Psi \in \mathcal{H}$ is an eigenstate of $\hat{F}$ iff $\hat{F}\Psi = \lambda\Psi$, for some (real) λ. Alternatively, by Eq. (3),

$$X_F|_\Psi = Y_{\hat{F}}|_\Psi = (\lambda/\hbar)\mathcal{J}|_\Psi. \tag{28}$$

That is, Ψ is an eigenstate of $\hat{F}$ if and only if the Hamiltonian vector field X_F is *vertical* (i.e., purely gauge) at Ψ. This will be the case if and only if X_f vanishes at $\pi(\Psi)$; $\pi(\Psi)$ is then a critical point of the function f. Evidently, the corresponding eigenvalue is exactly the (critical) value of f at $\pi(\Psi)$. In summary:

Definition 3.2.3 *Let $f : \mathcal{P} \to \mathbb{R}$ be an observable function. Critical points of f are called* eigenstates *of f. The corresponding critical values are called* eigenvalues.

We now consider the measurement of an observable f whose spectrum is discrete but possibly degenerate (of course, the "spectrum of f" coincides, by definition, with the spectrum of the corresponding operator $\hat{F}$). Let λ be a degenerate eigenvalue of $\hat{F}$, and denote by $\tilde{\mathcal{E}}_\lambda$ the associated eigenspace of $\mathcal{H}$. Associated to this eigenspace is a submanifold, $\mathcal{E}_\lambda$, of $\mathcal{P}$, which we shall call the *eigenmanifold* associated to λ. Suppose that the system is prepared in the state labeled by p_0 and let $\Psi_0 \in \pi^{-1}p_0$. The postulates of ordinary quantum mechanics assert that measurement of f will yield the value λ with probability $\langle P_\lambda\Psi_0, P_\lambda\Psi_0\rangle = \langle\Psi_0, P_\lambda\Psi_0\rangle = |\langle\Psi_0, P_\lambda\Psi_0/\|P_\lambda\Psi_0\|\rangle|^2$, where P_λ is the projector onto the relevant eigenspace of $\hat{F}$. From the above considerations, we know that this probability may be expressed in terms of the geodesic separation of the points p_0 and $\pi\left(P_\lambda\Psi_0/\|P_\lambda\Psi_0\|\right) \in \mathcal{E}_\lambda$. We will denote the latter point by $P_\lambda(p_0)$.

What sets the point $P_\lambda(p_0)$ apart from all other elements of $\mathcal{E}_\lambda$? We need only notice that for any point $\Phi \in \tilde{\mathcal{E}}_\lambda \cap S$,

$$|\langle\Psi_0, \Phi\rangle|^2 = |\langle\Psi_0, P_\lambda\Phi\rangle|^2 = |\langle P_\lambda\Psi_0, \Phi\rangle|^2 \le \|P_\lambda\Psi_0\|^2. \tag{29}$$

Therefore, of all elements $\Phi \in \tilde{\mathcal{E}}_\lambda$ with unit normalization, that which maximizes the quantity $|\langle\Psi_0, \Phi\rangle|^2$ is simply $P_\lambda\Psi_0/\|P_\lambda\Psi_0\|$, i.e., that to which the state Ψ_0 will "collapse" in the event that measurement of $\hat{F}$ yields the value λ. Therefore, by Theorem 3.2.2, $P_\lambda(p_0)$ is simply that point of $\mathcal{E}_\lambda$ which is *nearest* p_0!

We may now describe the measurement of an observable with discrete spectrum as follows. Suppose that immediately prior to measurement of f, the system is in the state $p_0 \in \mathcal{P}$. Denote by λ_i the critical values of f and by $\mathcal{E}_{\lambda_i}$ the corresponding eigenmanifolds. Interaction with the measurement device causes the system to be projected to one of the eigenmanifolds, say $\mathcal{E}_{\lambda_0}$. "Realizing that it collapsed" to the state $P_{\lambda_0}(p_0)$, the system returns what it knows to be the value of the observable under consideration, i.e., $f\left(P_{\lambda_0}(p_0)\right) = \lambda_0$. Of course, the probability that measurement causes reduction to $\mathcal{E}_{\lambda_0}$ is given by $\cos^2\left(\sigma(p_0, \mathcal{E}_{\lambda_0})/\sqrt{2\hbar}\right)$, where $\sigma(p_0, \mathcal{E}_{\lambda_0})$ denotes the *minimal* geodesic separation of p_0 and the sub-manifold $\mathcal{E}_{\lambda_0}$.

Now let us study the generic case. Let $\hat{F}$ be any observable operator on $\mathcal{H}$, the spectrum of which is allowed to be continuous. We first need a definition of the spectrum of $\hat{F}$ in terms of the corresponding observable function $f : \mathcal{P} \to \mathbb{R}$. Recall the standard definition [37]: λ is an element of the spectrum $sp(\hat{F})$ if and only if the operator $\hat{F} - \lambda\hat{1}$ is *not* invertible. Equivalently, $\lambda \in sp(\hat{F})$ iff given any positive $\varepsilon \in \mathbb{R}$, $\exists\, \Psi \in S$ such that $\|\hat{F}\Psi - \lambda\Psi\| < \varepsilon$. This condition guarantees that, to arbitrary precision, λ is an approximate eigenvalue of $\hat{F}$.

Using Eqs. (23) and (24), we may write the (square of the) above quantity as

$$\|\hat{F}\Psi - \lambda\Psi\|^2 = \{f-\lambda, f-\lambda\}_+|_{\pi(\Psi)} = \left[(\Delta f)^2 + (f-\lambda)^2\right]\Big|_{\pi(\Psi)}. \tag{30}$$

This equation allows us to define the spectrum of $\hat{F}$ in terms of the function $f : \mathcal{P} \to \mathbb{R}$;

Definition 3.2.4 *The* spectrum $sp(f)$ *of an observable* f *consists of all real numbers* λ *for which the function* $n_\lambda : \mathcal{P} \to \mathbb{R} \cup \{\infty\}$, $n_\lambda : p \mapsto \left[(\Delta f)^2(p) + (f(p) - \lambda)^2\right]^{-1}$ *is* unbounded.

Of course, a point at which $n_\lambda = \infty$ corresponds to an eigenstate of f.

The next step is a description of the spectral projection operators. Let Λ be a closed subset of the spectrum $sp(f)$ of f, and denote by $P_{\hat{F},\Lambda}$ the projection operator associated to $\hat{F}$ and Λ. In analogy with the above, put $\tilde{\mathcal{E}}_{\hat{F},\Lambda} = \left\{P_{\hat{F},\Lambda}\Psi \mid \Psi \in \mathcal{H} - \{\mathbf{0}\}\right\}$, and let $\mathcal{E}_{f,\Lambda}$ denote the projection of $\tilde{\mathcal{E}}_{\hat{F},\Lambda}$ to $\mathcal{P}$. Note that the set $\tilde{\mathcal{E}}_{\hat{F},\Lambda}$—the analogue of the eigenspace above—actually *is* the eigenspace of $P_{\hat{F},\Lambda}$ corresponding to the eigenvalue 1. Therefore, we have $\mathcal{E}_{f,\Lambda}$ consists of the critical points of an expectation value function, associated to the critical value 1. Unfortunately, this expectation value function is not directly expressible in terms of f and Λ. We must look for an alternative description of the submanifold $\mathcal{E}_{f,\Lambda}$.

In the representation defined by the operator $\hat{F}$, elements of $\tilde{\mathcal{E}}_{\hat{F},\Lambda}$ have support on Λ. Therefore, $\Psi \in \tilde{\mathcal{E}}_{\hat{F},\Lambda}$ iff $\langle\hat{F}\rangle_\Psi \in \Lambda^n\ \forall n > 0$, where Λ^n denotes the image of Λ under the map $\lambda \mapsto \lambda^n$. Recall that $\{f, f\}_+$ is the (projection to $\mathcal{P}$ of the) expectation value of $\hat{F}^2$. In general, the expectation value of $\hat{F}^n$ projects to the n-fold symmetric product $\{f, \{f, \{f, \ldots\}_+\}_+\}_+$. Therefore, we have

$$\mathcal{E}_{f,\Lambda} = \{q \in \mathcal{P} | \{f, \{f, \{f, \ldots\}_+\}_+\}_+|_q \in \Lambda^n\ \forall n > 0\}, \tag{31}$$

where there are n factors of f occurring above.

Having obtained a description of $\mathcal{E}_{f,\Lambda}$, we may now define the spectral projections in a manner intrinsic to the projective space. By precisely the same reasoning surrounding Eq. (29), $P_{f,\Lambda}$ maps a point $p \in \mathcal{P}$ to that element of $\mathcal{E}_{f,\Lambda}$ which is nearest p.

The measurement process may then be described as follows. Suppose the quantum system is in the state labeled by the point $p_0 \in \mathcal{P}$ at the instant an experimenter performs a measurement of the observable f. Following the rules of quantum mechanics, she "asks the system" whether the value of f lies in Λ—a closed subset of $sp(f)$, which she is free to choose. The experimental apparatus drives the system to one of two states—either $P_{f,\Lambda}(p_0)$ or $P_{f,\Lambda^c}(p_0)$, where Λ^c is the (closure of the) complement, in $sp(f)$, of Λ. The system is reduced to the former with probability

$$\cos^2\left(\frac{\sigma(p_0, P_{f,\Lambda}(p_0))}{\sqrt{2\hbar}}\right);$$

in this event, the experiment yields the positive result ($f \in \Lambda$). Having precisely prepared the system in the state p_0, the experimenter may then infer the value $f(P_{f,\Lambda}(p_0))$ of the observable f. The probability of reduction to the latter state is obtained by replacing Λ by Λ^c above.

Note that this description encompasses all measurement situations. In the event that the spectrum of f is discrete and nondegenerate, the experimenter may choose to let Λ_i contain the single eigenvalue λ_i. Moreover, she may measure all of the projections simultaneously. In this way, one recovers the first familiar description of the measurement process. Note also that, while the above discussion of the spectral projections may seem complicated and somewhat unnatural at first, the definition of the spectral projection operators on the Hilbert space has the same features. (Indeed, most textbooks simply skip this technical discussion.) This is simply one of the technical complications that the geometric formalism inherits from the Hilbert space framework.

To conclude, we wish to emphasize that the topic of our discussion has been ordinary quantum mechanics. We have just restated the well-known quantum mechanical formalism in a language intrinsic to the true space of states—the quantum phase space, $\mathcal{P}$; no new ingredients have been added to the physics. A particularly attractive feature of the formalism, however, is the fact that slight modifications of the standard picture naturally present themselves. For example, using many of these geometric ideas, Hughston [28] has explored a novel approach to the measurement problem in terms of stochastic evolution.

3.2.4 *The Postulates of Quantum Mechanics*

Let us collect the results obtained in the first three subsections.

We have formulated ordinary quantum mechanics in a language which is intrinsic to the true space of quantum states—the projective Hilbert space $\mathcal{P}$. As in classical mechanics, observables are smooth, real-valued functions which preserve the kinematic structure. Being a Kähler manifold, $\mathcal{P}$ is a symplectic manifold. The role of the quantum symplectic structure is precisely that of classical mechanics; it defines both the Lie algebraic structure on the space of observables and generates motions including the time evolution.

There are, however, two important features of quantum mechanics which are not shared by the classical description. First, the phase space is of a very particular nature; it is a Kähler manifold and, as we will see in Section 3.3.2, one of a rather special type—namely one of constant holomorphic sectional curvature.[3]

[3] In this sense, the quantum framework is actually a special case of the classical one!

The second difference lies in the probabilistic aspects of the formalism, which is itself intimately related to the presence of the Riemannian metric. More generally, this metric describes those quantum mechanical features which are absent in the classical theory—namely, the notions of uncertainty and state reduction. For example, the transition probabilities which arise in quantum mechanics are determined by a simple function of the geodesic distance between points of the phase space.

These results are most easily summarized by stating the postulates in the geometric language:

($\mathcal{H}$) *Physical states:* Physical states of the quantum system are in one-to-one correspondence with points of a Kähler manifold $\mathcal{P}$, which is a projective Hilbert space.[4]

($\mathcal{U}$) *Kähler evolution:* The evolution of the system is determined by a flow on $\mathcal{P}$, which preserves the Kähler structure. The generator of this flow is a densely defined vector field on $\mathcal{P}$.

($\mathcal{O}$) *Observables:* Physical observables are represented by real-valued, smooth functions f on $\mathcal{P}$ whose Hamiltonian vector fields X_f preserve the Kähler structure.

($\mathcal{P}$) *Probabilistic interpretation:* Let $\Lambda \subset \mathbb{R}$ be a closed subset of the spectrum of an observable f, and suppose the system is in the state corresponding to the point $p \in \mathcal{P}$. The probability that measurement of f will yield an element of Λ is given by

$$\delta_p(\Lambda) = \cos^2\left(\frac{\sigma(p,\ P_{f,\Lambda}(p))}{\sqrt{2\hbar}}\right), \tag{32}$$

where $P_{f,\Lambda}(p)$ is the point, closest to p, in the space $\mathcal{E}_{f,\Lambda}$, defined by Eq. (31).

($\mathcal{R}_D$) *Reduction, discrete spectrum:* Suppose the spectrum of an observable f is discrete. This spectrum provides the set of possible outcomes of the ideal measurement of f. If measurement of f yields the eigenvalue λ, the state of the system immediately after the measurement is given by the associated projection, $P_{f,\lambda}(p)$, of the initial state p.

($\mathcal{R}_C$) *Reduction, continuous spectrum:* A closed subset Λ of the spectrum of f determines an ideal measurement that may be performed on the system. This measurement corresponds to inquiring whether the value of f lies in Λ. Immediately after this measurement, the state of the system is given by $P_{f,\Lambda}(p)$ or $P_{f,\Lambda^c}(p)$, depending on whether the result of the measurement is positive or negative, respectively.

In the last postulate, Λ^c is the closure of the complement, in the spectrum of f of the set Λ. Although the first "reduction postulate" is a special case of

[4]We will see in Section 3.3.2 that quantum phase spaces $\mathcal{P}$ can be alternatively singled out as Kähler manifolds which admit maximal symmetries. This provides an intrinsic characterization without any reference to Hilbert spaces.

the second, both have been included for comparison with standard textbook presentations.

To conclude, although it is not obvious from textbook presentations, the postulates of quantum mechanics can be formulated in an intrinsically geometric fashion, without any reference to the Hilbert space. The Hilbert space and associated algebraic machinery provides convenient technical tools. But they are not essential. Mathematically, the situation is similar to the discussion of manifolds of constant curvature. In practice, one often establishes their properties by first embedding them in $\mathbb{R}^n$ (equipped with a flat metric of appropriate signature). However, the embedding is only for convenience; the object of interest is the manifold itself. There is also a potential analogy from physics, alluded to in the Introduction. Perhaps the habitual linear structures of quantum mechanics are analogous to the inertial rest frames in special relativity and the geometric description summarized here, analogous to Minkowski's reformulation of special relativity. Minkowski's description paved the way to general relativity. Could the geometric formulation of quantum mechanics lead to a more complete theory one day?

3.3 A Unified Framework for Generalizations of Quantum Mechanics

There are three basic elements of the quantum mechanical formalism which may be considered for generalization: the state space, the algebra of observables and the dynamics. The framework developed in Section 3.2 suggests avenues for each of these. First, while it is not obvious how one might "wiggle" a Hilbert space, one may generalize the quantum phase space by considering, say, the class of all Kähler manifolds $\{\mathcal{M}, g, \omega\}$. The geometric language also suggests an obvious generalization of the space of quantum observables: one might consider the space of *all* smooth, real-valued functions on the phase space. Finally, whether or not one chooses either of these extensions, one may consider generalized dynamics which, as in classical mechanics, preserves only the symplectic structure. Thus, one might require the dynamical flow to preserve only the symplectic structure, and not necessarily the metric.

While each of these structures may be extended separately, they are intimately related and construction of a complete, consistent framework is a highly nontrivial task. Thus, for example, if one allows *all* Kähler manifolds as possible quantum phase spaces, is seems very difficult to obtain consistent probabilistic predictions for outcomes of measurements. More generally, the problem of systematically analyzing viable, nontrivial generalizations of the the kinematic structure would be a major undertaking, although the payoff may well be exceptional. Modification of dynamics, on the other hand, is easier at least in principle. Therefore, we will first consider these in Section 3.3.1 and return to kinematics in Section 3.3.2.

3.3.1 Generalized Dynamics

Let us then suppose that we continue to use a projective Hilbert space for the quantum phase space, and let dynamics be generated by a preferred Hamiltonian function. However, let us only require that time evolution should preserve the symplectic structure ω (as in classical mechanics), and not necessarily the metric g. From the viewpoint of the geometric formulation, this is the simplest and most obvious generalization of the standard quantum dynamics.[5] The idea is reminiscent of the "nonlinear Schrödinger equations" that have been considered in the past. Therefore, it is natural to ask if these there is a relation between the two. We will see that the answer is in the affirmative. Furthermore, the geometric framework provides a unified treatment of these proposals and makes the relation between them transparent, thereby enabling one to correct a misconception.

Let us begin by defining the the class C_H of Hamiltonians we now wish to consider. C_H will consist of densely defined functions f on $\mathcal{P}$ satisfying the following properties: i) f is smooth on its domain of definition; and ii) the Hamiltonian vector field it defines generates a flow on all of $\mathcal{P}$. In particular, the Hamiltonian functions we considered in Section 3.2—expectation values of a possibly unbounded selfadjoint operator—belong to C_H, but they constitute only a "small subset" of C_H.

The existing proposals of nonlinear dynamics refer to flows in the full Hilbert space $\mathcal{H}$ rather than in the quantum phase space $\mathcal{P}$. To compare the two, we need to lift our flows to $\mathcal{H}$. Let us begin by recalling that it is natural to regard $\mathcal{P}$ as a reduced phase space, resulting from the first-class constraint $C(\Psi) := \langle\Psi, \Psi\rangle - 1 = 0$ on $\mathcal{H}$ (see Eq. (11)). Therefore, a function f on $\mathcal{P}$ admits a natural lift F to S, the unit sphere in $\mathcal{H}$. This function F on S is constant along the integral curves of the vector field $\mathcal{J}$ which generates phase rotations; $\pounds_{\mathcal{J}} f = 0$. Denote the space of these lifts by $\tilde{C}_H$.

To discuss dynamics on $\mathcal{H}$, we need to extend these functions[6] off S. From the reduced phase space viewpoint, the extension is completely arbitrary. For we can construct the Hamiltonian vector field on $\mathcal{H}$ generated by *any* extension F_{ext}. The restriction to S of this vector field does depend on the extension, but the the *horizontal part* of the restriction—i.e., the part orthogonal to $\mathcal{J}$—does not. Hence, the projection of the vector field to $\mathcal{P}$ agrees with X_f, irrespective of the choice of the initial extension. Thus, the generalized dynamics generated by a given

---

[5]Recall that the Hamiltonian need not be an observable function in the sense of Section 3.2.2. We could extend the kinematical setup as well and regard any smooth function on $\mathcal{P}$ as an observable function. (This would be analogous to Weinberg's [7] proposal which, however, was made at the level of the Hilbert space $\mathcal{H}$ rather than the quantum phase space $\mathcal{P}$.) We have refrained from doing this because the required extension of measurement theory is far from obvious.

[6]Strictly speaking, we only need to extend the dynamical flow off S. However, to compare our results with Weinberg's [7] we need to consider extensions of Hamiltonians. In discussions on generalized dynamics [7, 5], the issue of domains of definition of operators is generally ignored. Our treatment will be at the same level of rigor. In particular, we will ignore the fact that our Hamiltonian functions and vector fields are only densely defined.

Hamiltonian function f on $\mathcal{P}$ can be lifted to a whole family of flows on $\mathcal{H}$, all of which, however, evolve the physical quantum states—elements of $\mathcal{P}$—in the same way. Because of this, apparently distinct proposals for nonlinear evolutions on $\mathcal{H}$ can in fact be physically equivalent. This point is rather trivial from the viewpoint of geometric quantum mechanics. The reason for our elaboration is that—as we will see below—it has not been appreciated in the Hilbert space formulations.

While the extension off S of elements of $\tilde{\mathcal{C}}_H$ is completely arbitrary, one can use the standard quantum mechanical framework to select a specific rule. Consider, to begin with, a bounded, selfadjoint operator $\hat{F}$ on $\mathcal{H}$, and let F be the restriction to S of the corresponding expectation value function. There is then an obvious extension of F to all of $\mathcal{H}$: set $F_{\text{ext}}(\Psi) := \langle\Psi, \hat{F}\Psi\rangle$ (which we denoted by F in Section 3.2.1). One can restate this rule as

$$F_{\text{ext}}(\Psi) := \|\Psi\|^2 F(\Psi/\|\Psi\|). \tag{33}$$

The advantage is that this equation may now be used to extend *any* element F of $\tilde{\mathcal{C}}_H$ to all of $\mathcal{H}^\times = \mathcal{H} - \{\mathbf{0}\}$. Note that, with this preferred extension, the Hamiltonian vector field $X_{F_{\text{ext}}}$ is homogeneous of degree one on $\mathcal{H}^\times$:

$$X_{F_{ext}}(c\Psi) = cX_{F_{ext}}(\Psi) \quad \forall c \in \mathbb{C}. \tag{34}$$

Hence, the flow on $\mathcal{H}^\times$ which is generated by F_{ext} is homogeneous, but fails to be linear unless F is the restriction to S of the expectation value function defined by a self-adjoint operator. Next, it is easy to verify that these F_{ext} *strongly* commute with the constraint function $C(\Psi)$. Hence the flow along $X_{F_{\text{ext}}}$ preserves the constraint. Therefore, the specific extension considered above has the property that the corresponding flow preserves not only the symplectic structure Ω *but also the norm* on $\mathcal{H}^\times$. However, unless f is an observable function, it does not preserve the metric G.

Note that, even if we consider just these preferred extensions, the set of possible Hamiltonian functions on $\mathcal{H}^\times$ has been extended quite dramatically. To see this, consider the case when $\mathcal{H}$ is finite-dimensional. Then, the class of Hamiltonian functions on $\mathcal{H}$ allowed in standard quantum mechanics forms a *finite*-dimensional real vector space; it is just the space of expectation value functions constructed from selfadjoint operators. The space $\tilde{\mathcal{C}}_H$, on the other hand, is *infinite*-dimensional, since its elements are in one-to-one correspondence with smooth functions on $\mathcal{P}$. And each element of $\tilde{\mathcal{C}}_H$ admits an unique extension to $\mathcal{H}^\times$ via Eq. (33). It is natural to ask if one can do something "in-between." Can we impose more stringent requirements to select a class of potential Hamiltonians which is larger than that of observable functions of Section 3.2 but smaller than $\mathcal{C}_H$? For example, one might imagine looking for the class of functions on $\mathcal{H}^\times$ whose Hamiltonian flows preserve not just the norms but also the inner product. It turns out, however, that this class consists precisely of the expectation value functions defined by selfadjoint operators; there is no such thing as a nonlinear unitary flow on $\mathcal{H}$ [16]. Despite the magnitude of our generalization, it seems to be the only available choice.

We are now ready to discuss the relation between this generalization and those that have appeared in the literature. Note first that Eq. (33) implies that there is a one-to-one correspondence between elements of C_H and smooth functions on $\mathcal{H}^\times$ which are gauge invariant (i.e., insensitive to phase) and homogeneous of degree 2. (If $\mathcal{H}$ is viewed as a vector space over complex numbers, this corresponds to homogeneity of degree 1 in both Ψ and $\bar{\Psi}$.) It turns out that this is precisely the class of permissible Hamiltonians that Weinberg [7] was led to consider while looking for a general framework for nonlinear generalizations of quantum mechanics. Let us therefore call functions on $\mathcal{H}^\times$ satisfying Eq. (33) *Weinberg functions*, and denote the space of these functions by $\mathcal{O}_W$. Our discussion shows that there is a one-to-one correspondence between smooth functions on the projective Hilbert space $\mathcal{P}$ and Weinberg functions on the punctured Hilbert space $\mathcal{H}^\times$; the homogeneity restriction simply serves to eliminate the freedom in the extension of the function on $\mathcal{H}^\times$. Thus the extension of quantum dynamics that is immediately suggested by the geometrical framework reproduces key features of Weinberg's proposal.

There are, however, considerable differences in the motivations and general viewpoints of the two treatments. In particular, Weinberg works with the Hilbert space $\mathcal{H}$ (and, without explicitly saying so, sometimes with $\mathcal{H}^\times$). However, he does assume at the outset that elements Ψ and $c\Psi$ of $\mathcal{H}$ define the same physical state of the quantum system for all *complex* numbers c. Thus, although it is not explicitly stated, his space of physical states is also $\mathcal{P}$. Therefore, it is possible to translate his constructions to the the geometric language. As we will see below, the geometric viewpoint is often clarifying.

Next, let us consider two specific examples of nonlinear dynamics that have been considered in the literature. Each of these involves the nonrelativistic mechanics of a point particle moving in $\mathbb{R}^n$ and the Hilbert space consists of square-integrable functions on $\mathbb{R}^n$. Therefore, it will be useful to reinstate a complex notation for the remainder of this subsection. In this notation, an element of $\mathcal{O}_W$ is a real-valued function $F(\Psi, \bar{\Psi})$ of both Ψ and its conjugate, which is homogeneous of degree 1 in each argument. The Hamiltonian vector field generated by such a function corresponds to

$$X_F[\Psi](x) = \frac{1}{i\hbar}\frac{\delta F}{\delta \Psi^*}(x). \tag{35}$$

The simplest example of this type is provided by is the so-called "nonlinear Schrödinger equation." This equation is given by

$$i\hbar\frac{\partial \Psi}{\partial t}(x,t) = (\hat{H}_0\Psi)(x,t) + \epsilon|\Psi(x,t)|^2\Psi(x,t), \tag{36}$$

where $\hat{H}_0 = \hat{P}^2/2m + \hat{V}$ is the standard Hamiltonian operator describing a non-relativistic particle under the influence of a conservative force with potential $\hat{V}$. Note that the quantity $|\Psi(x,t)|$ is the modulus of $\Psi(x,t)$, *not* the norm $\|\Psi\|$ of the state-vector Ψ. It is easy to verify that (36) induces a flow on $\mathcal{P}$. This flow is Hamiltonian and the generating function is the projection to $\mathcal{P}$ of the function $\langle\Psi, H_0\Psi\rangle + H_\epsilon(\Psi)$ on $\mathcal{H}$, where

$$H_\epsilon(\Psi) := \frac{\epsilon}{2}\int d^n x\, \left[\Psi^*(x,t)\Psi(x,t)\right]^2. \tag{37}$$

Thus, Eq. (36) is indeed a specific example of our generalized dynamics.

Note, however, that the dynamical vector field X on $\mathcal{H}$ defined directly by the nonlinear Schrödinger equation is given by

$$X[\Psi](x) = X_{H_0}[\Psi](x) + X_\epsilon[\Psi](x), \tag{38}$$

where $X_\epsilon[\Psi](x) = (\epsilon/i\hbar)|\Psi(x,t)|^2\Psi(x,t)$. Clearly, it is *not* homogeneous in the sense of Eq. (34), and hence is not generated by a Weinberg function. Therefore, Weinberg was led to state that the "results obtained by the mathematical studies of this equation are unfortunately of no use" to the generalization he considered. However, we just saw that the nonlinear Schrödinger equation does correspond to generalized dynamics on $\mathcal{P}$ and is therefore of "Weinberg type." Hence, the statement in quotes is somewhat misleading.

Let us elaborate on this point. If we first focus on physical states, what matters is just the projected flow on $\mathcal{P}$. This in turn is completely determined by the restriction $H_\epsilon\big|_S$ of H_ϵ to S. Therefore, we may feel free to ignore the behavior of H_ϵ off of S. Of course, to construct a vector field on S which projects to the relevant one on $\mathcal{P}$, we can extend $H_\epsilon\big|_S$ *arbitrarily* and compute the associated Hamiltonian vector field. In particular, we may extend it in the way which Weinberg would suggest:

$$H'_\epsilon(\Psi) := \|\Psi\|^2 H_\epsilon\big|_S(\Psi/\|\Psi\|). \tag{39}$$

Thus, we have seen explicitly that the flow on $\mathcal{P}$ which is defined by H_ϵ may also be described by a Weinberg function! The emphasis on the true space of states $\mathcal{P}$ of the geometric treatment clarifies this point which seems rather confusing at first from the Hilbert space perspective.

The nonlinear Schrödinger equation is a fairly simple example, since the generating function H_ϵ is itself homogeneous (but of the "wrong" degree to be a Weinberg function). Let us now consider an example which is more sophisticated.

In an effort to address the problem of combining systems which are subject to a nonlinear equation of motion, Bialynicki-Birula and Mycielski were led to a *logarithmic* equation of motion [5]. They began with a general equation of motion of the form

$$i\hbar\frac{\partial\Psi}{\partial t}(x,t) = (\hat{H}_0\Psi)(x,t) + \alpha(|\Psi(x,t)|^2)\Psi(x,t), \tag{40}$$

and showed that physical considerations, particularly the requirement that α-term should not introduce interactions between otherwise noninteracting subsystems, imposes severe restrictions the functional form of α. The only possibility is to have $\alpha(\rho) = -b\ln(a^n\rho)$ for some constants a and b. (For details see [5].) Choosing units[7] with respect to which $a = 1$, the vector field along which the system evolves may be written as

$$X[\Psi](x) = X_{H_0}[\Psi](x) + X_1[\Psi](x), \tag{41}$$

[7] The constant a is a length scale of no physical significance, since it may be altered by addition of a constant to the Hamiltonian operator.

where $X_1[\Psi](x) = -(b/i\hbar)\ln(|\Psi|^2)\Psi(x)$. Again, the extra term may be seen to be Hamiltonian; $X_1 = X_{H_1}$, where

$$H_1(\Psi) = b \int d^n x \Psi^*(x)\Psi(x)\big[1 - \ln(\Psi^*(x)\Psi(x)\big]. \tag{42}$$

Since H_1 is phase invariant, this is also an example of our generalized dynamics.

Again, we can carry out the procedure used for the nonlinear Schrödinger equation to see that the corresponding motion on the projective space may be described by use of a Weinberg function. It is not difficult to show that the corresponding homogeneous function is given by

$$H_1'(\Psi) = H_1(\Psi) + b\|\Psi\|^2 \ln(\|\Psi\|^2). \tag{43}$$

Therefore, the logarithmic equation induces a flow on $\mathcal{P}$ which may be described by a function of the Weinberg type.

In retrospect, essentially any generalized dynamics of the form specified by Eq. (40) may be written as a Hamiltonian flow on $\mathcal{H}^\times$ by "integrating" the function F. The resulting Hamiltonian function is not likely to satisfy Weinberg's homogeneity condition. However, one may restrict it to the unit sphere, then extend this restriction as in Eq. (33). The resulting function will generate a Hamiltonian vector field which differs from the original one in general, but agrees with it on the unit sphere. Hence, both will generate the same flows on the projective space. This may be done with *any* Hamiltonian function on $\mathcal{H}^\times$ which preserves the unit sphere.

3.3.2 Characterization of the Standard Quantum Kinematics

Let us now turn to the issue of generalizations of the kinematic framework itself. As noted above, these extensions may well turn out to be profound. However, they also appear to be much more difficult to carry out. Therefore, as a first step, it is natural to ask what sets ordinary quantum mechanics apart from its natural generalizations. The purpose of this subsection is provide such a characterization of the standard kinematic framework.

Let us consider an *arbitrary* Kähler manifold $\{\mathcal{M}, g, \omega\}$, which is to represent the phase space of a (radically!) generalized quantum theory. (As before, $\mathcal{M}$ will be assumed to be a Hilbert manifold, and g and ω to be strongly nondegenerate.) The question is then, Are there natural conditions which will single out ordinary quantum mechanics from this class of generalizations? As we saw in Section 3.2, the observables of ordinary quantum mechanics are smooth functions on the phase space whose Hamiltonian vector fields satisfy Killing's equation; we therefore let the observables of the generalized framework consist of all smooth functions on $\mathcal{M}$ whose Hamiltonian flows are isometries. Denote this set as

$$\mathcal{O} := \{f : \mathcal{M} \mapsto \mathbb{R} \mid \pounds_{X_f} g = 0\}. \tag{44}$$

(Note that while we call elements of $\mathcal{O}$ *observables*, we do not claim to possess a complete and consistent formalism incorporating measurement. Indeed, this seems an impossible feat at the present level of generality.)

It should be emphasized that there are Kähler manifolds which do not admit a single observable function (other than constants); the torus is an example. Let us amplify this point briefly. Since the discussion leading to Definition 3.2.2 is valid for an arbitrary Kähler manifold, any observable is completely determined by its value and first two derivatives at a single point. As before, we will denote by $\mathcal{S}_p$ the set of symmetry data at the point $p \in \mathcal{M}$ (if λ, X and K are 0-, 1- and 2-forms at p for which $\omega_\alpha{}^\gamma K_{\gamma\beta}$ is symmetric, then $(\lambda, X, K) \in \mathcal{S}_p$). Given an arbitrary $p \in \mathcal{M}$, each observable function then determines an element of $\mathcal{S}_p$. According to Theorem 3.2.1, the converse holds for ordinary quantum mechanics: there, the algebra of symmetry data at each point p is *integrable*. In this sense, the phase space of standard quantum mechanics admits "as many observables as possible." This useful idea is captured in

Definition 3.3.1 *If for each $p \in \mathcal{M}$, every element $(\lambda, X, K) \in \mathcal{S}_p$ is integrable, we say that $\mathcal{O}$ is* maximal.

The set of observables of ordinary quantum mechanics is maximal. As we will see, it is essentially this property of the quantum observables which allows us to recover the standard formalism.

In order to illustrate the importance of the maximality of the space of observables, it is necessary to introduce one mathematical concept. It is fairly easy to verify that the Riemann curvature tensor of a projective Hilbert space is of the special form

$$R_{\alpha\beta\gamma\delta} = \frac{C}{2}\left[g_{\gamma[\alpha}g_{\beta]\delta} + \omega_{\alpha\beta}\omega_{\delta\gamma} - \omega_{\gamma[\alpha}\omega_{\beta]\delta}\right], \tag{45}$$

where $C = \hbar/2$ and $[\ldots]$ denotes skew-symmetrization. If the curvature tensor of our generalized phase space $\mathcal{M}$ satisfies this equation at a point p, it is said to be of *constant holomorphic sectional curvature* (CHSC) at p. In that case, the real number C is the value of the holomorphic sectional curvature at p. If for some real C, the Riemann tensor assumes the form written in Eq. (45) (for all $p \in \mathcal{M}$), $\mathcal{M}$ is called a manifold of CHSC $= C$. It may be useful to note that if a Kähler manifold $\mathcal{M}$ is of CHSC at each point, then it must be of overall CHSC; that is, if R satisfies Eq. (45) at each p, then C must be a constant [38].

Holomorphic sectional curvature, in the context of complex manifolds, is analogous to the scalar curvature of real manifolds. Since there is a very strong relationship between the number of independent Killing vector fields on a real manifold and the form of the Riemann curvature tensor [36], one may expect that the maximality of the space of quantum observables to be strongly related to Eq. (45). We will therefore use an approach analogous to that presented in [36] to consider more closely the interplay between the geometry of the phase space and the algebra of observables. The results in this section are stated without proof; for details, see [16].

Since the commutator of two Killing vector fields is another Killing vector field, the set of observables on an arbitrary Kähler manifold is closed under the Poisson bracket {, }. The Poisson bracket also satisfies the Jacobi identity; this equips $\mathcal{O}$

with the structure of a Lie algebra. On the entire function space, we may also define the commutative operation

$$\{f, k\}_+ := (f, k) + fk, \qquad (f, k) := \frac{\hbar}{2}(\nabla_\alpha f)g^{\alpha\beta}(\nabla_\beta g), \tag{46}$$

which we call the *symmetric bracket*. The symmetric bracket is not necessarily closed on the space of observables on an arbitrary Kähler manifold.

Suppose that f_1 and f_2 are any two observable functions on M, and let $f_3 = \{f_1, f_2\}$. Since each f_i generates a Killing vector field, it determines an element $(f_i, X_i, K_i)\big|_p \in \mathcal{S}_p$ for any $p \in \mathcal{M}$, where $X_i = X_{f_i}$ and $K_{i\alpha\beta} = \nabla_\alpha(X_{f_i})_\beta$. Of course, $f_3 = \omega(X_1, X_2)$ and $X_{3\alpha} = -g_{\alpha\beta}\left[X_{f_1}, X_{f_2}\right]^\beta\big|_p = X_2^\beta K_{1\beta\alpha} - X_1^\beta K_{2\beta\alpha}$, where [,] denotes the commutator of vector fields. It is straightforward to derive the corresponding expression for the 2-form K_3; the result is

$$K_{3\alpha\beta} = K_{2\alpha}{}^\gamma K_{1\gamma\beta} - K_{1\alpha}{}^\gamma K_{2\gamma\beta} + X_{1\mu} X_{2\nu} R_{\alpha\beta}{}^{\mu\nu}. \tag{47}$$

This fact suggests that we define the following operation on the set $\mathcal{S}_p$ of symmetry data:

$$\begin{aligned}&\left[(f_1, X_1, K_1), (f_2, X_2, K_2)\right]_p \\ &:= \left(\omega(X_1, X_2), X_2^\beta K_{1\beta\alpha} - X_1^\beta K_{2\beta\alpha}, K_{2\alpha}{}^\gamma K_{1\gamma\beta} - K_{1\alpha}{}^\gamma K_{2\gamma\beta} + X_{1\mu} X_{2\nu} R_{\alpha\beta}{}^{\mu\nu}\right).\end{aligned} \tag{48}$$

This bracket on $\mathcal{S}_p$ is defined only to mirror the Poisson bracket on $\mathcal{O}$. That is, to obtain the symmetry data corresponding to the Poisson bracket of two observables, one may simply apply the above "bracket" operation to the symmetry data of the initial observables.

For an arbitrary point p on an arbitrary Kähler manifold, $\mathcal{S}_p$ is closed under $[\,,\,]_p$. However, the Jacobi identity will, in general, fail. It is natural to ask for circumstances under which $\mathcal{S}_p$ forms a Lie algebra. The answer to this question is provided by

Lemma 3.3.1 $[\,,\,]_p$ *is a Lie bracket on* $\mathcal{S}_p$ *if and only if the Riemann tensor is of CHSC at* p.

If $\mathcal{O}$ is maximal, then $[,\,]_p$ is a Lie bracket on $\mathcal{S}_p$ for any $p \in \mathcal{M}$. As a consequence of Lemma 3.3.1, the Riemann tensor is of CHSC at each point of $\mathcal{M}$. Therefore $\mathcal{M}$ is a manifold of CHSC; this is summarized by

Corollary 3.3.1 *Suppose* $\mathcal{O}$ *is maximal. Then* $\mathcal{M}$ *is a manifold of CHSC.*

Let us make the analogous construction for the symmetric bracket. Again, let $f_1, f_2 \in \mathcal{O}$ and let f_4 denote the symmetric bracket, $f_4 = \{f_1, f_2\}_+ = f_1 f_2 + (\hbar/2)g(X_1, X_2)$. One can easily show the following:

$$\begin{aligned}\omega_\alpha{}^\beta \nabla_\beta f_4 &= f_1 X_{2\alpha} + f_2 X_{1\alpha} \\ &\quad + (\hbar/2)\omega_\alpha{}^\beta\left(K_{1\beta\gamma}X_2^\gamma + K_{2\beta\gamma}X_1^\gamma\right),\end{aligned} \tag{49}$$

$$\nabla_\alpha X_{4\beta} = f_1 K_{2\alpha\beta} + f_2 K_{1\alpha\beta} + \hbar K_{1\gamma[\alpha}\omega^{\gamma\delta} K_{2\beta]\delta} + X_1^\mu X_{2\nu}\left[\hbar R_{\alpha(\mu\nu)\gamma} - 2g_{\alpha(\mu}g_{\nu)\gamma}\right]\omega_\beta{}^\gamma . \tag{50}$$

Therefore it is natural to define the following commutative operation on $\mathcal{S}_p$:

$$\begin{aligned}
&((f_1, X_1, K_1), (f_2, X_2, K_2))_p \\
&:= \Big(f_1 f_2 + \hbar/2 g(X_1, X_2),\ f_1 X_{2\alpha} + f_2 X_{1\alpha} + (\hbar/2)\omega_\alpha{}^\beta (K_{1\beta\gamma} X_2^\gamma \\
&\quad + K_{2\beta\gamma} X_1^\gamma),\ f_1 K_{2\alpha\beta} + f_2 K_{1\alpha\beta} + \hbar K_{1\gamma[\alpha}\omega^{\gamma\delta} K_{2\beta]\delta} \\
&\quad + X_1^\mu X_{2\nu}\left[\hbar R_{\alpha(\mu\nu)\gamma} - 2g_{\alpha(\mu}g_{\nu)\gamma}\right]\omega_\beta{}^\gamma \Big).
\end{aligned} \tag{51}$$

We know that if $\mathcal{M}$ is a projective Hilbert space this operation produces the symmetry data determined by the symmetric (i.e., Jordan) bracket of the corresponding observables. For the generic case however, $\mathcal{O}$ will not be closed under the symmetric bracket; hence we do not expect $\mathcal{S}_p$ to be closed under its symmetric bracket. Therefore, the symmetric bracket defined by Eq. (51) should be viewed as an operation on the space of *all* triples (f, X, K), without the symmetry condition imposed on K.

The condition that $\mathcal{S}_p$ be closed under $(,)_p$ is even stronger than that found in Lemma 3.3.1:

Lemma 3.3.2 *The set $\mathcal{S}_p$ of symmetry data at p is closed under the symmetric bracket $(,)_p$ if and only if the Riemann tensor is of CHSC= $2/\hbar$ at p.*

The difference between this condition and that specified in Lemma 3.3.1 is that here the actual value of the holomorphic sectional curvature is determined by the coefficient appearing in the definition of the symmetric bracket. Lemma 3.3.1 states that the holomorphic sectional curvature be constant, but does not specify its value.

Note that each of these lemmas involves only a single point of $\mathcal{M}$. If the set of observables is maximal, then by definition its elements are in one-to-one correspondence with elements of $\mathcal{S}_p$ for any $p \in \mathcal{M}$. Therefore, the closure of a maximal set of observables under the symmetric bracket is equivalent to the closure of $\mathcal{S}_p$ under $(,)_p$ for any $p \in \mathcal{M}$. If we apply Lemma 3.3.2 to each point of $\mathcal{M}$, we immediately obtain

Corollary 3.3.2 *Suppose $\mathcal{O}$ is maximal. Then $\mathcal{O}$ is closed under the symmetric bracket if and only if the Riemann tensor is of CHSC= $2/\hbar$.*

Corollary 3.3.1 does not specify the value of the holomorphic sectional curvature, but merely states that the Riemann tensor assumes the special form written in Eq. (45) for some constant C. However, combining it with Lemma 3.3.2, one obtains

Theorem 3.3.1 *Suppose $\mathcal{O}$ is maximal and that $\mathcal{S}_p$ is closed under $(,)_p$ for a single point $p \in \mathcal{M}$. Then $\mathcal{M}$ is a manifold of CHSC= $2/\hbar$.*

Of course, we would like to go one step further to say that $\mathcal{M}$ must be a projective Hilbert space. Now, it is known that any two finite-dimensional Kähler manifolds which are complete, simply connected, and of CHSC= $2/\hbar$ are isomorphic [38]. Therefore, in the finite-dimensional case, we have obtained a characterization of the structure that picks out the standard quantum kinematics from possible generalized frameworks: If the generalized quantum phase space is a complete, simply connected Kähler manifold and the set of observables is maximal and closed under $\{,\}_+$, then one is dealing with the structure of ordinary quantum mechanics. This characterization should be useful in the search for genuine generalizations. In a generalized theory, one or more of the conditions must be violated. Thus, the characterization systematizes the search for generalizations and suggests concrete directions to proceed.

The case when $\mathcal{M}$ is infinite-dimensional is much more interesting physically. However, in this case, we do not know if we can conclude that a (complete) Kähler manifold on which $\mathcal{O}$ is maximal and closed under $\{,\}_+$ is isomorphic to a projective Hilbert space. Indeed there may exist *many* different infinite-dimensional Kähler manifolds (satisfying the above completeness requirements) of the same constant holomorphic sectional curvature. This is an important open problem. If the situation turns out to be the same as in the finite-dimensional case, we will again have a theorem characterizing ordinary quantum mechanics. If the situation is different and we have many such Kähler manifolds, it would be even more interesting. For, we would then be presented with viable generalizations of the standard quantum formalism. Such examples would be very interesting because they are likely to admit a consistent measurement theory and thus lead to physically complete generalizations of a rather subtle type.

3.4 Semiclassical Considerations

One of the most striking features of the geometric approach to quantum mechanics is its resemblance to the classical formalism. One might therefore expect that the geometric framework may shed some light on the relation between quantum and classical physics and, in particular, semiclassical approximations. We will see that this expectation is correct. In Section 3.4.1 we consider the relation between the two theories at a kinematical level and elucidate the special role played by coherent states. These results are then used in the remaining section to analyze dynamical issues. In 3.4.2 we consider systems such as harmonic oscillators and free fields; in 3.4.3, we discuss dynamics in the WKB approximation.

3.4.1 Kinematics

Let us now suppose that we are given a classical phase space (Γ, α), where Γ is assumed to be a finite-dimensional vector space and α is the symplectic form thereon. Let $(\mathcal{P}, g, \omega)$ be the quantum phase space which results from the application of the textbook quantization procedure to (Γ, α). Thus, on the quantum Hilbert space $\mathcal{H}$, there are position and momentum operators ($\hat{Q}_i$ and $\hat{P}_i$, $i = 1, \ldots, n$) corresponding to the classical position and momentum observables. We are all accustomed to this direction of construction. Our goal now is to do things in "reverse": Given the quantum phase space, we will provide a construction of the classical one.

We will continue with our previous convention and write the projections to $\mathcal{P}$ of the expectation values of these operators as q_i and p_i. It should be emphasized, however, that these functions are the elementary *quantum* observables, not the classical ones; they are functions on the infinite-dimensional space $\mathcal{P}$ and not on the finite-dimensional space Γ. Note that by standard construction the Poisson brackets of these elementary quantum observables satisfy the same "commutation relations" as do the classical variables:

$$\{q_i, p_j\} = \delta_{ij} \quad \text{and} \quad \{q_i, q_j\} = 0 = \{p_i, p_j\}. \tag{52}$$

One of the most common approaches to the classical limit comes from a simple theorem of Ehrenfest's which suggests that expectation values of quantum operators are to be approximated, in some sense, by the corresponding classical observables. The geometric formulation is particularly well suited to implement these ideas. To each quantum state $x \in \mathcal{P}$, let us associate the classical state $(q_i(x), p_j(x)) \in \Gamma$. This association defines the obvious mapping $\rho : \mathcal{P} \to \Gamma$. In fact, one might view this as the *definition* of the classical phase space. (The astute reader will object, correctly pointing out that one must specify the elementary quantum observables q_i and p_i before such a "definition" of the classical phase space can be made. Recall, however, that this is just the counterpart of the fact that elementary classical observables must be specified before construction of the quantum theory.)

For convenience, let us denote the generic elementary observables by f_r, $r = 1, \ldots, 2n$ (i.e., $\{f_r\} = \{q_i, p_i, i = 1, \ldots, n\}$). Now let us define an equivalence relation on $\mathcal{P}$: $x_1 \sim x_2 \Leftrightarrow f_r(x_1) = f_r(x_2)\ \forall r$. This equivalence relation fibrates the quantum phase space, which may now view as a bundle over the classical phase space, $\Gamma = \mathcal{P}/\sim$.

Associated to any bundle is a "vertical" distribution; i.e., a special class of (vertical) tangent vectors at each state $x \in \mathcal{P}$. That a tangent vector $v \in T_x\mathcal{P}$ is vertical simply means that $v(f_r) = 0\ \forall r = 1, \ldots, 2n$. Equivalently the vertical subspace may be defined as

$$\mathcal{V}_x := \left\{ v \in T_x\mathcal{P} \;\middle|\; \omega(X_{f_r}|_x,\ v) = 0 \quad \forall r = 1, \ldots, 2n \right\}. \tag{53}$$

The vertical directions are simply those in which the elementary quantum observables assume constant values.

Let $\mathcal{V}_x^{\perp}$ denote the ω-orthogonal complement of the vertical subspace at x. One can show [16] that each tangent space $T_x\mathcal{P}$ may be written as the sum $T_x\mathcal{P} = \mathcal{V}_x \oplus \mathcal{V}_x^{\perp}$. Let us therefore call elements of $\mathcal{V}_x^{\perp}$ *horizontal*. If Y is a vector *field* which is everywhere horizontal (vertical), we will simply write $Y \in \mathcal{V}^{\perp}$ ($Y \in \mathcal{V}$). Note in particular that any algebraic combination of the X_{f_r} is necessarily horizontal everywhere. Let us make three preliminary observations that will be used later.

First, the distribution $\mathcal{V}$ is *integrable*, since for any $v_1, v_2 \in \mathcal{V}$, $[v_1, v_2](f_r) = 0 \Rightarrow [v_1, v_2] \in \mathcal{V}$. Next, since $X_{f_r}(f_s) = \{f_s, f_r\}$, we must have $[X_{f_r}, v](f_s) = 0 \;\forall v \in \mathcal{V}$. Therefore, the X_{f_r} preserve the vertical distribution. Similarly, the Hamiltonian vector field of any algebraic function of the f_r also preserves the distribution. Lastly, since $[X_{f_r}, X_{f_s}] = 0 \;\forall r, s$ the horizontal spaces are also integrable. Therefore, there exist global horizontal sections of our quantum bundle over Γ.

We are now prepared to reconstruct the classical symplectic structure from the geometry of the quantum phase space. Let ξ and ζ be two vector fields on Γ and denote by $\tilde{\xi}$ and $\tilde{\zeta}$ their horizontal lifts to $\mathcal{P}$. Then the classical symplectic structure is defined by

$$\alpha(\xi, \zeta) := \omega(\tilde{\xi}, \tilde{\zeta}). \tag{54}$$

We have simply defined α as the "horizontal part" of the quantum symplectic structure. Since the classical phase space is linear, it is obvious that Eq. (54) correctly defines the classical symplectic structure. However, in order to allow more general constructions, let us see why this definition of α provides a symplectic structure on Γ. First of all, it is easy to show that $\omega(\tilde{\xi}, \tilde{\zeta})$ is constant along the fibres of $\mathcal{P}$. Therefore, the definition is self-consistent. To see that α is nondegenerate, one need only notice that $\alpha(\xi, \zeta) = 0 \;\forall \zeta \Rightarrow \omega(\tilde{\xi}, \tilde{\zeta}) = 0 \;\forall \tilde{\zeta} \in \mathcal{V}^{\perp} \Rightarrow \tilde{\zeta} \in \mathcal{V}$; by construction, ζ must therefore vanish. Finally, that α is closed is obvious, since it is the pull-back, via any horizontal section, of a closed form.

Let us summarize these results. Quantization of a classical theory with a *linear* phase space is canonical, given the elementary position and momentum observables. For this case, the quantum phase space may be naturally viewed as a bundle over the classical phase space. The classical phase space is simply the base space of this bundle. The symplectic structure on $\mathcal{P}$ naturally defines a notion of horizontality, and the classical symplectic structure is simply the horizontal part of ω.

Incidentally, note that our arguments go a short step toward the more general theory for which the classical phase space is not necessarily linear. Suppose that $\mathcal{V}$ is an integrable, symplectic distribution on $\mathcal{P}$ of finite co-dimension $2n$ which may be specified *locally* by the constancy of functions $f_r, r = 1, \ldots, 2n$. Suppose also that these functions may be chosen in such a way that their Poisson algebra closes (up to constants). Then the quotient of $\mathcal{P}$ by $\mathcal{V}$ inherits a natural symplectic structure. It is desirable to eliminate the requirement that the co-dimension of $\mathcal{V}$ be finite; otherwise treatment does not apply to, for example, quantum field theory. We believe that this assumption can be removed although a detailed analysis is yet to be carried out.

Note that our construction provides not only a projection from $\mathcal{P}$ to Γ but also a class of preferred embeddings of classical phase space into the quantum one. One may guess that these embeddings are related, in some way, to coherent states. After all, it is well known that spaces of coherent states are endowed with natural symplectic structures [39, 18].

Let us therefore examine the nature of the coherent state spaces from the geometric point of view. There are a number of different constructions of coherent states. We consider the attractive and fairly general approach introduced by Perelomov [19], Gilmore [20] and, to some extent much earlier, by Klauder [21]. Consider the the Heisenberg-Weyl group,[8] obtained by exponentiating the Lie algebra generated by $\hat{Q}_i$, $\hat{P}_j$, and the identity operator on $\mathcal{H}$. One defines Perelomov's space of generalized coherent states by the action of this group on an arbitrary element $\Psi_0 \in \mathcal{H}$. In this manner, one obtains generalized coherent states $\Psi_{(q_i', p_i')} := \exp[-\frac{i}{\hbar}\sum_i (q_i' \hat{P}_i - p_i' \hat{Q}_i)]\,\Psi_0$, labeled by pairs of parameters (q_i', p_i') with dimensions of position and momenta, respectively. If one chooses for Ψ_0 the ground state of the oscillator, one recovers the space of standard coherent states. However, one may apply the above construction using arbitrary Ψ_0 as the fiducial state. Hence, the notion of generalized coherent states is a viable one for the generic system.

Let us now return to the relation between the quantum and classical phase spaces. Given an arbitrary element $x_0 \in \mathcal{P}$ of the quantum phase space, one can obtain a submanifold of generalized coherent states as follows: Choose a state Ψ_0 which projects to x_0 and construct the generalized coherent states via the standard method described above. To begin with, these states constitute a genuinely nonlinear subspace of the Hilbert space. Nonetheless, we may project the entire space to $\mathcal{P}$. An understanding of the nature of the resulting submanifold of $\mathcal{P}$ is provided by the following facts. First, the expectation values of the basic operators at the coherent states are given by

$$q_i(x_{(q',p')}) = q_i(x_0) + q_i' \quad \text{and} \quad p_i(x_{(q',p')}) = p_i(x_0) + p_i' \tag{55}$$

(so that if we choose for Ψ_0 the ground state of the harmonic oscillator, the expectation values are precisely q' and p'). Second, any two coherent states generated by the same fiducial element x_0 possesses the same uncertainties. Therefore, the uncertainties in q_i and p_j are constant on the generalized coherent state space, and hence on the submanifold of $\mathcal{P}$ they define. Finally, note that the definition of the Heisenberg-Weyl group implies

$$\frac{\partial}{\partial q'}\Psi_{(q',p')} = \frac{1}{i\hbar}\hat{P}\Psi_{(q',p')}, \tag{56}$$

$$\frac{\partial}{\partial p'}\Psi_{(q',p')} = -\frac{1}{i\hbar}\hat{Q}\Psi_{(q',p')}. \tag{57}$$

Projecting this result to $\mathcal{P}$, one immediately verifies that each space of generalized coherent states is everywhere horizontal. Thus, our horizontal sections on $\mathcal{P}$ are precisely the generalized coherent state spaces! Hence, it follows in particular that the uncertainties in the elementary quantum observables are constant on the horizontal sections.

---

[8] Although this group is kinematical in origin, in the literature on coherent states, it is often referred to as the *dynamical group* associated with the simple harmonic oscillator.

Finally, it is interesting to note that our construction of the horizontal sections did *not* refer to the Heisenberg group. Given a general system, Perelomov's generalized coherent state spaces are constructed as above, but by use of a different ("dynamical"; see footnote 8) group; an entirely different set of generalized coherent state spaces will then result, which will *not* correspond to horizontal sections. This seems an a point worth further investigation.

3.4.2 Dynamics: Oscillators

Let us now consider dynamical issues. Since the horizontal spaces are integrable, the classical phase space may be embedded into $\mathcal{P}$ in infinitely many ways. Is there a *preferred* horizontal section? The answer to this question will involve the dynamics of the system. In general, we expect this issue to be far from trivial. The case of the harmonic oscillator is, as one may guess, fairly simple. In this subsection we will restrict ourselves to this case.

Recall the elementary treatment of the classical limit in terms of Ehrenfest's theorem. The rate of change of the expectation value $\langle \hat{F} \rangle$ of any observable operator $\hat{F}$ is given by

$$\frac{\mathrm{d}}{\mathrm{d}t}\langle \hat{F} \rangle = \frac{1}{i\hbar}\left\langle \left[\hat{F}, \hat{H}\right] \right\rangle, \tag{58}$$

where $\hat{H}$ is the Hamiltonian operator. As usual, let f and h denote the corresponding observable functions on $\mathcal{P}$. Then as a result of Eq. (14) and its subsequent discussion, Eq. (58) directly translates to

$$\frac{\mathrm{d}f}{\mathrm{d}t} = \{f, h\}. \tag{59}$$

The fact that this equation exactly mirrors the classical expression is, however, deceiving, because the functional form of the quantum Hamiltonian h is typically entirely different from that of the classical Hamiltonian. For example, in general, one may not even be able to express h in terms of q_i and p_i alone. In fact, this is the case already for the harmonic oscillator. At first, this seems puzzling because, as is well known, the q_i and p_i "follow the classical trajectories" if the Hamiltonian operator is a quadratic function of $\hat{Q}_i$ and $\hat{P}_i$.

Let us therefore pursue this a bit further. For simplicity, consider the one-dimensional case. First, we must obtain the form of the quantum Hamiltonian. This is easy. Since $\hat{H} = (1/2m)\hat{P}^2 + (m\omega^2/2)\hat{Q}^2 = (1/2m)\{\hat{P}, \hat{P}\}_+ + (m\omega^2/2)\{\hat{Q}, \hat{Q}\}_+$, the Hamiltonian function on $\mathcal{P}$ must be of the form

$$h = \frac{1}{2m}p^2 + \frac{m\omega^2}{2}q^2 + \frac{1}{2m}(\Delta p)^2 + \frac{m\omega^2}{2}(\Delta q)^2, \tag{60}$$

where we have used only the definition and properties of the symmetric bracket. Cross terms involving the uncertainties—and hence the total Hamiltonian h—can not be expressed in terms of q_i and p_i alone. More generally, such cross terms are by-products of the quantization process and provide one of the significant differences between the forms of the classical and quantum dynamics.

Notice that the Hamiltonian may be decomposed into two parts; hence, so may the corresponding Hamiltonian vector field. $X_h = X_{h_0} + X_{h_\Delta}$, where h_0 takes the form of the classical Hamiltonian and $h_\Delta = (1/2m)(\Delta p)^2 + (m\omega^2/2)(\Delta q)^2$. Now, recall that the uncertainties Δq and Δp are constant in the horizontal directions. As a consequence, X_{h_Δ} must be purely vertical. This should not be at all surprising since, as we already know, time dependence of q and p is the same as for the classical observables. Since X_{h_Δ} is everywhere vertical, the pull-back s^*h of the quantum Hamiltonian, via any horizontal section $s : \Gamma \to \mathcal{P}$, to the classical phase space is precisely the classical Hamiltonian, up to a physically irrelevant overall constant. This is essentially the reason why the evolution of the basic quantum observables agrees with the classical evolution. However, the quantum evolution is quite different from the classical, for it does not generally preserve the horizontal sections. A natural question arises: Is there a horizontal section which is preserved by the Hamiltonian evolution?

Note that we are asking for much more than a dynamical trajectory confined to one horizontal section. The question is whether there exists an *entire* horizontal section of $\mathcal{P}$ which is preserved by the quantum evolution. A section $s_0 : \Gamma \to \mathcal{P}$ is of this special nature if and only if X_{h_Δ} vanishes on the entire image of s_0. Equivalently, we may search for a section s_0 on which the uncertainty term h_Δ attains a local extremum. It seems a rather natural guess that this will be the case at states which saturate the uncertainty relation between q and p. This expectation is correct. One can see this by writing

$$(1/\omega\hbar)h_\Delta = \left[\Delta q\sqrt{m\omega/2\hbar} - \Delta p\sqrt{1/2m\omega\hbar}\right]^2 + (1/\hbar)\Delta p\Delta q \geq 1/2,$$

where the inequality is due to Heisenberg. Therefore, any state x at which $h_\Delta(x) = \omega\hbar/2$ extremizes the uncertainty term. It is now easy to see that the only values of Δp and Δq (subject, of course, to Heisenberg's uncertainty relation) which extremize h_Δ are given by $(\Delta q)^2 = \hbar/2m\omega$ and $(\Delta p)^2 = m\omega\hbar/2$. There is therefore a *single* horizontal section which is preserved by the Hamiltonian evolution; it is the section on which the quantum evolution is "most classical." As one might suspect, this preferred section corresponds to the standard coherent state space, generated by the oscillator's ground state. These results clearly hold for any finite-dimensional oscillator, and likely for the infinite-dimensional case (e.g., quantum Maxwell theory).

3.4.3 *Dynamics: WKB Approximation*

The previous discussion of dynamics has been, to some extent, in the context of the Ehrenfest approach to semiclassical dynamics. Let us also consider the problem from the point of view of Hamilton-Jacobi theory—the context in which one often introduces the WKB approximation. While the discussion of Section 3.4.2 has been limited to the special case of the oscillator, we now consider more general systems and obtain two main results. First, we will present interesting condition for the validity of the WKB approximation which, to our knowledge, has not been discussed in the literature. Second, we will show that the WKB equation actually corresponds to a Hamiltonian evolution on the projective space, and therefore defines a generalized quantum dynamics of the Weinberg type.

We will consider the dynamics of a nonrelativistic particle moving in $\mathbb{R}^n$ under the influence of a general conservative force (see footnote 6). The wave function is therefore assumed to satisfy the Schrödinger equation

$$i\hbar \frac{\partial}{\partial t}\Psi(x,t) = \left(-\frac{\hbar^2}{2m}\Delta + V(x)\right)\Psi(x,t), \tag{61}$$

where Δ is the Laplace operator on $\mathbb{R}^n$. In the spirit of our approach, we decompose the state vector into real and imaginary parts. This defines two real fields ϕ and π via $\Psi = (\phi + i\pi)/\sqrt{2\hbar}$. In terms of these fields, the metric and symplectic structure on $\mathcal{H}$ assume the forms

$$G((\phi_1,\pi_1),(\phi_2,\pi_2)) = \int \mathrm{d}^n x\,[\phi_1(x)\phi_2(x) + \pi_1(x)\pi_2(x)], \tag{62}$$

$$\Omega((\phi_1,\pi_1),(\phi_2,\pi_2)) = \int \mathrm{d}^n x\,[\phi_1(x)\pi_2(x) - \phi_2(x)\pi_1(x)]. \tag{63}$$

The second equation may look familiar from classical field theory. It implies that the fields ϕ and π are canonically conjugate; one may view them, respectively, as the "field" and "momentum density" of the classical field theory under consideration. This is not surprising because, after all, any quantum theory can be regarded as a field theory.

Indeed, one may calculate the expectation value of the above Hamiltonian operator,

$$H(\phi,\pi) = \frac{1}{2\hbar}\int \mathrm{d}^n x \left[\frac{\hbar^2}{2m}\left((\vec{\partial}\phi)^2 + (\vec{\partial}\pi)^2\right) + V(x)\left(\phi^2(x) + \pi^2(x)\right)\right],$$

and verify that the fields evolve according to the canonical equations of motion:

$$\frac{\partial\phi}{\partial t} = \frac{\delta H}{\delta\pi} \quad\text{and}\quad \frac{\partial\pi}{\partial t} = -\frac{\delta H}{\delta\phi}. \tag{64}$$

We are interested in the relationship between these equations of motion provided by quantum dynamics and the Hamilton-Jacobi equation.

For this it is convenient to express Ψ as $\Psi = \sqrt{\rho}\exp(\mathrm{i}S/\hbar)$ and rewrite Schrödinger dynamics as equations of motion for ρ and S:

$$\frac{\partial S}{\partial t} + \frac{1}{2m}(\vec{\partial}S)^2 + V(x) = \frac{\hbar^2}{2m}\frac{\Delta\sqrt{\rho}}{\sqrt{\rho}}, \tag{65}$$

$$m\frac{\partial\rho}{\partial t} + \vec{\partial}\cdot(\rho\vec{\partial}S) = 0. \tag{66}$$

The second equation is the conservation equation, $\partial\rho/\partial t + \vec{\partial}\cdot\vec{J} = 0$, where $\vec{J} = \rho\vec{\partial}S/m$ is the probability current density. The familiar observation is the fact that Eq. (65) becomes the Hamilton-Jacobi equation upon dropping the term involving $\hbar$; in a loose sense, one obtains the classical limit upon taking $\hbar \to 0$. (For details, see standard texts, for example, [40, 41].)

Let us explore the standard approach in which one simply drops the "quantum correction" to the Hamilton-Jacobi equation. To be concrete, let us refer to the corresponding evolution as *WKB evolution*. First, one may express the Hamiltonian function in terms of ρ and S:

$$H(\rho, S) = \int \mathrm{d}^n x \left[\frac{\hbar^2}{8m\rho(x)} (\vec{\partial}\rho)^2 + \frac{1}{2m}\rho(x)(\vec{\partial}S)^2 + \rho(x)V(x) \right]. \tag{67}$$

It may be seen that ρ and S are also canonically related and that their equations of motion may be obtained from Eqs. (64) by making the replacements $\phi \to \rho$ and $\pi \to S$. One may also verify that the WKB evolution corresponds to dropping the first term in Eq. (67). That is, if $H_\hbar$ denotes this first term, and $H_{\mathrm{WKB}} := H - H_\hbar$, then the WKB evolution corresponds to integration of the Hamiltonian vector field X_{WKB} generated by H_{WKB}.

Let us first examine the condition of validity of the the WKB approximation. We require conditions under which the Hamiltonian vector field generated by $H_\hbar$ is small compared to that generated by H. To this end, we compute the functional derivatives of $H_\hbar$ with respect to the fields ϕ and π. We obtain

$$\frac{\delta H_\hbar}{\delta\phi} = -\frac{1}{8m\hbar\rho^2}\left[2\hbar\rho\,(\phi\Delta\phi + \pi\Delta\pi) + \left(\phi\vec{\partial}\pi - \pi\vec{\partial}\phi\right)^2\right]\phi; \tag{68}$$

the corresponding expression for $\delta H/\delta\pi$ is obtained by making the replacement $\phi \leftrightarrow \pi$ above. This expression may be written in terms of quantities which are somewhat more physical. One need only notice that

$$\begin{aligned} \left\langle \hat{P} \right\rangle &= \int \mathrm{d}^n x\; m\vec{J}(x), & \vec{J} &= \frac{1}{2}(\phi\vec{\partial}\pi - \pi\vec{\partial}\phi), \\ \left\langle \hat{P}^2 \right\rangle &= \int \mathrm{d}^n x\; K(x), & K &= \frac{\hbar}{2}(\phi\Delta\phi + \pi\Delta\pi). \end{aligned} \tag{69}$$

The quantities $m\vec{J}$ and K which appear in Eq. (68) may be interpreted physically as the "density of momentum" and the "density of squared-momentum".

In terms of these quantities, the above functional derivative may be written as

$$\frac{\delta H}{\delta\phi} = -\frac{1}{2m\hbar\rho^2}\left[(m\vec{J})^2 - \rho K\right]\phi. \tag{70}$$

Therefore,

$$i\hbar\left(\frac{\delta H_\hbar}{\delta\pi} - i\frac{\delta H_\hbar}{\delta\phi}\right) = \frac{i}{2m\rho^2}\left((m\vec{J})^2 - \rho K\right)(\phi + \mathrm{i}\pi). \tag{71}$$

The WKB evolution is obtained by subtracting ($1/\sqrt{2\hbar}$ times) this term from the Schrödinger equation. The WKB equation of motion may therefore be written

$$i\hbar\frac{\partial}{\partial t}\Psi = \hat{H}\Psi + \frac{1}{2m\rho^2}\left[(m\vec{J})^2 - \rho K\right]\Psi, \tag{72}$$

where $\hat{H}$ is the unaltered Hamiltonian operator. Hence, the condition of validity of the WKB equation is that the second term on the right side of Eq. (72) be small; i.e., that the "density of squared-momentum" (weighted by the probability density) be comparable in magnitude to the square of the "density of momentum."

We have just arrived at an interesting form of the WKB equation of motion. One should compare it with the general form given by Eq. (40) of the nonlinear Schrödinger equation. Note that the additional term α appearing in that equation is a function of only the amplitude ρ. Hence, the WKB equation appears to be considerably more complicated than the explicit forms of generalized dynamics that are generally considered.

However, we have already seen (see the discussion following Eq. (67)) that the WKB evolution is generated by a Hamiltonian function on $\mathcal{H}$. Surprisingly, this evolution actually induces a Hamiltonian flow on the projective Hilbert space. To see this, in the light of Section 3.3.1, we need only show that X_{WKB} preserves the unit sphere. This is the case if and only if the Poisson bracket of H_{WKB} and the constraint function, $C = \int \mathrm{d}^n x \rho(x) - 1$ vanishes at each normalized state. This fact may be proven as follows: Since $H_\hbar$ is independent of S, $\{H_\hbar, C\} = 0$. Moreover, since H generates the Schrödinger evolution, which preserves the unit sphere, $\{H, C\} = 0$. Taking the difference between these two Poisson brackets, we obtain the desired result: $\{H_{\mathrm{WKB}}, C\} = 0$. This Poisson bracket actually vanishes *strongly*; therefore, the WKB evolution actually preserves the norm of *all* state-vectors (not just the unit vectors). Therefore, the dynamics of the WKB approximation is an example of generalized dynamics of the Weinberg type; in particular, it defines a Hamiltonian flow on the projective Hilbert space.

3.5 Discussion

Let us begin with a summary of the main results.

We first showed that ordinary quantum mechanics can be reformulated in a geometric language. In particular, as in classical mechanics, the space of physical states—the "quantum phase space"—is a symplectic manifold and dynamics is generated by a Hamiltonian vector field. However, unlike in classical mechanics, the quantum phase space is equipped also with a Kähler structure. As one might suspect, the Riemannian metric, which is absent in the classical description, governs uncertainty relations and state vector reduction which are hallmarks of quantum mechanics. The geometric formulation shows that the linear structure which is at the forefront in textbook treatments of quantum mechanics is, primarily, only a technical convenience, and the essential ingredients—the manifold of states, the symplectic structure and the Riemannian metric—do not share this linearity. Therefore, the framework can serve as a stepping stone, for nonlinear generalizations of quantum mechanics.

One can consider such generalizations in various directions. A "conservative" approach would retain the kinematical structure and generalize only the dynamics. This is the viewpoint that underlies the various nonlinear generalizations of the Schrödinger equation that have been considered in the literature. The strategy is also natural and easy to implement in the geometric formulation: While in the standard formulation, dynamics can be generated by a very restricted class of functions on the quantum phase space, we can generalize the framework by allowing the Hamiltonian flow to be generated by *any* smooth function on the quantum phase space. We saw that the known proposals of generalized dynamics fall in this class. Furthermore, by concentrating on the quantum phase space—rather than the fiducial Hilbert space—one can separate essential features of dynamics from inessential ones. In particular, this naturally led to a clarification of the relation between Weinberg's framework and nonlinear Schrödinger equations and corrected some misconceptions.

The kinematical generalizations are more difficult. However, we were able to streamline the search for such generalizations by providing a characterization of ordinary quantum mechanics using structures that have direct physical interpretation: In essence, what singles out quantum mechanics is that the underlying Kähler manifold has maximal symmetries. From the geometric viewpoint then, restricting oneself to ordinary quantum mechanics is rather analogous, in (pseudo-)Riemannian geometry, to working *only* with manifolds of constant curvature. Now, in (pseudo-)Riemannian geometry, a rich theory remains even if one drops the restriction of maximal symmetries. Indeed, in the context of general relativity, the manifolds of constant curvature are only the "vacua," and in most physically interesting situations, there are significant departures from these geometries. Is the situation perhaps similar in the case of quantum mechanics? Have we been restricting our attention only to the most elementary of viable theories?

The geometrical formulation is also useful to probe semiclassical issues. We saw in particular that the quantum phase space has a natural bundle structure and the horizontal cross sections correspond precisely to families of generalized coherent states. It also provides a succinct and clear condition for validity of the WKB approximation. Furthermore, it turned out that dynamics in the WKB approximation yields a well-defined flow on the quantum phase space which corresponds to a "generalized quantum dynamics" in the sense of Weinberg.

The quantum (Hilbert space and hence) phase space is finite-dimensional only in exceptional cases, such as spin systems. Most work in the literature in the area of geometric formulations of quantum mechanics deals only with this case. By working with Hilbert manifolds, we were able to treat the generic case—such as particles moving in $\mathbb{R}^n$—where the quantum phase space is infinite-dimensional. Finally, most of our results go through also in quantum field theory (although the measurement postulates are, as is usual, geared to nonrelativistic quantum mechanics). Indeed, the geometric treatment sheds new light on the second quantization procedure. Because the space of quantum states is itself a symplectic manifold equipped with a Kähler structure, it turns out that one can use (a natural infinite-dimensional generalization of) the machinery of geometric quantization to carry out quantization again. The resulting theory is precisely the second quantized one. Thus, second quantization is indeed $(\text{quantization})^2$! (For details, see [16].)

We will conclude by listing a few of the important open problems. First, we have given an intrinsically geometric formulation of the five postulates of quantum mechanics that deal with kinematics, unitary evolution, and measurements of observables (possibly with continuous spectra). However, we did not include the spin-statistics postulate. The reason is that we do not have a succinct formulation of this postulate which refers only to the essential geometric structures. Obtaining such a formulation is a key open problem. The remaining issues deal with generalizations of the standard framework. We saw that it is rather straightforward to extend dynamics by allowing the Hamiltonian to be any densely defined function on the quantum phase space. However, unless it is an observable function in the sense of Definition 3.2.1, we may not have a consistent measurement theory for it. Whether this is a problem is not so clear. Indeed, this feature arises already for the nonlinear Schrödinger equations considered in the literature and there it is generally not perceived as a problem. However, a systematic analysis of this issue should be carried out. Next, as indicated in Section 3.3.2, there is possibility that there exist infinite-dimensional Kähler manifolds with constant holomorphic sectional curvature ($= 2/\hbar$) which are *not* isomorphic to a projective Hilbert space. If this does turn out to be the case, we will obtain viable, nontrivial generalizations of quantum kinematics for which even the measurement theory could be developed in detail. Therefore, it is important to settle this issue. Also, even in the finite-dimensional case, we do not know if there exist *any* Kähler manifolds other than projective Hilbert spaces for which a satisfactory measurement theory can be developed. Even isolated examples of such manifolds would be very illuminating. The final issue stems from the fact that the space of quantum states shares several features with the classical phase space. Is there then a quantization procedure to arrive directly at $(\mathcal{P}, \omega, g)$ without having to pass through the Hilbert space? Not only may the answer be in the affirmative but the new procedure may even be useful in cases when the standard quantization procedure encounters difficulties. That is, such a procedure may itself suggest generalizations of quantum mechanics.

Acknowledgments

We would like to thank Lane Hughston for correspondence, Ted Newman for suggesting that we examine the WKB approximation, and Domenico Guilini for pointing out some early references. This work was supported in part by the NSF grants PHY93-96246 and PHY95-14240 and by the Eberly fund of The Pennsylvania State University.

References

[1] G. C. Ghirardi, A. Rimini, and T. Weber, "Unified dynamics for microscopic and macroscopic systems," *Phys. Rev.* **D34**, 470–491 (1986).

[2] P. Pearle, "Combining stochastic dynamical state reduction with spontaneous localization," *Phys. Rev.* **A39**, 2277–2289 (1989).

[3] G. C. Ghirardi, R. Grassi and A. Rimini, "Continuous spontaneous reduction involving gravity," *Phys. Rev.* **A42**, 1057–1064 (1990).

[4] J. S. Bell, "Introduction to the hidden-variable question," in *Proceedings of the international school of physics Enrico Fermi, course IL: Foundations of quantum mechanics*, (Academic Press, New York, 1971).

[5] I. Bialynicki-Birula and J. Mycielski, "Nonlinear wave mechanics," *Ann. Phys.* New York **100**, 62–93 (1976).

[6] P. Pearle, "Reduction of a state vector by a non-linear Schrödinger equation," *Phys. Rev.* **D13**, 857–868 (1976).

[7] S. Weinberg, "Testing quantum mechanics," *Ann. Phys.* New York **194**, 336–386 (1989).

[8] P. Pearle, "Towards a relativistic theory of state vector reduction," in *Sixty-two years of uncertainty*, Ed A. I. Miller (Plenum Press, New York, 1990).

[9] J. S. Bell, *Speakable and unspeakable in quantum mechanics* (Cambridge University Press, Cambridge, 1987).

[10] A. Ashtekar and J. Stachel, *Conceptual problems of quantum gravity* (Birkhäuser, Boston, 1991).

[11] K. V. Kuchař, "Time and interpretations of quantum gravity," in *Proceedings of the 4th Canadian conference on general relativity and relativistic astrophysics*, Eds. G. Kunstatter, D. Vincent, and J. Williams (World Scientific, Singapore, 1992).

[12] C. J. Isham, "Canonical quantum gravity and the problem of time," in *Integrable systems, quantum groups and quantum field theory*, Eds. L. A. Ibart and M. A. Rodrigues (Kluwer, Dordrecht, 1992).

[13] R. Penrose, *Emperor's new mind: Concerning computers, minds and the laws of physics* (Oxford University Press, Oxford 1989).

[14] M. Gell-Mann and J. B. Hartle, "Classical equations for quantum systems," *Phys. Rev.* **D47**, 3345–3382 (1993).

[15] C. J. Isham, "Quantum logic and the histories approaches to quantum theory," *J. Math. Phys.* **35**, 2157–2185 (1994).

[16] Troy A. Schilling, *Geometry of quantum mechanics*, doctoral thesis (The Pennsylvania State University 1996).

[17] T. W. B. Kibble, "Geometrization of Quantum Mechanics," *Commun. Math. Phys.* **65**, 189–201 (1979).

[18] A. M. Perelomov, *Generalized coherent states and their applications* (Springer-Verlag, New York, 1986).

[19] A. M. Perelomov, *Commun. Math. Phys.* **26**, 222 (1972).

[20] R. Gilmore, *Ann. Phys.* New York **74**, 391 (1972).

[21] J. R. Klauder, *J. Math. Phys.* **4**, 1055 (1963), *J. Math. Phys.* **4**, 1058 (1963).

[22] R. Penrose and W. Rindler, *Spinors and space-time, vol 1* (Cambridge University Press, Cambridge, 1985).

[23] A. Ashtekar, G. T. Horowitz, and A. Magnon-Ashtekar, "A generalization of tensor calculus and its applications to physics," *Gen. Rel. Grav.* **14**, 411–428 (1982).

[24] A. Heslot, "Quantum mechanics as a classical theory," *Phys, Rev.* **D31**, 1341–1348 (1985).

[25] J. Anandan and Y. Aharonov, "Geometry of quantum evolution," *Phys. Rev. Lett.* **65**, 1697–1700 (1990).

[26] G. W. Gibbons, "Typical states and density matrices," *Jour. Geom. Phys.* **8**, 147–162 (1992).

[27] L. P. Hughston, "Geometric aspects of quantum mechanics," in *Twistor Theory*, Ed. S. A. Huggett, (Marcel Dekker, New York, 1995).

[28] L. P. Hughston, "Geometry of stochastic state vector reduction," *Proc. R. Soc. Lond.* **A452**, 953–979 (1996).

[29] R. Cirelli, A. Manià, and L. Pizzocchero, "Quantum mechanics as an infinite-dimensional Hamiltonian system with uncertainty structure: Part I," *J. Math. Phys.* **31**, 2891–2897 (1990).

[30] R. Cirelli, A. Manià, and L. Pizzocchero, "Quantum mechanics as an infinite-dimensional Hamiltonian system with uncertainty structure: Part II," *J. Math. Phys.* **31**, 2898–2903 (1990).

[31] P. R. Chernoff, J. E. Marsden, *Properties of infinite-dimensional Hamiltonian systems* (Springer-Verlag, Berlin, 1974).

[32] R. Shankar, *Principles of quantum mechanics*, Chapter 9 (Plenum Press, New York, 1980).

[33] P. A. M. Dirac, *Lectures in quantum mechanics* (Yeshiva University Press, New York, 1964).

[34] R. Geroch, "A method for generating solutions of Einstein's equations," *J. Math. Phys.* **12**, 918–924 (1971).

[35] R. M. Wald, *General relativity* (The University of Chicago Press, Chicago, 1984).

[36] A. Ashtekar and A. Magnon-Ashtekar, "A technique for analyzing the structure of isometries," *J. Math. Phys.* **19**, 1567–1572 (1978).

[37] M. Reed and B. Simon, *Methods of modern mathematical physics*, v. I, *Functional analysis* (Academic Press, 1980).

[38] K. Yano, *Structures on manifolds* (World Scientific, Singapore, 1984).

[39] W.-M. Zhang, "Coherent states: Theory and some applications," *Rev. Mod. Phys.* **62**, 867–927 (1990).

[40] H. Goldstein, *Classical Mechanics* (Addison-Wesley, Reading, MA, 1981).

[41] C. Lanczos, *The variational principles of mechanics* (Dover Publications, New York, 1970).

4

General Covariance is Bose-Einstein Statistics

James Baugh
David Ritz Finkelstein
Heinrich Saller
Zhong Tang

ABSTRACT Points of spacetime, like bosons, are indistinguishable. This may be a macroscopic quantum effect. Sets are fermionic, however, in that double occupation is excluded. Representing spacetime as a point set ("bosons" as "fermions") introduces many unphysical degrees of freedom. Instead we represent a spacetime as a network of bosonic links $\iota = \{\cdots\}$ generalizing and quantizing Peano's theory of the natural numbers. His ι becomes the fermionic quantizer but is itself bosonic. Immediate causal succession is the membership relation $\in$ derived from ι. The causal relation is the transitive relation $\in^* = \in \cdots \in$ derived from $\in$. We use ι to express a simple time axis and then Minkowski spacetime as quantum logic networks. While ι is used infinitely often in the mathematical foundations of differential geometry, and merely once or twice as quantization during quantum theory construction, in this theory it occurs at a rate of $\sim 10^{120}\ \mathrm{s}^{-4}$ in the vacuum. A Compton limit to the precision of spacetime coordinates, 16 orders of magnitude above the usual Planck limit, sets this rate.

4.1 Quantum Relativity

The usual quantum theory is not fully quantum in that it assumes a classical spacetime structure, ignoring the dynamical variations in this structure that one expects today; and it uses a quantum predicate algebra for the system level only, reverting to classical set theory for every other computational level, including the metalevel.

Nor is general relativity fully general. When Einstein first spoke of "arbitrary" transformations and "general" covariance, he explicitly excluded classical tearings, and he could hardly include quantum superpositions, as they had not yet been discovered.[1]

---

[1]In his later work, well into the quantum era, he held to his original classical position, though he more than once suggested quantum spacetime ("the application of Heisenberg's

Thus the desert between the modern quantum and general relativity theories limits both. To cross it we must simultaneously quantize spacetime and extend quantum theory to new levels.

In this paper, dedicated to Engelbert Schucking with admiration and affection, we correct a residual objectivization of spacetime points that marred our earlier efforts in these directions [Finkelstein (1996)]. Then we assembled an absolute spacetime out of given fermionic points. Here we make a relational spacetime out of replicas of one bosonic link, generating spacetime points as needed.

To estimate the scale time of the link, we point out fundamental limits to the precision of coordinates that set in at Compton sizes, 16 orders of magnitude closer to experiment than the Planck size.

We propose that the key symmetry of general relativity, general covariance, is an expression of a more fundamental bosonic statistics, and is a macroscopic quantum effect. Two of the best-known symmetries of twentieth-century physics may be one. Perhaps we accommodated to the vacuum's amazing properties so long before the quantum theory was discovered that we have been blind to its quantum nature.

4.2 General Covariance and Bosonic Statistics

It is amazing that laws of geometry worked out on the soils of the Nile and the sands of the Aegean work as well as they do in the tangent spaces of spacetime, considering that particles of soil and sand are visible and distinguishable while points of spacetime are not. We distinguish two electrons (say) by their positions. What distinguishes their positions? Not the electrons that occupy them, evidently.

Heisenberg taught us how to use an algebra of operations to overcome some of the mismatch between physical systems, which do not have absolute states, and their paper descriptions, which effectively do. Here we discuss the mismatch between physical spacetime regions, which do not have absolutely distinguishable points, and their paper or set-theoretic descriptions, which effectively do.

The usual formulation of general relativity gives quite separate kinds of theory for the points and the connections of spacetime. The point set of the manifold is given in the kinematics and not subject to variation in the dynamical principle. The causal connections in the manifold are defined by the metric tensor $g_{\nu\mu}(x)$ and dynamical, subject to variation.

This field/point distinction seems unphysical. The discovery of black hole singularities makes points dynamical too. We can annihilate a point of spacetime if we choose by focusing a black hole singularity on it. This unnatural distinction also forfeits a major goal, the unity of spacetime and matter-field in one entity that has been called space-time-matter [Weyl (1922, 1993)].

---

algebraic method to spacetime") as the next thing to try if his asymmetric or complex metric did not succeed.

Yet such a split is inevitable in a classical theory, for discontinuous creation and annihilation of points cannot be coupled deterministically to continuous variation of fields. In the quantum theory, however, the continuous and the discontinuous merge into the quantum, and it is not clearly impossible to couple points and fields, or even to eliminate spacetime points in favor of their causal connections. Points are as nonobjective as quanta.[2] Nevertheless we did not completely eliminate the field/point split from our earlier attempts at a quantum spacetime [Finkelstein (1996)], because we did not describe how the points of our spacetime models were generated from their mutual relations. We do both here.

In experience, spacetime points are distinguished solely by their metrical relations to each other and the other fields they carry. They do not carry names or addresses besides these fields. When we determine the radar (null) coordinates of an event, for example, we are simply measuring electromagnetic fields propagating in a gravitational field.

Field theory distinguishes points twice: once by their coordinates and once by their fields, including the metric field. Experiment distinguishes them only once. Field theory uses a redundant description. One then cancels this redundancy by postulating general covariance: diffeomorphisms interchange points with no experimental consequence. Points are identical.

A classical example of such identical but plural entities is the cube $\phi\phi\phi$ of a mapping ϕ. Here are three identical mappings ϕ in a line. These ϕ's are identical, yet we can still say that one ϕ is first, another ϕ second, and another ϕ third, assigning them coordinates 1, 2, 3 by their mutual relations only.

We construct here spacetime topologies with a similar cellular relational structure, not linear but reticular, and not classical but quantum, and hence not fixed *a priori* but dynamically varying.

4.3 The Two-Point Paradox

Especially for a quantum theory of spacetime, where actions are primary, and unfeasible actions should not be postulated, it seems necessary to dispense at the start with the fiction that points of spacetime are distinguishable, just as one dispenses with the less transparent fiction that systems have objective states. To stress that this is a radical change in theory, we formulate it as another pair-paradox to accompany the two-slit and the twin paradoxes.

Two points of spacetime are identical, and yet they are two.

The two-slit and twin paradoxes bring out unexpected path dependences, hence noncommutativities, and were found in this century. The two-point paradox was apparent to Leibniz if not Aristotle. The effect is normal in a relational theory of spacetime. It counts as paradox only because today absolute spacetime is orthodox. Here we express it as an unexpected commutativity, of point annihilation operators.

---

[2] Pointless spacetime is studied without benefit of quantum theory by [Pandres (1997)].

Points would have to store at least 100 bits to distinguish themselves intrinsically. But this relational distinction requires no internal structure of the point at all.

Because spacetime is ordinarily represented by a set of distinguishable points with additional structure, our earlier models represented a quantum spacetime and a point both by quantum sets or qusets. But sets which are identical are one. Sets that differ in their relations to other sets must differ in their elements. So the representation of spacetime as a set of points forces the redundancy already mentioned. Namely, it distinguishes points from each other twice: by their elements as sets and by their relations to each other.

Point sets are fundamentally fermionic. A set cannot contain an element twice (as can a sequence, for example). Set theory was not designed to represent bosons. Using them to do so introduces unphysical redundancy.

We may see general covariance as a compensating error. Lorentz-Fitzgerald contraction compensates for the error of absolute time and space; state-vector collapse, for the error of absolute states; general covariance, for the error of intrinsically distinguishable spacetime points.

Classical mechanics started with systems like planets and pendulums in modes of high quantum number and so behaving quite classically, and took up quantum effects later. Spacetime physics begins with a ground mode, the vacuum, where quantum effects like this bosonic property dominate.

4.4 The Quantum Causal Relation

We turn now to what truly separates and connects spacetime points: their causal relation. Einstein's metric tensor field $g_{\mu\nu}(x)$ tells us who knows who in spacetime and how often, by the light cone $g_{\mu\nu}(x)\,dx^{\mu}\,dx^{\nu} = 0$ and by the volume element $\sqrt{-g}$, respectively. Quantum theory factors "knowing" (that is, projection operators) into creating and annihilating (kets and bras). Therefore we propose that quantum space-time-matter is a multilevel pattern of creation and annihilation acts on itself. What one space-time-matter act "knows" can only be another.

To describe such reflexive acts, we use a Clifford algebra $\mathcal{Q}$ given earlier [Finkelstein (1996)].[3]

Classical finite set theory generates its sets with

- a constant $\downarrow$, the null set .
- a monadic operation ι, $\iota(X) = \{X\}$ being the unit set containing X.
- a dyadic operation $\vee$, the disjoint union.

When X represents the system and $\iota(X)$ represents a class or predicate about the system, an element of the metasystem, the unitization ι connects system to metasystem.

[3] $\mathcal{Q}$ was wrongly assigned an indefinite metric, corrected here.

Present formal theories of the real axis, and so of space and time, are still based ultimately on the numbers, which are generated by an infinite iteration of ι. Indeed, the world of classical mathematics is built entirely out of ι's. This provides some slight clue to building the world of quantum physics out of the quantum ι.

4.5 Quantum Is Simpler

Quantum theory refers essentially to the system-metasystem[4] interaction, making statements like, "If we prepare the system in this way then we can register it in that." The system-metasystem interaction has become part of our theory of nature.

It is certainly possible that bringing the system-metasystem interaction into the theory makes quantum theory fundamentally more complicated than classical. But since the quantum theory is simpler than the classical in every other respect, and since this interaction is actually there in any physical theory, implicit or explicit, it is more likely that making it explicit will simplify the theory. The most one could hope for in this vein is to reduce all interactions to the one expressed by ι. Such a quantum theory would be simpler than classical theories and present quantum theories. We have been attempting it for some time.

If Peano's theory of the natural numbers is regarded as a theory of a toy time axis, ι is a triunity of quantification, quantization, and succession. In the Peano model of linear time, the points of time are created in the course of time. Their interconnection is their generation. Topology is genealogy.[5]

We seek a similar triune model of physical spacetime.

In the corresponding construction of quantum set theory, the disjoint union $\vee$ becomes the Grassmann product, the null set $\downarrow$ becomes the number 1, ι becomes the (fermionic) quantizer, a module endomorphism from the one-system $*$ module to the many-system $*$ ring, and the dyadic operations of quantum superposition $\pm$ are appended. If ψ is a one-particle mode extensor, then $\iota(\psi)$ is a creator of a quantum in the mode ψ and so a vector in a higher-level $*$ module than ψ, forgetting the ring multiplication. We can therefore apply ι again. The result $\iota^2(X)$ is a creator-creator.

In quantum theory iterating ι means iterating quantization. To represent quantum time in the way that Peano represented classical time is to quantize many times,

---

[4]In general we use the prefix "meta-" to mean "related to higher logical levels," as in "metalanguage" and "metalogic." A metapredicate of a system is a predicate about a predicate of the system. We define the predicates of the metasystem to be the metapredicates of the system.

The metasystem in this sense mediates between the experimenter and the system, as a simplified abstraction of that which directly interacts with the system.

[5]To put it more pithily: "All relatedness has its foundation in the relatedness of actualities; and such relatedness is wholly concerned with appropriation of the dead by the living—that is to say, with 'objective immortality' whereby what is divested of its own living immediacy becomes a real component in other living immediacies of becoming." [Whitehead (1929)]

forming a quasi-continuum of levels of quantization, instead of the mere first or second quantization of the usual quantum theories.

The limit of infinitely many metalevels is a ring $\mathcal{Q}$ of quantum sets, defined as the least Clifford $*$ ring (here over $\mathbb{Z}$) that includes the trivial $*$ ring $\mathbb{Z}$ (representing the null or identity act) and is closed under the functor Meta := End* Grass*.[6] The adjoint $*$ on $\mathcal{Q}$ ultimately derives from that on the $*$ ring $\mathbb{Z}$ and is likewise definite. We call elements of $\mathcal{Q}$ extensors because $\mathcal{Q}$ is a Grassmann algebra among other things. To use an evocative term of Rota [Rota (1997)], $\mathcal{Q}$ is an "algebra over itself."

Irony! The operation of fermionic quantization is itself bosonic:

$$\iota^*\iota - \iota\iota^* = 1, \qquad \iota^*(\downarrow) = 0. \tag{1}$$

Building spacetime from ι's makes spacetime points less basic than spacetime links.

$\mathcal{Q}$ is a quantum homologue of the class $\mathcal{S}$ of finite pure sets. The operator ι may be considered as an ideal element of $\mathcal{Q}$, yet enters into its construction.

The number operator $N = \iota\iota^*$ for this boson has as its eigenvalue for any extensor ϕ the length of the longest chain of ι's in the factorization $\phi = \iota^n\psi$ with $\iota^*\psi = 0$. We call the subextensor ψ the *mooring* of ϕ and the quantum variable N the *scope*[7] of the quset.

We describe any act on a quset sharply by an element in $\mathcal{Q}$.

4.6 Limits to Spacetime

Our goal is to make a stick-figure cartoon of the vacuum as one does of a molecule or crystal, and of particles as of crystal defects. There are some inevitable differences between the spacetime hypercrystal and atomic crystals besides their dimensionalities.

- We add and subtract these cartoons to show quantum superposition.
- We put arrowheads on the sticks to show the sense of time.
- We nest cartoons within cartoons to show the multilevel hierarchy of reflexive quantum acts.

Atoms of a crystal are so massive that often we can ignore their statistics. We can never treat spacetime points in the vacuum hypercrystal as distinguishable.

We describe the space-time-matter under study as a quantum, relativistic, hierarchic algorithm; Al for short. Language, computer, and neural network are also workable metaphors. Each subelement of space-time-matter defines a qubit in Al, with the value 1 for presence and 0 for absence.

[6] Where Grass* forms the Grassmann $*$ module of a given $*$ module, and End* forms the endomorphism $*$ ring of its argument, a $*$ module. Two forgetful functors in Meta are themselves forgotten in this definition.

[7] Scope in the nautical sense: the length of the chain leading to the mooring.

A vacuum is any highly regular, even hypercrystalline, mode of Al that we use as a background for less regular modes, excitations of the vacuum. Matter is not a tenant of the spacetime but a variant of the vacuum computation or message.

In classical set theory every set can be reduced to a disjoint union of unit sets. Qusets, however, may entangle by superposition to defy such reduction. This simplifies the theory. Classical theory needs uncountably many cells in order to be relativistically invariant. In the quantum theory, even a handful of quantum cells suffices for this with entanglement. Relativistic invariance of a quantum lattice is thus simpler than that of a classical one. The duration of the unit cell of this cellular vacuum provides a fundamental unit of time that we designate by Π.

4.7 The Elephantine Chronon

For consistency with present physics, Π cannot be much greater than the present-day limit to the precision with which spacetime coordinates can be measured. This is usually set at the Planck time $T_P \sim 10^{-45}$ s, based solely on quantum and gravitational effects.

At very short times and high energies, however, gravity fuses with all the other forces, and they also must enter into this limit. To measure four null coordinates of an event at once, we have the event emit a short light pulse and time its arrival as a blip at four remote receivers. The times of the four blips are the coordinates.

How short can a pulse be? To be sure that a photon reaches each receiver with a sharp arrival time, we must make a pulse of many photons. Since they fit into in a small hypervolume they have high energy. Therefore they collide with each other and produce pairs of every kind of quantum. These quanta Compton scatter the original photons, and re-annihilate each other in pairs to produce new photons, broadening the original sharp pulse to about the Compton time for an average quantum.

So the operative limit to the precision of spacetime coordinates is not the Planck time, associated with an energy of $\sim 10^{16}$ TeV, but some Compton time, associated with a rest energy E of a mere TeV or less, $\Pi \sim T_C = \hbar/E \sim\sim 10^{-30}$ s, roughly 16 orders of magnitude closer to experimental detection. Quantization goes on in the vacuum at a rate[8] $\sim \Pi^{-4} \sim 10^{120}\ \mathrm{s}^{-4} \sim 10^{96}\ \mathrm{s}^{-1}\,\mathrm{m}^{-3}$. To see its effects in laboratory experiments requires not still higher energy experiments but a deeper theory of the voluminous data already in hand.

4.8 Topological Nature of Gauge

This quantum lattice also provides a nonintegrable gauge transport from cell to neighboring cell, much as a crystal topology does in the classic defect theories of

---

[8] We thank Theodore Jacobson for a helpful correction.

Volterra and Burgers. Some have proposed that all gauge is a consequence of higher dimensions. We propose that all gauge results from quantum fine structure of the usual dimensions at the scale of 1 TeV or less [Finkelstein (1978), Madore (1995), Selesnick (1995)].

When we transport a unit cell of the spacetime hypercrystal from one place to another, its path threads between defect strings, so the transport depends on the path. Such a path-dependent transport underlies every gauge theory. Crystalline defect strings manifest at the level of present-day physics as gauge flux tubes, and their ends as particles. Thus the quanta of particle physics, for example, should ultimately be defined as hypercrystalline defects solely by the internal topology that replaces the "good" hypercrystal of the vacuum. The usual field theory is a continuum limit of the network theory.

Topological defects in continuum theories [Vilenkin and Shellard (1994)] describe singularities in a smoothed distribution of the defects considered here.

There is thus an abundance of data to determine the spacetime code (the dynamics of Al). In the limit of classical spacetime ($\hbar \to 0$) it must imply a distribution of defect sheets that near the vacuum mode obeys the known gauge field equations of gravity and the standard model, with sheet edges obeying the Dirac equations for the leptons and quarks.

Such a space-time-matter unification is implicit in varying degree in many inspiring works:

- The crystalline ether of Isaac Newton.[9]
- The superconducting vacuum of the especially influential work of Nambu and Yona-Lasinio [Nambu and Yona-Lasinio (1961)].
- The spin networks of Roger Penrose [R. Penrose (1971)].[10]

---

[9]Newton proposed in his *Opticks* that space was filled with a crystal, "aether," whose atoms vibrated faster than light. Transverse vibrational waves in this crystal guided polarized light-corpuscles at critical decision points. But light is polarized just because the electromagnetic field is a massless gauge vector field. Our inference of the hypercrystalline vacuum structure from the data of gauge physics merely updates Newton's inference of a crystalline ether from light polarization, and eliminates its underlying spacetime continuum in the process.

Did Newton nowhere mention how planets and people could move as freely as they do through his hot crystalline ether? He must have considered the possibility that ponderable matter is not a foreign body in the ether but a disturbance of the ether itself; for Descartes had said just that in his celebrated vortex theory of matter. When Newton crystallized Descartes' fluid ether he presumably considered the immediate conceptual consequences.

In some measure, a quantum ether is a synthesis of the continuous one of Descartes and the discrete one of Newton. Our quantum hypercrystalline vacuum too might be opaque to some of its own excitations, transparent to others, and supermobile to others, like a quantum superfluid or superconductor.

[10]The spacetime-code program started from Penrose's spin networks. Penrose used Schwinger's bosonic algebra over a two-dimensional spinor space to represent the unit sphere, the geometry of directions at one point in three-dimensional space. The bosonic algebra became fermionic for a time and here reverts to bosonic, the little spinor space has

- The skeletal spacetime of Regge [Regge (1961)].[11]
- The quantized spacetime of Hartland Snyder [Snyder (1947)].
- The solid spacetime of Sakharov.
- The ur theory of Weizsäcker [Weizsäcker (1943)].
- The pre-geometry of Misner, Thorne, and Wheeler [Misner et al. (1973)].
- The nonlocal cellular spacetime of Yukawa [Yukawa (1966)].
- And the noncommutative spacetime geometry of Connes [Connes (1994)], among others.

The reflexive multilevel quantum structure we use seems to be foreshadowed by Weizsäcker's suggestion of multiquantization, and by cosmological writings of Whitehead [Whitehead (1929)] and other philosophers.

Standard field theory associates both coordinates and fields, two significantly different kinds of information, with each point of spacetime. Both must have topological meaning. Evidently the spacetime coordinates, which in the classical limit define the classical manifold topology, express long-range topological relations of a spacetime-matter cell to the entire system, while the fields express short-range topological structure of the cell and its vicinity. We suppose that the internal degrees of freedom of the standard model arise not from new external dimensions but from such truly internal structure, omitted in the low-energy continuum approximation because it cannot be resolved by long waves.

4.9 Paradox Lost

For a concrete and elementary example, we depict a time line T_3 that is three Π long. We do this in three ways: simple, simpler, and simplest.

The most common and least simple way to represent the natural numbers, used for example by C. S. Peirce and J. Von Neuman, defines the number n as a set of cardinality n. We imitate their theory for a piece of the time axis.

In this model

$$\begin{aligned} 1 &= \iota(\downarrow) \\ 2 &= 1 \vee \iota(1), \\ 3 &= 2 \vee \iota(2), \\ T_3 &= \iota(1) \vee \iota(2) \vee \iota(3)\,. \end{aligned} \tag{2}$$

---

become an inductively generated infinite-dimensional space, and the generator ι has been added to reach other points, but spacetime is still a smoothed tensor diagram.

[11] A quset may be regarded as a generalized Regge skeleton. Instead of simplicial faces, each carrying a continuous measure, it has cells of arbitrary dimension, each carrying one quantum bit of information (qubit). The link is either present or absent as its qubit has the value 1 or 0. The usual continuous gradation from shorter edges to longer of the Regge theory presumably arises from concatenations and quantum superpositions of such binary alternatives.

Combining these definitions, we find

$$T_3 = \iota(\iota(\downarrow)) \vee \iota(\iota(\downarrow) \vee \iota(\iota(\downarrow))) \vee \iota(\iota(\downarrow) \vee \iota(\iota(\downarrow)) \vee \iota(\iota(\downarrow) \vee \iota(\iota(\downarrow)))) \qquad (3)$$

Each number appears 2 times in its successor, 2^2 times in its second successor, and so forth. All these repetitions are then accumulated in the time axis T. In this set of only three distinguishable points, $\iota(\downarrow)$ appears $1+2+4=7$ times. In the limit of long times, all the redundancies diverge exponentially.

The simpler model used by Peano and Finsler [Booth and Ziegler (1996)] associates with the number n a set of height n and cardinality only 1:

$$\begin{aligned} 1 &= \iota(\downarrow), \\ 2 &= \iota(1), \\ 3 &= \iota(2), \\ T &= \iota(1) \vee \iota(2) \vee \iota(3)\,. \end{aligned} \qquad (4)$$

Now

$$T_3 = \iota(\iota(\downarrow)) \vee \iota(\iota(\iota(\downarrow))) \vee \iota(\iota(\iota(\iota(\downarrow)))). \qquad (5)$$

Each point reappears only once in each successor. In this product $\iota(1)$ appears only $1+1+1=3$ times. In the limit of infinite duration, the redundancy diverges only linearly. In the quantum theory this still becomes a Grassmann product of infinite grade with infinite redundancy.

We propose to model this toy time axis simply as

$$T_3 = \iota^3 = \quad \leftarrow \Box \leftarrow \Box \leftarrow . \qquad (6)$$

Now $\iota(1)$ appears only once. There is no redundancy. But this time axis is not a set of time points, even in this classical model. It is a mapping of the past to the future. It does not include or contain any distinguishable points of the number axis. The propagators can be rearranged without changing the result because they are identical though three.

Only a certain tradition, not experience, leads one to represent history as a set composed of states or points at successive times. What we need to represent is only our actions on the system. We do not need states or points to do this, but an algebra (in a suitably reticulated sense) of the actions themselves.

In the first two models, we represent time translation (the action of waiting) through one Π by the point-mapping $1 \to 2$ and $2 \to 3$. These map points of time into points of time.

In the third model we represent the same time translation by the abstract mapping ι itself. Instead of permuting points of time, ι increases computational depth.

The bosonic element in this description is ι. There are three identical factors of ι in T_3 and nevertheless $T_3 \neq 0$, so ι is not fermionic. We may regard these factors ι as representing three moments of this time axis.

In the quantum version the simplest model of T_3 (or any T_n for that matter) is a Clifford extensor of total Grassmann degree 2 (one initial index and one final index) and depth 3 (or n). The products in (6) are convolutions of transition amplitudes.

This representation is much "thinner" than the two previous representations, whose Grassmann degrees diverge with n, and no deeper.

The extension of this theory to the line of any length is obvious. Our earlier construction of the time coordinate t and generator ∂_t [Finkelstein (1996)] is unaffected by this change in interpretation, but the plethora of such operators is greatly reduced.

In this 1-dimensional toy, ι represents a causal link relating an act and its immediate successor, and is equivalent to the associated membership relation $\in$. since all the sets used are unit sets. The ancestral membership relation usually designated by $\in^*$ (this $*$ is not an adjoint but stands for an undetermined exponent) then represents the causal relation. In the most common set theory this representation excludes time loops, unlike general relativity.[12]

In the most obvious extensions to the 4-dimensional Minkowski vacuum, the vertices have 4 inputs and 4 outputs instead of 1 and form a hypercubic network instead of a linear one. ι is still the causal link between event pairs, but $\in$ (now distinct from the ι relation) represents the successor relation, while $\in^*$ still represents the causal relation. We show a 2-dimensional analogue and section: A piece of simple quantum plane, a 2-dimensional generalization of the quantum line (6), is

$\mathrm{qN}^2 =$

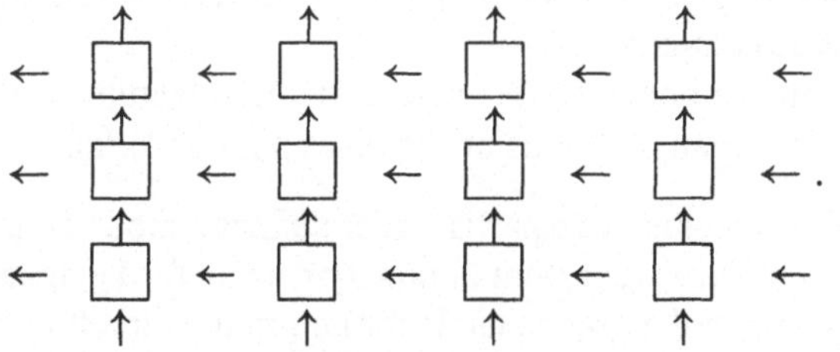

(7)

This net, like (6), when provided with initial and final extensor data in $\mathcal{Q}$ on its entire boundary, is to produce a transition amplitude.

The first question is what the many crossroads that appear in this figure all stand for. In some earlier models [Finkelstein (1996)], each vertex was a Lorentz-invariant Grassmann element and the net was their Grassmann product.

In the simplest and most plausible nonlineal generalization of (6), the net is the directed graph of an arbitrary partial order, and each box is the disjoint union or $\vee$-product of the *unitizations* of its inputs; and is therefore dually the de-unitized intersection of its outputs. In the hypercubic case this vertex is still invariant under the SL(4) group on its 4 inputs, and its 4 outputs.

Evidently the SL(4) symmetry of these networks is excessive. A slightly deeper microtopology for the unit cell of the hypercubic lattice that naturally reduces SL(4) to the Lorentz group SO(1, 3) will be described elsewhere.

The immediate development then goes much as before [Finkelstein (1996)] but with less duplication. We keep the same definitions expressing coordinates x^μ and derivatives ∂_μ in terms of discrete translations along each axis, relative to such a

---

[12] Finsler set theory [Booth and Ziegler (1996)] allows time loops, like general relativity.

spacetime net, and so also the same definitions for the generators of the diffeomorphism covariance group, the continuous Poincaré invariance group, and the internal unitary groups. Where before each point had its own family of operators, now one such family serves for the whole net.

Networks like (6), (7), and (to account for gauge) such defective nets as the dislocation

$\mathrm{q}\mathbf{N}^2 =$

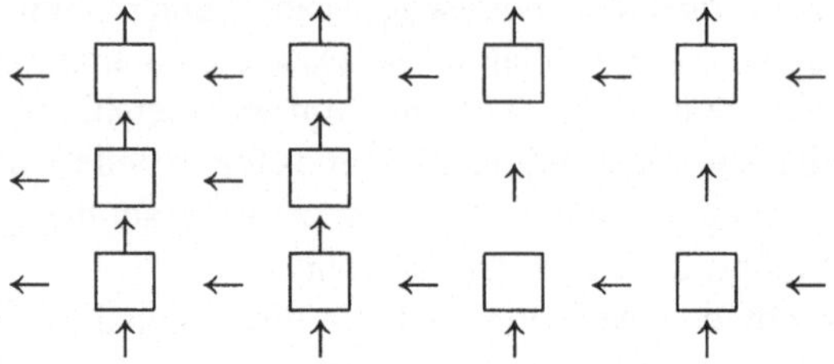

, (8)

and disclinations are extensors in $\mathcal{Q}$ expressed in terms of ι. The link ι itself is the infinite sum

$$\iota \sim \sum_{\Psi \in B} [\iota(\Psi)]^* \vee \Psi \quad (?) \tag{9}$$

where B is a classical basis for $\mathcal{Q}$. We count each ι link as a single simple element of spacetime structure.

We diagram elements of $\mathcal{Q}$, modes of the quantum spacetime network, by generalized Feynman diagrams or Penrose spin networks:

- Any directed graph of a partial order defines a mode of the quantum network.
- External vertices represent boundary data as for Feynman diagrams.
- Each arrow in the graph stands for an ι in the direction of the arrow or an ι^* in the opposite direction (with one index at each end).
- Each vertex represents a disjoint union (antisymmetrizing all the indices at that vertex).
- Any superposition (formal sum or difference) of modes is a mode.

It is then straightforward to compute the transition amplitude of this network for given boundary data.

Causal loops and completely closed diagrams have amplitude 0.

The question of action principles arises anew for these new and fewer variables, and is now more manageable. Now that ι is an element of the network it is natural to vary it in the action principle.

We still must effectively enumerate a basis for these quantum nets. Baugh provides an elegant and effective way to enumerate the pure sets, the special case with no boundary data [Baugh (1997)]. Assign to each set s a natural number $B(s)$ (that we call its *Baugh number*) by the recursive rules

$$B(\downarrow) = 0, \tag{10}$$

$$B(s \vee t) = B(s) + B(t), \tag{11}$$

$$B(\iota(s)) = 2^{B(s)}. \tag{12}$$

It is easy to prove that B enumerates and well orders the finite pure sets.

4.10 Summation

The following ideas are tested in this development:

- A physical system is defined by a semigroup of modes of action, not a succession of states of being.
- Spacetime is a smoothing of space-time-matter.
- Space-time-matter is a physical system akin to a neural network.
- Quantization is a form of quantification.
- Causal succession is another instance of quantification.

General covariance indicates that the local elements of space-time-matter are identical. We infer that they are identical in the sense of Bose-Einstein statistics. General covariance is then a macroscopic quantum effect.

We take the local element to be a causal link represented by the operator ι of fermionic quantization. which is itself bosonic.

This moves us out of the absolute spacetime used in physics from Newton through Einstein to today into an operational, hence relational, hence Leibnizian, theory of spacetime.[13]

Acknowledgments

James Baugh and David Ritz Finkelstein gratefully acknowledge support by the Institute for Scientific Interchange, Turin. Zhong Tang gratefully acknowledges support by the M. & H. Ferst Foundation, Atlanta.

References

[Baugh (1997)] Private communication.

[Booth and Ziegler (1996)] Booth, D. and R. Ziegler (eds.). *Finsler Set Theory: Platonism and Circularity*. Translation of Paul Finsler's papers on set theory with introductory comments (Birkhäuser, Basel).

---

[13] *Note added in proof.* In this paper we still assume that the event-permutation operators are in the center of the algebra of actions, implying that the event statistics must be fermionic or bosonic. In Finkelstein (1996) we proposed a quantum group of event permutations and experimented with non-central two-valued exchange operators for the special case of four events. Since then F. Wilczek has pointed out a beautiful new domain of two-valued non-central statistics; see F. Wilczek, hep-th/9710135, 9806228, of which the first has appeared in Nuclear Physics Supplement. We currently explore the possibility that the statistics of space-time-matter events is non-central, not merely projective but quantum, and yields the central bosonic and fermionic statistics only as low-energy approximations.

[Connes (1994)] Connes, A. *Non-Commutative Geometry* (Academic Press, San Diego.)

[Finkelstein (1969)] Finkelstein, D. (1969). Space-Time code. *Physical Review* **184**, 1261.

[Finkelstein (1978)] Finkelstein, D. Beneath time. In J. T. Fraser, N. Lawrence, and D. Park (eds.), *The Study of Time*, III. Springer, Berlin.

[Finkelstein (1996)] . Finkelstein, D. (1996). *Quantum Relativity* (Springer, Heidelberg). And references cited there.

[Finkelstein et al. (1974)] Finkelstein, D., G. Frye, and L. Susskind (1974). Space-time code V, *Physical Review* **D 9**, 2231.

[Haag (1990)] Haag, R. *Communications on Mathematical Physics* **132**, 245–251.

[Landau (1960)] Landau, E. *Grundlagen der Analysis* Chelsea, New York.

[Madore (1995)] Madore, J. *Non-commutative differential geometry and its physical application.* Cambridge University Press, Cambridge.

[Misner et al. (1973)] Misner, C., K. Thorne, and J. A. Wheeler., *Gravitation.* W. H. Freeman, San Francisco.

[Nambu and Yona-Lasinio (1961)] Nambu, Y. & G. Yona-Lasinio. Dynamical model of elementary particles based on an analogy with superconductivity. I, II. *Physical Review* **122**, 345; **124**, 246.

[Pandres (1997)] Pandres, D. Gravitational and electroweak interactions. *International Journal of Theoretical Physics*, in press.

[R. Penrose (1971)] Penrose, R. Angular momentum: an approach to combinatorial space-time. In T. Bastin (ed.) *Quantum Theory and Beyond.* Cambridge University Press, Cambridge.

[Regge (1961)] Regge, T. (1961). General relativity without coordinates. *Nuovo Cimento* **19** 558–571.

[Rota (1997)] Rota, G.-C. Private communication.

[Saller (1997)] Saller, H. (1997). The analysis of time-space translations in quantum fields. *International Journal of Theoretical Physics* (in press).

[Selesnick (1995)] Selesnick, S. (1995). Gauge fields on the quantum net. *J. Math. Phys.* **36**, 5465–5479.

[Snyder (1947)] Snyder, H. P. Quantized space-time. *Physical Review* **71**, 38.

[Vilenkin and Shellard (1994)] Vilenkin, A. and E. P. S. Shepard, *Cosmic Strings and other Topological Defects.* Cambridge University Press, Cambridge.

[Weizsäcker (1943)] Weizsäcker, C. F. v. (1943). *Zum Weltbild der Physik.* Hirzel Verlag, Stuttgart (1943).

[Weyl (1922, 1993)] Weyl, H. *Raum-Zeit-Materie.* 1st edn. 1922; 8th edn., with J. Ehlers, 1993. Springer, Berlin. Translation: *Space-Time-Matter.* Dover, New York (1985).

[Whitehead (1929)] Whitehead, A. N. *Process and Reality: an Essay in Cosmology.* Cambridge University Press. Corrected edn. (D. R. Griffin and D. W. Sherburne, eds.) Free Press, New York (1978).

[Yukawa (1966)] Yukawa, H. (1966). Atomistics and the divisibility of space and time. *Prog. Theor. Phys.*, Supplements no. 37 and 38, 512–523.

5

The Split and Propagation of Light Rays in Relativity

Stanisław L. Bażański

ABSTRACT A geometrical formalism is developed that describes in a relativistic way physical effects in which a split of light rays occurs that is next followed by reconvergence of rays. Besides a general geometric description that uses the approach of geometric optics, a procedure has been worked out of how to compute the difference of the proper times of arrivals of the two light beams to a measuring apparatus, as well as frequency shifts of each of the beams taken separately. The formalism is applicable to both the special and the general theory of relativity, and it can be used equally well when the light split is produced by a man-made optical device or when it is caused by the gravity field itself. The geometric description used in the formalism is independent of the physical origin of the frequency shift; i.e., or whether it is a Doppler, gravitational, or cosmological frequency shift effect.

5.1 Introduction

In the theory of relativity, there exists a unified approach to the description of the frequency shift of light propagating in a spacetime. Several situations—which in some previous approaches were considered to be a consequence of different physical effects, like the Doppler shift caused by the relative motion of the source and observer, the gravitational shift in a stationary space-time, and the cosmological redshift in nonstationary cosmological models—are in this description reduced to always the same simple fact from space-time geometry.

The objective of this paper is to derive an analogous unified geometric description of the frequency shift accompanying a propagation of light in a relativistic space-time, where light initially suffers a split which is next followed by a reconvergence of rays, regardless of the physical mechanism responsible for processes of such a kind.

5.2 Traditional Approach

From the derivation of the classical frequency shift formula it follows that the limits of applicability of the formula are slightly wider than those within which it has been usually considered. For this end, let us examine here an outline of such a derivation.

In the approximation of geometric optics the propagation of light in any pseudo-Riemmanian spacetime is represented by particles with null geodesic world lines. In a system of coordinates $\{x^\alpha\}$, a null geodesic can be described by equations of the form

$$x^\alpha = \xi^\alpha(p), \tag{1}$$

where p is an affine parameter, and the functions ξ^α satisfy the set of equations

$$k^\alpha = \frac{d\xi^\alpha}{dp}, \qquad \frac{Dk^\alpha}{dp} = 0, \qquad k_\alpha k^\alpha = 0. \tag{2}$$

Let us now consider two time like world lines in a space-time: S, the source, and O, the observer, both lying in a star-shaped neighborhood of each other, and let us take a 2-dimensional null surface Σ generated by a continuous family of null geodesics outgoing from S and incoming to O in the neighborhood considered. The generating null geodesics define on Σ a null vector field k^α, which due to Eqs. (2) satisfies all over Σ the equations $k_{[\alpha,\beta]}k^\beta = 0$, where comma stands for ordinary partial differentiation, and the square brackets denote antisymmetrization. As a consequence of these equations, after integrating the 1-form $\kappa = k_\alpha dx^\alpha$ over any closed region C of Σ with the boundary ∂C and using the Stokes theorem (see

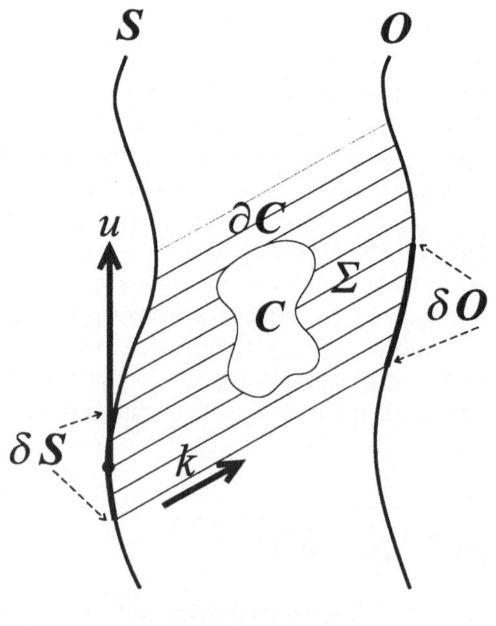

Figure 5.1

Fig. 5.1), one obtains the relation

$$\oint_{\partial C} k_\alpha \, dx^\alpha = 0, \tag{3}$$

which was derived for the first time by Ehlers and Sachs [1].

In the case when the light sent out by the source S is monochromatic, to derive the frequency shift formula one must chose for the region C on Σ the region bounded by two null geodesics γ_1 and γ_2 taken from those which generate Σ, and by the world lines S and O (see Fig. 5.1), and apply the relation (3) to its boundary. Because the integrals along γ_1 and γ_2 of $k_\alpha dx^\alpha$ vanish, one obtains from (3) the equality of the two integrals

$$\int_{\delta S} k_\alpha \, dx^\alpha = \int_{\delta O} k_\alpha \, dx^\alpha$$

along the segments δS and δO cut out of S and O by γ_1 and γ_2 respectively. Taking into account that along S and O $dx^\alpha = u^\alpha \, ds$, where ds is the proper time line element along the respective geodesic and u^α is its tangent vector normalized to unity, after making use of the mean value theorem, one obtains

$$\overline{(k_\alpha u^\alpha)}\Big|_{\delta S} \Delta s_s = \overline{(k_\alpha u^\alpha)}\Big|_{\delta O} \Delta s_o, \tag{4}$$

where the bar stands for the mean value along the segments δS and δO respectively, and Δs_s, Δs_o are the proper time lengths of the corresponding segments. Solving equation (4) for the ratio of the proper time increments and going to the limit with $\gamma_2 \to \gamma_1 = \gamma$, we obtain

$$\frac{\nu_o}{\nu_s} = \frac{(k_\alpha u^\alpha)_{P_o}}{(k_\alpha u^\alpha)_{P_s}}, \tag{5}$$

where P_s and P_o are the points of intersection of the null geodesic γ with the world lines S and O respectively, while ν_s is the frequency measured at S of the light signal emitted there, and ν_o is the frequency of the same light signal received and measured at O. The formula above can also be written in the form

$$\nu_o = \nu_s \frac{(k_\alpha u^\alpha)_{P_o}}{(k_\alpha u_{\parallel}^\alpha)_{P_o}}, \tag{6}$$

in which all quantities on the right-hand side are evaluated at the point P_o, and where $u_{\parallel}^\alpha$ is the tangent vector to S at P_s transported parallel to P_o. Formula (6) was derived for the first time by Schrödinger [2].

The procedure just presented can easily be applied to a more general situation when a light ray sent by S and received by O meets along its way a number, say n, of timelike world lines $\mathcal{R}_i$, $i = 1, 2, \ldots, n$ at each of which a change of the light propagation direction takes place. Physically, such lines may be interpreted as representing optical devices like mirrors, prisms, etc. As a result, the light rays are now only piecewise differentiable, though still continuous. This does not, however, affect the derivation of equation (5), due to the properties of integrals that were used for this purpose. Also the parallel transport appearing in Eq. (6) can obviously be generalized to piecewise differentiable world lines by a requirement that at the points where the light propagation direction is discontinuous, the result of the parallel propagation along the previous differentiable section of the ray's world line be taken as the initial value for the parallel propagation along the next such section. Summing up, we arrive at the following conclusion:

Lemma 5.2.1. The relativistic redshift formula (6) is applicable not only in the case when the world lines which connect the source and the observer are null geodesics, but also when they are piecewise null geodesics.

Let us turn now our attention to the case in which the timelike world lines S and O, which represent a source and an observer respectively, are completely lying in a timelike 2-dimensional world sheet Σ that can be interpreted as a history of an optical fibre connecting S and O. The history of a light ray that propagates through the optical fibre from S to O is represented by a null line γ lying in Σ. In general, γ is not a null geodesic any longer. The frequency shift of the light transmitted by the fibre is, however, still given by an equation analogous to equation (6).

In order to find the corresponding redshift formula, we introduce in Σ a number n of timelike world lines $\mathcal{R}_i$, $i = 1, 2, \ldots, n$ which are lying between S and O in the direction of proceeding along any null line γ, completely lying in Σ, which connects S and O. Let us now consider a piecewise null geodesic γ_n which connects the world lines S, $\mathcal{R}_1, \ldots, \mathcal{R}_n$, O. Obviously, γ_n does not lie in Σ, although its vertices do lie there. Due to Lemma 6.2.1, equation (2.5) is still valid at both ends of γ_n. For a sufficiently smooth Σ, this equation preserves its validity in the limits of $n \to \infty$, so that the maximal distance between any intermediate lines $\mathcal{R}_{i-1}$ and $\mathcal{R}_i$ will tend to zero, and consequently γ_n will tend to γ. When we would, however, like to obtain a formula analogous to (6), in which all quantities on the left-hand side are evaluated along the world line O, we should replace (6) by

$$\nu_o = \nu_s \frac{(k_\alpha u^\alpha)_{P_o}}{(k_{\alpha_\parallel} u^\alpha_\parallel)_{P_o}}, \tag{7}$$

where $u^\alpha_\parallel$ and $k_{\alpha_\parallel}$ are obtained by a parallel transport of vectors tangent to S and γ, respectively, from the point P_s of intersection of S and γ to the point P_o of intersection of O and γ. Since γ is not a geodesic line, the vectors $k_{\alpha_\parallel}$ and k_α at P_o are in general not equal to each other.

The result just obtained can be summarised in the following form:

Lemma 6.2.2. The relativistic redshift formula can be adapted to the description of the transmission of light through an optical fibre whose history is represented by a family of null lines joining the two world lines of the source and the observer, respectively.

5.3 Modified Approach

Formula (6) defines in a geometric way a construction of a function $\phi(s_o)$ of the proper time s_o along O's world line that enables us to write (2.6) in the form $\nu_s = \nu_o\phi(s_o)$. Another function of the same type is defined by equation (7). It is also possible, after performing the parallel transport not in the denominator but in the numerator of (2.5), to define a function $\psi(s_s)$, of the proper time s_s along S's world line instead, such that a relation of the form $\nu_s = \nu_o\psi(s_s)$ will take place.

There is also another geometric approach that may be used to define the functions ϕ and ψ, which turns out to be more convenient for applications of the type considered here. Its description is the subject of the present section.

Let in a coordinate system x^α the world line S of the source be analytically described by four equations $x^\alpha = \xi^\alpha(s_s)$ in terms of the proper time parameter s_s along S, and similarly let the observer's world line O be represented by $x^\alpha = \eta^\alpha(s_o)$; see Fig. 5.2. Consider the light cone emanating from a point P with the coordinates $\xi^\alpha(s_s)$ lying on S. In the accepted coordinate system the light cone is described by four equations of the form

$$x^\alpha = \kappa^\alpha\big(p, z_1, z_2, \xi^\mu(s_s)\big), \tag{8}$$

where p, z_1, and z_2 are three parameters labeling the points on the surface of the cone. In particular, we take for p the affine parameter distance from the world line S, and for z_A, $A = 1, 2$, any two coordinates that can be used for parametrizing the "celestial sphere."

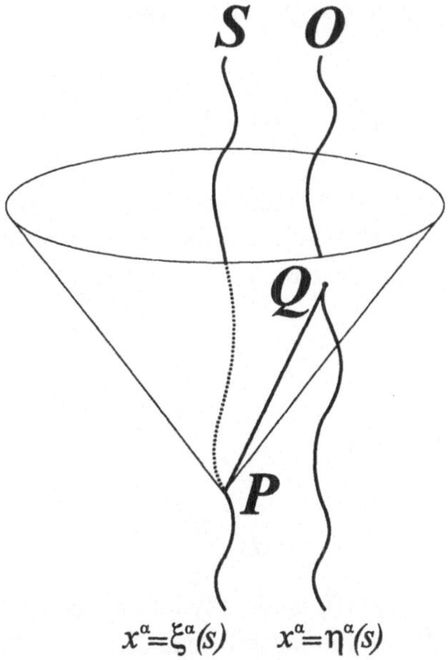

Figure 5.2

Suppose that the world lines S and O as well as the light cone with its vertex at a fixed point P of S are given; i.e., both the two functions ξ, η, and the function κ in (3.1) are known. The question is, What are the coordinates of the point Q at which the world line O will intersect the light cone? To find them, we must consider the system of the four algebraic equations

$$\eta^{\alpha}(s_o) = \kappa^{\alpha}\big(p, z_1, z_2, \xi^{\mu}(s_s)\big). \tag{9}$$

After eliminating the parameters p, z_A, $A = 1, 2$ from these equations, one obtains a single equation that, in general, admits a solution of the form

$$s_o = f(s_s), \tag{10}$$

and the coordinates of Q are then equal to $\eta^{\alpha}(s_o)$.

If the point Q belongs to a star-shaped neighborhood of P, there will be a single solution (10) of equations (9). Multiple solutions of the form (10) will occur if there is a number larger than one of null geodesic outgoing from the point P that reach the world line O, after refocusing or forming caustics on their way before intersecting O. There will be no solution (10) at all if the point P is separated from O by an event horizon.

Although the procedure of deriving (10) makes use of a coordinate system, the result (10) itself is coordinate independent, for it represents a mapping of two invariants, under which a proper time segment along S is mapped onto a proper time segment along O. In the case when the corresponding two segments of the world lines S and O belong to the intersection of the star-shaped neighborhoods of their points, the proper time segment Δs_o can be expressed in terms of Δs_s in the form

$$\Delta s_o = \frac{df}{ds_s}(s_s + \theta\,\Delta s_s)\Delta s_s, \tag{11}$$

where $0 \le \theta \le 1$. Relation (11) is just another way of writing equation (4), and can be used to derive a redshift formula that is an alternative to (6). In the high frequency limit, when both Δs_o and $\Delta s_s \to 0$, from (11) we obtain

$$\frac{\nu_s}{\nu_o} = \frac{df}{ds_s}(s_s), \tag{12}$$

and the redshift parameter z is

$$z = \frac{df}{ds_s}(s_s) - 1. \tag{13}$$

When the function $f(s_s)$ in (10) is already known, equations (3.5) and (3.6) are more suitable for calculating the redshift than (2.6). On the other hand, the latter equation reveals a geometric construction of the inverse function f^{-1}, which in the first paragraph of the present section was denoted by ϕ.

5.4 Sagnac-Like Effects

Let us now turn over to the case when equations (9) admit multiple solutions of the form (10). Instead of the single function f we had in (10), we may now have a larger number of functions. In what follows, a particular case of two functions, say f_+ and f_-, will be considered such that

$$s_{1o} = f_+(s_s), \qquad s_{2o} = f_-(s_s) \tag{14}$$

are two different solutions of equation (9), which to the same value of s_s assign two different values s_{1o} and s_{2o} of the parameter s_o. From the physical point of view, such a situation may occur both in the special and the general theory of relativity.

In the first case, this may only happen if between the world lines S and O an optical apparatus is inserted that causes a split of the light rays, which are now represented by two sequences of piecewise null geodesics that form two paths of light signals. With a null geodesic being a segment of each of the paths a function f_i, $i = 1, 2, \ldots, n$ of the type (14) is associated that maps the proper time along a time-like world line at one end of the segment into a proper time at the other end. The final mapping of the proper time along S into that along O is for every of the two paths realized by a function f which is a superposition $f_1 \circ f_2 \circ \cdots \circ f_n$ of all the functions f_i. Examples of two constructions of this type are presented in Section 6.5. A continuous limit of such a construction must be taken if the source S and the observer O are connected by two different optical fibres. A limiting case of such a kind will be presented elsewhere.

In the second case, in general relativity, in addition to the effects enumerated above, a split may occur that is caused by the gravitational field itself, due to the gravitational lensing effect [3]. In this case, in consequence of refocusing or due to forming caustics, two different null geodesics will join the world lines S and O, and equation (9) will have two solutions of the form (14).

In the general case, the two light rays, sent out from the same point P on the line S, with the coordinates $\xi^\alpha(s_s)$, will intersect the observer's world line O at two different points which correspond to two values, $s_{1o} = s_o$ and $s_{2o} = s_o + \Delta_o$, of the proper time parameter s_o along O, as is schematically shown in Fig. 5.3. In terms of the mappings f_+ and f_- introduced in (14), we have

$$\Delta_o = f_-(s_s) - f_+(s_s). \tag{15}$$

The equation above is the definition of the difference Δ_o of arrival times of the slower light signal represented by f_- and the faster one described by f_+. Regardless of the mechanism which causes the split, all effects of the kind considered now will be called Sagnac-like effects, and, correspondingly, the quantity Δ_o defined by (15) will be called a (generalized) Sagnac time (difference).

Formula (15) defines the Sagnac time Δ_o, measured by O, as a function of the proper time parameter s_s along S. It is, however, more convenient to regard Δ_o as a function of the proper time parameter $s_o = f_+(s_s)$ along O,

$$\Delta_o(s_o) = f_-(f_+^{-1}(s_o)) - s_o, \tag{16}$$

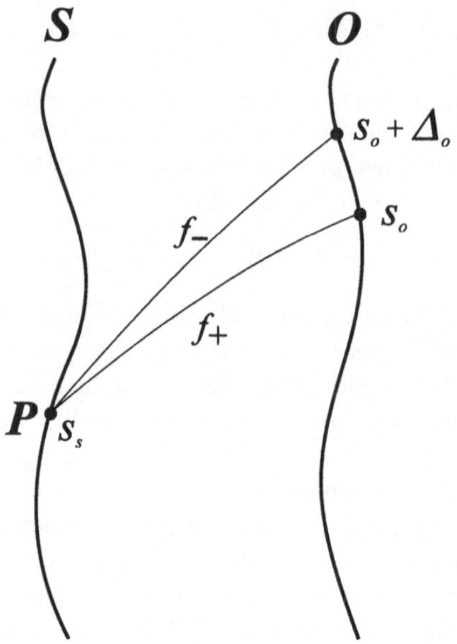

Figure 5.3

where, in accordance with the normal abuse of the language in physics, $\Delta_o(s_o)$ has been used to denote $\Delta_o(f_+^{-1}(s_o))$. Differentiating (4.3) with respect to s_o, one obtains

$$\begin{aligned} \frac{d\Delta_o}{ds_o} + 1 &= \frac{d}{ds_o}(f_+^{-1}(s_o)) \frac{df_-}{ds_s}\bigg|_{s_s = f_+^{-1}(s_o)} \\ &= \frac{f'_-(s_s)}{f'_+(s_s)}\bigg|_{s_s = f_+^{-1}(s_o)}. \end{aligned} \tag{17}$$

In practice, the two signals, described respectively by f_- and f_+, arrive to O from different directions. It is then possible to shutter one of them off and observe the other one. In this way the two frequencies ν_- and ν_+ of the respective signals can be measured independently as functions of s_o along O's world line. Due to (12),

$$\nu_-(s_o + \Delta_o) = \frac{\nu_s}{f'_-(f_+^{-1}(s_o))}, \qquad \nu_+(s_o) = \frac{\nu_s}{f'_+(f_+^{-1}(s_o))},$$

and (17) can be rewritten in the form

$$\frac{d\Delta_o}{ds_o} + 1 = \frac{\nu_+(s_o)}{\nu_-(s_o + \Delta_o)}. \tag{18}$$

Formula (18) gives a relation between three measurable quantities defined along the observer's world line O. Its immediate consequence is the following observation.

Lemma 6.4.1. The Sagnac difference of times $\Delta_o(s_o)$ remains constant along the observer's world line O if and only if for any value of the proper time parameter s_o along O the equality $\nu_+(s_o) = \nu_-(s_o + \Delta_o)$ is taking place.

From the mathematical point of view, this lemma means that, provided $\Delta_o =$ constant, the two frequencies ν_- and ν_+ are determined by the same function $\nu(s_o)$ of the parameter s_o, their values being only shifted by a constant term Δ_o, $\nu_+(s_o) = \nu(s_o)$ and $\nu_-(s_o) = \nu(s_o - \Delta_o)$.

In other words, if the source emits a frequency modulated signal, then the observer will, in general, receive a signal with a distorted modulation. In case, however, the signal suffers on its way a split which is followed next by a refocusing of the rays, the two received signals will maintain *the same* distorted modulation, with a relative time delay of one against the other if and only if $\Delta_o =$ constant.

5.5 Examples

The first of the two examples that will be discussed here is the classical Sagnac effect considered in the framework of special relativity. It was primarily my attempt to understand this effect in terms of space-time geometry that initiated the generalization presented in the previous sections. This circumstance also explains why I have started to call all the effects which may be described within the generalized approach Sagnac-like effects, although I admit that this may be regarded as too narrow a name, especially when one takes into account that, for example, effects within the classical Michelson interferometer also belong to this class.

In the Minkowski space-time with the signature -2, let us take into account n $(n > 2)$ timelike world lines O_k, $k = 1, 2, \ldots, n$, described by the equations $x = O_k(s_k)$, which in a coordinate frame take the form

$$\begin{aligned} x_k^0 &= \gamma s_k, \\ x_k^1 &= r\cos\Big[\frac{\omega\gamma}{c}s_k + \frac{2\pi}{n}(k-1)\Big], \\ x_k^2 &= r\sin\Big[\frac{\omega\gamma}{c}s_k + \frac{2\pi}{n}(k-1)\Big], \\ x_k^3 &= 0, \end{aligned} \tag{19}$$

where r, ω, and

$$\gamma = \frac{1}{\sqrt{1 - \omega^2 r^2/c^2}}$$

are constant, and s_k denotes the proper time taken along O_k for $k = 1, 2, \ldots, n$.

In the Minkowski space-time, we shall consider two piecewise null geodesics f_- and f_+, both starting from the same point $O_1(s_1)$, which form two loops.

The first loop, denoted by f_-, is described by a light signal that is by every observer who receives it instantly retransmitted to his nearest neighbor in front of him, that is, in accordance with the pattern

$$O_1 \to O_2 \to \cdots \to O_k \to O_{k+1} \to \ldots \to O_n \to O_1. \tag{20}$$

As a result, the light signal f_-, which starts from the observer O_1 at the instant s_1 of the local proper time returns, after being consecutively retransmitted by all the other observers, to the same world line O_1 at a value s_1^- of the proper time there. The piece of its path between every two neighboring world lines O_k and O_{k+1} is just a segment of the generator of the future light cone with its vertex at $O_k(s_k)$, joining the vertex with the point $O_{k+1}(s_{k+1})$ on the world line O_{k+1}. From the equation of the Minkowski null cone, by making use of (19) we find that the proper times s_k and s_{k+1}, for $k = 1, 2, \ldots, k, 1$, satisfy the transcedental equation

$$\frac{\gamma}{2r}(s_{k+1} - s_k) - \sin\left[\frac{\omega\gamma}{2c}(s_{k+1} - s_k) + \frac{\pi}{n}\right] = 0, \tag{21}$$

which, after introducing a dimensionless variable $x = (s_{k+1} - s_k)\gamma/(2r)$ and putting $\beta = \omega r/c$, is equivalent to

$$x - \sin\left(\beta x + \frac{\pi}{n}\right) = 0. \tag{22}$$

Here, for $0 < \beta < 1$, the left-hand side is a monotonic increasing function of x, and thus in the interval $(0, \pi(n-1)/\beta n)$, equation (22) has a single root $x = F_-(\beta, n)$ which is a function of the indicated parameters only. Therefore

$$s_{k+1} - s_k = \frac{2r}{\gamma} F_-(\beta, n),$$

and the consecutive proper times at which the light signal from the loop (20) meets the timelike world lines O_k, for $k = 1, 2, \ldots, k, 1$, form an arithmetic sequence. Hence, for the last term in this sequence we obtain

$$s_1^- = s_1 + \frac{2rn}{\gamma} F_-(\beta, n). \tag{23}$$

The equation above is just an example of the second of the relations (14): $s_1^- = f_-(s_1)$.

The second of the loops, f_+, is formed by an analogous light signal, which is being sequentially retransmitted in the opposite direction:

$$O_1 \to O_n \to \cdots \to O_{k+1} \to O_k \to \cdots \to O_2 \to O_1. \tag{24}$$

The analysis of the Minkowski light cone, now with its vertex on O_{k+1}, leads to the equation

$$\frac{\gamma}{2r}(s_k - s_{k+1}) + \sin\left[\frac{\omega\gamma}{2c}(s_k - s_{k+1}) - \frac{\pi}{n}\right] = 0.$$

After introducing the variable $y = (s_k - s_{k+1})\gamma/(2r)$, this equation takes the form

$$y + \sin(\beta y - \frac{\pi}{n}) = 0. \tag{25}$$

For $0 < \beta < 1$, equation (25) also admits a single root only, which will be denoted by $y = F_+(\beta, n)$. In terms of this root, one can express the proper time s_1^+ of the return to the world line O_1 of a light signal f_+ which was emitted by O_1 at the proper time instant s_1:

$$s_1^+ = s_1 + \frac{2rn}{\gamma} F_+(\beta, n). \tag{26}$$

This equation defines the first of the functions (14), $s_1^+ = f_+(s_1)$, for the case considered now.

From (23), (26), and (12) it follows that there is no frequency shift during any of the transmissions of the light signals considered in this section up to now.

Since for $\beta \neq 0$ the proper times of arrivals of the two light signals, f_- and f_+, to O_1 are different from one another, in accordance with (15), a nonvanishing Sagnac difference Δ_o of the proper times will arise,

$$\Delta_o = \frac{2rn}{\gamma}\Big[F_-(\beta, n) - F_+(\beta, n)\Big]. \tag{27}$$

The solutions F_- and F_+ of equations (22) and (25), respectively, can be found either by means of an approximation procedure or numerically.

If Newton's approximation method is applied to (22), with $x = 0$ as the starting point, in the first approximation one obtains

$$F_-(\beta, n) = \frac{\sin \pi/n}{1 - \beta \cos \pi/n},$$

and equation (25) leads similarly to

$$F_+(\beta, n) = \frac{\sin \pi/n}{1 + \beta \cos \pi/n}.$$

Thus, in this approximation, the Sagnac difference of times, due to (27), (24), and (26), is

$$\Delta_o = \frac{2\,\omega\, r^2 \sin 2\pi/n}{c} \frac{\sqrt{1 - \omega^2 r^2/c^2}}{1 - \omega^2 r^2/c^2 \cos^2 \pi/n}, \tag{28}$$

which, when the terms containing the ultra relativistic expression $(\omega^2 r^2)/c^2$ in the second factor above are neglected, is identical with the classical formula for the Sagnac effect: $\Delta_o = 2\beta r\, \sin(2\pi/n)$.

As the second example, let us now consider a family of n, $n > 2$, straight, timelike world lines O_k, $k = 1, 2, \ldots, n$, which in Minkowski space-time are described by the equations $x = O_k(s_k)$ that in a coordinate frame take the form

$$\begin{aligned} x_k^0 &= \gamma s_k, \\ x_k^1 &= -\beta\gamma s_k \sin\phi_k + R\cos\phi_k, \\ x_k^2 &= \beta\gamma s_k \cos\phi_k + R\sin\phi_k, \\ x_k^3 &= 0, \end{aligned} \tag{29}$$

where $\beta = v/c$, $\gamma^{-2} = 1 - \beta^2$ are constant, s_k is the proper time along the k-th straight line O_k, and $\phi_k = 2\pi(k-1)/n$.

Let us now assume that the first observer, O_1, sends out a light signal at an instant s_1 of his proper time, and that the signal is instantly after its reception retransmitted by every next observer. As a result, the light returns to the first observer after being propagated along two loops, f_- given by (20), and f_+ described by (24).

Let us take into account the segment of the null geodesic which corresponds to the piece of the path f_- between two neighboring world lines O_k and O_{k+1}. To find the proper time s_{k+1} at which O_{k+1} receives the signal retransmitted by O_k at his proper time s_k, we must write the equation of the light cone with the vertex at $O_k(s_k)$. From this equation, after some algebra, we obtain

$$(s_{k+1} - s_k)^2 - 4\beta\gamma \sin\frac{\pi}{n}\left[\beta\gamma \sin\frac{\pi}{n}\, s_k + R\cos\frac{\pi}{n}\right](s_{k+1} - s_k) \\ -4\sin^2\frac{\pi}{n}\left[\beta^2\gamma^2 s_k^2 + R^2\right] = 0.$$

If we solve this equation for $(s_{k+1} - s_k)$ and take into account its positive root, which corresponds to the future null cone, we can express s_{k+1} in terms of s_k

$$\begin{aligned} s_{k+1} &= F_-(s_k) \\ &= s_k + 2\sin\frac{\pi}{n}\left[\beta\gamma\left(\beta\gamma \sin\frac{\pi}{n}\, s_k + R\cos\frac{\pi}{n}\right)\right. \\ &\quad \left. + \sqrt{\beta^2\gamma^2\left(\beta\gamma\sin\frac{\pi}{n} + R\cos\frac{\pi}{n}\right)^2 + \beta^2\gamma^2 s_k + R^2}\,\right]. \end{aligned} \tag{30}$$

In an analogous way, for the path (24), the equation of the null cone with the vertex at $O_{k+1}(s_{k+1})$ leads to

$$(s_k - s_{k+1})^2 - 4\beta\gamma \sin\frac{\pi}{n}\left[\beta\gamma \sin\frac{\pi}{n}\, s_{k+1} - R\cos\frac{\pi}{n}\right](s_k - s_{k+1}) \\ -\sin^2\frac{\pi}{n}\left[\beta^2\gamma^2 s_{k+1}^2 + R^2\right] = 0.$$

This equation determines the proper time s_k in terms of s_{k+1} for any two neighboring observers along f_+. Solving it for $(s_k - s_{k+1})$, and taking the positive root, we obtain

$$\begin{aligned} s_k &= F_+(s_{k+1}) \\ &= s_{k+1} + 2\sin\frac{\pi}{n}\left[\beta\gamma\left(\beta\gamma \sin\frac{\pi}{n}\, s_{k+1} - R\cos\frac{\pi}{n}\right)\right. \\ &\quad \left. + \sqrt{\beta^2\gamma^2\left(\beta\gamma\sin\frac{\pi}{n} - R\cos\frac{\pi}{n}\right)^2 + \beta^2\gamma^2 s_{k+1} + R^2}\,\right]. \end{aligned} \tag{31}$$

Since both in (30) and (31) the corresponding differences $s_{k+1} - s_k$ are functions of the proper time, the frequencies of the two signals, after their travel between O_k and O_{k+1}, will be shifted. To compute the proper times s_1^- and s_1^+ at which the two corresponding signals f_- and f_+ will return to the starting world line O_1, one must perform an n-fold composition of the respective functions F, that is,

$$s_1^- = f_-(s_1) = F_-(\ldots F_-(F_-(s_1))\ldots),$$

and

$$s_1^+ = f_+(s_1) = F_+(\ldots F_+(F_+(s_1))\ldots),$$

Using equations (12) and (14), one could find from here the global frequency shifts along the two loops f_- and f_+. Also, in accordance with (15), the Sagnac difference of times does not vanish here. It is equal to

$$\begin{aligned}\Delta_o &= f_-(s_1) - f_+(s_1)\\ &= F_-(\ldots F_-(F_-(s_1))\ldots) - F_+(\ldots F_+(F_+(s_1))\ldots).\end{aligned} \tag{32}$$

Of course, even for such simple solutions like (30) and (31), an exact application of formula (32) would be quite tedious. If, however, one expands (30) and (31) into power series with respect to β, keeping under control the linear terms only, then one can easily obtain the leading term for the "Sagnac effect" produced by a family (29) of rectilinear motions with constant velocities of equal magnitude:

$$\Delta_o = 2\beta\gamma Rn \sin\frac{2\pi}{n} + O(\beta^3), \tag{33}$$

where the neglected terms of higher order in β depend on s_1. This indicates, in accordance with Lemma 6.4.1, that the frequency shifts along the two loops will be different.

Sometimes one can encounter an opinion that the classical Sagnac effect generated within a family of motions like (19) is an optical counterpart of the Foucault pendulum, which is producing a measure of how the motions in a system under consideration deviate from the inertial motion. The example of the family (29) indicates that such an opinion is not quite correct. It is interesting to observe that the leading terms which in formulae (28) and (33) are proportional to β are in the two cases of the same order of magnitude.

This work has been supported by the Polish Research Program KBN, contract no PB 1371/P03/97/12, registration no 2 P03B 017 12.

References

[1] Ehlers, J. & Sachs, R. K. 1959, Erhaltungsätze für die Wirkung in elektromagnetischen und gravischen Strahlungsfeldern, *Z. Physik*, **155**, 498.

[2] Schrödinger, E. 1956, *Expanding Universes*, Cambridge University Press, Cambridge.

[3] Schneider, P., Ehlers, J. & Falco, E. E. 1993, *Gravitational Lenses*, Springer-Verlag, New York.

6

How to Define a Unique Vacuum in Cosmology

Lluís Bel

ABSTRACT We propose a distinguished set of positive and negative energy modes of the Klein-Gordon equation as a time-independent definition of the vacuum state of a quantized scalar field.

6.1 Klein–Gordon Equation

Given any spacetime with line element

$$ds^2 = g_{\alpha\beta}(x^\rho)\,dx^\alpha\,dx^\beta, \qquad \alpha, \beta, \cdots = 0, 1, 2, 3, \tag{1}$$

the Klein-Gordon equation for a classical field $\psi(x^\rho)$ reads, using a system of units such that $c = 1$:

$$\left(\Box^2 - \frac{m^2}{\hbar^2}\right)\psi = \left(g^{\alpha\beta}\partial_{\alpha\beta} - \Gamma^\alpha\partial_\alpha - \frac{m^2}{\hbar^2}\right)\psi = 0, \tag{2}$$

where

$$\Gamma^\alpha = g^{\lambda\mu}\Gamma^\alpha_{\lambda\mu}. \tag{3}$$

If ψ_1 and ψ_2 are two, in general complex, solutions, then the current

$$J^\alpha(\psi_1, \psi_2) = -i\hbar g^{\alpha\beta}(\psi_1^*\partial_\beta\psi_2 - \psi_2\partial_\beta\psi_1^*) \tag{4}$$

is conserved:

$$\nabla_\alpha J^\alpha = \frac{1}{\sqrt{-g}}\partial_\alpha(\sqrt{-g}\,J^\alpha) = 0, \qquad g = \det(g_{\alpha\beta}), \tag{5}$$

and this allows us to define the invariant scalar product (ψ_1, ψ_2) of two well-behaved solutions as the flux of the preceding current across any space-like hypersurface Σ:

$$(\psi_1, \psi_2) = \int_\Sigma J^\alpha\,d\Sigma_\alpha \tag{6}$$

6.2 Quantization of a Scalar Field

The canonical quantization of a scalar field is a two-step process. The first step consists in selecting a distinguished set of modes of the Klein-Gordon equation to define what in the jargon of quantum field theory is called the *vacuum state*. The second step consists of implementing the so-called canonical commutation relations to be satisfied by the field operator and its conjugate momentum. Only the first step raises new problems in general relativity, and this paper is entirely dedicated to it. A common belief[1] is that in general, and in particular in cosmology, there is no unique way of choosing a unique vacuum state and therefore that there is an inevitable spontaneous creation of particles. We claim in this paper that for Robertson-Walker models with flat space-sections it is possible to distinguish a preferred set of modes, with no time mixing of positive and negative modes, thus defining a unique vacuum state and suppressing spontaneous particle creation from the vacuum state.

In Minkowski spacetime and in a Galilean frame of reference the vacuum state is defined by the set of modes ($\epsilon = \pm$)

$$\varphi_\epsilon(x^\alpha, \vec{k}) = (2\pi\hbar)^{-3/2} u_\epsilon(t, \vec{k}) \exp(\frac{i}{\hbar}\vec{k}\vec{x}), \qquad \varphi_{-\epsilon}(x^\alpha, \vec{k}) = \varphi^*_\epsilon(x^\alpha, -\vec{k}), \quad (7)$$

with

$$u_+(t, \vec{k}) = (2\omega)^{-1/2} \exp(-\frac{i}{\hbar}\omega t) \quad \omega(\vec{k}) = +(\vec{k}^2 + m^2)^{1/2} \quad (8)$$

where $\vec{k}$ is a constant index vector. The modes φ_+ are by definition the positive energy modes and φ_- the negative energy modes. This set of modes can be characterized by the following conditions:

i) The set of modes, having the general form (7), must be a complete orthonormal set of particular solutions of the Klein-Gordon equation. Orthonormality here means that the scalar product of two modes is

$$(\varphi_{\epsilon_1}(x^\alpha, \vec{k}_1), \varphi_{\epsilon_2}(x^\alpha, \vec{k}_2)) = \frac{1}{2}(\epsilon_1 + \epsilon_2)\delta(\vec{k}_1 - \vec{k}_2), \quad (9)$$

and completeness means that any solution of the Klein-Gordon equation that can be written as a Fourier transform on the flat space-sections t =const.:

$$\psi(t, \vec{x}) = \frac{1}{(2\pi\hbar)^{3/2}} \int c(t, \vec{k}) \exp(\frac{i}{\hbar}\vec{k}\vec{x})\, d^3\vec{k} \quad (10)$$

can also be written as

$$\psi(t, \vec{x}) = \int (a_+(\vec{k})\varphi_+(x^\alpha, \vec{k}) + a_-(\vec{k})\varphi_-(x^\alpha, \vec{k}))\, d^3\vec{k} \quad (11)$$

[1] See for example [1], [2]

ii) The functions u_ϵ are solutions of the first-order differential equation

$$i\hbar\dot{u}_\pm = \pm\omega u_\pm, \qquad \dot{u} = \frac{du}{dt} \tag{12}$$

It is this condition that guarantees that there will not be time mixing of positive and negative modes. Generalizing this second condition to Roberson-Walker spacetimes with flat space-sections is the main contribution of this paper.

6.3 Robertson–Walker Models

The line element of a Robertson-Walker cosmological model with flat space-sections is

$$ds^2 = -dt^2 + e^{2\sigma(t)}\delta_{ij}dx^i dx^j, \qquad i, j, \cdots = 1, 2, 3, \tag{13}$$

and the Klein-Gordon equation reads

$$\left(-\partial_t^2 - 3\dot{\sigma}\,\partial_t + e^{-2\sigma}\Delta - \frac{m^2}{\hbar^2}\right)\psi = 0, \tag{14}$$

where

$$\dot{\sigma} = \frac{d\sigma}{dt}, \qquad \Delta = \delta^{ij}\partial_{ij}. \tag{15}$$

The scalar product of two solutions can then be written using as hypersurface Σ any space-section t =const.:

$$(\psi_1, \psi_2) = i\hbar e^{3\sigma}\int_t (\psi_1^*\partial_t\psi_2 - \psi_2\partial_t\psi_1^*)\, d^3\vec{x}. \tag{16}$$

6.4 Modes

We shall define a mode, as it is usual in this case, as a solution of the following form:

$$\varphi(x^\alpha, \vec{k}) = \frac{1}{(2\pi\hbar)^{3/2}}u(t, \vec{k})\exp(\frac{i}{\hbar}\vec{k}\vec{x}), \tag{17}$$

where $\vec{k}$, the index of the mode, is a constant vector and where u must therefore be a solution of the following evolution second-order differential equation:

$$\hbar^2\ddot{u} + 3\hbar^2\dot{\sigma}\dot{u} + \omega^2 u = 0, \qquad \omega^2 = e^{-2\sigma}\vec{k}^2 + m^2 \tag{18}$$

The scalar product of two modes corresponding to any two vector indices is

$$(\varphi(x^\alpha, \vec{k}_1), \varphi(x^\alpha, \vec{k}_2)) = .i\hbar e^{3\sigma}(u^*(t, \vec{k}_1)\dot{u}(t, \vec{k}_2) - u(t, \vec{k}_2)\dot{u}^*(t, \vec{k}_1))\delta(\vec{k}_1 - \vec{k}_2). \tag{19}$$

6.5 Reduction of the Evolution Equation

Let us consider for each index $\vec{k}$ the following first-order[2] differential equation:

$$i\hbar\dot{u}(t,\vec{k}) = f(t,\vec{k})u(t,\vec{k}). \tag{20}$$

We say that this equation is a reduction of the corresponding evolution equation (18) if the function f is such that every solution of (20) is also a solution of (18). Or equivalently, if f is such that

$$u(t,\vec{k}) = A(t_0,\vec{k})\exp(-\frac{i}{\hbar}\int_{t_0}^{t} f(s,\vec{k})\,ds), \tag{21}$$

where A, a constant which will depend on a normalization condition and on the lower limit of integration that has been chosen, is a solution of (18).

Taking the time derivative of Eq. (20), multiplying by $i\hbar$, and taking into account Eq. (20) itself we get

$$-\hbar^2\ddot{u} = i\hbar\dot{f}u + f^2 u, \tag{22}$$

and using Eq. (18) and dividing by u,

$$i\hbar\dot{f} + f^2 + 3i\hbar\dot{\sigma} f - \omega^2 = 0. \tag{23}$$

This is a Riccati equation that f has to satisfy if Eq. (18) is to be a reduction of Eq. (20).[3] Any particular values of the initial time t_0 and $f(t_0)$ define a particular solution of this equation and therefore a set of modes. We consider the problem of deciding whether among all solutions there are some that are distinguished in some particular sense. Since for $\hbar = 0$ this equation becomes

$$f_0^2 = \omega^2, \tag{24}$$

we shall consider as distinguished those particular solutions, if they exist, which considered as functions of $\hbar$ become

$$f_0^+ = +\omega, \quad \text{or} \quad f_0^- = -\omega, \quad \omega > 0 \tag{25}$$

when $\hbar = 0$. We shall note them f_+ and f_-, and we shall refer to them as the *regular solutions*. Only two such solutions can exist. In fact from the theory of the Riccati equation it follows that the general solution can be written as

$$f = f_+ + \frac{f_- - f_+}{1 + C(f_- - f_+)\exp(-(i/\hbar)\int(2f_+ + 3i\hbar\dot{\sigma})\,dt)}. \tag{26}$$

If the arbitrary constant C is zero, then $f = f_-$. If it has infinite modulus, then $f = f_+$. Since f_+ and f_- are regular, it follows from the expression above that any

---

[2]The concept of order reduction has been used widely in many contexts. The use of this concept here is elementary. Another, nonelementary, application to cosmology can be seen in [3].

[3]The general relationship between the generalized Riccati equation and the linear differential equation of second order can be seen in [4].

other solution f will have an essential singularity for $\hbar = 0$. Clearly the regular solutions are distinguished.

A rigorous proof of the existence of regular solutions should require one to make precise assumptions about the function σ, i.e., about the cosmological model being considered. No attempt in this direction will be made here, but we can offer for $m = 0$ two simple though nontrivial examples for which regular solutions exist. The first one is

$$\sigma = \frac{1}{2}\ln(Ht) \tag{27}$$

where H is a constant and describes a radiation-dominated model near $t = 0$. The two regular solutions are in this case

$$f_\epsilon = \epsilon k(Ht)^{-1/2} - i\hbar(2t)^{-1}, \quad k = |\,\vec{k}\,|, \tag{28}$$

The modes corresponding to these solutions are:

$$u_\epsilon = A(t_0, \vec{k})(t_0/t)^{1/2}\exp(-2i\epsilon k/\hbar)H^{-1/2}t^{1/2}, \tag{29}$$

A being a normalization constant.

The second example is the de Sitter model.[4] Then

$$\sigma = Ht. \tag{30}$$

The regular solutions of Eq. (23) in this case, as can be readily checked, are

$$f_\epsilon = \frac{\epsilon k e^{-Ht}}{1 + \epsilon i\hbar H k^{-1} e^{Ht}}. \tag{31}$$

The modes corresponding to these solutions are

$$u_\epsilon = A(k^{-2}\hbar^2 H^2 + e^{-2Ht})^{1/2}\exp[\epsilon i(pe^{-Ht} - \tan^{-1} pe^{-Ht})], \tag{32}$$

where $p = k/(\hbar H)$.

6.6 Approximations to the Regular Solutions

Formal approximations to the regular solutions can be obtained assuming that they could be developed as power series of $\hbar$ starting with either one of the two values 25:

$$f = \sum_{n=0}^{\infty}(i\hbar)^n f_n. \tag{33}$$

Substituting (33) into (23), the subsequent terms could then be obtained recursively from an equation of the following type:

$$f_{2n+1} = -\frac{1}{2f_0}(\dot{f}_{2n} + 3\dot{\sigma} f_{2n} + 2f_1 f_{2n} + \cdots + 2f_n f_{n+1}), \quad n \geq 0 \tag{34}$$

---

[4]In 1 + 1 dimensions the existence of a single distinguished vacuum in this case was already pointed out in [5].

or

$$f_{2n} = -\frac{1}{2f_0}(\dot{f}_{2n-1}+3\dot{\sigma} f_{2n-1}+2f_1 f_{2n-1}+\cdots+2f_{n-1}f_{n+1}+f_n^2), \quad n>0. \quad (35)$$

In particular we have

$$f_1 = -\frac{1}{2}(3\dot{\sigma} + f_0^{-1}\dot{f}_0), \quad (36)$$

and therefore the behavior of (21) when $\hbar \to 0$ is

$$u(t,\vec{k}) \to Ae^{3/2(\sigma(t_0)-\sigma(t))}(f_0(t_0)/f_0(t))^{1/2}\exp(-\frac{i}{\hbar}\int_{t_0}^{t} f_0(s,\vec{k})\,ds) \quad (37)$$

For the case (27) it is of course trivial to derive the exact solution (28) using the algorithm above. For the de Sitter example considered in (30), one can find by induction the expression of the general term of the power series (33); namely,

$$f_n = \epsilon k(-i\epsilon\hbar Hk^{-1})^n e^{(n-1)Ht}, \qquad n \geq 0. \quad (38)$$

The series (33) can be summed in the interval defined by $\hbar Hk^{-1}\exp(Ht) < 1$ where it is convergent to yield the result that we anticipated in (31). This latter result, though, is valid also in the interval where the series diverge.

The relationship between the approximation method presented in this section and the usual WKB method can be seen as follows. Let us separate the real and the imaginary parts of f by writing

$$f = f_x + if_y. \quad (39)$$

The Riccati equation (23) then becomes an ordinary system of two differential equations:

$$\hbar(\dot{f}_x + 3\dot{\sigma} f_x) + 2f_x f_y = 0, \quad (40)$$

$$\hbar(\dot{f}_y + 3\dot{\sigma} f_y) + f_y^2 - f_x^2 + \omega^2 = 0. \quad (41)$$

From (40) we derive

$$f_y = -\frac{\hbar}{2}(3\dot{\sigma} + f_x^{-1}\dot{f}_x), \quad (42)$$

and substituting the above expression into (41), we obtain

$$\hbar^2(2f_x^{-1}\ddot{f}_x + 6\ddot{\sigma} - 3f_x^{-2}\dot{f}^2 + 9\dot{\sigma}^2) + 4f_x^2 - 4\omega^2 = 0. \quad (43)$$

On the other hand, using (42) in (21), we get for each mode

$$u(t) = A(t_0)e^{3/2(\sigma(t_0)-\sigma(t))}(f_x(t_0)/f_x(t))^{1/2}\exp(\frac{-i}{\hbar}\int_{t_0}^{t} f_x(s)\,ds) \quad (44)$$

We see then that the standard WKB method consists of using the real equation (43) to derive recurrence relations to calculate the coefficients of a formal expansion of the real part f_x in powers of $\hbar^2$, while we used a simpler Riccati equation to calculate the coefficients of f in a formal expansion in powers of $i\hbar$.

6.7 Critical Points at $t = \infty$

There is a second point of view that suggests the existence of two distinguished solutions. Let us assume that both σ and $\dot{\sigma}$ have a limit, say σ_∞ and $\dot{\sigma}_\infty$, when $t \to \infty$. In this regime Eq. (23), or equivalently the dynamical system (40)–(41), can be considered as an autonomous dynamical system, and as such it has two critical points, namely the two solutions of the second degree equation:

$$f^2 + 3i\hbar\dot{\sigma}_\infty f - \omega_\infty^2 = 0, \tag{45}$$

which are

$$f_\infty^\pm = \frac{1}{2}\left(-3i\hbar\dot{\sigma}_\infty \pm \sqrt{4\omega_\infty^2 - 9\hbar^2\dot{\sigma}_\infty^2}\right). \tag{46}$$

If these two critical points are not *attractors*, as a phase-portrait analysis shows in many cases, then there will be only one solution going through each of the critical points—namely the solutions that we would obtain choosing one critical point as initial condition, so to speak, for $t = \infty$ and integrating backwards in time. We shall call them the *critical solutions*.

Let us consider again as an example the case of the de Sitter model as in (30), but with arbitrary mass m. We have then $\sigma_\infty = \infty$, $\dot{\sigma} = H$, and the two critical points (46) are

$$f_\infty^\pm = \frac{1}{2}\left(-3i\hbar H \pm \sqrt{4m^2 - 9\hbar^2 H^2}\right). \tag{47}$$

We have to distinguish two cases: i) If $2m > 3\hbar H$, then

$$\mathrm{Re}(f_\infty^\pm) = \pm\frac{1}{2}\sqrt{4m^2 - 9\hbar^2 H^2}, \qquad \mathrm{Im}(f_\infty^\pm) = -\frac{3}{2}\hbar H. \tag{48}$$

A phase-portrait analysis shows that these two critical points for $t \to \infty$ are *center* points. ii) If on the contrary $2m < 3\hbar H$, then the two critical points are

$$\mathrm{Re}(f_\infty^\pm) = 0, \qquad \mathrm{Im}(f_\infty^\pm) = \frac{1}{2}(-3\hbar H \pm \sqrt{9\hbar^2 H^2 - 4m^2}) \tag{49}$$

A phase-portrait analysis shows that f_∞^+ is an *attractor* and that f_∞^- is a *repellor*. Notice that the analytic solution (31) goes through the *attractor* f_∞^+ but it shares this property with an infinity of other solutions and therefore it is not distinguished as a critical solution.

Since for $\hbar = 0$ the two critical values coincide with the values of $f_0^\pm$ for $t = \infty$, we may conjecture that when both the regular and the critical solutions exist, then they coincide. If this conjecture could be proven true, then the regular solutions $f_\pm$ could be approximated by a numerical calculation of the critical solutions.

6.8 Special Cases

Among the variety of things[5] that may happen to the series (33), we may consider the case where they converge in two disjoint time intervals—for instance for $t < t_0$ and $t > t_1$ with $t_1 > t_0$. Let $f_\pm^{(0)}$ and $f_\pm^{(1)}$ be the corresponding regular solutions supposed to be different. Either we suppose this case to be just a particular one among the more general class of cosmological models considered in the preceding section, or we consider that it is a special case deserving a special treatment. In the first case the appropriate choice to distinguish two particular solutions of the Riccati equation (23) is to choose $f_\pm^{(1)}$. In the second case a possible choice could be to choose $1/2(f_\pm^{(1)} + f_\pm^{(0)})$.

Let us suppose as a final example that in some time interval $\Delta T = [t_0, t_1]$ the function σ is almost constant. If one considers that the concept of mode is a local one and for the purpose of defining it a spacetime which has almost zero curvature on ΔT can be identified globally to Minkowski's spacetime, then one might consider the possibility of choosing as distinguished solutions those which reduce to $\exp(\pm(i\omega/\hbar)t)$ on ΔT. On the other hand, if one considers, as we do, that the concept of mode is a global one and must be a characteristic of the spacetime considered as a whole, then the preceding choice becomes unjustified and one needs some other criterion to select distinguished solutions of the Riccati equation (23).

6.9 Positive and Negative Energy Modes

The complex conjugate of Eq. (23) can be written as

$$i\hbar\frac{d}{dt}(-f^*) + (-f^*)^2 + 3i\hbar\dot{\sigma}(-f^*) - \omega^2 = 0, \tag{50}$$

which proves that if f is a solution of (23), then $-f^*$ is also a solution. Since $f_0^- = -(f_0^+)^*$ because ω is real, and since each of these initial terms characterizes the corresponding solutions f_- and f_+, it follows that

$$f_- = -f_+^*. \tag{51}$$

Notice that since ω in (18) is a function of $\vec{k}^2$, both functions $f_\pm$ are even functions of $\vec{k}$:

$$f(t, \vec{k}) = f(t, -\vec{k}). \tag{52}$$

We shall define the positive (negative) energy modes u_+ (u_-) as those modes for which u is a solution of (20), f being f_+ (f_-):

$$i\hbar\dot{u}_\pm(t, \vec{k}) = f_\pm(t, \vec{k})u_\pm(t, \vec{k}), \tag{53}$$

[5]The very special case where $\sigma(t)$ is a periodic function was considered in [6].

or

$$u_{\pm}(t,\vec{k}) = A_{\pm}(t_0,\vec{k})\exp(-\frac{i}{\hbar}\int_{t_0}^{t} f_{\pm}(s,\vec{k})\,ds), \tag{54}$$

Let $c(t,\vec{k})$ be a solution of the second-order equation (18) corresponding to initial conditions $c(t_0,\vec{k})$ and $\dot{c}(t_0,\vec{k})$ on some space-section $t = t_0$. If u_+ and u_- are two particular energy modes, one positive and the other negative, there will exist two constants $a_{\pm}(k)$ such that

$$c(t_0,\vec{k}) = a_+(\vec{k})u_+(t_0,\vec{k}) + a_-(\vec{k})u_-(t_0,\vec{k}), \tag{55}$$

and from (20) we shall have also

$$i\hbar\dot{c}(t_0,\vec{k}) = a_+(\vec{k})f_+(t_0,\vec{k})u_+(t_0,\vec{k}). + a_-(\vec{k})f_-(t_0,\vec{k})u_-(t_0,\vec{k}) \tag{56}$$

Solving for $a_{\pm}$, we obtain

$$a_{\pm}(\vec{k}) = \frac{i\hbar\dot{c}(t_0,\vec{k}) - f_{\mp}(t_0,\vec{k})c(t_0,\vec{k})}{u_{\pm}(t_0,\vec{k})(f_{\pm}(t_0,\vec{k}) - f_{\mp}(t_0,\vec{k}))}. \tag{57}$$

If c is itself a positive (negative) energy mode, then $a_- = 0$ ($a_+ = 0$), demonstrating explicitly that an energy mode is positive (negative) independently of the choice of the space-section and initial condition on it, as far as it satisfies the appropriate first-order equation (53).

From (20) and (51) we have

$$i\hbar\frac{d}{dt}(u_+ - u_-^*) = f_+(u_+ - u_-^*), \tag{58}$$

and therefore we shall have

$$u_-(t,\vec{k}) = u_+^*(t,\vec{k}), \tag{59}$$

provided that we choose initial conditions which satisfy this condition on some arbitrary space-section. From (52) we shall have

$$u_{\pm}(t,\vec{k}) = u_{\pm}(t,-\vec{k}). \tag{60}$$

Let us consider two modes with energy condition $\epsilon_1 = \pm$ and $\epsilon_2 = \pm$. From (19), and (20) and its complex conjugate, we have

$$\Big[(\varphi_{\epsilon_1}(x^\alpha,\vec{k}_1),\varphi_{\epsilon_2}(x^\alpha,\vec{k}_2)) = e^{3\sigma}(u_{\epsilon_1}^*(t,\vec{k}_1)u_{\epsilon_2}(t,\vec{k}_2)(f_{\epsilon_2}(t,\vec{k}_2) + f_{\epsilon_1}^*(t,\vec{k}_1))\delta(\vec{k}_1 - \vec{k}_2))\Big] \tag{61}$$

From this result we can see that the scalar product of two different modes is zero. If $\vec{k}_1 \neq \vec{k}_2$, then this follows from the properties of the Dirac δ function. If $\vec{k}_1 = \vec{k}_2$ and $\epsilon_1 = -\epsilon_2$, then this is a consequence of (51). To normalize the modes we shall require the generalized modes to satisfy again conditions (9). This is equivalent to requiring the constant $A_{\pm}(t_0,\vec{k})$ in (21) to satisfy

$$A_{\pm}A_{\pm}^*(t_0\vec{k}) = e^{-3\sigma(t_0)}\Big|f_{\pm}(t_0,\vec{k}) + f_{\pm}^*(t_0,\vec{k})\Big|^{-1}. \tag{62}$$

This fixes the norm of $A_\pm$ but not its phase. Since we want to have everywhere the relation (59), we may require that $A_\pm$ satisfy the relation

$$A_-(t_0,\vec{k}) = A_+^*(t_0,\vec{k}). \tag{63}$$

Let ψ be any solution of the Klein-Gordon equation (14) which can be expressed as a Fourier integral on each space-section $t = constant$:

$$\psi(t,\vec{x}) = \frac{1}{(2\pi\hbar)^{3/2}}\int c(t,\vec{k})\exp(\frac{i}{\hbar}\vec{k}\vec{x})\, d^3\vec{k}. \tag{64}$$

Since the c's must be, for each $\vec{k}$, a solution of (18), they will be a linear combination of u_+ and u_- as in (55). Substituting in (64), using (60), and after some rewriting we get

$$\psi(x^\alpha) = \int a_+(\vec{k})\varphi_+(x^\alpha,\vec{k}) + a_-(\vec{k})\varphi_-(x^\alpha,\vec{k})\, d^3\vec{k}, \tag{65}$$

where

$$\varphi_-(x^\alpha,\vec{k}) = \varphi_+^*(x^\alpha,-\vec{k}). \tag{66}$$

We have thus proved that the set of modes defined above provides a straightforward generalization of the modes (7) associated with the Galilean frames of reference of Minkowski spacetime. The vacuum state that they define does not depend on time.

6.10 Concluding Remarks

We have assumed that the space-sections t =constant were flat. This is not an essential restriction. Models with nonflat space-sections can be dealt with using the eigenstates of the Laplacian of a constant curvature 3-dimensional Riemannian metric.[6]

Also, the main idea of this paper, which consisted in reducing the Klein-Gordon equation to two complex conjugate first-order equations with respect to time, can be generalized to more general spacetimes and frames of reference. This work will be published elsewhere.

Acknowledgments

It is a pleasure to acknowledge many stimulating and useful discussions with Jérôme Martin and Pierre Teyssandier.

---

[6] See for instance [7].

REFERENCES

[1] Birrell, N. D. and Davies, P. C. (1982). *Quantum fields in curved spacetime*, Cambridge University Press, Cambridge,

[2] Fulling, S. A. (1989).*Aspects of quantum field theory in curved spacetime*, Cambridge University Press, Cambridge.

[3] Bel, L. l. and Sirousse, Zia H. (1985). *Phys. Rev. D*, **32**, 12, 3128–3135.

[4] Davis, H. T. (1962). *Introduction to Nonlinear Differential and Integral Equations*, Dover Publications, New York.

[5] Floreanini, R., Hill, C. T. and Jackiw, R. (1987). *Annals of Physics*, **175**, 345.

[6] Droz-Vincent, Ph. (1996). *Letters in Mathematical Physics*, **36**, 277–290

[7] Lifshitz, E. M. and Khalatnikov, I. M. (1963). *Adv. Phys.*, **12**, Appendix J.

7

EIH Theory and Noether's Theorem

Peter G. Bergmann[1]

7.1 Introduction

Classical field theory separates the field, which obeys partial differential equations (the field equations), from the sources, which satisfy ordinary (ponderomotive) differential equations, at least when the sources are conceived to be point masses. That the sources of the field are affected by a field's presence is a fact beyond question. But as point-like sources carry a field with them that tends to infinity at the location of each source, the total field must be split into an "incident field" and a "self-field" in order to obtain a finite field that determines the behavior of the source.

This procedure is open to criticisms on two scores. First, given the total field, its incident part depends on the choice of self-field, which is far from unique. Second, the whole splitting procedure depends on the linearity of the field laws. As the experimental physicist can only measure the total field, the splitting prescription appears to go against the spirit of a field theory.

General relativity postulates as field equations of the gravitational field partial differential equations that approximate linearity only for weak fields. The equations of motion of binary stars presented a particularly serious conceptual problem, in that each component of the binary system was to travel on geodesics determined by the other only. This choice cannot be justified as the first step of an approximation. At each component's location the weaker contribution to the total field is adopted as the important part, and the stronger (self-)field discarded.

The Einstein-Infeld-Hoffmann theory avoids these shortcomings. The field is accepted as a whole, and the dynamic laws are given the form of conservation laws. The conserved quantities appear as integrals over domains that do not include the locations of the field sources. The existence of these integrals depends on invariance properties of the theory being examined.

---

[1]Dedicated to Engelbert Schucking on the occasion of his seventieth birthday.

7.2 Invariance Group and Noether's Theorem

The notions of *invariant transformations* and of *invariance group* play key roles. An invariant transformation is one that leaves the form of the dynamical law unchanged, whether that is the Lagrangian or the Hamiltonian. The invariant transformations of a dynamical system form a group, the invariance group. The invariance group need not be connected. If it is at least locally connected, there are infinitesimal invariant transformations, and one can construct a canonical (Hamiltonian) formalism.

An infinitesimal invariant transformation has a generator. Noether's theorem states that that generator is a constant of the motion. Loosely, Noether's theorem is a special case of the fact that in a Poisson bracket $[A, B] = C$ either component generates an infinitesimal transformation that changes the other component by an amount C. Noether's theorem results from one component of the Poisson bracket being the Hamiltonian, the other the generator of an infinitesimal invariant transformation, so that $C = O$.

7.3 An Example

Return now to the EIH theory and consider as an example the gauge properties of the electromagnetic potentials, $\vec{A}$ and Φ, and the invariance of the electromagnetic charge, the source of the electromagnetic field. If the charges present are all within a compact 3-dimensional domain, V, then Q, the total charge inside V, equals the 2-dimensional integral over S, the boundary of V,

$$Q = \oint \vec{E} \cdot d\vec{S}. \tag{1}$$

The canonical momentum densities of the field are $\vec{\pi} = -\frac{1}{c}\frac{\partial \vec{A}}{\partial t}$ and π_o. An infinitesimal gauge transformation leaves the momentum densities unchanged, but affects the configuration variables.

$$\delta \vec{A} = \nabla \Psi, \qquad \delta \Phi = -\frac{1}{c}\frac{\partial \Psi}{\partial t}, \tag{2}$$

where Ψ is an arbitrary (smooth) function of the space and time coordinates.

The generator of the gauge transformation is

$$\Gamma = \int \left(\vec{\pi} \cdot \nabla \Psi - \frac{1}{c} \pi_o \frac{\partial \Psi}{\partial t} \right) d^3x. \tag{3}$$

By Noether's theorem the generator Γ is constant. As Ψ and $\frac{\partial \Psi}{\partial t}$ are arbitrary functions, their coefficients must be zero,

$$\nabla \cdot \vec{E} = 0, \qquad \pi_o = 0. \tag{4}$$

If the generators of infinitesimal invariant transformations vanish, Poisson brackets between them must be zero as well. In Dirac's terminology, the generators must not only vanish (they are *constraints*), but they are *first-class constraints*.

That Q does not change in time is proven by way of the Maxwell-Lorentz equations governing the electromagnetic field away from sources. As

$$\frac{dQ}{dt} \equiv \oint \frac{\partial \vec{E}}{\partial t} \cdot d\vec{S} \tag{5}$$

and

$$\frac{\partial \vec{E}}{\partial t} = c\nabla \times \vec{B}, \tag{6}$$

Stokes's theorem guarantees the vanishing of the integral (5). Hence the total charge, defined by Eq. (1), is a constant of the motion.

In a somewhat more general situation than in the conservation of electric charge, a two-dimensional integral need not be a constant of the motion, but changes in accordance with an expression that represents a "flux." Consider a field that obeys Euler-Lagrange equations outside a compact domain, the site of the sources. I shall designate the algebraically independent field variables by y^A, where the superscript A runs from 1 to n. The summation convention is to apply to such superscripts and subscripts as well. Coordinate indices are to be identified by lowercase Greek letters.

The field equations outside the source domain are

$$\frac{\delta L}{\delta y^A} \equiv \frac{\partial L}{\partial y^A} - \left(\frac{\partial L}{\partial y^A{}_{,\rho}}\right)_{,\rho} = 0. \tag{7}$$

A "symmetry" of the field equations is indicated where a mapping of the field on itself leaves the form of the Lagrangian, i.e., the function

$$L(y^A, y^A{}_{,\rho})$$

unchanged. These are the invariant transformations defined earlier. I assume that there are symmetries that include the neighborhood of the identity mapping. The existence of other symmetries is not excluded. An infinitesimal transformation of the field variable y^A will be written as Ψ^A. The change in the form of L resulting from the infinitesimal transformation will be

$$\begin{aligned} \delta L &= -\left(\frac{\partial L}{\partial y^A}\Psi^A + \frac{\partial L}{\partial y^A{}_{,\rho}}\Psi^A{}_{,\rho} + Q^\rho{}_{,\rho}\right) \\ &= -\left(\Psi^A \frac{\partial L}{\partial y^A} + C^\mu{}_{,\mu}\right), \\ C^\mu &= \Psi^A \frac{\partial L}{\partial y^A{}_{,\mu}} + Q^\mu. \end{aligned} \tag{8}$$

7.4 The Generalized EIH Theory

Q^ρ (and therefore also C^ρ) is a function to be chosen arbitrarily, as it makes no contribution to the Euler-Lagrange equations. The transformation Ψ^A will be invariant if C^μ can be chosen so that δL vanishes:

$$\Psi^A \frac{\delta L}{\delta y^A} + C^\mu{}_{,\mu} = 0. \tag{9}$$

Such a choice is

$$C^\mu = \Psi^A \frac{\partial L}{\partial y^A{}_{,\mu}}. \tag{10}$$

Other choices are obtained by adding to C^μ expressions of the form $U^{\mu\nu}{}_{,\nu}$ where $U^{\mu\nu}$ must be skew-symmetric,

$$U^{\mu\nu} + U^{\nu\mu} = 0, \tag{11}$$

but is otherwise arbitrary.

The conservation of electric charge provides a hint how one may formulate dynamical laws for the sources of a field without integrating across the sources themselves. A volume integral can be converted into a surface integral, whose dimension may be reduced once more by the applications of Stokes's theorem. By Gauss's theorem a 4-dimensional integral turns into a 3-dimensional integral taken over the boundary of the original domain of integration. If the original integrand, $C^\mu{}_{,\mu}$, vanishes, then the corresponding 3-dimensional integral,

$$H_B = \oint C^\mu d^3\Sigma_\mu = 0, \tag{12}$$

where

$$d^3\Sigma_\mu \equiv \frac{1}{6}\delta_{\mu\rho\sigma\tau} dx^\rho \wedge dx^\sigma \wedge dx^\tau,$$

vanishes as well.

As the next step we cut the 3-dimensional integral domain in two. The cutting surface will be 2-dimensional, and the two 3-dimensional pieces have their boundaries in common. Obviously, as the two resulting 3-dimensional pieces, after adjusting their orientations, are equal,

$$H = \int_M C^\mu d^3\Sigma_\mu = \int_N C^\mu d^3\Sigma_\mu, \tag{13}$$

their individual magnitudes depend only on the common 2-dimensional boundary,

$$H = \frac{1}{2}\oint W^{\mu\nu} d\Sigma_{\mu\nu}, \tag{14}$$

$$d^3\Sigma_{\mu\nu} \equiv \frac{1}{2}\delta_{\mu\nu\rho\sigma} dx^\rho \wedge dx^\sigma,$$

$$W^{\mu\nu} + W^{\nu\mu} = 0,$$

$$W^{\mu\rho}{}_{,\rho} = C^\mu.$$

The integrand $W^{\mu\nu}$ is not uniquely determined: Given one solution compatible with (14), another solution is obtained by adding a term

$$W^{\mu\nu\prime} = W^{\mu\nu} + T^{\mu\nu\sigma}{}_{,\sigma}, \tag{15}$$

with

$$T^{\mu\nu\sigma} = \delta^{\mu\nu\sigma\tau} V_{\tau}. \tag{16}$$

V_τ is arbitrary.

The EIH theory is based on the fact that the 2-dimensional integral (14) equals the 3-dimensional integral (13). Neither domain of integration includes the locations of the sources.

The 3-dimensional integral (13) has the topology of a sphere S^2 multiplied by a finite curve segment. It surrounds the domain containing the sources, $S^2 \times R$. Its boundary, the domain of integration (14), will consist of two spheres, S^2.

After adjusting the orientation of the two pieces of the boundary, the relationship between Eqs. (13) and (14) can be given the form

$$H_N = H_M + \int C^{\mu} d^3\Sigma_{\mu}. \tag{17}$$

The change in the value of the 2-dimensional integral going from M to N is determined by the "flux," C^{μ}. If that flux vanishes, then $H_N = H_N$.

7.5 Concluding Remarks

In this presentation I have avoided introducing the components of the metric tensor as the field variables, without ruling out that choice. To this extent the sources of the field need not be components of a tensor. Directions of line elements may, but need not be labeled as spacelike and timelike.

I have not used symbols for covariant differentiation. All symbols for differentiation indicate partial derivatives.

REFERENCES

[1] Einstein, A., Infeld, L., and Hoffmann, B., *Ann. Math.* **39**, 65 (1938).
[2] Noether, E., *Goett. Nachr.* **37**, 235 (1918).

8

The Static Cylinder in General Relativity

W.B. Bonnor

ABSTRACT For a certain range of a parameter m the Levi-Civita vacuum solution describes the exterior field of a massive static cylinder. When $m = 0$, the spacetime is flat, and when m is small and positive, the gravitational field can be reconciled with the corresponding Newtonian one. When $m = \frac{1}{4}$, the circular orbits of test particles become null, and for $m > \frac{1}{4}$, all circular orbits are space like. Increasing m still further, one finds that when $m = \frac{1}{2}$ spacetime again becomes flat. It was formerly thought that for $m > \frac{1}{4}$ the Levi-Civita spacetime did not represent the field of a cylinder. Recently, however, it has been matched with interior cylindrical solutions throughout the range $0 \leq m < \frac{1}{2}$. The strange behavior for $\frac{1}{4} \leq m \leq \frac{1}{2}$ therefore requires explanation. In this paper I examine one of these interior solutions from this point of view. It seems that for $\frac{1}{4} \leq m < \frac{1}{2}$ the Levi-Civita solution does refer to a cylinder, but of increasing radius. As m approaches $\frac{1}{2}$, the radius tends to infinity, suggesting that when $m = \frac{1}{2}$, the bounding spatial surface is plane.

8.1 Introduction

It is a pleasure to dedicate this paper to Engelbert Schucking, who once showed me not only the red light district of Hamburg but also many insights into cosmology.

The Levi-Civita (hereafter LC) spacetime continues to puzzle relativists. It is a solution of the vacuum Einstein equations with a metric whose general form is

$$ds^2 = -(r+b)^{8m^2-4m}(K^2dr^2 + L^2dz^2) - P^2(r+b)^{2-4m}d\phi^2 + Q^2(r+b)^{4m}dt^2,$$

where b, m, K, L, P, Q are real constants. To represent cylindrical symmetry, one may impose the following ranges on the coordinates

$$r > 0, \qquad -\infty < z < \infty, \qquad 0 \leq \phi \leq 2\pi, \qquad -\infty < t < \infty, \tag{1}$$

and identify $\phi = 0$ and $\phi = 2\pi$. If one locates the curvature singularity on the axis of symmetry one puts $b = 0$; then by scale transformations in r, z, t the constants K, L, Q can be removed. However, P cannot be got rid of by such a transformation if one wishes to preserve the range of ϕ in (1). From this point of view the LC metric has two essential constants m and P and may be written

$$ds^2 = -r^{8m^2-4m}(dr^2 + dz^2) - P^2 r^{2-4m} d\phi^2 + r^{4m} dt^2. \qquad (2)$$

(For a discussion of the number of constants in cylindrical spacetimes see [1] and [16].)

The effect of P is well known in the theory of cosmic strings, and it clearly has to do with the topology of spacetime. The real challenge is to give a physical interpretation of m.

The difficulties of interpreting m have been well documented [2, 3, 4, 5], and there is no need to dwell on them here. Mathematically, m can assume any real value, and there seems little doubt that its interpretation is not the same throughout the entire range. In this paper I shall concentrate on the limited range $0 \le m \le \frac{1}{2}$. When m is zero the spacetime is flat, and for small m the metric describes the exterior field of a static cylinder of mass per unit length m [2]. One familiar property of such a field is the existence of circular test particle orbits, but one finds that if $m = \frac{1}{4}$ the circular orbits of (1) become null, and for $m > \frac{1}{4}$ they are spacelike. Stranger still, when $m = \frac{1}{2}$ the spacetime is again flat!

The absence of circular test particle orbits led me and others [2, 3] to suggest that if $m > \frac{1}{4}$, (1) does not describe the field of a cylinder. This conclusion seemed to be strengthened by the fact that the Kretschmann scalar *diminishes monotonically* in $\frac{1}{4} \le m \le \frac{1}{2}$, whereas one would expect the gravitational field to become stronger with increasing mass per unit length [3].

However, as has been pointed out by Philbin [5], these arguments are by no means conclusive. Regarding the first, in Newtonian theory circular orbits are possible for any mass per unit length because, since velocity is unlimited, a centrifugal force is always available to balance the gravitational attraction however strong; but in relativity there is a maximum centifugal force which occurs when the test particle reaches the speed of light, and this maximum is reached when $m = \frac{1}{4}$. Assuming that for $m > \frac{1}{4}$ the gravitational field is stronger, one therefore finds a natural explanation for the absence of timelike circular orbits. Turning to the second of the above arguments, one must admit that the Kretschmann scalar, which is determined by tidal forces, may not be the best guide to the strength of the gravitational field. A better, and simpler, guide is the force required to keep a test particle at rest at a given proper distance from the centre of the cylinder. This, of course, is measured by the acceleration of the test particle in the metric (1). One finds that this acceleration *increases monotonically* in $\frac{1}{4} \le m \le \frac{1}{2}$.

These considerations indicate that (1) may after all represent the exterior of a cylinder if $\frac{1}{4} < m < \frac{1}{2}$. Moreover they have been confirmed by the discovery recently of perfect fluid interior solutions which can by matched to (1) on a boundary r =constant [5, 6, 7]. If we assume, as now seems reasonable, that increasing m means increasing mass per unit length, it remains to explain why, when m actually reaches $\frac{1}{2}$, the spacetime becomes flat, indicating the absence of any source at all. This is the question I try to answer in this paper, and to do so I shall use one of the exact interior solutions which represent sources of (1). My method is to examine what happens to this as $m \to \frac{1}{2}$, though the limit cannot actually be reached since the solution does not apply when $m = \frac{1}{2}$. It turns out that the curvature of the bounding 2-surface of the source tends to zero, so that the cylinder is turning into a plane. Speaking loosely, one can say that when m actually reaches $\frac{1}{2}$ there remains only a plane wall and this produces a uniform gravitational field, that is, a flat spacetime.

The interior solution I choose is one recently published by Haggag and Desokey [7], hereafter referred to as the HD solution. This is particularly simple and is described in Section 2. In Section 3 it is matched to the Levi-Civita metric, and the global solution is interpreted physically in Section 4. Section 5 examines the mass per unit length of the cylinder by the use of Whittaker's formula. There is a concluding Section 6.

8.2 The HD Solution

This solution is case (i) of [7]; I present it in a slightly modified form, and omit an inessential constant (denoted by Q in [7]) that can be removed by a scale coordinate transformation:

$$ds^2 = -(1 - A^2r^2)^{-1}e^{a^2r^2}dr^2 - (1 - A^2r^2)\,dz^2 - r^2d\phi^2 + e^{a^2r^2}dt^2, \tag{3}$$

where $A^2 := a^2(1 + 2k^2)^{-1}$ and a, k are positive arbitrary constants, This metric and the ranges of ϕ, t are as in (1). It is Minkowskian on the symmetry axis $r = 0$ and has no angular defect there. The pressure and density are

$$p = (4\pi)^{-1}A^2(k^2 - a^2r^2)e^{-a^2r^2}, \tag{4}$$

$$\mu = (4\pi)^{-1}A^2(2 + k^2 - a^2r^2)e^{-a^2r^2}, \tag{5}$$

so we can take the zero pressure surface

$$r_0 = k/a \tag{6}$$

as the surface S of the cylinder. Then inside the cylinder $A^2r^2 < 1$ and the interior is nonsingular. p and ρ satisfy

$$\mu > p \geq 0,$$

so the interior matter is reasonable.

8.3 Matching to LC Spacetime

The metric (3) has to be matched to the LC spacetime on S. Rather than use (2) as the exterior metric, I prefer to take the equivalent Kasner form

$$ds^2 = -dR^2 - FR^{2n_1}dz^2 - GR^{2n_2}d\phi^2 + HR^{2n_3}dt^2, \tag{7}$$

where F, G, H are positive arbitrary constants, and n_i satisfy

$$n_1 + n_2 + n_3 = 1, \qquad n_1^2 + n_2^2 + n_3^2 = 1. \tag{8}$$

The ranges of z, ϕ, and t are as in (1). Metric (7) is a transform of (1) and n_i can be parametrised in terms of m as follows:

$$\begin{aligned} n_1 &= -2m(1-2m)N^{-1}, \\ n_2 &= (1-2m)N^{-1}, \\ n_3 &= 2mN^{-1}, \\ N &= 4m^2 - 2m + 1, \end{aligned} \tag{9}$$

which satisfy (8). It should be noted that m here is *half* the m used in [7].

In (3) and (7) the coordinates z, ϕ, t are taken to be the same, but the radial coordinates are not necessarily continuous on S. Indeed they do not need to be if we use the Darmois junction conditions [8, 9] to match the two metrics. Let S be given by (6) in the interior and by $R = R_0$ in the exterior, and let $x^2 = z, x^3 = \phi, x^4 = t$ be coordinates on S. We require that the first and second fundamental forms be the same on S when calculated from the interior and exterior metrics. This leads to six equations:

$$g_{22} = -(1+k^2)(1+2k^2)^{-1} = -FR_0^{2n_1}, \tag{10}$$

$$g_{33} = -(k/a)^2 = -GR_0^{2n_2}, \tag{11}$$

$$g_{44} = e^{k^2} = HR_0^{2n_3}, \tag{12}$$

$$\Omega_{22} = -n_1 FR_0^{2n_1-1} = akB(1+2k^2)^{-1}e^{-k^2/2}, \tag{13}$$

$$\Omega_{33} = -n_2 GR_0^{2n_2-1} = (k/a)Be^{-k^2/2}, \tag{14}$$

$$\Omega_{44} = n_3 HR_0^{2n_3-1} = akBe^{k^2/2}, \tag{15}$$

where $B = \mid [(1+k^2)/(1+2k^2)]^{1/2} \mid$. Here Ω_{ik} is the second fundamental form, given on each line for the exterior and interior regions.

If from the above relations we form Ω_{22}/g_{22} etc, we get three formulae for the normal curvatures of the coordinate lines on S:

$$\kappa_z = n_1/R_0 = -ak(1+k^2)^{-1}Be^{-k^2/2}, \tag{16}$$

$$\kappa_\phi = n_2/R_0 = (a/k)Be^{-k^2/2}, \tag{17}$$

$$\kappa_t = n_3/R_0 = akBe^{-k^2/2}. \tag{18}$$

These formulae will play a role later in the limiting process by which $m \to \frac{1}{2}$.

The equations (10)–(15) and (8) comprise eight equations for nine unknowns $k, a, n_1, n_2, n_3, R_0, F, G, H$. However, the equations are not all independent, and there are two independent parameters, k and a, in terms of which the other unknowns can be expressed; in particular

$$\begin{aligned} n_1 &= -k^2 K, \\ n_2 &= (1+k^2)K, \qquad n_3 = k^2(1+k^2)K, \end{aligned} \tag{19}$$

$$R_0 = (k/a)(1+2k^2)Be^{k^2/2}K, \tag{20}$$

where $K := (1 + k^2 + k^4)^{-1}$. From (9) and (19) we obtain

$$m = \frac{k^2}{2(1 + k^2)}. \tag{21}$$

8.4 Physical Interpretation

We now have a global, regular solution comprising a perfect fluid interior matching an LC exterior, valid for $0 \leq m < \frac{1}{2}$, and depending on two parameters k and a. Comparing with [7], one finds that though the interior and exterior metrics are equivalent, some differences arise in the relations between the constants, presumably owing to different treatments of the matching process. HD's global solution seems to depend on only one parameter. However, the basic relation (21) appears in [7].

As m runs from 0 to $\frac{1}{2}$, k goes from 0 to ∞. This march of k produces no spectacular changes in the interior solution. The pressure and density, given by (4) and (5), are well behaved and decrease outward. When $k = 1$, corresponding to $m = \frac{1}{4}$, the ratio p/μ on the axis $r = 0$ is $\frac{1}{3}$, which would correspond to the equation of state of incoherent radiation, but away from the axis the ratio is less. As $k \to \infty$, p tends to equality with μ throughout the cylinder.

Consider now the normal curvatures of the coordinate lines on the bounding surface S, given by (16)–(18). These all tend to zero as $k \to \infty$. Corresponding to this, the proper radius of the cylinder

$$\int_0^{k/a} (1 - A^2 r^2)^{-1/2} e^{a^2 r^2/2}\, dr$$

tends to infinity. We can interpret this as meaning that the surface of the cylinder is becoming more and more like a plane. The HD solution does not permit us to proceed to the limit $k = \infty$, or $m = \frac{1}{2}$, but *it strongly suggests that the exterior solution with $m = \frac{1}{2}$ has a plane source.* This would be very reasonable because, according to Newtonian theory, such a source produces a uniform gravitational field. Using now the principle of equivalence, this is equivalent to an accelerated coordinate system and therefore a flat spacetime; which is what the LC solution is when $m = \frac{1}{2}$.

8.5 The Whittaker Mass per Unit Length

When m is small, it corresponds to the Newtonian mass per unit length of an infinite line mass. Its precise meaning as its value proceeds to $\frac{1}{2}$ is unclear, although, as stated in the introduction, it seems likely that the increase is accompanied by a strengthening of the gravitational field. In this section I shall derive a formula for m in terms of the mass per unit length M determined by the formula of Whittaker [10].

The Whittaker formula for the active gravitational mass of a static distribution of perfect fluid of density μ and pressure p is

$$M = \int_V (\mu + 3p)\sqrt{(-g)}\, dv \tag{22}$$

where g is the determinant of the 4-dimensional metric, V is the 3-dimensional spatial volume containing the matter, and dv is the volume element. Since we are taking a 3-dimensional integral and using a 4-dimensional determinant, M is not an invariant. Indeed by making a scale transformation of the time $t \to \alpha t$, we can give it an arbitrary value. The dangers in using the Whittaker formula have been discussed in [11, 12, 13]. Nevertheless the formula has often been used successfully for isolated matter distributions by ensuring that the metric is Minkowskian at infinity. We cannot do that here because the LC metric cannot be made Minkowskian at infinity. However, as we have a nonsingular interior, we can make the metric Minkowskian on the symmetry axis $r = 0$, which calibrates the time coordinate, so to speak. This has already been arranged in (3), so we shall substitute (3) into (22) to calculate M for the HD solution.[1]

The calculation of M is easy. Substituting into (22) from (3), (4), and (5), we find

$$M = (4\pi)^{-1} \int_0^{k/a} \int_0^1 \int_0^{2\pi} A^2(2 + 4k^2 - 4a^2r^2) r\, d\phi\, dz dr$$

and, inserting the value of A,

$$M = \frac{k^2(1 + k^2)}{2(1 + 2k^2)}. \tag{23}$$

Using (21), we obtain a relation between M and m:

$$M = \frac{m}{1 - 4m^2}, \tag{24}$$

so as $m \to 1/2$, $M \to \infty$. This is consistent with the result, obtained in the previous section, that as m approaches $\frac{1}{2}$ the radius of the cylinder tends to infinity.

It should be noted that M as obtained above is the mass per unit *coordinate length* in the z direction. m, on the other hand, is presumably related to the mass per unit *proper length*, since it is invariant to a change in scale of z. However, it is remarkable that the same formula (24) has been obtained previously for two different cylinders. The first of these is described in [14], and the second one is that in [6], for which (24) holds [15].

---

[1] Whittaker proved his formula for an isolated system, which could be enclosed in a two-dimensional surface in vacuum. This is not so for our infinite cylinder, so the result must be treated with reserve.

8.6 Conclusion

I have been studying a global solution for an infinite static cylinder with a perfect fluid interior, matched at a boundary to the Levi-Civita vacuum metric. The solution is physically reasonable if the parameter m satisfies $0 \leq m < \frac{1}{2}$. As $m \to \frac{1}{2}$, the curvature of the bounding surface tends to zero, suggesting that the case $m = \frac{1}{2}$ refers to a plane and not a cylinder. This is consistent with the fact that when $m = \frac{1}{2}$ the exterior is described by a flat uniformly accelerated metric. A similar conclusion was reached in [2] by a different method.

Finally I return to a remark made at the beginning of the paper that the LC metric contains two arbitrary constants. It makes no difference that we take it in the form (7) rather than (1). The constants F and H can evidently be made unity by scale transformations of z and t without affecting their ranges in (2). However, the constant G cannot be removed in this way if we wish to preserve the range $0 \leq \phi \leq 2\pi$ for ϕ; and this range must be preserved to ensure a match over the whole boundary S, since in the interior ϕ certainly has this range. Thus the exterior depends on two arbitrary constants G and m, which are both determined by the interior structure constants.

References

[1] da Silva, M. F. A., Herrera, L., Paiva, F. M. and Santos, N. O. 1995. *Gen. Rel. Grav.* **27** 859.

[2] Gautreau, R. and Hoffman, R. B. 1969. *Nuovo Cimento* **B61** 411.

[3] Bonnor, W. B. and Martins, M. A. P. 1991. *Class. Quantum Grav.* **8** 727.

[4] Lathrop, J. D. and Orsene, M. S. 1980. *J. Math. Phys.* **21** 152.

[5] Philbin, T. G. 1996. *Class. Quantum Grav.* **13** 1217.

[6] Bonnor, W. B. and Davidson, W. 1992. *Class. Quantum Grav.* **9** 2065.

[7] Haggag, S. and Desokey, F. 1996. *Class. Quantum Grav.* **13** 3221.

[8] Darmois, G. 1927. *Mémorial des Sciences Mathematiques* (Gautiers Villars, Paris) Fasc. 25.

[9] Bonnor, W. B. and Vickers, P. A. 1981. *Gen. Rel. Grav.* **13** 29.

[10] Whittaker, E. T. 1935. *Proc. R. Soc.* A**149** 384.

[11] Cooperstock, F. I. and Sarracino, R. S. 1978. *J. Phys. A: Math. Gen.* **11** 877.

[12] Cooperstock, F. I., Sarracino, R. S. and Bayin, S. S. 1981. *J. Phys. A: Math. Gen.* **14** 181.

[13] Devitt, J. and Florides, P. S. 1989. *Gen. Rel. Grav.* **21** 585.

[14] Bonnor, W. B. 1979. *J. Phys. A: Math. Gen.* **12** 847.

[15] Bonnor, W. B. and Davidson, W. 1992. (unpublished).

[16] MacCallum, M. A. H. 1998. *Gen. Rel. Grav.* **30** 131.

9

Gravity and the Tenacious Scalar Field

Carl H. Brans

ABSTRACT Scalar fields have had a long and controversial life in gravity theories, having progressed through many deaths and resurrections. The first scientific gravity theory, Newton's, was that of a scalar potential field, so it was natural for Einstein and others to consider the possibility of incorporating gravity into special relativity as a scalar theory. This effort, though fruitless in its original intent, nevertheless was useful in leading the way to Einstein's general relativity, a purely two-tensor field theory. However, a universally coupled scalar field again appeared, both in the context of Dirac's large number hypothesis and in 5-dimensional unified field theories as studied by Fierz, Jordan, and others. While later experimentation seems to indicate that if such a scalar exists its influence on solar system–size interactions is negligible, other reincarnations have been proposed under the guise of dilatons in string theory and inflatons in cosmology. This paper presents a brief overview of this history.

9.1 Scalar Gravity?

After the conceptual foundations of special relativity had been laid by Einstein and the natural 4-dimensional formalism for space-time had been clarified by him, Minkowski, and others, it was natural to consider how field theories should fit into the new framework. Of course, since it lay at the foundations of special relativity, Maxwell's electromagnetic theory translated beautifully using the 4-vector potential and 2-form field formalism. The other classical field theory was gravity, so the question of incorporating gravity into the new relativity arose next. The standard textbook introductions to the subject naturally emphasize the logical path to Einstein's resolution of this, his general theory of relativity. Such pedagogical treatments can even seduce the reader into believing that Einstein's general relativity is the logically necessary consequence of special relativity and the gravitational principle of equivalence. The actual history was of course not so linear, and the logical path not so obvious to the participants. The first, apparently most simple, approach is to generalize Newton's scalar gravitational theory to a special relativistic scalar one. In fact, this was precisely what was tried. In the following we will sketch a brief overview of the physics of this period, 1907 to 1915. This a is fascinating and highly instructive story. Fortunately, John Norton has provided an excellent, very readable, review of the history of this subject [1].

Let us start with the natural 4-force generalization of Newtonian mechanics,

$$\frac{d}{d\tau}\left(m\frac{dx^\mu}{d\tau}\right) = \mathcal{F}^\mu, \tag{1}$$

where the right side is the 4-force vector. Implicit in the physics of this equation is that the path parameter, τ, must be proper time, so the auxiliary condition

$$\eta_{\mu\nu}\frac{dx^\mu}{d\tau}\frac{dx^\nu}{d\tau} = -1, \tag{2}$$

must be satisfied. In modern terms this fixing of the path parameter would be described as a "conformal gauge-fixing condition." At any rate, (1), together with the assumption that m is constant in the (2) gauge, implies that

$$\eta_{\mu\nu}\mathcal{F}^\mu\frac{dx^\nu}{d\tau} = 0. \tag{3}$$

As mentioned earlier, Maxwell's electromagnetism fits into special relativity quite naturally, since electromagnetism was the theory whose consistency triggered the reinvestigation of spacetime that ultimately led to it. For electromagnetism the force in (1) is

$$\mathcal{F}^\mu_{em} = F^{\mu\nu}\eta_{\nu\rho}\frac{dx^\rho}{d\tau}, \tag{4}$$

and (3) follows neatly from the antisymmetry of $F^{\mu\nu}$.

Now, what of gravity? In Newtonian-Galilean theory, the field can be described by a scalar potential, ϕ, with

$$\nabla^2\phi = \frac{\kappa}{2}\rho, \tag{5}$$

where ρ is mass density, $\kappa \equiv 8\pi G$, and G the usual Newton constant. Using Galilean 3-vector notation,

$$\mathbf{E}_g = -\nabla\phi, \tag{6}$$

the equations of motion become

$$\frac{d}{dt}\left(m\frac{d\mathbf{r}}{dt}\right) = m\mathbf{E}_g. \tag{7}$$

Note that this potential has units of velocity squared, so that in the standard relativistic choice used in this paper, $c = 1$, ϕ is dimensionless.

The natural, perhaps logically minimal, approach to including gravity in special relativity might thus seem to be "relativizing" (5), (6), and (7), simply extending them from three to four dimensions,

$$\Box^2\phi = \frac{\kappa}{2}\rho, \tag{8}$$

$$\mathcal{F}^\mu_g = -m\phi^{,\nu}, \tag{9}$$

and (1). However, (3) applied to (9) results in

$$\frac{dx^\mu}{d\tau}\frac{\partial\phi}{\partial x^\mu} = \frac{d\phi}{d\tau} = 0, \tag{10}$$

along the particle's path. That is, the potential is constant along the path of every particle, so the gravitational force must necessarily be zero on every particle! Although the historical details are not entirely clear, it seems likely that this is the problem to which Einstein referred in 1907 in discounting the appropriateness of a scalar special relativistic theory of gravitation without allowing "... the inert mass of a body to depend on the gravitational potential"[2]. In the context of the time, this seemed unacceptable, and so the problem of incorporating gravity into special relativity was the root of a great deal of concern on Einstein's part. It also provided fuel for criticism of the entire structure of special relativity by others, notably Abraham [1].

For our purposes, the next significant contribution was by Nordström [3], who attacked the question of the possible dependence of mass on gravitational field directly, using

$$m = m_0 e^\phi. \tag{11}$$

This is a direct precursor to the idea that the time component of the metric, or perhaps the proper time itself (gauge dependence of mass) depends on the gravitational potential.

In current formalism we guarantee consistency by starting from an action approach. For a single point particle of mass m, the field variable can be regarded as the path functions, $z^\mu(\tau)$, and the single particle action is

$$A_p = -\int m\sqrt{-\dot{z}^\mu \dot{z}_\mu}\, d\tau, \tag{12}$$

where $\dot{z}^\mu \equiv dz^\mu/d\tau$. To fit into the volume integration needed for field theories, this can be written

$$A_p = -\int \left(\int m\sqrt{-\dot{z}^\mu \dot{z}_\mu}\,\delta^4(x^\mu - z^\mu(\tau))d\tau\right) d^4x. \tag{13}$$

As a model for particle-field interaction consider electromagnetism, with total field plus interaction actions,

$$A_{em} + A_I = \frac{-1}{16\pi}\int (A_{\mu,\nu} - A_{\nu,\mu})(A^{\mu,\nu} - A^{\nu,\mu})\, d^4x$$
$$+q\int\left(\int \dot{z}^\mu(\tau) A_\mu \delta^4(x^\nu - z^\nu(\tau))\, d\tau\right) d^4x. \tag{14}$$

Using $A_p + A_{em} + A_I$ as the total action results in the Maxwell field equations with current density source $J^\mu(x^\nu) = q\int \dot{z}^\mu(\tau)\delta^4(x^\nu - z^\nu(\tau))\, d\tau$, together with the correct equation of motion for the particles. Note that the inertial mass, m, in A_p is independent of the coupling constant, q in A_I.

By analogy, the total action for scalar gravity would be

$$\begin{aligned} A_p + A_I + A_\phi = & -\int \Big(\int m\sqrt{-\dot{z}^\mu \dot{z}_\mu}\,\delta^4(x^\mu - z^\mu(\tau))\,d\tau \Big) d^4x \\ & -\int \phi \Big(\int m\sqrt{-\dot{z}^\mu \dot{z}_\mu}\,\delta^4(x^\mu - z^\mu(\tau))\,d\tau \Big)\, d^4x \\ & -\frac{1}{\kappa} \int \phi_{,\mu}\phi^{,\mu}\, d^4x. \end{aligned} \tag{15}$$

Clearly the field variation results in (8) with $\rho(x^\mu) = m \int \delta^4(x^\mu - z^\mu(\tau))d\tau$ in the conformal gauge, (2). However, the variation over the particle's variables, $z^\mu(\tau)$, $\dot{z}^\mu(\tau)$ results in something quite different from (1) and (9), namely,

$$\frac{d}{d\tau}\Big(m(1+\phi)\dot{z}^\mu(\tau)\Big) = -m\phi^{,\mu}. \tag{16}$$

This can of course be identified with Nordström's (11), with $\exp(\phi) \approx 1 + \phi$, to first order in ϕ.

While this provides a self-consistent and thus potentially viable scalar gravitational theory, it was not an attractive approach to Einstein, who seemed to have some strong intuitive drive to incorporate the principle of equivalence at the basis for the correct relativistic gravitational theory, even though he apparently was not yet aware of the Eötvös experimental work on this matter.

Einstein proceeded to criticize scalar special relativistic gravitational theories as being specifically inconsistent with the equivalence principle. When the four-vector equations of motion, say (1) or (16), are expressed in terms of local coordinate time, it is clear that local coordinate acceleration of a particle will depend on the the particle's full kinetic energy, thus violating the equal acceleration principle. For example, replacing the two $d\tau$'s in the denominator of the left side of (16) with their coordinate expressions gives

$$\frac{d^2\mathbf{r}}{dt^2} \sim -(1 - v^2)\nabla\phi. \tag{17}$$

Thus, for example, spinning bodies would have smaller accelerations in a gravitational field than nonspinning identical ones, hot bodies than cold, etc. Nordström pointed out that this effect would be too small to be measured by contemporary technology, but nevertheless it was regarded by Einstein as a serious obstacle to the consideration of such theories.

By a coincidence, von Laue was investigating the influence of special relativity on the theory of stress in bodies, and in so doing discovered what we now call the 4-dimensional stress-energy tensor, with T^{00} identified with energy density, and $T^{ij} = p^{ij}$ the components of the spatial stresses on the body, $p^{ij} = p\delta^{ij}$, for an isotropic fluid. Clearly a Lorentz velocity transformation would then mix the purely spatial stress components into the energy density, so that the energy density of a moving body would depend on its stress and quadratically on speed.

At this point it is appropriate to take a close look at the concepts of active and passive gravitational mass, as well as inertial mass. A complete study of these questions is beyond our scope here. In fact, the extent to which Newton's laws are laws as opposed to simply definitions of force and mass is not an easy one to settle. For a review of the ideas of Mach and others on these questions, see the book by Ray [4]. Here we content ourselves with the following somewhat superficial review. First assume some adequate operational definition has been given for Newtonian force. Then gravitational force turns out to be proportional to a single scalar parameter for a given particle at a particular spacetime point. This parameter is called "*passive gravitational mass,* m_{gp}," so

$$\mathbf{F} = m_{gp}\mathbf{E_g}. \tag{18}$$

On the other hand, the gravitational field is determined by field equations with source density proportional to "active gravitational mass, m_{ga}." It is now easy to see however, that these two parameters cannot be independent of other without violation of Newton's third law and thus conservation of momentum. Thus, for a pair of particles, the force of particle 1 on particle 2 is proportional to the $m^1_{ga}m^2_{gp}$, while the reaction force is proportional to $m^2_{ga}m^1_{gp}$, so that Newton's third law requires

$$\frac{m_{ga}}{m_{gp}} = \text{universal constant.} \tag{19}$$

Clearly this constant can be chosen to be one and active and passive masses assumed to be equal, unless Newton's third law is to be relaxed. The remaining mass is "*inertial mass,* m_i," with

$$\mathbf{F} = m_i\mathbf{a}. \tag{20}$$

One form of the equivalence principle, the universality of gravitational acceleration at a fixed spacetime point, is then simply the statement that

$$\frac{m_g}{m_i} = \text{universal constant.} \tag{21}$$

Returning to the historical matter of a scalar gravitational mass, the discovery by von Laue and others that special relativity would imply that stresses contribute to mass, the question was how to fit this into a gravitational theory. Nordström had already taken into account the contribution of the gravitational potential to mass, in (11), or in more modern formalism, (16), but still needed to account for the questions raised by von Laue's study of $T^{\mu\nu}$. Specifically, what should be the ρ on the right side of (8)? Nordström first considered T^{00}, evaluated in the rest frame of the particle, or invariantly $T^{\mu\nu}u_\mu u_\nu$, where u_μ are the components of the body's 4-velocity. Einstein suggested that the trace of the tensor, $T = T^\mu_\mu$, explicitly including stresses, would be more appropriate. However, what of stress-energy tensors defined by null fields, such as electromagnetism, for which $T = 0$? Einstein pointed out that if such fields are to be treated as localized bodies, they must be contained and their contribution to the gravitational field might be accounted for in terms of the stresses the confined fields exert on the container.

At this point, the prospects for a scalar special relativistic gravitational theory looked good. However, a critical thought experiment remained. Consider a closed cycle in which a stressed body (a rod) is lowered in a gravitational field, then unstressed and raised again. Clearly energy is not conserved in this cycle. The net gain in energy is associated with the lack of energy associated with a pure stress (no strain). Nordström and Einstein were then able to show that energy conservation could be restored if they assumed that movement through the gravitational field was also associated with a change in length of the rod along the direction of the stress. Thus work is done and energy released. In other words,

Gravity and Geometry? *A full working through of the implications of a special relativistic scalar gravitational theory leads to the suggestions that the lengths of rods, and rates of clocks, might depend on their location in a gravitational field.*

In modern notation Nordström's theory would be expressed by saying that the metric is conformally flat,

$$ds^2 = e^{2\phi}(dx^2 + dy^2 + dz^2 - dt^2), \tag{22}$$

with the gravitational field ϕ determined by

$$\Box^2 \phi = \frac{\kappa}{2} T, \tag{23}$$

where the exact form of the source field equations was developed through several trial and error stages. The crucial step of associating gravity with a distortion of spacetime measurements, thus with empirical spacetime geometry had been taken. Einstein and Fokker [5] built on this work, exploring the full geometric implications of Nordström's ideas. Einstein noted that Nordström's geometric form, (22), conformally flat, was too specialized and eventually generalized to an arbitrary metric form, leading to the full Einstein equations, with metric tensor playing the role of gravitational potential. Apparently this finally put an end to the search for a purely scalar special relativistic gravitational theory. Nevertheless, a universally coupled scalar field pops up in yet another context!

9.2 Kaluza–Klein Theories

Very early in the development of relativity, searches began for a unification of gravity with electromagnetism. One of the most enduring of such attempts is that associated with the names of Kaluza and Klein. The collection of papers on this subject edited by Appelquist et al [6] provides a valuable resource for this subject, both classically and in more modern contexts.

Briefly, the Kaluza-Klein idea is to embed the electromagnetic 4-potential into the metric by enlarging the dimension of the space to 5, and inserting the 4-potential as

$$\gamma_{AB} = \begin{pmatrix} V^2 & V^2 A_\beta \\ V^2 A_\alpha & g_{\alpha\beta} + V^2 A_\alpha A_\beta \end{pmatrix}. \tag{24}$$

The vector tangent to the extra dimension is assumed to be a Killing vector, so all variables depend only on the spacetime coordinates x^α. The existence of this Killing vector gives rise to a natural foliation of the 5-space of codimension 1, with local expressions as in (24). Maintaining this foliation and simply counting variables, we see the usual 10 components of the spacetime metric, $g_{\alpha\beta}$, the four A_α, a 4-vector, and a 15th, surprise quantity, a 4-scalar, V. In the 1921 paper of Kaluza [7], this field is described as "noch ungedeutete," and left at that. In 1948 Thiry [6] wrote out the full field equations 5-Ricci, $\mathcal{R}_{AB}$, equals zero, explicitly, in effect taking this new scalar seriously as a possible universally coupled scalar field. The 5-dimensional vacuum Einstein tensor reduces to a scalar part,

$$\mathcal{S}_{55} = \frac{3V^2 F_{\alpha\beta}F^{\alpha\beta}}{8} - \frac{R}{2}, \tag{25}$$

a 4-vector set,

$$\mathcal{S}_{5\beta} = \frac{3}{2}V_{,\alpha}F_\beta{}^\alpha/2 + \frac{V}{2}F_\beta{}^\alpha{}_{;\alpha}, \tag{26}$$

and the 4-tensor part,

$$\mathcal{S}_{\alpha\beta} = S_{\alpha\beta} + \frac{V^2}{2}\left(F_{\alpha\mu}F^\mu{}_\beta + \frac{\eta_{\alpha\beta}}{4}F_{\mu\nu}F^{\mu\nu}\right) - \left(\frac{V_{,\alpha;\beta}}{V} - \frac{\eta_{\alpha\beta}\Box^2 V}{V}\right). \tag{27}$$

Here

$$dA = \frac{1}{2}(A_{\alpha|\beta} - A_{\beta|\alpha})\sigma^\beta \wedge \sigma^\alpha = \frac{1}{2}F_{\beta\alpha}\sigma^\beta \wedge \sigma^\alpha. \tag{28}$$

The full 5-dimensional analog of the Einstein equations are

$$\mathcal{S}_{AB} = 0, \tag{29}$$

derived from

$$\delta\int \mathcal{R}\sqrt{|g^{(5)}|}d^5x. \tag{30}$$

However, the resulting (25), (26),and (27) contain the new field, V, apparently unrelated to either gravity or electromagnetism. With the *ad hoc* choice

$$V^2 = \frac{\kappa}{2\pi} = \text{const.} \tag{31}$$

the result is standard vacuum Einstein-Maxwell,

$$F_{[\alpha\beta,\mu]} = 0, \tag{32}$$

from (28);

$$F_\beta{}^\alpha{}_{;\alpha} = 0, \tag{33}$$

from (26); and

$$S_{\alpha\beta} = \frac{\kappa}{4\pi}\left(F_{\alpha\mu}F_\beta{}^\mu - \frac{g_{\alpha\beta}}{4}F_{\mu\nu}F^{\mu\nu}\right), \tag{34}$$

from (27) plus one additional condition,

$$F_{\mu\nu}F^{\mu\nu} = 0, \tag{35}$$

which follows from $\mathcal{S}_{55} = 0$. Of course, this last condition is not a part of Maxwell's theory. To eliminate it and reproduce the full Maxwell theory, the variational principle can be replaced by

$$\delta \int \sqrt{|g^{(5)}|}\left(\mathcal{R} + \lambda\left(g_{55} - \frac{\kappa}{2\pi}\right)\right) d^5x = 0. \tag{36}$$

From this the combined vacuum Maxwell-Einstein field equations are recovered. However, there are some residual difficulties in the interpretation of the five-geodesic as the path of a particle subject to both gravity and electromagnetic forces, so this approach has never been widely accepted as fully satisfactory in its original intent. Nevertheless, the basic ideas of the Kaluza-Klein approach continue to be used in the wider sense of modern gauge theories [6].

From the viewpoint of this paper, however, Kaluza-Klein is important for introducing the new scalar V. In fact, from the identification required in (31), this new scalar can be associated with the Newtonian gravitational constant.

9.3 Dirac's Numbers

Meanwhile Dirac [8], building on the work of Eddington and Milne, pointed out some remarkable clustering of dimensionless numbers composed from observed values of certain fundamental constants. Dirac's starting point was the value of present cosmological age of the universe, T_u, as defined by the best available value of the Hubble constant in 1938. Of course, the value of this number depends on arbitrary choice of units and so, of itself, cannot have any particular physical significance. Thus, another natural time unit is needed. A natural choice is provided by any of a number of "atomic" time scales, such as e^2/m, or $\hbar/m$, where e is the electronic charge and m is some natural mass, such as that of the electron, or of a nucleon. Clearly, the range of numbers available for this choice is 10^3, and this arbitrariness will not affect the agreement to orders of magnitude that follow. Dirac then noticed that this fundamental time ratio results in

$$t \equiv \frac{T_u}{T_a} \approx 10^{40}, \tag{37}$$

where T_a is one of the natural atomic time units. Next Dirac decided to look at the dimensionless ratio of two of the fundamental forces, electrical and gravitational, on some standard atomic particle:

$$\gamma \equiv \frac{e^2}{\kappa m^2} \approx 10^{40}. \tag{38}$$

Another number to consider is the ratio of the present observed mass of the universe to the standard atomic mass,

$$\mu \equiv M_u/m \approx 10^{80}. \tag{39}$$

These empirical numbers were first discussed by Eddington, and are generally known as Eddington numbers. From Dirac's perspective, the clustering of these natural, dimensionless constants into groups of widely varying magnitude 10^{40}, 10^{80} is remarkable indeed and led Dirac to speculate that this clustering might well have some causal basis in some physical theory. He proceeded to develop a cosmological model in which

$$\mu \approx t^2, \tag{40}$$

$$\gamma \approx t, \tag{41}$$

so that the quantities μ, γ change with the age of the universe, the "epoch" in Dirac's terms. Dirac's cosmology was later reconsidered in more detail by Canuto and others [9]. At this point, Dirac's clustering leads to a number

$$\frac{\mu}{t\gamma} \approx 10^0, \tag{42}$$

or

$$\frac{\kappa M_u}{R} \approx 10^0. \tag{43}$$

Later, as a preface to inflationary cosmological models, these "large number coincidences" will play an important role. This will be discussed in the section on "inflatons" below. However, at this point they raise the question of whether κ is a truly universal constant, or if it would change in circumstances with different values for M/R. In other words, (43) raises the possibility that κ is determined by the mass distribution in the universe.

9.4 Scalar–Tensor Theories

Pascual Jordan [10] was intrigued by the occurrence of the new scalar field in the Kaluza-Klein type of theories, and especially its possible role as a generalized gravitational constant in the spirit of Dirac's hypothesis, (43). Building on this idea Jordan and his colleagues, including Englebert Schücking, began an investigation of Kaluza-Klein theories with special concern for the idea that the new 5-dimensional metric component, a spacetime scalar, might play the role of a varying gravitational "constant." The resulting 4-dimensional form of the field equations can be so interpreted. However, Jordan and his colleagues took the next step of separating the scalar field from the original 5-dimensional unified gravitational-electromagnetic context. Later Brans and Dicke [11] independently arrived at a similar proposal. Brans and Dicke were especially motivated by Mach's ideas on inertial induction. Sciama [12] had proposed a model theory of inertial induction, that is, a theoretical mechanism for generating the inertial forces felt during acceleration of a reference frame. These forces were hypothesized to be of gravitational origin, occurring only during acceleration relative to the "fixed stars." In this model the ratio of inertial to gravitational mass will depend on the average distribution of mass in the universe, in effect making κ a function of the mass distribution in the universe.

In commonly used notation such theories introduce a scalar field ϕ, which will (locally and approximately) play the role of reciprocal Newtonian gravitational constant κ. One obvious motivation for this choice of field quantity (rather than κ itself) is Dirac's large number hypothesis in the form

$$1/\kappa \approx M/R. \tag{44}$$

From this, rather than (43), we see the possibility that $1/\kappa$ itself might be a field variable and satisfy a field equation with mass as a source, something like

$$\Box \frac{1}{\kappa} = \rho. \tag{45}$$

Of course, the usual Lagrangian for Einstein theory including matter has κ directly multiplying the matter contributions. In this form, changes would be made to the local behavior of matter, the local equations of motion, as a result of variations in ϕ. Consequently, in order to incorporate a Mach's principle by way of a variable gravitational "constant," some modifications must be made to standard general relativity. Start by writing the standard Einstein action as

$$\delta \int \sqrt{-g}(R + \kappa L_m) d^4x = 0, \tag{46}$$

where L_m is the "usual" matter Lagrangian, presumably derived from some classical or quantum model.

At this point it is useful to consider several aspects of the famous "principle of equivalence." First is the statement that all bodies at the same spacetime point in a given gravitational field will undergo the same acceleration. We will refer to this as the "weak" equivalence principle, WEP. A stronger statement, on which standard Einstein's general relativity is built, is that the *only* influence of gravity is through the metric, and can thus (apart from tidal effects) be locally, approximately transformed away, by going to an appropriately accelerated reference frame. This is the "strong" principle, SEP. If we start from an action of the form in (46) with variable κ, we will be risking the geodesic equation for test particles, and thus, possibly the WEP, and even mass conservation. However, we are allowing for a possible violation of the SEP, since gravity, the universal interaction of mass, will influence local physics by changing the local κ. As Dicke noted, the Eötvös experiment verifies only the WEP and not the SEP, so in the 1960s it was reasonable to consider such alternatives.

Returning to the form of the action, let us then isolate κ from matter in the original (46) by dividing by it,

$$\delta \int \sqrt{-g}(\phi R + L_m)d^4x = 0, \tag{47}$$

where $\phi \equiv 1/\kappa$. While we seem to have thus saved the geodesic equations for test particles, it is now known, of course, that the motion of composite bodies is more complex. It turns out that with refined observation techniques, even the coupling of ϕ directly to the gravitational field gives rise to observable effects for matter configurations to which gravitational energy contributes significantly. This is now known as the "Dicke-Nordtvedt" effect and has been investigated in the earth-moon system with the lunar laser reflector.

Nevertheless, let us proceed to see what follows from (46). We are anticipating field equations for ϕ, so some action for this new field must be supplied,

$$\delta \int \sqrt{-g}(\phi R + L_m + L_\phi)d^4x = 0. \tag{48}$$

The usual requirement that the field equations be second order leads to

$$L_\phi = L(\phi, \phi_{,\mu}). \tag{49}$$

Apart from this, there seem to be few *a priori* restrictions on L_ϕ. At first glance, the standard choice for a scalar field,

$$L_\phi = -\omega \phi_{,\mu}\phi_{,\nu}g^{\mu\nu}, \tag{50}$$

leading to a wave equation for ϕ with R as source would seem to be natural. However, the coupling constant ω would itself then need to have the same dimensions as the gravitational κ that the new field is to replace! It would at least seem reasonable to require that any new coupling constant be dimensionless for various reasons, so a natural minimal choice is

$$L_\phi = -\omega \phi_{,\mu}\phi_{,\nu}g^{\mu\nu}/\phi, \tag{51}$$

in which ϕ has dimensions of inverse gravitational constant,

$$[\phi] = [\kappa^{-1}]. \tag{52}$$

The form (51) leads to an action which is often referred to as the "Jordan-Brans-Dicke," JBD, action,

$$\delta \int \sqrt{-g}(\phi R + L_m - \frac{\omega}{\phi}\phi_{,\mu}\phi_{,\nu}g^{\mu\nu})d^4x = 0. \tag{53}$$

The variational principle, with standard topological and surface term assumptions, results in

$$\delta_m \int \sqrt{-g}L_m dx^4 = 0, \tag{54}$$

$$\phi S_{\alpha\beta} = T_{(m)\alpha\beta} + \phi_{;\alpha;\beta} - g_{\alpha\beta}\Box\phi + \frac{\omega}{\phi}\left(\phi_{,\alpha}\phi_{,\beta} - \frac{1}{2}g_{\alpha\beta}\phi_{,\lambda}\phi^{,\lambda}\right), \tag{55}$$

$$\omega\left(\frac{2\Box\phi}{\phi}-\frac{\phi_{,\lambda}\phi^{,\lambda}}{\phi^2}\right)=-R. \tag{56}$$

The first of these, (54), is the standard variational principle for matter, which follows the same equations as in Einstein theory, thus (apparently) satisfying the weak equivalence principle. For test particles, (54), results in the geodesic equations. However, for extended, or composite, particles, this is may no longer be true, even in standard general relativity. The second-order interaction of matter by way of the scalar-metric coupling gives rise to violations of the weak equivalence principle, so that bodies of different mass may have different gravitational accelerations in identical gravitational fields. Of course, because of the free standing L_m in (53), the energy tensor for matter is still conserved,

$$T_{(m)\alpha\ ;\beta}^{\ \ \ \ \ \beta}=0. \tag{57}$$

Taking the trace of (55) and solving for R leads to another form for (56),

$$\Box\phi=\frac{1}{(2\omega+3)}T_{(m)}, \tag{58}$$

in which $T_{(m)}$ is the trace of the ordinary matter tensor. In a weak field model situation, within a static spherical shell of mass M, radius R, and otherwise empty universe, this equation produces

$$\phi\approx\phi_\infty+\frac{1}{4\pi(2\omega+3)}\frac{M}{R}. \tag{59}$$

If ϕ can be identified with the local reciprocal gravitational constant, and ϕ_∞ is set zero as a default asymptotic condition, then this equation is seen to be consistent with the Dirac coincidence (44). Another natural approximation to (58) is to consider the effect of local matter over some background ϕ_0 equal to the present observed value,

$$\phi\approx\phi_0+\frac{1}{4\pi}\sum_{\text{local matter}}\frac{m}{r}. \tag{60}$$

In equation (55) $T_{(m)\alpha\beta}$ are the components of the stress-energy tensor for matter derived from the matter Lagrangian L_m in the standard fashion. This equation, which describes the sources of the gravitational field, can be rewritten

$$S_{\alpha\beta}=(1/\phi)(T_{(m)\alpha\beta}+T_{(\phi)\alpha\beta}). \tag{61}$$

This form clearly suggests that $(1/\phi)$ does indeed act as a generalized gravitational "constant," with both ordinary matter and the field ϕ itself serving as sources for the metric. However, it turns out that the presence of the $\Box\phi$ term on the right hand side of (55), together with (58), results in two occurrences of the matter tensor as a source, effectively producing a constant renormalization of ϕ as "gravitational constant."

The earliest serious investigations of this theory were by Jordan and his group, prominent among whom was Englebert Schücking. Heckman gave the first nontrivial exact vacuum solution, the generalization of the Schwarzschild solution of standard Einstein theory. Later, Schücking [13] investigated the natural question of a Birkhoff-type theorem for scalar-tensor equations. He was able to show that again the most general spherically symmetric solution must be static, if the scalar field is assumed to be static, or to have a lightlike gradient. However, if ϕ is allowed to be a function of time, more general solutions can exist, of course. A class of such solutions was also presented in Schücking's paper, and opened the way for studies of spherically symmetric, nonstatic, phenomena occurring in scalar-tensor but not standard Einstein theory.

Early on questions of the choice of "conformal gauge" for the metric were considered. In other words, replacing the metric, $g_{\mu\nu} \to \bar{g}_{\mu\nu} = \psi g_{\mu\nu}$, leads to a replacement of the action (53) discarding the surface (topological) part, by

$$\delta \int d^4x \sqrt{-\bar{g}} \left(\frac{\phi}{\psi} \bar{R} + \frac{3\phi}{2} \frac{|\bar{\nabla}\psi|^2}{\psi^3} - 3\bar{\nabla}\psi \cdot \bar{\nabla}\phi/\psi^2 + L_m/\psi^2 - \frac{\omega}{\phi\psi} |\bar{\nabla}\phi|^2 \right) = 0. \tag{62}$$

In particular, if ψ is chosen to be ϕ, (62) becomes

$$\delta \int d^4x \sqrt{-\bar{g}} \left(\bar{R} - \left(\omega + \frac{3}{2} \right) |\bar{\nabla}\alpha|^2 + e^{-2\alpha} L_m(\bar{g}) \right) = 0, \tag{63}$$

where $\phi = e^{\alpha}$. This variational principle is of course just the Einstein one for a massless scalar field (dimensionless), α, but universally coupled to all other matter through the $e^{-2\alpha}$ factor. Regarding conformal rescalings of the metric as a "gauge," (63) is an expression of the theory in the "Einstein gauge," as opposed to the original (53), the "Jordan" gauge. However, it should be clear that there is more to the conformal scaling than merely the formal expression of the equations. In fact, the universal coupling of α to all matter in (63) means that in this metric test particles will not follow geodesics, nor have conserved inertial mass, etc., in the Einstein gauge. In effect, the identification of the Einstein metric used in the formulation (63) as the "physical" metric leads to significant and observable violations of mass conservation and the WEP. Nevertheless, the choice of various conformal gauges continues to be studied.

The investigation of such scalar-tensor generalizations of Einstein theory was strongly influenced by the work of Dicke. In fact, the 1960s and 1970s saw an explosion of interest in relativity and gravitational theories prompted at least in part by the presence of theoretically viable alternatives to standard Einstein theory, and Dicke's energetic promotion of them. By fortuitous coincidence this was also the time when NASA was coming of age and searching for space-related experiments of fundamental importance. Simultaneously, Nordtvedt, Will, and others [14] were led to provide rigorous underpinnings to the operational significance of various theories, especially in solar system context, developing the parameterized post-Newtonian (PPN) formalism as a theoretical standard for expressing the predictions of relativistic gravitational theories in terms which could be directly related to experimental observations. The equations of scalar-tensor theory approach those of standard Einstein theory as ϕ approaches a constant. From (59) this would seem to occur in the limit of large ω. In fact, it is generally true that the predictions of scalar-tensor approach those of Einstein for large ω, although there are interesting questions to be considered in general [15].

The ultimate outcome of these efforts was to set limits on the value of the parameter ω so large as to make the predictions of this theory essentially equivalent to those of standard Einstein theory. In other words, solar system experimentation led to the conclusion that scalar-tensor modifications of standard Einstein theory would necessarily differ insignificantly from the standard, leading many workers to regard such theories as irrelevant. Nevertheless, as the next two sections show, universally coupled, thus gravitational, scalar fields continue to play important roles in contemporary physics.

9.5 Dilatons

In the preceding discussion the scalar field was universally coupled to all matter and played a role determining the locally measured Newtonian gravitational "constant." Of course, scalar fields occur throughout physics, especially as quantum fields. Investigations of internal spaces for particle symmetries directly involve gauge theories with internal symmetry spaces occupied by families of fields which, while having interesting transformation properties from the internal gauge group viewpoint, are nonetheless spacetime scalars. Some of the earliest are the $SO(N)$ bosons of the dual model, the Nambu-Goldstone bosons, and the famous Higgs fields. Certainly, these scalars, as quantum fields, are based on different motivations than those leading to the scalar field in scalar-tensor theories. Nevertheless, the formalism, and perhaps macroscopic manifestations may turn out to be not too different.

Historically, quantum dual models led to string theory and later superstring theory. In this process, a scalar field referred to as a "dilaton" appears quite naturally. This field couples to the trace of the 2-dimensional string stress tensor. It thus manifestly breaks the Weyl conformal (dilation) symmetry of the string. Nevertheless it is precisely what is needed to balance the quantum anomalies of this tensor by way of beta functionals of this tensor. Along the way, the Einstein equations can be derived as the beta functions related to some external spacetime metric. The two volumes of Green, Schwartz, and Witten [16] provide useful description of the origin and role of dilatons.

This is clearly a long and complicated subject, which we only summarize here. Consider a string action as a natural generalization of a point particle action. For a background metric, $g_{\alpha\beta}$, an obvious action choice is

$$S_1 = \frac{-1}{4\pi\alpha'} \int \sqrt{|h|} h^{ab} \partial_a X^\alpha \partial_b X^\beta g_{\alpha\beta}(X^c) d^2\sigma, \tag{64}$$

with internal coordinate area $d^2\sigma$, internal string metric, h_{ab}, $a, b \ldots = 1, 2$, and α' a tension-related coupling parameter. If S_1 is compared to the relativistic point particle action, the role of point particle parameter has been replaced by the intrinsic surface metric, h_{ab}. As in the point particle case, the derived physics should be independent of the internal parameterization, and in particular, the choice of string metric. Thus, $\delta S_1/\delta h_{ab} = 0$, or

$$T_{ab} = \partial_a X \cdot \partial_b X = 0. \tag{65}$$

Trivially, this implies that the trace of T vanishes, but this would also be predicted by the conformal invariance of the string metric in (64). Actually, an even stronger result obtains: Any 2-dimensional metric is conformally flat (but only locally, in general!),

$$h_{ab} = \phi \eta_{ab}, \tag{66}$$

with constant η_{ab}. Thus, the surface element appearing in (64) reduces to the flat one,

$$d^2\sigma \sqrt{|h|} h^{ab} = d^2\sigma \eta^{ab}. \tag{67}$$

The independence of the classical action from the choice of string metric is thus equivalent to invariance under Weyl (conformal) transformation internal to the string surface. In addition to S_1, it is natural to consider two additional terms, with associated fields derived from the string quantities. The first contains a spacetime 2-form field, $B_{\alpha\beta}$, derived from the intrinsic volume 2-form in the string,

$$S_2 = \frac{-1}{4\pi\alpha'} \int \epsilon_{ab} \partial_a X^\alpha \partial_b X_\beta B_{\alpha\beta}(X^c) d^2\sigma. \tag{68}$$

The second introduces the geometry of the string,

$$\chi = \frac{1}{4\pi} \int \sqrt{|h|} R^{(2)} d^2\sigma. \tag{69}$$

However, one of the first discoveries relating geometry and topology was that this integral depends only on the topology of the string surface, and not the particular geometry. This is in fact the first Chern class for two dimensions. The value for χ is the Euler number of the surface, and cannot be a dynamical variable. The 2-geometry of the string can be introduced nontrivially by adding to (67) a scalar field, the "dilaton," Φ, giving

$$S_3 = \frac{1}{4\pi} \int \sqrt{|h|} \Phi(X^c) R^{(2)} d^2\sigma. \tag{70}$$

This term apparently breaks with the conformal invariance classically, thus violating the desired invariance at the classical level. However, paradoxically, it is precisely this term which can restore this invariance after quantization. Thus, when the action $S = S_1 + S_2 + S_3$ is quantized, conformal invariance is broken (an anomaly) unless the external fields satisfy three equations, as described in detail in GSW, volume 1, page 180. For brevity, we drop the antisymmetric field, setting $B_{\alpha\beta} = 0$, and get (in the magical string dimension 26!) Einstein-like equation,

$$0 = R_{\alpha\beta} - 2\Phi_{;\alpha;\beta}, \tag{71}$$

$$0 = 4\Phi_{,\alpha}\Phi^{,\alpha} - 4\Phi^{;\alpha}_{;\alpha} + R. \tag{72}$$

Equivalently, these background field conditions can be derived from an "effective action,"

$$\delta \int e^{-2\Phi}(R - 4\Phi_{,\alpha}\Phi^{,\alpha})d^D X. \tag{73}$$

It is easy to verify then that this action is a special case of the vacuum scalar-tensor one, (53), with $-2\Phi = \ln\phi$ and $\omega = 1$. Nevertheless, it is difficult not to notice the close parallel between the universally coupled scalar of the old scalar-tensor theories and the new dilaton.

9.6 Inflatons

Cosmological models in standard general relativity have long been known to contain serious conceptual difficulties. In particular, using standard general relativistic models, initial conditions must be fantastically fine-tuned in order to result in the universe as we now see it some 10^{10} years later. See for example Peebles [17], Linde [18]. Consider the standard Robertson-Walker isotropic homogeneous metric model,

$$ds^2 = -dt^2 + R(t)^2 d\sigma_\epsilon^2, \tag{74}$$

where the three-space metric, $d\sigma_\epsilon^2$, is hyperbolic, flat, or spherical depending on whether ϵ is -1, 0, or +1. The Einstein equations result in

$$\left(\frac{\dot{R}}{R}\right)^2 = \frac{\kappa\rho}{3} + \frac{\epsilon}{R(t)^2} + \frac{\Lambda}{3}. \tag{75}$$

Defining the Hubble variable as usual, this can be rewritten,

$$1 = \Omega + \epsilon\Omega_R + \Omega_\Lambda, \tag{76}$$

where

$$\Omega \equiv \frac{\kappa\rho}{3H^2}, \tag{77}$$

$$\Omega_R \equiv \frac{1}{(RH)^2}, \tag{78}$$

and

$$\Omega_\Lambda \equiv \frac{\Lambda}{3H^2}. \tag{79}$$

Present data certainly gives values for these three quantities each in the ballpark of one. In fact,

$$\Omega(now) \approx \frac{\kappa M}{R} \approx 10^0, \tag{80}$$

which is one of Dirac's large number coincidences which was so instrumental in leading to the scalar-tensor theories. Now, however, we note it in the context of a universe evolving from earlier ("initial") data drastically different from that at present. For example, in the present matter-dominated era the equation of state leads to

$$\rho R^3 = M \approx \text{const.}, \tag{81}$$

whereas in an earlier radiation-dominated state

$$\rho R^4 \approx \text{const.} \tag{82}$$

An analysis of the time evolution of these quantities under drastically different regimes show that an extremely small variation of the values of the Ω's at early times would result in drastically different values now. But this is not the only conceptual problem. For example, there are questions of how the universe could have homogenized itself from random early data (the "horizon" problem), and others [17], [18].

Guth [19] pointed out that this myriad of difficulties could be at least partially resolved if the early stages of evolution were "inflationary," that is,

$$R(t) = R(0)e^{Ht}, \tag{83}$$

with constant H. Such a model is consistent with (75) for $\rho = \epsilon = 0,\ \Lambda \neq 0$. Of course, this is not consistent with present data, so something other than a cosmological constant is needed. One way to achieve it is to introduce a new massless scalar field was the "inflaton," ϕ, with Lagrangian density,

$$\mathcal{L} = g^{\alpha\beta}\phi_\alpha\phi_\beta - V(\phi). \tag{84}$$

The resulting stress tensor produces an effective mass density and pressure given by

$$\rho_\phi = \dot{\phi}^2/2 + V, \qquad p_\phi = \dot{\phi}^2/2 - V. \tag{85}$$

By "fine-tuning" the potential, V, at least some, but certainly not all, of the problems discussed above can be resolved. In some versions, the inflaton has a dilaton-like nature, in others it is reminiscent of the ϕ in the old scalar-tensor theories, with ω so large as to make the deviations from general relativity insignificant in contemporary solar system physics, but very significant in earlier cosmological contexts. At present, it seems likely that more than one scalar field will be required. The entire field of inflationary models is a very active one at present, with many competing models. However, the role of scalar fields such as ϕ is prominent in many of them.

9.7 Conclusion

Universally coupled, thus gravitational, scalar fields are still active players in contemporary theoretical physics. So, what is the relationship between the scalar of scalar-tensor theories, the dilaton, and the inflaton? Clearly this is an unanswered and important question. The scalar field is still alive and active, if not always well, in current gravity research.

Acknowledgments

Bob Dicke led the way for me in the late 1950s, starting with inertial forces, Mach's principle, all the way through the ideas of scalar-tensor modifications of Einstein's general relativity. His death in March of 1997 deprives all of us in relativity of a source of ideas, and perhaps more importantly, a driving spirit of enthusiasm for understanding the mysteries of gravity.

John Norton was of great help in teaching me some of the interesting early history of scalar gravity and its role in the development of general relativity.

References

[1] John D. Norton, *Archive for History of Exact Science*, **45**, 17 (1992).
[2] Albert Einstein, quotation excerpted from Norton's article, preceding reference.
[3] Gunnar Nordström, *Annalen der Physik*, **40**, 856 (1913).
[4] Christopher Ray, *The Evolution of Relativity* (Adam Hilger, Bristol and Philadelphia, 1987).
[5] Albert Einstein and Adriaan Fokker, *Annalen der Physik*, **44**, 321 (1914).
[6] Thomas Appelquist, Alan Chodos, Peter G. O. Freund, *Modern Kaluza-Klein Theories* (Addison-Wesley, Reading, MA, 1987).
[7] Th. Kaluza, *Sitz. d. Preuss. Akad. d. Wiss., Physik.-Mat. Klasse*, 966 (1921).
[8] P. A. M. Dirac, *Proc. Roy. Soc.* **A165**, 199 (1938).
[9] V. Canuto, P. J. Adams, S. H. Hsieh, E. Tsiang, *Phys. Rev.* **D16**, 1643 (1977).
[10] Pascual Jordan, *Schwerkraft und Weltall*, (Vieweg Braunschweig, 1955).
[11] C. H. Brans and R. H. Dicke, *Phys. Rev.* **124** 925 (1961).
[12] Dennis Sciama, *Mon. Not. Roy. Ast. Soc.* **113**, 34 (1953).
[13] Engelbert Schücking, *Zeit. f. Physik* **148**, 72 (1957).
[14] Clifford M. Will, *Theory and Experiment in Gravitational Physics*, rev. ed. (Cambridge University Press, Cambridge, 1993).
[15] T. Damour and K Nordtvedt, *Phys. Rev.* **D48**, 3436 (1993).
[16] Michael B. Green, John H. Schwarz, Edward Witten, *Superstring Theory* (Cambridge University Press, Cambridge, 1987).
[17] P. J. E. Peebles, *Physical Cosmology* (Princeton University Press, Princeton, 1993).
[18] A. Linde, *Particle Physics and Inflationary Cosmology* (Harwood, London, 1990).
[19] A. Guth, *Phys. Rev.* **D23** 347 (1981).

10

The Cavendish Experiment in General Relativity

Dieter Brill

ABSTRACT Solutions of Einstein's equations are discussed in which the "gravitational force" is balanced by an electrical force, and which can serve as models for the Cavendish experiment.

10.1 Introduction

One of many useful lessons one can learn from Engelbert is the appreciation of simple situations and examples that nonetheless can teach us valuable physics. For me, such an Engelbert lesson was an introduction to the Bertotti-Robinson universe (which, as Engelbert added with characteristic precision that extends also to the history of physics, was first discovered by Levi-Civita), and its relation to extremal solutions [1]. Below is a bit of physics that we can learn from extremal solutions to Einstein's equations.

In general relativity there is a well-defined sense in which the equations of motion for particles follow from the field equations. This is well known but not easily checked out,[1] for the necessary manipulations are rather formidable. It has been remarked [2] that when predictions of general relativity are based on particle equations of motion, they appear to lack the transparency and cogency that we appreciate in Newtonian physics and in some alternative theories; that even the outcome of the Cavendish experiment has not been derived in a way that is both simple and rigorous; and that, at least in the case of 2-dimensional ("planar") translational symmetry, there exist "anti-Cavendish" solutions of Einstein's field equations, describing slabs that do not attract each other. (In these solutions there are no interactions other than gravity between the slabs, but the stress-energy of the matter is "exotic.")

---

[1] I really mean *nachvollziehbar*, a fashionable German word that seems to have no good English equivalent.

In the present contribution we examine a question suggested by these considerations,[2] namely whether there are simple models in general relativity that are relevant to the Cavendish experiment. We will construct one such model that is easily analyzed and whose predictions agree with the expected experimental outcome. These models are not confined to the plane symmetric case for which they were first discussed, and they have no connection with the "exotic" slab solutions. Nevertheless we begin with a few elementary remarks about the special status of planar symmetry in general relativity as compared to other field theories.

10.2 Planar Symmetry

In electrostatics, problems with planar symmetry (such as two parallel charged plates) are among the simplest to treat. The translational and rotational symmetry of the physical setup prevents dependence of physical quantities on the transverse (y, z) directions. The problem therefore becomes 1-dimensional. Physically one cannot, of course, realize strict translational symmetry, because the system's total charge and mass would be infinite. However, one can approximate the 1-dimensional situation by systems whose properties are independent of y and z out to some large distance D, when one considers only longitudinal distances x small compared to D. In the limit $D \to \infty$, the 1-dimensional approximation becomes arbitrarily accurate, and reasonable physical quantities have finite limits. These include the electric field, the force per area, and the acceleration of the plates. (However, when the total charge on the plates is nonzero, the electrostatic potential does not have a finite limit if it is normalized to zero at infinity.)

Another simple feature of translationally symmetric electrostatics is the uniqueness of the relative acceleration between the plates. (The electric field is unique up to an additive constant.) If one has *any* solution with the appropriate symmetry, it is the *correct* solution.[3] This simplicity makes the (approximately) parallel plate geometry so useful in both pedagogy and practice.

In Newtonian gravitation the situation is essentially identical; everything one knows about electrostatics can be taken over (with the appropriate sign of the force), except that there is no arbitrary charge to mass ratio—the (strong) principle of equivalence fixes the ratio of gravitational to inertial mass to be a positive constant. One might expect that the simplicity of the parallel plate geometry will also carry over in general relativity.

---

[2] I thank Prof. C. Alley (University of Maryland) for numerous discussions which called attention to the status of the Cavendish experiment *vis a vis* general relativity, and in which he supplied the experimental ideas mentioned in passing below.

[3] This is true provided the plates are indeed static. In a typical experiment one balances the electric force between plates by an elastic force, and measures how much elastic force is needed to keep the plates static. If plates of finite size and non-negligible charge were allowed to accelerate, they would of course radiate. The radiation reaction would affect the plates' net acceleration, and this would depend on radiation conditions imposed at infinity.

There are of course important differences between these theories, which can destroy the analogy. One relevant difference that is usually cited is the role of the potential. In electrostatics and in Newtonian gravity the potential has no direct physical meaning separate from the fields. In general relativity the analogous quantity is the metric, and it measures directly the physically meaningful space-time distances. When the size D of the system increases (with constant mass density), the Newtonian potential, normalized to zero at infinity, typically diverges. This is not a serious problem in electrostatics or in Newtonian gravity, because another normalization can be chosen with impunity. However, a diverging metric offers more serious problems, and is certainly not allowed in an asymptotically flat space-time. On the other hand, it is not obvious that this divergence cannot be undone in the limit by suitable gauge changes; and in any case one can take the view that if translationally symmetric solutions exist, they should (approximately) describe physically realistic parallel plates, since the general relativity solutions for finite plates presumably exist.

Static solutions with plane symmetry have in fact been studied in general relativity, for example by Taub [3], for matter with a fluid equation of state. Unfortunately they do not readily lend themselves to physical interpretation of the type sought here. (One can however show on the basis of this work that, as expected, no solution exists with vanishing pressure and positive mass density.) Also, the relation of these solutions to any description of finite parallel plates with proper (asymptotically flat) behavior at infinity is not transparent.

10.3 Exact, Static Solutions

Static solutions would not seem to present a very versatile arena for exploring the features of the gravitational interaction. They do however correspond to a possible physical arrangement that would reasonably be used in a sensitive experiment to measure the strength of the gravitational interaction (G). In the usual Cavendish experiment,[4] even if initially the proof mass is in free fall, the long-time behavior is typically governed by an interplay of gravitational interaction and torsion fiber reaction. The initial acceleration can generally not be measured as accurately as the final displacement, which one can model as masses with constant separation—in other words, a static situation.

---

[4]Prof. Alley (private communication) points out that this could be modified to realize the plane-symmetric geometry by replacing the usual masses with parallel plates, one of which is suspended (for example, by means of the traditional torsion fiber) so that the total force on it can be monitored. This geometry has several advantages, for example that the distance between the plates does not have to be known with great accuracy, and that many of the devices used in a parallel plate electrostatic measurement to increase the accuracy, such as "guard rings" to make the field more uniform, could be adapted to the gravitational experiment. However, I do not know of any attempt to obtain a more accurate measurement of G in this way.

The gravitational interaction is then measured by the force necessary to keep the masses apart, and the basic nature of this force is electrical. (One could replace the force of the torsion fiber by the explicitly electrical force obtained, for example in the parallel plate version of the experiment, by putting equal charges on the plates.) We model this force by assuming that each volume also carries a net charge, proportional to the mass of that volume, and all of the same sign. For a suitable choice of the constant charge/mass ratio, the attractive gravitational and repulsive electrical forces will then balance in the Newtonian description.

How do we describe this situation in general relativity? Because mass and charge are present, we must solve the Einstein and the Maxwell equations, as well as the equations of motion of the matter. The source in the Einstein equations is the stress-energy of the electric field and that of the matter; the source in the Maxwell equations is the charge density of the matter; and these equations imply the matter equation of motion, at least for the simplest kind of matter, "charged dust."[5] If it is indeed possible to balance the forces in detail—an expectation discouraged by the nonlinear nature of Einstein gravitation, but encouraged by the absence of interaction energies in the the corresponding Newtonian situation—then there should be a static solution. It is a remarkable theorem that such solutions not only exist [4], but that the fields have a unique form under these conditions [5], the Majumdar-Papapetrou form [6] that is well-known when gravity is generated not by matter but only by charged black holes (and the stress-energy of their electric field). In the latter case the geometry and field can represent any static arrangement of a finite number of black holes with an *extremal* charge.

The Majumdar-Papapetrou ansatz for the metric and field can be written as

$$ds^2 = -V^{-2}dt^2 + V^2\left(dx^2 + dy^2 + dz^2\right), \qquad A^\mu = V\delta_t^\mu . \tag{1}$$

By explicit computation[6] of the Einstein tensor $G_{\alpha\beta}$ and the Maxwell stress-energy tensor $T_{\alpha\beta}^{\text{EM}}$, one finds agreement of most of the components of the two, for example

$$\begin{aligned} G_{xx} &= V^{-2}\left(-V_x^2 + V_y^2 + V_z^2\right) = T_{xx}^{\text{EM}}, \\ G_{xy} &= -2V_xV_y = T_{xy}^{\text{EM}}, \\ &\text{etc.} \end{aligned}$$

[5]The static sources in these solutions are unstressed, due to the detailed balance between electric and gravitational forces. So we can imagine that these are elastic bodies rather than dust, but with vanishing stress and strain. The stress-energy tensor is then the same as that of dust, and the solution still applies. It is clear that in this model the stress-energy tensor satisfies all energy conditions one might reasonably want to impose.

[6]I am grateful to the group of Prof. C. Alley for providing me with many of the results cited below, as obtained by their computer calculations. The conclusions drawn from these calculations are my own and have not been fully discussed with Prof. Alley.

The exception is $(\alpha, \beta) = (t, t)$. Similarly one finds that the A^μ of Eq. (1) satisfy most of the components of the vacuum Maxwell equations,

$$F_{\alpha\ ;\beta}^{\beta} = 0 \quad \text{except for} \quad \alpha = t.$$

This structure of the field equations is appropriate for a static dust source, since that type of source contributes only to the components that are excepted above. For these one has the condition (in units where the gravitational and electromagnetic coupling is unity)

$$\begin{aligned} G_{tt} - T_{tt}^{\mathrm{EM}} &= -2V^{-5}\nabla^2 V = T_{tt}^{\mathrm{matter}} \\ F_{t\ ;\beta}^{\beta} &= -V^{-4}\nabla^2 V = J_t^{\mathrm{matter}}. \end{aligned} \tag{2}$$

Here the Laplacian ∇^2 is to be evaluated in the flat 3-dimensional background metric $dx^2 + dy^2 + dz^2$.

All the equations are satisfied if charged dust can supply both sources in the equations (2). For this type of matter, with mass density ρ and charge density σ, we have

$$\begin{aligned} T_{\alpha\beta}^{\mathrm{matter}} &= \rho u_\alpha u_\beta, \\ J_\alpha^{\mathrm{matter}} &= \sigma u_\alpha. \end{aligned}$$

From the metric (1) we find that the unit 4-velocity for the static matter has the form $u_\alpha = V^{-1}\delta_\alpha^t$. Thus we see that equations (2) are satisfied if we choose

$$\rho = \sigma = -V^{-3}\nabla^2 V. \tag{3}$$

The equation relating the "potential" V and the source ρ is very similar to the Newtonian equation; in the vacuum region they are identical. One way to make a correspondence between the two is the following: Given any Newtonian potential V_N (vanishing at infinity) and source ρ_N, one finds a solution of equation (3) by

$$V = 1 + V_N, \qquad \rho = \rho_N / V^3.$$

Thus ρ has the same support as ρ_N, and the two differ only slightly if the gravitational fields are weak, $|V_N| \ll 1$.

This solution to the equations of general relativity has all the physically reasonable properties one expects; but could there be other solutions to the same problem with different properties? Suppose any static solution to the Einstein-Maxwell equations is given. For simplicity, confine attention to the region outside the matter. Let the electrostatic potential A^t of the solution be V. Let g_{ij} be the spacelike metric on the 3-dimensional hypersurfaces that are orthogonal to the timelike Killing vector. One can then show that when one modifies the metric by a conformal factor V^{-2}, its Ricci tensor vanishes, $R_{ij}[V^{-2}g_{ij}] = 0$. The 3-dimensional modified metric must therefore be flat, and hence the original metric and field must have the form of equations (1). In this sense, then, the solutions given above are unique.

10.4 Test Particle Motion

It is not necessary to verify separately that the equations of motion for the matter are satisfied by the solution given by equations (1, 3), because the matter motion is a consequence of the field equations. However, as a further check that this solution is reasonable, we derive the equation of motion for a test particle in the fields of this solution.

As in Newtonian physics, the general relativistic motion of test particles in the general metric (1) (even with V harmonic) is not integrable, but there is always an energy integral. The energy integral is enough to find the motion if we know that it is confined to one coordinate line. This is the case for example when there is planar (x, y) or axial (about the z-axis) symmetry. We therefore confine attention to these cases where the energy integral yields the essential information about the motion.

In the case of uncharged particles we can apply the usual theorems about geodesics, that any Killing vector like $\partial/\partial t$ yields a conservation law of the corresponding covariant component of the 4-velocity u, $u_t = -E$. From the metric (1) we therefore have, with $\tau =$ proper time,

$$E = -g_{tt}\frac{dt}{d\tau} = \frac{1}{V^2}\frac{dt}{d\tau}.$$

We also know that setting the length of u to unity is always an integral of the equation of motion,

$$u \cdot u = -1 = -\frac{1}{V^2}\left(\frac{dt}{d\tau}\right)^2 + V^2\left(\frac{dx}{d\tau}\right)^2. \tag{4}$$

Elimination of $dt/d\tau$ yields

$$\left(\frac{dx}{d\tau}\right)^2 + \frac{1}{V^2} = E^2. \tag{5}$$

So for geodesic motion the quantity $1/V^2$ acts as an effective potential, and the particle will be deflected from the Killing orbit $(x, y, z) =$ const. by an amount proportional to ∇V, as in the Newtonian description.

For weak fields we have $V \approx 1 - \Phi$, where Φ is the Newtonian potential (for a spherically symmetric mass M, $\Phi = -M/r$), hence

$$\frac{1}{2}\left(\frac{dx}{d\tau}\right)^2 + \Phi \approx \frac{1}{2}(E^2 - 1),$$

which is (essentially) the usual energy integral for potential motion, yielding the usual motion corresponding to attractive gravity.

When the particle is charged (with charge/mass q), the corresponding conservation laws are derived from the variational principle

$$\delta \int (u \cdot u + 2qu \cdot A)\, d\tau = 0.$$

If there is a Killing vector that leaves the metric and the potential A invariant, there is a conserved quantity (Noether's theorem); for the Killing vector $\partial/\partial t$ the conserved quantity is the momentum conjugate to t, $-E = u_t + qA_t$, or

$$E = \frac{1}{V^2}\frac{dt}{d\tau} - \frac{q}{V}.$$

We substitute this in (4), eliminate $dt/d\tau$, and find

$$\left(\frac{dx}{d\tau}\right)^2 + \frac{1-q^2}{V^2} + \frac{2Eq}{V} = E^2.$$

We see that the attractive gravitational potential of the comparable equation (5) is reduced, and becomes repulsive for $q^2 > 1$. There is also a contribution to the potential that is proportional to E.

In the special ("extremal"[7]) case $q^2 = 1$ and $E = 0$, an initially static test particle remains at rest at any position (as does the matter that produces these fields). This is the case where attractive gravity and repulsive electrostatics are in perfect balance. If the particle is initially not quite at rest, then Eq must be slightly negative (since $V > 0$), hence the term $2Eq/V$ represents an attractive potential. It may be interpreted as a response of the particle's increased "relativistic mass" to gravity, with no compensating increase in the particle's charge.

10.5 Conclusion

We have exhibited the unique static solutions to the Einstein-Maxwell matter field equations that represent an arbitrary distribution of extremally charged matter in the form of dust. In particular, these solutions can be a good approximation to the geometry of a Cavendish experiment. Because all the charges have the same sign, the electric interaction is repulsive between two volumes of matter. The constancy in time of the physical distance between the masses implies, in ordinary language, a balancing attractive gravitational interaction. In this sense we have shown that in general relativity, as in Newtonian gravity, the gravitational interaction between the bodies is nonzero and attractive. Because the solution is valid only when the charge has the extremal value, such a balancing Cavendish experiment could be used to find the extremal charge value for a given mass.[8] Measurement of this extremal charge-to-mass ratio is equivalent to measuring the gravitational constant G.

---

[7] In this context the term "extremal" is somewhat misleading: it is the maximum charge that a black hole of the given mass could have, but is not a large charge at all for that mass of ordinary matter to carry.

[8] Results from an electrically balancing Cavendish experiment have recently been reported [7]; however, that experiment did not measure the extremal charge value because (for good reasons) the attracting "large mass" was not the same as the repelling electrode.

REFERENCES

[1] E. Schucking, personal communication.
[2] C. Alley, in *Fundamental Problems in Quantum Theory* (Ann. New York Acad. Sci. **755**, 1995), p.464; *Frontiers of Fundamental Physics*, Eds., M. Barone and F. Selleri, (Plenum Press, New York, 1994), p.125.
[3] A. H. Taub, Phys. Rev. **103**, 454 (1956); *General Relativity*, Ed., L. O'Raifeartaigh (Oxford Univ. Press, Oxford, 1972) p. 133.
[4] W. B. Bonnor, *Gen. Rel. Grav.* **12**, 453 (1980) and the references cited there.
[5] P. Chruściel and N. S. Nadirashvili, *Class. Quant. Grav.* **12**, L17 (1995); also see P. Chruściel, *Contemporary Mathematics* **170**, 23 (1994) and the references cited there.
[6] S. D. Majumdar, *Phys. Rev.* **72**, 390 (1947); A. Papapetrou, *Proc. Roy. Irish Acad.* **A51**, 191 (1947).
[7] M. F. Fitzgerald and T. R. Armstrong, *IEEE Trans. on Instrumentation and Measurement* **44**, 494 (1995).

11

Wave Maps in General Relativity

Yvonne Choquet-Bruhat

Dedicated to Engelbert Schucking with esteem and affection.

11.1 Introduction

Wave maps from a pseudo-Riemannian manifold of hyperbolic (Lorentzian) signature (V, g) into a pseudo-Riemannian manifold are the generalization of the usual wave equations for scalar functions on (V, g). They are the counterpart in hyperbolic signature of the harmonic mappings between properly Riemannian manifolds. The first wave maps to be considered in physics were the σ-models, e.g., the mapping from the Minkowski spacetime into the 3-sphere which models the classical dynamics of 4-meson fields linked by the relation

$$\sum_{a=1}^{4} | f^a |^2 = 1.$$

Wave maps appear in various areas of physics (see for instance Nutku (1974), Misner (1978)). Stoeger (1983) remarks that Rosen's bi-metric theory of gravity gives a wave map equation. But wave maps play an important role in general relativity itself; for example, harmonic coordinates can be considered as a condition for the identity map from $(U, g), U \subset V$, into an open set of Euclidean space (U, e) to be a harmonic map. Wave maps between (V, g) into $(V, \hat{e})$, $\hat{e}$ a given metric, gives a global harmonic gauge condition on (V, g).

Einstein, or Einstein-Maxwell, equations for metrics possessing a 1-parameter spacelike isometry group can be written as a coupled system of a wave map equation from a manifold of dimension 3 and an elliptic, time-dependent, system of partial differential equations on a 2-dimensional manifold, together with ordinary differential equations for the Teichmüller parameters (Moncrief (1986), Choquet-Bruhat and Moncrief (1995)).

The natural problem for wave maps is the Cauchy problem. It is a nonlinear problem, complicated by the fact that the unknowns do not take their values in a vector space, but in a manifold. Gu Chaohao (1980) has proven global existence of smooth wave maps from the 2-dimensional Minkowski spacetime into a complete Riemannian manifold by using the Riemann method of characteristics. Ginibre and Velo (1982) have proven a local-in-time existence theorem for wave maps from a Minkowski spacetime of arbitrary dimensions into the compact Riemannian manifolds $O(N)$, $CP(N)$, $GC(N, p)$ by semigroup methods. They prove global existence on 2-dimensional Minkowski spacetimes. These local and global results have been extended to arbitrary regularly hyperbolic sources and complete Riemannian targets in Choquet-Bruhat (1987), which proves also global existence for small data on $n+1$ dimensional Minkowski spacetime, $n \geq 3$ and odd. This last result has been proved to hold for $n = 2$ by Choquet-Bruhat and Gu Chaohao (1989) if the target is a symmetric space.

In this article we survey some of these and more recent results which we think pertinent to general relativity, and indicate applications.

11.2 Definitions

Let (V, g) and (M, h) be two smooth pseudo-Riemanian manifolds of arbitrary signature $n+1$ and dimension d. Let u be a mapping from V into M:

$$u: \qquad V \rightarrow M .$$

In an open set ω of V with local coordinates (x^α), with ω sufficiently small for the mapping u to take its value in a coordinate chart (y^A) of M, the mapping u is represented by d functions u^A of the $n+1$ variables x^α:

$$(x^\alpha) \quad \mapsto \quad y^A = u^A(x^\alpha).$$

The mapping u is differentiable at $x \in V$ if the functions u^A are differentiable. The differential $du(x)$ at x is a linear map between the tangent space at x to V and the tangent space to M at $u(x)$. It is therefore an element of the tensor product of the cotangent space to V at x by the tangent space to M at $u(x)$:

$$du(x)\, T_x V \rightarrow T_{u(x)}M, \text{ i.e., } du(x) \in T^*_x V \otimes T_{u(x)}M. \tag{1}$$

The differential itself is a section of the vector bundle E with base V and fiber E_x at x the above tensor product.

The metrics g on V and h on M together with the mapping u endow E_x with a scalar product (we still denote by g the contravariant tensor associated with g)

$$\mathrm{G(x)} \equiv g(x) \otimes h(u(x))$$

We denote by ∇ the linear connexion on the vector bundle E. It is such that, for any tangent vector v to V, any section t of T^*V (or tensor product of such sections) and section of the vector bundle with base V and fibre $T_{u(x)}$ at x (or tensor product of such sections) this covariant derivative in the direction of v, ∇_v is given by

$$\nabla_v(t \otimes s) \equiv (\nabla_v(g)t) \otimes s + t \otimes (u^* \nabla_{duv}(h)s),$$

where $\nabla(g)$ and $\nabla(h)$ are the Riemannian covariant derivatives with respect to g and h. This covariant derivative leaves invariant the scalar product G.
If f is a section of E represented in local coordinates by the $(n+1) \times d$ functions f_α^A, we have

$$\nabla_\alpha f_\beta^A(x) \equiv \delta_\alpha f_\beta^A(x) - \Gamma_{\alpha\beta}^\mu(x) f_\mu^A(x) + \delta_\alpha u^B(x) \Gamma_{BC}^A(u(x)) f_\beta^C(x),$$

where $\Gamma\,^\mu_{\alpha\beta}$ and Γ_{BC}^A denote the components of the Riemannian connections of g and h.
Analogous formulas give the covariant derivatives of sections of bundles over V with fiber $\otimes^p T_x^* V \otimes T_{u(x)} M$. In particular, by the very definition of the Riemann curvature tensors:

$$(\nabla_\alpha \nabla_\beta - \nabla_\beta \nabla_\alpha) f_\lambda^A = R_{\alpha\beta\lambda}{}^\mu(x) f_\mu^A(x) + \delta_\alpha u^C \delta_\beta u^B R_{CB}{}^A{}_D f_\mu^D$$

Definition *A mapping $u: (V, g) \to (M, h)$ is called a "harmonic" map if it satisfies the following second-order partial differential equation taking its values in TM:*

$$g.\nabla^2 u = 0.$$

This equation reads in local coordinates on V and M:

$$g^{\alpha\beta} \nabla_\alpha \partial_\beta u^A \equiv g^{\alpha\beta} (\partial_{\alpha\beta}^2 u^A - \Gamma_{\alpha\beta}^\lambda \partial_\lambda u^A + \Gamma_{BC}^A \partial_\alpha u^B \partial_\beta u^C) = 0.$$

The harmonic map equation is invariant under isometries of (V, g) and (M, h): Let u be a harmonic map from (V, g) into (M, h), let f and F be diffeomorphisms of V and M respectively; then $F \circ u \circ f$ is a harmonic map from $(f^{-1}(V), f_* g)$ into $(F(M), dFh)$.

Harmonic maps are critical points of the Dirichlet integral:

$$\int_M g^{\alpha\beta} \mu_g \partial_\alpha u \cdot \partial_\beta u,$$

with $\partial_\alpha u \cdot \partial_\beta u \equiv h_{AB} \partial_\alpha u^A \partial_\beta u^B$ and μ_g the volume element of g. The Dirichlet integral is positive if g and h are properly Riemannian; it is then the energy of the mapping u. A harmonic map will eventually be an extremum of this energy.

We consider from now on the case where the metric g is of hyperbolic signature, the corresponding harmonic maps are now called wave maps; the Dirichlet integral is no more their energy. In physics, when the source is the Minkowski spacetime and the target the sphere S^3, they are known as σ-models.

11.3 Wave Maps—the Cauchy Problem

Throughout this paper we stipulate the following conditions on the manifold $(V, \mathbf{g})$: The manifold V is of the type $S \times \mathbb{R}$, S an n-dimensional oriented smooth manifold, the metric $\mathbf{g}$ is pseudo-Riemannian with signature $(-,+,\ldots,+)$, each submanifold $S_t \equiv S \times \{t\}$ is spacelike. The metric then reads in a moving frame with time axis orthogonal to the S_t's:

$$\mathbf{g} = -N^2 dt^2 + g_{ij}\theta^i\theta^j, \qquad \theta^i \equiv dx^i + \beta^i dt.$$

The scalar function N, called lapse, is strictly positive; the vector β is called the shift; the induced metric on each S_t, $g_t = g_{ij}dx^i dx^j$, is properly Riemannian. We denote, respectively, by ∇_α and ∇_i the covariant derivatives in $\mathbf{g}$ on V and in g_t on S_t. (We will suppress the boldface notation when Greek indices make clear the spacetime nature of an object; we will suppress the t index in space objects with Latin coordinates, or when the context makes clear the manifold S_t in which it lies.)

The first natural problem to solve for a wave map is the *Cauchy problem*, i.e., the construction of a wave map taking together with its first derivative given values on a submanifold S_0.

We will in this article suppose that $\mathbf{g}$ is C^1 and $(V, \mathbf{g})$ is globally hyperbolic with S_0 a Cauchy surface, that is, the set of timelike or null paths joining two points of V is compact in the set of paths (Leray (1953)), each timelike or null path without end point cuts S_0 once (Geroch (1970)). The Cauchy problem for a linear wave equation on $(V, \mathbf{g})$ has then a global solution whose support is contained in the future of the support of its Cauchy data. The solution is C^∞ if $\mathbf{g}$ is C^∞ as well as the Cauchy data.

Remark. A necessary and sufficient condition for $(V, \mathbf{g})$ to be globally hyperbolic is that it is a product $S \times \mathbb{R}$ with S_0 a Cauchy surface (Geroch (1970)). A sufficient condition is that it is regularly hyperbolic (see definition later) with g_t uniformly equivalent to a complete metric e (see Choquet-Bruhat (1967)).

Theorem 1. *Suppose* $\mathbf{g}$ *and* h *are smooth. Let* $u(0, x) = \varphi$ *and* $\partial_t u(0, x) = \psi$ *be smooth Cauchy data, with* φ *uniformly continuous for some complete Riemannian metric topologies given to* M *and* S*. Then there exists a neighborhood* Ω *of* S_0 *in* V *and a smooth wave map* $u\colon (\Omega, \mathbf{g}) \to (M, h)$*, such that* u *takes the given Cauchy data on* S_0*.*

Sketch of proof. Becasue the mapping $\varphi : S \to M$ is continuous, there exist atlases on S and M such that each chart of M is mapped by φ and all sufficiently nearby maps into a chart of M. The wave map equation reads in coordinates charts on $S \times \mathbb{R}$ and M as a quasi-diagonal quasi-linear system of second-order hyperbolic equations. The theorem is proved by using the classical existence, uniqueness, and domain of dependence results for such systems.

We will give a more complete proof, directly global in space, after defining relevant functional spaces.

Definition. A metric $\mathbf{g}$ on $V \equiv S \times \mathbf{I}$, $\mathbf{I} \subset \mathbb{R}$ is said to be regularly hyperbolic if
1. There exist numbers B_1, B_2 *such that on* $S \times \mathbf{I}$

$$0 < B_1 \leq N \leq B_2.$$

2. The metrics g_t, $t \in \mathbf{I}$, *induced by* $\mathbf{g}$ *on* S_t, *are uniformly equivalent to a given Riemannian metric* e *on* S, *that is, there exist strictly positive numbers* A_1 *and* A_2 *which for any vector field* ξ *on* S *and* $t \in \mathbf{I}$ *satisfy on* S

$$A_1 e(\xi, \xi) \leq g_t(\xi, \xi) \leq A_2 e(\xi, \xi).$$

The Sobolev spaces W_s^p of tensors on S are defined through the metric e. We set $W_s^2 = H_s$. The covariant derivative in the metric e of a tensor on S (which may be time dependent) is denoted by D. We suppose (S, e) is such that the usual imbedding and multiplication properties hold, as well as the density in W_s^p of the space of C^∞ tensors with compact support, denoted C_0^∞: it will be the case if (S, e) is a smooth complete Riemannian manifold with curvature bounded as well as all its derivatives.
We denote by $E_s^p(T)$ the space of tensors f on $V_T \equiv S \times [0, T]$ such that

$$f \in E_s^p(T) \quad \rightleftarrows \quad f \in C_b^0(V_T), Df, \partial_0 f \in C^k([0, T], W_{s-k-1}^p), 0 \leq k \leq s-1,$$

where C_b^0 denotes continuous and bounded tensors.
We denote by E_s^p the space of tensors on V which are in $E_s^p(T)$ for any finite T. We set $E_s \equiv E_s^2$.

Theorem 2a. *Suppose M is diffeomorphic to $\mathbb{R}^m$. A mapping $u: V \to M$ is then globally defined by a set of m scalar functions u^A on V.*
Suppose the metric h is smooth. Suppose **g** *is smooth with uniformly bounded derivatives of arbitrary order with respect to t and e, i.e., N, β, g possess these properties. Let $\varphi^A \in C_b^0$, $D\varphi^A \in H_{s-1}$, and $\psi^A \in H_{s-1}$ be Cauchy data such that $s > \frac{n}{2} + 1$.*
There exists T> 0 and a wave map $u = (u^A)$, $u^A \in E_s(T)$, taking the given data. The interval T of existence for any s is equal to the interval corresponding to $s = s_0$, the smallest integer greater than $\frac{n}{2} + 1$. The solution is unique and depends continuously on the data. A solution in $E_s(T)$, $s > \frac{n}{2} + 1$ can be approximated by smooth solutions.

<u>Corollary 1</u>. *The existence in $E_s(T)$, uniqueness, and continuous dependence on the data of the solution still hold if the metric* **g**, *i.e., N, β, and g belong to E_s.*

<u>Corollary 2</u>. *An existence theorem in a neighborhood of S can be proven in local spaces, i.e., by replacing the spaces C_b^0 and H_s on S by spaces of functions which are in C_b^0 or H_s in each open relatively compact subset of S. The neighborhood will be of the form $S \times [0, T]$ only under some further hypothesis, for instance either the uniform bound of the norms in some locally finite covering of S, or the uniform bound on R^m of the Christoffel symbols and curvature, with enough of its derivatives, of the metric h.*

<u>*Proof*</u>. The wave map equation reads as a semilinear quasi-diagonal hyperbolic system of m scalar second-order equations on V. The Leray theory, as completed by Dionne (1962), Choquet-Bruhat (1971), Y.C.B. Christodoulou-Francaviglia (1979), gives the results. One uses the fact that E_{s-1} is an algebra when $s - 1 > \frac{n}{2}$

<u>*Remark 1*</u>. It is possible to prove an analogous theorem with less time regularity of the metric. One then obtains less time regularity of the solution.
<u>*Remark 2.*</u> The hypothesis made on the metric imply in all cases that D**g** is uniformly bounded on V_T. They do not necessarily imply that it is Lipshitzian: the geodesics between two nearby points may not be unique.

An analogue of Theorem 2a can be proved when the target M is not diffeomorphic to $\mathbb{R}^m$ by embedding (M, h) isometrically into some space $(\mathbb{R}^N, q)$ for instance a Euclidean space, which is always possible.

Definition. *The manifold (M, h) is said to be regularly embedded in $(\mathbb{R}^N, q)$ if it is defined by p smooth scalar equations $\Phi^{(I)} = 0$ on R^N, of rank p on M, and h is the pullback of q under this embedding. We denote by $\nu^{(I)}$ the normal to M in R^N defined by the gradient of $\Phi^{(I)}$, and we set*

$$\nu_{(I)} = m_{IJ}\nu^{(J)},$$

where m_{IJ} is the inverse of the matrix with elements $(\nu^{(I)}, \nu^{(J)})_q$, scalar products in the metric q. We denote by $\nabla_{(q)}$ the covariant derivative in **g** *and q.*

Lemma. *A mapping U: $(V_T, \mathbf{g}) \to (\mathbb{R}^N, q)$ will be a wave map u: $(V_T, \mathbf{g}) \to (M, h)$ if and only if $U = (U^A)$ is such that*
a. It satisfies the system of N scalar semilinear equations

$$g^{\alpha\beta}\{\nabla_{(q)\beta}\partial_\alpha U^A + \partial_\alpha U^B \partial_\beta U^C \nu^A_{(I)}(\nabla_{(q)B}\nu^{(I)}_C)\} = 0.$$

b. Its initial data $U^A{}_{|S_0} = \varphi^A$ take their values in M while $(\partial_t U^A{}_{|S_0}) = \psi^A$ take their values in the tangent space to M at φ, i.e., $\Phi^{(I)}(\varphi) = 0$ and $\partial_A\Phi^{(I)}(\varphi)\psi^A =$ 0.

Proof. (Choquet-Bruhat (1987): for special cases see Ginibre and Velo (1982)) A mapping u: $(V, \mathbf{g}) \to (M, h)$ is a wave map if and only if it is a critical point of the Dirichlet integral:

$$\mathcal{E}(\mathrm{u}) \equiv \int_V (\mathbf{g} \otimes h)(\nabla u, \nabla u)\mu_g.$$

The Dirichlet integral $\mathcal{E}(U)$ constructed with **g**, q and $U = i(u)$, i, the identity embedding of M into R^N, is equal to $\mathcal{E}(u)$ when $h = i^*q$ while u is restricted to take its values in $M \subset \mathbb{R}^N$. Hence a critical point of $\mathcal{E}(u)$ is a critical point of $\mathcal{E}(U)$ with the contraints $\Phi^{(I)}(U) = 0$. The classical variational calculus with Lagrange multipliers gives the Euler equation satisfied by such a U:

$$g^{\alpha\beta}\nabla_{(q)\beta}\partial_\alpha U^A + \lambda_{(I)}q^{AB}\partial_B\Phi^{(I)} = 0,$$

with $\lambda_{(I)}$ determined by the compatibility of the above equations with the p scalar conditions which must be satisfied for U to take its values in M:

$$g^{\alpha\beta}\nabla_\alpha\partial_\beta(\Phi^{(I)}(U)) = 0,$$

that is,

$$g^{\alpha\beta}\{\nu^{(I)}_A\nabla_{(q)\beta}\partial_\alpha U^A + \partial_\alpha U^B\partial_\beta U^C \nabla_{(q)B}\nu^{(I)}_C)\} = 0.$$

We deduce the value of $\lambda_{(I)}$ and the given equations from the above formulas. Conversely a mapping U satisfying the given equation can be identified with a mapping u which will be a critical point of $\mathcal{E}(u)$, hence a wave map $(V, \mathbf{g}) \to (M, h)$, if U takes its values in M.

By the computation of $\lambda_{(I)}$, it is immediate that the equation satisfied by U implies the previously written linear equation for the scalar function $\Phi^{(I)}(U)$ on $S \times [0, T]$. It follows from the uniqueness theorem for such an equation that U takes its values in M, since $\Phi^{(I)}(U)$ and $\partial_t \Phi^{(I)}(U)$ have been supposed to vanish on S_0.

We deduce from the above lemma and the Leray theory the following theorem:

Theorem 2b. Theorem 2a and its corollaries apply to a target manifold (M, h), which can be regularly embedded in a smooth manifold $(\mathbb{R}^N, q)$.

11.4 Harmonic Gauges in General Relativity

As an example of application of the local existence theorem for wave maps, we will recall the proof of the local-in-time, global-in-space uniqueness theorem for vacuum Einsteinian spacetimes.

Two spacetimes of general relativity $(V, \mathbf{g})$ and $(V', \mathbf{g}')$ are considered as equivalent if they are isometric, i.e., if there exists a diffeomorphism u from V onto V′ such that $\mathbf{g}$ is the reciprocal image by u of $\mathbf{g}'$, i.e.

$$\mathbf{g} = u_* \mathbf{g}'$$

One chooses a representative $\mathbf{g}$ in this equivalence class, i.e., chooses on V a gauge for the metric, by generalizing the harmonic coordinate condition. The generalized condition has the advantage of turning the Ricci tensor of $\mathbf{g}$ into a globally defined quasi-linear, quasi-diagonal second-order operator.

Let us consider on V a given metric $\hat{e}$.

Definition. One says that $\mathbf{g}$ is in harmonic gauge with respect to $\hat{e}$ if the identity map $u : V \to V$ is a wave map from $(V, \mathbf{g})$ onto $(V, \hat{e})$.

When u is the identity, the equation for a wave map from $(V, \mathbf{g})$ into $(V, \hat{e})$ reduces to the following relation between the connections of $\mathbf{g}$ and $\hat{e}$:

$$\hat{F}^\lambda \equiv g^{\alpha\beta}(\Gamma^\lambda_{\alpha\beta} - \hat{\Gamma}^\lambda_{\alpha\beta}) = 0,$$

where Γ and $\hat{\Gamma}$ are respectively the connections of $\mathbf{g}$ and $\hat{e}$. This condition is tensorial, i.e., coordinate independent (the difference of two connections is a tensor) and defined on the whole of V if it is so of $\mathbf{g}$ and $\hat{e}$. It reduces to the usual harmonic coordinate condition if $\hat{e}$ is the flat metric in rectilinear coordinates on a manifold V diffeomorphic to $\mathbb{R}^{n+1}$.

The Ricci tensor of a metric $\mathbf{g}$ in $\hat{e}$ harmonic gauge is a symmetric 2-tensor, which we denote $\mathrm{Ricc}^{(\hat{e})}(\mathbf{g})$. It is a quasi-linear, quasi-diagonal operator on $\mathbf{g}$. Its contravariant components are (Choquet-Bruhat, Christodoulou, Francaviglia, (1979))

$$\mathrm{R}^{\alpha\beta}_{(\hat{e})} \equiv \frac{1}{2} g^{\lambda\mu} D_\lambda D_\mu g^{\alpha\beta} + Q^{\alpha\beta\rho\sigma}_{\gamma\delta\lambda\mu}(\mathbf{g}) D_\rho g^{\gamma\delta} D_\sigma g^{\lambda\mu}$$

$$-\frac{1}{2}g^{\lambda\mu}\{g^{\alpha\rho}\hat{R}^{\beta}{}_{\mu\lambda\rho}+g^{\beta\rho}\hat{R}^{\alpha}{}_{\mu\lambda\rho}\},$$

where D denotes here the covariant derivative in the metric $\hat{e}$ and Q is an analytic function of $\mathbf{g}$. The tensor $Ricc^{(\hat{e})}(\mathbf{g})$ is linked with the full Ricci tensor of $\mathbf{g}$ by the relation

$$\mathrm{R}^{(\hat{e})}_{\alpha\beta}\equiv\mathrm{R}_{\alpha\beta}-\frac{1}{2}\,(\mathrm{g}_{\alpha\lambda}D_{\beta}\hat{F}^{\lambda}+\mathrm{g}_{\beta\lambda}D_{\alpha}\hat{F}^{\lambda})$$

11.4.1 Existence of a Solution of the Vacuum Einstein Equations

The Cauchy data for $\mathbf{g}$, namely $\mathbf{g}|_S$ and $\partial_t\mathbf{g}|_S$, are determined by the geometrical data $g\equiv g_0$ and K (the second fundamental form) on $S\equiv S_0$ and an arbitrary choice of lapse and shift, for instance 1 and 0. Classical local existence and uniqueness theorems for a solution of the Cauchy problem apply to the tensor valued hyperbolic system

$$Ricc^{(\hat{e})}(\mathbf{g})=0.$$

The best known generic result in the case $n=3$ is for Cauchy data such that $\mathbf{g}_{|S}\in C^0_b$, $D\mathbf{g}_{|S}\in H_2$. The solution exists then in V_T for some small enough T. It belongs to $E_3(T)$.

One shows by using the Bianchi identities, as in the case of the harmonic coordinate choice, that this solution satisfies the full Einstein equations if the initial data satisfy the constraints and the initial harmonic gauge condition. Indeed for a solution of $Ricc^{(\hat{e})}(\mathbf{g})=0$, the Einstein tensor $\mathbf{S}$ reduces to

$$S_{\alpha\beta}=\frac{1}{2}(g_{\alpha\lambda}D_{\beta}\hat{F}^{\lambda}+g_{\beta\lambda}D_{\alpha}\hat{F}^{\lambda}-g_{\alpha\beta}D_{\lambda}\hat{F}^{\lambda}).$$

The Bianchi identities show that the harmonicity vector $\hat{F}$ satisfies a quasi-diagonal linear system which can be written

$$\mathrm{g}^{\alpha\lambda}D_{\alpha}D_{\lambda}\hat{F}^{\beta}+\hat{A}^{\beta\lambda}_{\alpha}D_{\lambda}\hat{F}^{\alpha}=0\,,$$

where $\hat{A}$ is a 3-tensor, known when $\mathbf{g}$ is known, linear in $D\mathbf{g}$ and analytic in $\mathbf{g}$ when $\mathrm{Det}\,\mathbf{g}\neq 0$ (which is a consequence of the properties of $\mathbf{g}$). This system has zero Cauchy data if the geometric data $(g,\ K)$ satisfy the constraints while the lapse and shift are chosen so as to satisfy the initial harmonic gauge condition. The uniqueness theorem for linear hyperbolic systems applies and shows that the $\mathbf{g}$ so obtained satisfies the original Einstein equations.

11.4.2 Uniqueness Theorem

We will have shown (local in time) uniqueness, up to isometries, of a solution $\mathbf{g}$ of the vacuum Einstein equations with given Cauchy data g and K on S if we can show that there exists a neighbourhood Ω of S in V and a diffeomorphism u of Ω to a neighborhood Ω' of S in V reducing to the identity on S and such that the reciprocal image $\mathbf{g}'$ of $\mathbf{g}$ by u, i.e., $\mathbf{g}' = u_*\mathbf{g}$, is in $\hat{e}$-harmonic gauge on $\Omega' \equiv u^{-1}(\Omega)$ and if the Cauchy data g′ and K′ for $\mathbf{g}'$ are equal to the given Cauchy data g and K while the lapse N′ and the shift β' take arbitrary preassigned values. Suppose there exists a wave map u, diffeomorphism from $(\Omega, \mathbf{g})$ onto $(\Omega', \hat{e})$, which reduces to the identity on S, while $\partial_t \mathrm{u}\ |_S = \psi$ is given and non-vanishing. Then $u \circ u^{-1}$ is the identity map from Ω' onto itself, and it is a wave map from $(\Omega', u_*\mathbf{g})$ onto (Ω',$\hat{e}$). The metric $u_*\mathbf{g}$ is therefore in $\hat{e}$-harmonic gauge in Ω'. Its Cauchy data are determined by the Cauchy data g and K and the initial values φ and ψ for the wave map u, hence uniquely determined.

When S is compact, we can take Ω of the form $V_T \equiv S \times [0, T]$. The existence theorem of the harmonic map $u \in E_3(T)$, for eventually some smaller T when $\mathbf{g} \in E_3(T)$ is a consequence of the local existence theorems. The mapping u is C^1 and is a diffeomorphism for small enough T, due to the choice of its initial values. However, the new metric $u_*\mathbf{g}$ is then only in $E_2(T)$, and the known uniqueness theorem does not apply. We must therefore suppose $\mathbf{g} \in E_4(T)$ then $u \in E_3(T)$ as well as $u_*\mathbf{g}$, and the uniqueness theorem applies. Note that an $E_4(T)$ metric is in $C^2(T)$. Its geodesics are locally unique. Perhaps this regularity is reasonable to ask from generic metrics.
When S is not compact one must use Corollary 2 of Theorems 2a and 2b, since the identity map of S is then only locally bounded and its derivative only locally square integrable.

11.5 Global Problem—the First Energy Estimate

To study global problems for wave maps one must use their special geometric properties, as for other fundamental equations of physics. In the sections that follow we suppose that the target (M, h) is a properly Riemannian manifold regularly embedded in a space R^N.
The *stress energy tensor* of the mapping u: $(V, \mathbf{g}) \to (M, h)$ is the covariant 2-tensor on M given by

$$T \equiv (h \circ u)(\partial u, \partial u) - \frac{1}{2}\mathbf{g}\{\mathbf{g} \otimes (h \circ u)\}.\{\partial u \otimes \partial u\},$$

that is

$$T_{\alpha\beta} = h_{AB}(u)\partial_\alpha u^A \partial_\beta u^B - \frac{1}{2} g_{\alpha\beta} g^{\lambda\mu} h_{AB}(u) \partial_\lambda u^A \partial_\mu u^B,$$

which we will usually write

$$T_{\alpha\beta} \equiv \partial_\alpha u \cdot \partial_\beta u - \frac{1}{2} g_{\alpha\beta} \partial_\lambda u \cdot \partial^\lambda u$$

Indices are raised with $\mathbf{g}$, and a dot denotes the scalar product in the metric h of the target space.

Lemma. *The stress-energy tensor of a wave map has zero divergence.*

Proof. A straightforward calculation gives

$$\nabla_\alpha T^\alpha_\lambda \equiv \partial_\lambda u \cdot g^{\alpha\beta}\nabla_\alpha \partial_\beta u \equiv h_{AB}(u)\partial_\lambda u^A g^{\alpha\beta}\nabla_\alpha \partial_\beta u^B.$$

Therefore $\nabla \cdot T = 0$ if u is a wave map.

The *energy-momentum vector* of u with respect to a vector X on V is:

$$\mathcal{P}^\alpha \equiv T^\alpha_\beta X^\beta$$

Lemma. *If X is timelike or null, then $\mathcal{P}$ is timelike or null, X and $\mathcal{P}$ have opposite time orientation.*

Proof. We have

$$\mathcal{P}^{\alpha} \equiv \partial^{\alpha} u \cdot (X^{\beta}\partial_{\beta} \mathrm{u}) - \frac{1}{2} \mathrm{X}^\alpha \, (\partial_\lambda u \cdot \partial^{\lambda} u)\,;$$

therefore

$$\mathcal{P}^{\alpha}\mathcal{P}_{\alpha} \equiv \frac{1}{4}(X^\alpha X_\alpha)(\partial_\beta u \cdot \partial^\beta \mathrm{u})^2 \le 0 \quad \text{if} \quad \mathrm{X}^\alpha \mathrm{X}_\alpha \le 0\,.$$

The time orientations of $\mathcal{P}$ and X are opposite, because one checks that

$$\mathcal{P}^{\alpha} X_\alpha \ge 0.$$

From the identity satisfied by a wave map, we deduce

$$\nabla_\alpha \mathcal{P}^\alpha = \frac{1}{2} T^{\alpha\beta}(L_X \mathbf{g})_{\alpha\beta}, \qquad (L_X \mathbf{g})_{\alpha\beta} \equiv \nabla_\alpha X_\beta + \nabla_\beta X_\alpha.$$

The symmetric 2-tensor $\pi \equiv L_X \mathbf{g}$ is the Lie derivative of the spacetime metric with respect to X. If there exists a timelike vector field with respect to which the metric has special properties, for instance is stationary, it may be appropriate to choose this vector field to define an energy-momentum vector. Otherwise it is convenient, in order to introduce only geometric quantities linked with the $n+1$ decomposition of the metric and their space derivatives, to choose for X the past-oriented unit normal, ν, to M, i.e., in the coframe θ^{α} :

$$\mathrm{X}_i = 0, \qquad X_0 = N;$$

the corresponding energy-momentum vector is timelike or null and future directed. We denote by K_{ij} the extrinsic curvature of S imbedded in $(V, \mathbf{g})$:

$$K_{ij} = -\tfrac{1}{2N}(\partial_t g_{ij} + \nabla_i \beta_j + \nabla_j \beta_i)$$

A straightforward computation shows that for such a choice of X we have

$$(L_X \mathbf{g})_{0i} = -\partial_i N, \qquad (L_X \mathbf{g})_{00} = 0, \qquad (L_X \mathbf{g})_{ij} = -2\,\omega^{\,0}_{\,ij} N = K_{ij}.$$

Lemma. For the given choice of X we have

$$\mathcal{P}^{0} N \equiv T^{00} N^2 \equiv \frac{1}{2}(\mid N^{-1}\partial_0 u \mid_h^2 + \mid Du \mid_{g,h}^2).$$

μ_t is the volume element of g on M_t, and we denote by $\mid\ \mid_{g,h}$ the norm both in g and h.

Proof: straightforward.

$\mathcal{P}^0 N$ is by definition the *energy density* of u at time t, with respect to the chosen X. We have

$$\mathcal{P}^0 N \equiv \frac{1}{2}\{|Du|^2_{g,h} + |N^{-1}\partial_0 u|^2_h\} \equiv \frac{1}{2}\{g^{ij}\partial_i u \cdot \partial_j u + N^{-2}\partial_0 u \cdot \partial_0 u\}$$

The integral of the energy density on a subset ω_t of S_t is, by definition, the *energy* $e_{\omega_t}(u)$ of u on this subset. We set $e_t(u) \equiv e_{S_t}(u)$.

Theorem. (Energy inequality) Let u be a solution of the wave map equation on a manifold $V = S \times \mathbb{R}$ with a C^1 regularly hyperbolic metric **g** such that DN and NK are uniformly bounded in g norm on each S_t. Suppose that $u \in C^2(T) \cap E_1(T)$. Then u satisfies for $t \in [0, T]$ the fundamental energy inequality

$$e_t(u) \le e_0(u) \exp C\{\textstyle\int_0^t \operatorname{Sup}_{M_\tau}(|DN|_g + 2|NK|_g) d\tau\}.$$

with C a number depending only on n.

Proof:
The integration of the divergence equation satisfied by $\mathcal{P}$ gives

$$\textstyle\int_{S_t} \mathcal{P}^0 N \mu_t = \int_{S_0} \mathcal{P}^0 N \mu_0 + \int_0^t \int_{S_\tau} N^{-1}\partial_i N \partial^i u \cdot \partial_0 u + NK^{ij} T_{ij} \mu_\tau d\tau.$$

This integral equality, together with the inequality satisfied by scalar products, implies the inequality

$$e_t(u) \le e_0(u) + C \textstyle\int_0^t \operatorname{Sup}_{S_\tau}(|DN|_g + 2|NK|_g) e_\tau(u) d\tau.$$

This inequality implies the theorem by the Gromwall lemma.

Remark 1. In the case where X is a Killing vector field of **g** and we use it to define the energy density, the energy inequality becomes an equality expressing the conservation of energy of the mapping u. We have chosen here for X the unit normal to S. It is a Killing field if $DN = 0$ and $K = 0$ the corresponding energy $e_t(u)$ is then conserved .
Remark 2. We have, if (M, h) is regularly embedded in $(\mathbb{R}^N, \delta)$, where δ is the Euclidean metric:

$$|Du|^2_{g,h} = g^{ij} h_{AB} \partial_i u^A \partial_j u^B = g^{ij} \delta_{IJ} \partial_i U^I \partial_j U^J .$$

We deduce from the hypothesis made on **g** that $|Du|^2_{g,h}$ is uniformly equivalent to $|Du|^2$. An analogous property holds for $|\partial_0 u|^2_h$; therefore $e_t(u)$ is uniformly equivalent to the sum of these norms, that is there exist positive numbers $C_{\mathbf{g}}$ depending only on the bounds on **g** such that,

$$C_{\mathbf{g}} e_t(u) \le \| \partial_0 u(.,t) \|_{L^2} + \| Du(.,t) \|_{L^2} \le C'_{\mathbf{g}} e_t(u)$$

Remark 3. The energy inequality gives only an estimate of ∂u. An estimate of u, as a mapping in $\mathbb{R}^N$, can be obtained from its initial data by the formula

$$u^I(.,t) = \varphi^I + \textstyle\int_0^t \partial_t u^I(., \tau) d\tau.$$

11.6 Second Energy Inequality

To bound the second derivatives of u, we consider the differentiated equations. A straightforward use of the Ricci formula for the commutation of covariant derivatives leads to the equations

$$\nabla^\lambda \nabla_\lambda \partial_\alpha u - R^\beta_\alpha \partial_\beta u + R_{A\dot{B}\ C}\ \partial_\alpha u^A \partial_\beta u^B \partial^\beta u^C = 0,$$

where $R_{\alpha\beta}$ is the Ricci tensor of $\mathbf{g}$ and R_{ABCD} the Riemann tensor of h.
We define a positive definite quadratic form E on V by setting

$$E^{00} = N^{-2}, \qquad E^{i0} = 0, \qquad E^{ij} = g^{ij}.$$

We introduce a stress-energy tensor for the first derivatives of u given by

$$T^{(1)}_{\alpha\beta} \equiv E^{\lambda\mu} T^{(1)}_{\alpha\beta,\lambda\mu}, \qquad T^{(1)}_{\alpha\beta,\lambda\mu} \equiv \{\nabla_\alpha \partial_\lambda u \cdot \nabla_\beta \partial_\mu u - \tfrac{1}{2} g_{\alpha\beta} \nabla^\rho \partial_\lambda u \cdot \nabla_\rho \partial_\mu u\}$$

and use again the Ricci identity to obtain

$$\begin{aligned} \nabla_\alpha T^{\alpha\beta}_{(1)} \equiv E^{\lambda\mu} \{ & \nabla_\alpha \nabla^\alpha \partial_\lambda u \nabla^\beta \partial_\mu u \\ & + (R_\alpha{}^\beta{}_\mu{}^\rho \partial_\rho u - R_{CD\ B} \partial_\mu u^B \partial_\alpha u^C \partial^\beta u^D) \nabla^\alpha \partial_\lambda u \} \\ & + (\nabla_\alpha E^{\lambda\mu}) T^{\alpha\beta}_{(1),\lambda\mu} \equiv J^\beta . \end{aligned}$$

with

$$\nabla_\alpha \nabla^\alpha \partial_\lambda u \equiv \nabla_\alpha \nabla_\lambda \partial^\alpha u = \nabla_\lambda \nabla_\alpha \partial^\alpha u + R^\mu_\lambda \partial_\mu u + R_{A\dot{B}\ C} \partial_\lambda u^B \partial_\alpha u^A \partial^\alpha u^C .$$

When u is a wave map, we have $\nabla_\lambda \nabla_\alpha \partial^\alpha u = 0$.
The energy-momentum vector of ∂u is defined by

$$\mathcal{P}^\alpha_{(1)} \equiv T^{\alpha\beta}_{(1)} X_\beta$$

We have

$$\nabla_\alpha \mathcal{P}^\alpha_{(1)} \equiv \frac{1}{2} T^{\alpha\beta}_{(1)} (L_X \mathbf{g})_{\alpha\beta} + J^\beta X_\beta$$

We take for X as before the past oriented unit normal to M_t, and we integrate on $M \times [0,\ t]$ the above divergence. The density of energy at time t is now found to be, using the definition of E

$$N \mathcal{P}^0_{(1)} \equiv \frac{1}{2} |\nabla^2 u|^2_{E,h} \equiv \frac{1}{2} \{ |N^{-2} \nabla_0 \partial_0 u|^2 + 2 |N^{-1} \nabla \partial_0 u|^2_g + |\nabla\nabla u|^2_g,$$

In the expression containing the norm in g only the spatial components ∇_i of the derivatives ∇ (which are here taken in the metrics $\mathbf{g}$ and h) are involved. Note that the space components of covariant derivatives in $\mathbf{g}$ and covariant derivatives in the space metric g_t are linked by the relation

$$\nabla_i \nabla_j u \equiv \nabla_i \nabla_j u - N K_{ij} \partial_0 u.$$

We set

$$e^{(1)}_t(u) \equiv \textstyle\int_{M_t} N \mathcal{P}^0_{(1)} \mu_t.$$

Integration of the divergence identity and use of the value of $L_X \mathbf{g}$ gives

$$e_t^{(1)}(u) = e_0^{(1)}(u) + \int_0^t \int_{M_\tau} N\{-\partial_i N T^{i0}_{(1)} + K_{ij} T^{ij}_{(1)} + N J^0\}\mu_\tau \, d\tau.$$

We use the equation satisfied by ∂u and the following values of $\nabla_\alpha E^{\lambda\mu}$:

$$\nabla_i E^{00} = 0, \quad \nabla_\alpha E^{ab} = 0, \quad \nabla_i E^{j0} = -2N^{-1} K_i^j, \quad \nabla_0 E^{i0} = 2N^{-1}\partial^i N.$$

We then find

$$N J^0 = N E^{\lambda\mu}\{(R_\lambda^\rho \partial_\rho u - R_{AB\ C}\partial_\lambda u^A \partial_\rho u^B \partial^\rho u^C)\cdot \nabla^0 \partial_\mu u + R_{\alpha\ \mu}^{\ 0\ \rho}\partial_\rho u . \nabla^\alpha \partial_\lambda u$$
$$-R_{AB\ C}\partial_\alpha u^A \partial^0 u^B \partial_\lambda u^C \cdot \nabla^\alpha \partial_\mu u\} + 2\,\partial^i N T^{00}_{(1),i0} - 2\,K_i^j (T^{i0}_{(1)0j} + T^{i0}_{(1),j0})]$$

with

$$T^{00}_{(1),i0} = T^{00}_{(1),0i} = \frac{1}{2} N^{-4}\nabla_0\partial_0 u \cdot \nabla_0 \partial_i u + N^{-2}\nabla_j \partial_0 u . \nabla^j \partial_i u,$$

$$T^{i0}_{(1),0j} = -N^{-2}\nabla^i \partial_0 u \cdot \nabla_0 \partial_j u, \quad T^{io}_{(1),j0} = -N^{-2}\nabla^i \partial_j u \cdot \nabla_0 \partial_0 u$$

Remark. When the metric h has constant curvature the cubic terms in ∂u take a particularly simple expression. Indeed we have then, with C a constant,

$$R_{AB.C} = C(h_{A.}h_{BC} - h_{AC}h_{B.})$$

Therefore:

$$R_{ABC}\partial_\alpha u^A \partial_\beta u^B \partial_\gamma u^C = C\{(\partial_\beta u \cdot \partial_\gamma u)\,\partial_\alpha u - (\partial_\alpha u \cdot \partial_\gamma u)\partial_\beta u\}$$

We deduce from the expression of $T^{(1)}_{\alpha\beta,\lambda\mu}$ that

$$N T^{i0}_{(1)} = -N^{-3}\nabla^i \partial_0 u . \nabla_0 \partial_0 u - N^{-1} g^{jh} \nabla^i \partial_j u \cdot \nabla_0 \partial_h u,$$

$$T^{ij}_{(1)} = N^{-2}\nabla^i \partial_0 u \cdot \nabla^j \partial_0 u + g^{hk}\nabla^i \partial_h u . \nabla^j \partial_k u - \frac{1}{2} g^{ij}(-|N^{-2}\nabla_0\partial_0 u|_h^2 + |\nabla \partial u|^2_{g,h})$$

We deduce frome these expressions the inequality

$$e_t^{(1)}(u) \le e_0^{(1)}(u) + \int_0^t \int_{M_\tau} \{|DN|_g(|N^{-1}\nabla\partial_0 u|_{g,h}|\nabla_0\partial_0 u|_h + |\nabla^2 u|_g |N^{-1}\nabla \partial_0 u|_g) +$$
$$|NK|_g(|N^{-1}\nabla\partial_0 u|^2_{g,h} + |\nabla D u|^2_{g,h}(1 - \tfrac{n}{2}) + \tfrac{1}{2}|N^{-2}\nabla_0\partial_0 u|_h^2) + |N J^0|\}\mu_\tau \, d\tau$$

We deduce then from the expression of NJ^0 that there exists a number C depending only on n such that the following inequality is satisfied:

$$e_t^{(1)}(u) \le e_0^{(1)}(u) + C\int_0^t \operatorname{Sup}_{M_\tau}(|DN|_g + 2N|K|_g) e_\tau^{(1)}(u) d\tau$$
$$+ \int_0^t \{\int_{M_\tau} (L(\partial u)^2 + Q(\partial u)^2)\mu_\tau\}^{1/2}\{e_\tau^{(1)}\}^{1/2}\, d\tau$$

where $L(\partial u)$ is an homogeneous linear form of the scalars $|N^{-1}\partial_0 u|_h$, $|Du|_{g,h}$ whose coefficients depend linearly on $|\text{Riemann}(g)|_E$, while $Q(\partial u)$ is a homogeneous polynomial of degree 3 in these same scalars whose coefficients depend linearly on $|\text{Riemann}(h)|_h$. There exists therefore a positive number C_h depending only on $|\text{Riemann}(h)|_h$ such that

$$\{\int_{M_\tau} Q(\partial u)^2 \mu_\tau\}^{1/2} \le C_h\{\|\, |N^{-1}\partial_0 u|_h\, \|^3_{L^6(M_\tau,g)} + \|\, |Du|_{g,h}\|^3_{L^6(M_\tau,g)}\}.$$

By the Sobolev embedding theorem on (S, e) we have for a scalar function f on S:

$$\| f \|_{L^p} \leq C\{\| f \|_{W^q_m}\} \qquad \text{if } 1 \leq q \leq p \leq nq/(n-mq)$$

in particular $L^6 \subset H_1$ if $n \leq 3$. By the hypothesis on g the L^p norms in g and e on S are equivalent. However there is some problem in applying directly the inequality above to the function $|Du|_{g,h}$ because this function is not differentiable when $Du = 0$.

We proceed as follows: We set

$$\mathrm{f} \equiv |\mathrm{Du}|^2_{g,h}; \text{ then } \| \, |Du|_{g,h} \, \|^2_{L^6} = \| \, |Du|^2_{g,h} \, \|_{L^3} \equiv \| f \|_{L^3}$$

On the other hand (note that for a scalar function $Df = \nabla f$)

$$|Df| \leq C_g |\nabla f|_g$$

and since the metric h has vanishing covariant derivative

$$|\nabla f|_g = 2|\nabla u . \nabla^2 u|_g$$

Finally

$$|Df| \leq C_g |\nabla u|_{g,h} |\nabla^2 u|_{g,h}$$

Hence by the Hölder inequality

$$\| Df \|_{L^q} \leq C_g \, \| \, |\nabla u|_{g,h} \, \|_{L^{2p}} \| \, |\nabla^2 u|_{g,h} \, \|_{L^2}, \qquad \frac{1}{q} = \frac{1}{2} + \frac{1}{2p}.$$

In particular Df is integrable when $|\nabla u|_{g,h}$ and $|\nabla^2 u|_{g,h}$ are square integrable. We have

$$\| Df \|_{L^1} \leq C_g (\| f \|_{L^1})^{1/2} \, \| \, |\nabla^2 u|_{g,h} \, \|_{L^2} \leq C_g e_t(u)^{1/2} e_t^{(1)}(u)^{1/2}$$

A. Suppose $n = 1$. The Sobolev embedding theorem says that there exists a constant C depending only on (S, e) such that

$$\| f \|_{C^0_b} \leq C\{\| f \|_{L^1} + \| Df \|_{L^1}\},$$

therefore

$$\| f \|_{L^3} \leq \{\| f \|^2_{C^0_b} \| f \|_{L^1}\}^{1/3} \leq C_g\{(\lambda^2 + \lambda\mu)^2 \lambda^2\}^{1/3},$$

with

$$\lambda \equiv e_t(u)^{1/2}, \qquad \mu \equiv e_t^{(1)}(u)^{1/2},$$

i.e.,

$$\| f \|_{L^3} \leq C_g\{\lambda^2 + \lambda^{4/3}\mu^{2/3}\}.$$

B. Suppose $n > 1$. The Sobolev embedding theorem gives

$$\| f \|_{L^p} \leq \mathrm{C}\{\| f \|_{L^1} + \| Df \|_{L^1}\}, \qquad 1 \leq p \leq n/n-1.$$

When $f \in L^p$ we have

$$Df \in L^q, \quad \frac{1}{q} = \frac{1}{2} + \frac{1}{2p} \qquad 1 \leq q \leq 2n/2n-1$$

We see that f and Df are in $L^{2n/2n-1}$, hence $f \in L^p, 1 \leq p \leq \frac{2n}{2n-3}$. This result shows that $f \in L^4$ if $n = 2$, result sufficient for our estimation of $Q(\partial u)$ in this case. In the general case we see that the iteration gives a sequence of values of q and p given by the induction formula

$$p_1 = 1, \quad q_m = \frac{2p_m}{p_m + 1}, \qquad p_{m+1} = \frac{nq_m}{n - q_m}, \qquad \text{hence}\, p_{m+1} = \frac{2np_m}{(n-2)p_m + n}$$

The p sequence is increasing and tends to a limit, finite if $n > 2$, $p = \frac{n}{n-2}$, $p = 3$ if $n = 3$, as necessary for tour estimation of $Q(\partial u)$. As it was foreseen, the estimate of $e_t(u)$ and $e_t^{(1)}(u)$ is not sufficient to estimate $Q(\partial u)$ if $n > 3$.

CASE $n = 1$: We have seen that in the case $n = 1$ there exists a constant C_g depending only on (S, e) and the bounds on g such that

$$\| \, |Du|_{g,h} \, \|^3_{L^6(S_t,g)} \leq C_g\{e_t(u)^{3/2} + e_t(u)e_t^{(1)}(u)^{1/2}\}.$$

An analogous inequality valid for $|N^{-1}\partial_0 u|_h$ gives the same type of inequality for $Q(\partial u)$, with a constant $C_{g,h}$ depending now on the bounds on **g** and the h norm of Riemann(h). We find thus:

$$e_t^{(1)}(u) \leq e_0^{(1)}(u) + C\int_0^t \operatorname{Sup}_{S_\tau}(|DN| + 2|NK|)e_\tau^{(1)}(u)d\tau$$
$$+ \int_0^t C_{Ricc(\mathbf{g})}e_\tau(u)^{1/2}e_\tau^{(1)}(u)^{1/2}d\tau + C_{g,h}\int_0^t e_\tau(u)\{e_\tau(u)^{1/2}e_\tau^{(1)}(u)^{1/2}$$
$$+ e_\tau^{(1)}(u)\}\, d\tau$$

where $C_{Ricc(g)}$ is the supremum on S_τ of the E norm of the Ricci tensor of **g**. Variants of this second term can be obtained by using previous results and for instance the L^3 norm instead of the supremum norm for Ricc(**g**).

In the above integral inequality $e_t^{(1)}(u)$ appears on the right-hand side only with powers 1 and $\frac{1}{2}$.

__Lemma.__ *When $n = 1$, under the hypothesis made on* **g** *and h on $S \times [0, T]$, the second energy $e_t^{(1)}(u)$ is uniformly bounded on $[0,T]$.*

__*Proof.*__ We already know that the fundamental energy $e_t(u)$ is uniformly bounded on $[0, T]$. The integral inequality satisfied by $e_t^{(1)}(u)$ implies that it is bounded by a solution $y(t)$ of the corresponding integral equality, i.e., by a solution taking the same initial value of the associated differential equation, which is of the type

$$y' = 2ay + 2by^{1/2}$$

where a and b are continuous functions on $[0, T]$. The solution of this equation equals to y_0 for $t = 0$ is the continuous function on $[0, T]$:

$$y(t) = \{y_0 + \textstyle\int_0^t b(\tau)exp(-\int_0^\tau a(\theta)d\theta)d\tau\} \exp \int_0^t a(\tau)d\tau.$$

Theorem. *(global existence) Under the hypothesis made on $(V,$ **g**$)$ and (M, h) there exists for any $T > 0$ a wave map u: $(S \times [0, T],$ **g**$) \to (M, h)$ taking on S_0 the initial values*

$$u(\cdot,0)=\varphi\in C_b^0,\qquad D\varphi\in H_1,\qquad \partial_0 u(\cdot,0)=\psi\in H_1.$$

Proof. Recall that the notations mean that after the isometric embedding of (M, h) into the Euclidean space $\mathbb{R}^N$ the components φ^A and ψ^A belong to the indicated Sobolev spaces on (S, e), e the given Sobolev regular metric on S. The local existence theorem says that there exists a number $\ell > 0$ depending only on the norms of the Cauchy data such that there exists a wave map $(S, \times[0,\ell]) \to (M, h)$. Let T be the smallest number such that the wave map exists for $\ell < T$. One will have proven that T cannot be finite if one obtains an a priori bound of the relevant norms of $u(\cdot,\ell)$ and $\partial_0 u(\cdot,\ell)$ for $\ell < T$. This bound is deduced as follows from the energy estimates.

1. We have proven that $e_t(u) \le C(T)$ with $C(T)$ a continuous function of T, hence bounded for all finite T. We find, using the hypothesis on $\mathbf{g}$ and the isometric embedding $(M, h) \to (\mathbb{R}^N, q)$, the inequalities:

$$|N^{-1}\partial_0 u|_h^2 \equiv h_{AB}N^{-2}\partial_0 u^A\partial_0 u^B \equiv q_{IJ}N^{-2}\partial_0 u^I\partial_0 u^J \le C_N|\partial_0 u|^2$$

$$|Du|_{g,h}^2 \equiv g^{ij}h_{AB}\partial_i u^A\partial_j u^B \le C_g|Du|^2,$$

The L^2 norm of the "Cauchy data" $D\varphi_t$ and ψ_t at time t satisfy therefore the estimate found for $e_t(u)$.

2. To deduce the estimate of the H_1 norms from the a priori bound of $e_t^{(1)}(u)$, we first recall (cf. C.B-D.M Problems V.11 and V.12) see if i is an imbedding $(M, h) \to (Q, q)$ and $U = i\circ u$, $u : (V, \mathbf{g}) \to (M, h)$ then

$$\nabla_{(q)}\partial U \equiv \partial i\cdot\nabla\partial u + \nabla_{(q)}\partial i\cdot(\partial u\otimes\partial u),$$

i.e.,

$$\nabla_\alpha\partial_\beta U^I \equiv \partial_A i^I\nabla_\alpha\partial_\beta u^A + \nabla_A\partial_B i^I\partial_\alpha u^A\partial_\beta u^B.$$

When i is an isometric embedding, $\nabla\partial i$ is the second fundamental form L of M as a submanifold of (Q, q). We deduce from this relation the inequality

$$|\nabla^2 U|_{E,q} \le |\nabla^2 u|_{E,h} + |L|_{h,q}|\partial u|_{E,h}^2.$$

When (Q, q) is a euclidean space (R^N, δ) we have, with $\mathbf{D} = (D_\alpha)$ the covariant derivative in the metric $e + dt^2$ and U^I considered as a scalar function:

$$\nabla_\alpha\partial_\beta U^I \equiv D_\alpha\partial_\beta U^I + S^\lambda_{\alpha\beta}\partial_\lambda U^I,$$

where S is the tensor difference of the connexions of $\mathbf{g}$ and $e + dt^2$.
We deduce from these relations the inequality for each component of U:

$$|\mathbf{D}^2 U| \le C_{\mathbf{g}}\{|\nabla^2 u|_{E,h} + C_{\partial\mathbf{g}}|\partial u|_{E,h} + |L|_{h,q}|\partial u|_{E,h}^2\}$$

Therefore for each t, using the previous estimate of $\partial u(\cdot,t)$ in L^4:

$$\|\,\mathbf{D}^2 U(.,t)\,\|_{L^2} \le C_{\mathbf{g}}\{e_t^{(1)}(u)^{1/2} + C_{\partial\mathbf{g}}e_t(u)^{1/2} + |L|_{h,q}(e_t(u) + e_t(u)^{3/2}e_t^{(1}(u)^{1/2})\}.$$

We deduce from this inequality that the "t-Cauchy data" $D\varphi_t$, ψ_t are uniformly bounded in H_1 norm. The proof is complete.

11.7 Wave Map from the Outside of a Black Hole

The physical world is, at least at the classical macroscopic level, 4-dimensional. However examples of wave maps from manifolds of fewer dimensions occur in space times having a spacelike isometry group.

The existence of spherically symmetric wave maps on Minkowski spacetime has been proved by Christodoulou and Talvilar-Zadeh. The proof is delicate, due to the singularity which appears in the equation at the center of symmetry $r = 0$. We will avoid this problem by considering the outside of spherically symmetric black holes. However, we will be confronted with another problem: the corresponding spacetime is globally hyperbolic but not regularly hyperbolic, since the space metric is not complete. We will solve this difficulty by using the Regge-Wheeler coordinate and by turning the equation for a wave map on the Schwarzchild spacetime into an equation whose principal term is a wave map on 2-dimensional Minkowski spacetime, and has manageable lower order terms. Nonlinear Klein-Gordon equations on the outside of a black hole have been considered by Nicolas (1995); for previous work on the electromagnetic field on black hole metrics, see Bachelot (1991).

We consider the manifold $(V, \mathbf{g})$ with $V \equiv \Omega \times \mathbb{R}$, Ω exterior $r > 2m$ of a ball in $\mathbb{R}^3$, and $\mathbf{g}$ the Schwarzchild metric:

$$\mathbf{g} \equiv -(1 - \tfrac{2m}{r})dt^2 + (1 - \tfrac{2m}{r})^{-1}dr^2 + r^2(d\theta^2 + sin^2\theta d\varphi^2)$$

The wave map equation from the manifold $(V, \mathbf{g})$ into a Riemannian manifold (M, h) reads when u is a spherically symmetric map (i.e., depends only on r and t):

$$-\partial_t\{(1 - \frac{2m}{r})^{-1}\partial_t u\} + r^{-2}\partial_r\{r^2(1 - \frac{2m}{r})\partial_r u\}$$
$$+\Gamma^{\cdot}_{AB}\{-(1 - \frac{2m}{r})^{-1}\partial_t u^A \partial_t u^B + (1 - \frac{2m}{r})\partial_r u^A \partial_r u^B\}$$

We introduce on V the Regge-Wheeler coordinate:

$$\rho \equiv r + 2m\log(r - 2m)$$

With this new coordinate the Schwarzhild manifold $(V, \mathbf{g})$ is:

$$\mathbf{g} = (1 - 2m/r)(-dt^2 + d\rho^2) + r^2(d\theta^2 + \sin^2\theta\, d\varphi^2), \qquad -\infty < \rho < +\infty,$$

where r is a C^∞ function of ρ, increasing from $2m$ to $+\infty$.

A straightforward calculation shows that the above wave map equation reads in these coordinates

$$g^{\alpha\beta}\nabla_{\alpha\beta}u \equiv (1 - \tfrac{2m}{r})^{-1}\{-\partial^2 u/\partial t^2 + \partial^2 u/\partial\rho^2 + \tfrac{2}{r}(1 - \tfrac{2m}{r})\tfrac{\partial u}{\partial\rho}$$
$$+\Gamma^{\cdot}_{AB}(-\tfrac{\partial u^A}{\partial t}\tfrac{\partial u^B}{\partial t} + \tfrac{\partial u^A}{\partial\rho}\tfrac{\partial u^B}{\partial\rho})\} = 0$$

We see on this formula that the wave map equation for u on the 4-dimensional manifold (V, g) is equivalent to a wave map equation on the 2-dimensional Minkowski spacetime (R^2, η) with an added linear term with smooth coefficient (recall that we have $r > 2m$ for $-\infty < \rho < \infty$):

$$-\nabla_t \partial_t u + \nabla_\rho \partial_\rho u + \ell_\rho \partial_\rho u = 0, \quad \text{with } \ell_\rho \equiv 2r^{-1}(1 - 2mr^{-1},)$$

where ∇ denotes now the covariant derivative leaving invariant η and h.

A simple computation shows also that the above vector valued equation satisfied by a map whose target is regularly embedded in R^N reduces to a regular system of quasi-diagonal semilinear second-order hyperbolic equations on $\mathbb{R}^2$ which admits a local-in-time solution for Cauchy data $\rho \mapsto (\varphi(\rho), \psi(\rho))$ with $\varphi \in C_b^0$, $D\varphi$, and $\psi \in H_1$.

To prove global existence, we will use energy estimates of the mapping $u : (R^2, \eta) \to (M, h)$.

<u>First energy estimate</u>. The energy of the mapping $u : (R^2, -dt^2 + d\rho^2) \to (M, h)$ and the vector field $X = (-1, 0)$ is

$$\epsilon_t(u) \equiv \tfrac{1}{2} \int_{-\infty}^{\infty} \{\partial_t u \cdot \partial_t u + \partial_\rho u \cdot \partial_\rho u\} \, \partial\rho.$$

If u satisfies the modified wave map equation, this energy satisfies the equality

$$\epsilon_t(u) = \epsilon_0(u) + \int_0^t \int_{-\infty}^{\infty} \partial_\tau u \cdot \ell_\rho \partial_\rho u d\rho d\tau,$$

which implies when $r > 2m$, since then $0 < 2r^{-1}(1 - 2mr^{-1}) \le 1/4m$:

$$\epsilon_t(u) \le \epsilon_0(u) + \tfrac{1}{4m} \int_0^t \epsilon_\tau(u) d\tau.$$

We deduce then from the Gromwall lemma that $\epsilon_t(u)$ is bounded for all finite t, namely

$$\epsilon_t u) \le \epsilon_0(u) \exp(t/4m).$$

<u>Second energy estimate</u>. The second energy of u: $(R^2, -dt^2 + d\rho^2) \to (M, h)$ reads:

$$\epsilon_t^{(1)}(u) \equiv \tfrac{1}{2} \int_{-\infty}^{\infty} \{|\nabla_t \partial_t u|_h^2 + 2|\nabla_\rho \partial_t u|_h^2 + |\nabla_\rho \partial_\rho u|_h^2\} d\rho.$$

The computations done in the section on second energy inequalities apply (N is here equal to 1) but to the current term J^0 given in that section we must now, since $\nabla_\lambda \nabla^\alpha \partial_\alpha u = \nabla_\lambda \ell_\rho(u) \not\equiv 0$, add the term

$$-E^{\lambda\mu} \nabla_\lambda \ell_\rho \partial_\rho u \cdot \nabla^0 \partial_\mu u \equiv \ell_\rho \{|\nabla_t \partial_t u|_h^2 + \nabla_\rho \partial_\rho u|_h^2\} + \partial_\rho \ell_\rho \partial_\rho u . \nabla_t \partial_t u$$

with

$$\partial_\rho \ell_\rho \equiv 2r^{-2}(1 - 2mr^{-1})(-1 + 4mr^{-1}).$$

The function $\partial_\rho \ell_\rho$ is uniformly bounded on $r > 2m > 0$ (it vanishes for $r = 1/4m$, tends to zero when r tends to $2m$ or $+\infty$; it admits one negative minimum between $2m$ and $4m$ and one positive maximum between $4m$ and $+\infty$). denote by M the greatest absolute value of these extrema. It is in fact the value of the positive maximum and equal to Cm^{-2}, with C an absolute number, $C = (9 + \sqrt{17})^3(3 + \sqrt{17})2^{-9}$.

The Riemann tensor of the Minkowski metric is zero. We find for the second energy an inequality of the form

$$\epsilon_t^{(1}(u) \le \epsilon_0^{(1)}(u) + \int_0^t \{\int_{-\infty}^{\infty} Q(\partial u)^2) \, d\rho\}^{1/2} \{\epsilon_\tau^{(1)}(u)\}^{1/2} \, d\tau$$

$$+\tfrac{1}{4}\,\mathrm{m}^{-1}\int_0^t \epsilon_\tau^{(1)}\,d\tau + Cm^{-2}\int_0^t \{\epsilon_\tau(u)\}^{1/2}\{\epsilon_\tau^{(1)}\}^{1/2}\,d\tau.$$

We estimate the term in Q like we did in the general case $n = 1$, and we obain a bound for all finite t of $\epsilon_t^{(1)}(u)$, using the bound found for the fundamental energy. To enunciate the theorem in the original variable r we introduce the following definition:

Definition. A tensor field f on $\mathbb{R}$ is said to belong to $H_{1,\delta}^{\mathrm{Sch}}$ if it admits a generalized derivative and the following integral is finite:

$$\| f \|_{H_{1,\delta}^{Sch}} \equiv \left(\int_{2m}^{\infty}\{(1-2mr^{-1})^{2\delta}|f|^2 + (1-2mr^{-1})^{2\delta+2}|\partial_r f|^2\}\,dr\right)^{1/2}$$

Theorem. Let $(M,\ h)$ be a Riemannian manifold regularly embedded into a Euclidean space. If the Cauchy data φ is continuous and takes its values in M while $\partial_r\varphi \in H_{1,1/2}^{\mathrm{Sch}}$ and $\psi \in H_{1,-1/2}^{\mathrm{Sch}}$, with ψ taking its values in $T_\varphi M$, then there exists a global spherically symmetric wave map from the exterior of a Schwarzchild black hole into $(M,\ h)$ taking these Cauchy data.

Remark. The proof of global existence can also be made using the energies associated with the original Schwarzchild metric. Because this metric is static, it is natural to define these energies with the corresponding past-oriented Killing vector, whose coordinates are $X^0 = -1$, $X^i = 0$. The first energy is then

$$e_t(u) \equiv \int_{2m}^{+\infty}(1-\tfrac{2m}{r})^{-1}\partial_t u\cdot\partial_t u + (1-\tfrac{2m}{r})\partial_r u.\partial_r u\}(1-\tfrac{2m}{r})^{1/2}r^2\,dr,$$

that is,

$$e_t(u) \equiv \int_{-\infty}^{+\infty}\{\partial_t u\cdot\partial_t u + \partial_\rho u\cdot\partial_\rho u\}(1-\tfrac{2m}{r})^{1/2}r^2\,d\rho.$$

This energy is conserved, and gives bounds independent of t of weighted L^2 norms of $\partial_t u$ and $\partial_\rho u$. The difference with the previous is in the behaviour of u at infinity introduced by the weight r^2.

Third estimate.

To bound the third derivatives of an arbitrary wave map $u : (V,\ \mathbf{g}) \to (M,\ h)$, we derivate once more the wave map equation and obtain an equation of the form

$$\mathrm{I} + \mathrm{II} = 0,$$

with

$$I \equiv \nabla_\beta(\nabla^\lambda\nabla_\lambda\partial_\alpha u) \equiv \nabla^\lambda\nabla_\lambda\partial_\beta\nabla_\alpha u + \nabla^\lambda\{R_{\lambda\beta\alpha}{}^{\rho}\partial_\rho u + R_{cd}{}^{\cdot}{}_{b}\partial_\lambda u^d\partial_\beta u^c\partial_\alpha u^b\}$$
$$- R_\beta^{\ \rho}\nabla_\rho\partial_\alpha u + R_{\beta\ \alpha}^{\ \lambda\ \rho}\nabla_\lambda\partial_\rho u + R_{c\ b}^{\ d\cdot}\partial_\beta u^c\partial^\lambda u^d\nabla_\lambda\partial_\alpha u^b$$
$$II \equiv \nabla_\beta(-R_\alpha^\rho\partial_\rho u + R_{cd}{}^{\cdot}{}_{b}\partial_\alpha u^c\partial_\rho u^d\partial^\rho u^b).$$

We see that the covariant second derivatives of u satisfy an equation of the form

$$\nabla^\lambda\nabla_\lambda\nabla^2 u = F(u,\ \partial u,\ \nabla^2 u),$$

with

$$F(u,\ \partial u,\ \nabla^2 u) \equiv \nabla\,\mathrm{Riem}(\mathbf{g})\partial u + \mathrm{Riem}(\mathbf{g})\nabla^2 u + \mathrm{Riem}(h)\partial u\partial u\nabla^2 u$$
$$+\nabla\,\mathrm{Riem}(h)\partial u\partial u\partial u\partial u$$

One defines a stress-energy tensor for $\nabla^2 u$ by setting

$$T^{(2)}_{\rho\sigma} = E^{\alpha\alpha'} E^{\beta\beta'} T^{(2)}_{\rho\sigma,\alpha\alpha',\beta\beta'},$$

with

$$T^{(2)}_{\rho\sigma,\alpha\alpha',\beta\beta'} \equiv \nabla_\rho \nabla_\beta \partial_\alpha u . \nabla_\sigma \nabla_{\beta'} \partial_{\alpha'} u - \tfrac{1}{2} g_{\rho\sigma} \nabla^\lambda \nabla_\beta \partial_\alpha u . \nabla_\lambda \nabla_{\beta'} \partial_{\alpha'} u$$

The use of this stress-energy tensor and of the Ricci identity leads along the same lines as in the previous sections to an inequality for the energy of $\nabla^2 u$ at time t defined by

$$e^{(2)}_t(u) \equiv \int_{S_t} N^2 T^{00}_{(2)} \mu_t .$$

This inequality is of the form

$$e^{(2)}_t(u) \le e^{(2)}_0(u) + C \int_0^t \operatorname{Sup}_{S_\tau}(|\nabla N| + 2|NK|) e^{(2)}_t(u)\, d\tau .$$

$$\int_0^t \{\int_{S_\tau} |F(u, \partial u, \nabla^2 u|^2_{g,h} \mu_\tau\}^{1/2} e^{(2)}_\tau(u)^{1/2} .$$

where F is not identical to the expression previously denoted F, some other terms in curvatures come in computing the divergence of the stress-energy tensor $T^{(2)}$, but it has the same general form. To estimate its L^2 norm when the Riemann tensor of h, Riem(h) and ∇ Riem(h) are bounded we must estimate the L^2 norm of $|\nabla u|^4$ and $|\nabla u|^2 |\nabla^2 u|$. This will be possible interms of the energies e, $e^{(1)}$, $e^{(2)}$ in the cases $n = 2$ or 3, as we will show in the next section.

CASES $n = 2$ or 3. The case $n = 3$ is relevant for wave maps on a given physical spacetime.
The case $n = 2$ is important for Einsteinian spacetimes with a one-parameter space like isometry group, since the vacuum Einstein equations reduce then to a coupled system of elliptic equations, a wave map equation, and ordinary differential equations for the Teichmüller parameters (Moncrief (1986), Choquet-Bruhat and Moncrief (1994)). We show below that for $n = 2$ or 3 the local existence theorem is still valid for $s = 2$, and not only $s = 3$ as says the general theorem 2 of section 2.
In the case $n = 2$ we have seen that $f \equiv |\nabla u|^2_{g,h}$ belongs to L^p for any $p > 1$ if it belongs to L^1 as well as $|\nabla^2 u|^2_{g,h}$. In particular

$$\| \, |\nabla u|^4_{g,h} \, \|_{L^2} \equiv \| \, |\nabla u|^2_{g,h} \, \|^2_{L^4} \le C_g \{ e_t(u)^2 + e^{(1)}_t(u)^2 \},$$

On the other hand

$$\| \, |\nabla u|^2_{g,h} |\nabla^2 u|_{g,h} \, \|_{L^2} \le C_g \{ e_t(u) + e^{(1)}_t(u) \} \{ e^{(1)}_t(u)^{1/2} + e^{(2)}_t(u) \}^{1/2} .$$

Writing analogous inequalities for the other second derivatives of u, we obtain therefore for $e^{(2)}_t(u)$ an inequality which is <u>*linear*</u> in this quantity. In the case $n = 3$, we bound as follows the L^2 norms of $|\nabla u|^4$ and $|\nabla u|^2 |\nabla^2 u|$.

$$\| \, |\nabla u|^4 \, \|_{L^2} \le \| \, |\nabla u| \, \|_{C^0_b} \| \, |\nabla u| \, \|^3_{L^6},$$

By the Sobolev embedding theorem we have

$$\| \, |\nabla u| \, \|_{C^0_b} = \{ \| \, |\nabla u|^2 \, \|_{C^0_b} \}^{1/2} \le C \{ \| \, |\nabla u|^2 \, \|_{H^2} \}^{1/2},$$

and by previous estimates:

$$\{\| \, |\nabla \mathrm{u}|^2 \, \|_{H_2}\}^{1/2} \leq C_{\mathbf{g}} C(t)\{1 + e_t^{(2)}(u)\}^{1/2},$$

where $C_{\mathbf{g}}$ depends now on the C^2 norms of $\mathbf{g}$ (more refined hypothesis can be made) while $C(t)$ is a continuous function of the energies $e_t(u)$ and $e_t^{(1)}(u)$.
On the other hand, using again Schwarz and Sobolev inequalities and previous estimates, we find also

$$\| \, |\nabla u|^2 |\nabla^2 u| \, \|_{L^2} \leq \| \, |\nabla u|^2 \, \|_{L^3} \| \, |\nabla^2 u| \, \|_{L^6} \leq C_{\mathbf{g}} C(t)\{1 + e_t^{(2)}(u)\}^{1/2}.$$

We find, as in the case $n = 2$, an inequality for $e_t^{(2)}(u)$ which is linear when $e_t(u)$ and $e_t^{(1)}(u)$ are known.
We deduce from these properties the following lemma:

Lemma. *When $n = 2$ or 3 the energy $e_t^{(2)}(u)$ is bounded on any interval $[0, T]$ on which $e_t(u)$ and $e_t^{(1)}(u)$ are bounded.*

We will deduce from this lemma a local existence theorem for data such that $\varphi \in C_b^0, \nabla\varphi, \psi \in H_1$.

Theorem. *Let $\mathbf{g}$ and h satisfy the hypothesis of theorem 2. Let $\varphi \in C_b^0$, $\nabla\varphi, \psi \in H_1$. There exists a number $\ell > 0$ and a wave map u: $(S \times [0, \ell], \mathbf{g}) \to (N, h)$ taking these Cauchy data.*

Proof. The local-in-time existence for $\nabla\varphi, \psi \in H_2$ is already known. We first prove that the interval $[0, \ell]$ of existence depends in fact only on the H_1 norms of these quantities, because the energy $e_t(u)$ is bounded for all finite t and $e_t^{(1)}(u)$ is bounded for $t < T$, number depending on $e_0^{(1}(u)$, and then $e_t^{(2)}(u)$ is also bounded for $t < T$ (though this bound depends on $e_0^{(2)}(u)$). By the same procedure as one proves global existence from bounds of the H_1 norms in the case $n = 1$, one proves the existence of the wave map u for $t < T$.

To prove the existence of for $t < T$ when $\nabla\varphi, \psi \in H_1$, we approximate these quantities by quantities $\nabla\varphi_{(m)}, \psi_{(m)} \in H_2$, converging to $\nabla\varphi, \psi$ in H_1 norm. The corresponding solution $u_{(m)}$ exists on $S \times [0, T)$, T independant of m. We show that $u_{(m)}$ converges strongly on $S \times [0, T)$ to a mapping u taking the given Cauchy data by considering the associated mapping $U_{(m)} : S \times [0, \ell] \to \mathbb{R}^N$ and estimating by methods analogous to the previous ones the difference $\| \mathbf{D}U_{(m)} - \mathbf{D}U_{(n)} \|_{L^2}$ after taking the difference of the equations they satisfy.

The local existence theorem, and also the bound of $e_t^{(2)}(u)$ in terms of the previous energies, shows that a priori bounds of the energy of ∂u, $e_t^{(1)}(u)$, is sufficient to obtain global existence of the wave map u. This global existence does not hold for arbitrary regularly hyperbolic manifolds and arbitrary initial data.

The global existence of wave maps with small initial data from the Minkowski spacetime into an arbitrary Riemannian manifold has been proved (Choquet-Bruhat (1984)) when $n = 3$ (the wave map equation is then conformally regular) and when $n = 2$ (Choquet-Bruhat and Gu Chaohao (1989); in that case the target is restricted to be a symmetric space.

Recently a global existence theorem for initial data in the energy space, i.e., with bounded first energy, of weak solutions on Minkowski spacetime has been proved by compacity methods by Muller and Struwe ((1996) in the case $n = 2$. There is no uniqueness theorem.

Classes of smooth initial data for which the wave map equation from Minkowski spacetime $n = 3$ into a convex target manifold has a solution which blows up in a finite time have been constructed by Shatah (1988) and Shatah and Tahvildar-Zadeh (1996).

A global existence theorem for small initial data of classical solutions on expanding universes has been proved when $n = 2$ or 3 (Choquet-Bruhat, 1998).

REFERENCES

[1] Bachelot, A. Gravitational scattering of electromagnetic field by a Schwarzchild black hole, *Ann. Inst. H. Poincaré* **54** (1991) 261, 320.

[2] Chaohao, Gu. On the Cauchy problem for harmonic maps on two-dimensional Minkowski space, *Comm. Pure and App. Math* **33** (1980) 727–737.

[3] Choquet-Bruhat, Y. Partial Differential Equations on a Manifold, *Batelle Rencontres* (1967), eds. C. DeWitt and J. Wheeler, Benjamin.

[4] Choquet-Bruhat, Y. C^∞ solutions of nonlinear hyperbolic equations, *Gen. Rel. Grav.* **2** (1972) 359–362.

[5] Choquet-Bruhat, Y. and C. Gilain, G. Relativité Générale: difféomorphismes harmoniques et unicité, *C. R. Ac. Sci., Paris* **279** (1974) 827–830.

[6] Choquet-Bruhat, Y. Christodoulou, D. and Francaviglia, M. Cauchy data on a manifold, *Ann. Inst. Poincaré* **XXXI** no. 4 (1979) 399–414.

[7] Choquet-Bruhat, Y. Hyperbolic harmonic maps, *Ann Inst. Poincaré* **46**, 97–111.

[8] Choquet-Bruhat, Y. and DeWitt-Morette, C. *Analysis, Manifolds and Physics*, 2nd ed. North-Holland Publishing Co., Amsterdam (1989).

[9] Choquet-Bruhat, Y. and Moncrief, V. Existence theorem for Einstein's equations with one parameter isometry group, *Proc. Symp. Amer. Math. Soc.* **59**, Brezis and Segal ed. (1996) 61-80.

[10] Christodoulou, D. and Talvilar-Zadeh, A. On the regularity of spherically symmetric wave maps, *Comm. Pure Ap. Math.* **46** (1993) 1041–1091.

[11] Chrusciel, P. On uniqueness in the large of solutions of Einstein Equations, *Proc. Center for Math.*, Australian Nat. Univ. **27** (1991).

[12] Dionne, P. Sur les problèmes hyperboliques bien posés *J. Anal. Math.* Jerusalem 1 (1962) 1–90.

[13] Eells, J. and Sampson, H. Harmonic maps, *Amer. J. Math.* **86** (1964) 109–160.

[14] Geroch, R. Domains of dependence, *J. Math. Phys.* **11** (1970) 437–449.

[15] Ginibre, J. and Velo, G. The Cauchy problem for the O(N), CP(N-1) and GC(N,p) models, *Ann. Phys.* **142**, no. 2 (1982) 393–415.

[16] Hawking, S. and Ellis, G. F. R. *The Global Structure of Spacetime*, Cambridge Univ. Press (1973).

[17] Klainerman, S. and Machedon, M. Spacetime estimates for null forms and the local existence theorem, *Com. Pure App. Math.* **46** (1993) 1221–1268.

[18] Leray, J. *Hyperbolic differential equations* Inst. Adv. Stud., Princeton (1952).

[19] Lichnerowicz, A. Applications harmoniques et variétés Kahleriennes, *Symposia Matematica* **III** (1970) 341–402.

[20] Misner, C. W. Harmonic maps as models of physical theories, Phys. Rev. D (1978) 4510–4524.

[21] Moncrief, V. Reduction of Einstein's equations for vacuum spacetimes with spacelike U(1) isometry groups, *Ann. Phys.* **167** no. 1 (1986) 118–142.

[22] Muller, S. and Struwe, M. Global existence of wave maps in 2+1 dimensions with finite energy data, *Topological Methods in Nonlinear Analysis* **7** no. 2 (1996) 245–261.

[23] Nicolas, J. P. Non-linear Klein Gordon equations on Schwarzchild like metrics, *J. Math. Pures App.* **74** (1995) 35–58.

[24] Nutku, Y. Harmonic maps in physics, *Ann. Inst. H. Poincaré* **A 21** (1974) 175–183.

[25] Shatah, J. Weak solutions and development of singularities in the SU(2) σ-model, *Comm. Pure App. Math,* **41** (1988) 459–469.

[26] Shatah, J. and Talvilar-Zadeh, A. Non-uniqueness and development of singularities for the harmonic maps of the Minkowski space, preprint.

[27] Sogge, C. On local existence for wave equations satisfying variable coefficients null conditions, *Comm. Part. Diff. Eq.* **18** (1993) 1795–1821.

[28] Stoeger, W. R. Rosen bimetric theory of gravity as a harmonic map, Proc. Third M. Grossman Meeting ed. Hu Ning, North Holland (1983) 921–925.

[29] Choquet-Bruhat, Y. Applications d'ondes sur un univers en expansion, *C.R. Acad. Sci Paris* **326** (1998) 1175–1180.

12

General Relativity and Experiment

Thibault Damour

ABSTRACT The experimental tests of general relativity, notably binary pulsar ones, are reviewed, and their theoretical significance is discussed. Experiment and theory agree at the 10^{-3} level. All the basic structures of Einstein's theory (coupling of gravity to matter; propagation and self-interaction of the gravitational field, including in strong-field conditions) have been verified. However, some recent theoretical findings (cosmological relaxation toward zero scalar couplings) suggest that the present agreement between general relativity and experiment might be naturally compatible with the existence of a long-range scalar contribution to gravity (such as the dilaton, or a moduli field of string theory). This provides a new theoretical paradigm, and new motivations for improving the experimental tests of gravity. Ultra-high precision tests of the equivalence principle appear as the most sensitive way to look for possible long-range deviations from general relativity: they might open a low-energy window on string-scale physics.[1]

12.1 Introduction

Considered as a classical field theory, general relativity is defined by two postulates. One postulate states that the action functional describing the propagation and self-interaction of the gravitational field is

$$S_{\text{gravitation}}[g_{\mu\nu}] = \frac{c^4}{16\pi G} \int \frac{\sqrt{g}}{c} R(g) d^4x. \qquad (1)$$

A second postulate states that the action functional describing the behavior of "matter" in presence of a gravitational field is the minimal deformation of the special relativistic action functional used by particle physicists (the so-called "Standard Model") obtained by replacing everywhere the flat Minkowski metric $f_{\mu\nu} = \text{diag}(-1, +1, +1, +1)$ by $g_{\mu\nu}(x^\lambda)$ and the partial derivatives $\partial_\mu \equiv \partial/\partial x^\mu$ by g-covariant derivatives ∇_μ. [With the usual subtlety that one must also introduce a field of orthonormal frames, a "vierbein," for writing down the fermionic

[1]Updated write-up of review talks presented at various conferences.

terms.] Schematically, one has

$$S_{\text{matter}}\,[\psi, A, H, g] = \int \frac{\sqrt{g}}{c} \mathcal{L}_{\text{matter}} d^4x, \tag{2}$$

$$\begin{aligned} \mathcal{L}_{\text{matter}} = & -\frac{1}{4}\sum \frac{1}{g_*^2}\,\text{tr}(F_{\mu\nu}\,F^{\mu\nu}) - \sum \overline{\psi}\gamma^\mu D_\mu \psi \\ & -\frac{1}{2}\,|D_\mu\, H|^2 - V(H) - \sum y \overline{\psi} H \psi, \end{aligned} \tag{3}$$

where $F_{\mu\nu}$ denotes the curvature of a $U(1)$, $SU(2)$, or $SU(3)$ Yang-Mills connection A_μ (electro-weak and strong interactions), $F^{\mu\nu} = g^{\mu\alpha} g^{\nu\beta} F_{\alpha\beta}$, g_* being a (bare) gauge coupling constant; $D_\mu \equiv \nabla_\mu + A_\mu$; ψ denotes a fermion field (lepton or quark, coming in various flavors and three generations); γ^μ denotes four Dirac matrices such that $\gamma^\mu\gamma^\nu + \gamma^\nu\gamma^\mu = 2g^{\mu\nu}\,\mathbb{I}_4$, and H denotes the Higgs doublet of scalar fields, with y some (bare Yukawa) coupling constants.

Einstein's theory of gravitation is then defined by extremizing the total action functional,

$$S_{\text{tot}}\,[g, \psi, A, H] = S_{\text{gravitation}}\,[g] + S_{\text{matter}}\,[\psi, A, H, g]. \tag{4}$$

Seen within the framework of quantum field theory (with its basic constraints on locality, causality, absence of negative energy excitations,...), the postulates (1) and (2) are not independent but follow from the unique requirement that the gravitational interaction be mediated only by massless spin-2 excitations [1]. However, we shall retain the classical decomposition in two postulates, which is convenient for discussing the theoretical significance of various tests of general relativity. Let us discuss in turn the experimental tests of the coupling of matter to gravity (postulate (2)), and the experimental tests of the dynamics of the gravitational field (postulate (1)). For more details and references we refer the reader to [2] or [3].

12.2 Experimental Tests of the Coupling Between Matter and Gravity

The fact that the matter Lagrangian (3) depends only on a symmetric tensor $g_{\mu\nu}(x)$ and its first derivatives (i.e., the postulate of a "metric coupling" between matter and gravity) is a strong assumption (often referred to as the "equivalence principle") which has many observable consequences for the behaviour of localized test systems embedded in given external gravitational fields. Indeed, using a theorem of Fermi and Cartan [4] (stating the existence of coordinate systems such that, along any given timelike curve, the metric components can be set to their Minkowski values, and their first derivatives made to vanish), one derives from the postulate (2) the following observable consequences:

C$_1$: Constancy of the "constants": the outcome of local nongravitational experiments, referred to local standards, depends only on the values of the coupling constants and mass scales entering the Standard Model. [In particular, the cosmological evolution of the universe at large has no influence on local experiments.]

$\mathbf{C}_2$: Local Lorentz invariance: local nongravitational experiments exhibit no preferred directions in spacetime [i.e., neither spacelike ones (isotropy), nor timelike ones (boost invariance)].

$\mathbf{C}_3$: "Principle of geodesics" and universality of free fall: small, electrically neutral, non-self-gravitating bodies follow geodesics of the external spacetime $g_{\mu\nu}(x^\lambda)$. In particular, two test bodies dropped at the same location and with the same velocity in an external gravitational field fall in the same way, independently of their masses and compositions.

$\mathbf{C}_4$: Universality of gravitational redshift: when intercompared by means of electromagnetic signals, two identically constructed clocks located at two different positions in a static external Newtonian potential $U(\mathbf{x})$ exhibit, independently of their nature and constitution, the (apparent) difference in clock rate:

$$\frac{\tau_1}{\tau_2} = \frac{\nu_2}{\nu_1} = 1 + \frac{1}{c^2}\left[U(\mathbf{x}_1) - U(\mathbf{x}_2)\right] + O\left(\frac{1}{c^4}\right). \tag{5}$$

Many experiments or observations have tested the observable consequences $C_1 - C_4$ and found them to hold within the experimental errors. Many sorts of data (from spectral lines in distant galaxies to a natural fission reactor phenomenon which took place at Oklo, Gabon, two billion years ago) have been used to set limits on a possible time variation of the basic coupling constants of the Standard Model. The best results concern the electromagnetic coupling, i.e., the fine-structure constant α_{em}. A recent reanalysis of the Oklo phenomenon gives a conservative upper bound [5]

$$-6.7 \times 10^{-17}\,\text{yr}^{-1} < \frac{\dot{\alpha}_{\text{em}}}{\alpha_{\text{em}}} < 5.0 \times 10^{-17}\,\text{yr}^{-1}, \tag{6}$$

which is much smaller than the cosmological time scale $\sim 10^{-10}\,\text{yr}^{-1}$. By comparison, direct laboratory measurements comparing clocks based on atomic transitions having different dependences on α_{em} give the limit $|\dot{\alpha}_{\text{em}}/\alpha_{\text{em}}| < 3.7 \times 10^{-14}\,\text{yr}^{-1}$ [6]. [See references below for astronomical limits on the variability of α_{em}.]

Any "isotropy of space" having a direct effect on the energy levels of atomic nuclei has been constrained to the impressive 10^{-27} level [7]. The universality of free fall has been verified at the 10^{-12} level both for laboratory bodies [8], e.g. (from the last reference in [8])

$$\left(\frac{\Delta a}{a}\right)_{\text{Be Cu}} = (-1.9 \pm 2.5) \times 10^{-12}, \tag{7}$$

and for the gravitational accelerations of the Moon and the Earth toward the Sun [9],

$$\left(\frac{\Delta a}{a}\right)_{\text{Moon Earth}} = (-3.2 \pm 4.6) \times 10^{-13}. \tag{8}$$

The "gravitational redshift" of clock rates given by Eq. (5) has been verified at the 10^{-4} level by comparing a hydrogen-maser clock flying on a rocket up to an altitude ~ 10000 km to a similar clock on the ground [10].

In conclusion, the main observable consequences of the Einsteinian postulate (2) concerning the coupling between matter and gravity ("equivalence principle") have been verified with high precision by all experiments to date. The traditional paradigm (first put forward by Fierz [11]) is that the extremely high precision of free fall experiments (10^{-12} level) strongly suggests that the coupling between matter and gravity is exactly of the "metric" form (2), but leaves open possibilities more general than Eq. (1) for the spin-content and dynamics of the fields mediating the gravitational interaction. We shall provisionally adopt this paradigm to discuss the tests of the other Einsteinian postulate, Eq. (1). However, we shall emphasize at the end that recent theoretical findings suggest a new paradigm.

12.3 Tests of the Dynamics of the Gravitational Field in the Weak-Field Regime

Let us now consider the experimental tests of the dynamics of the gravitational field, defined in general relativity by the action functional (1). Following first the traditional paradigm, it is convenient to enlarge our framework by embedding general relativity within the class of the most natural relativistic theories of gravitation which satisfy exactly the matter-coupling tests discussed above while differing in the description of the degrees of freedom of the gravitational field. This class of theories are the metrically coupled tensor-scalar theories, first introduced by Fierz [11] in a work where he noticed that the class of non-metrically-coupled tensor-scalar theories previously introduced by Jordan [12] would generically entail unacceptably large violations of the consequence C_1. [The fact that it would, by the same token, entail even larger violations of the consequence C_3 was first emphasized by Dicke in subsequent work.] The metrically coupled (or equivalence-principle respecting) tensor-scalar theories are defined by keeping the postulate (2), but replacing the postulate (1) by demanding that the "physical" metric $g_{\mu\nu}$ (coupled to ordinary matter) be a composite object of the form

$$g_{\mu\nu} = A^2(\varphi)\, g^*_{\mu\nu}, \tag{9}$$

where the dynamics of the "Einstein" metric $g^*_{\mu\nu}$ is defined by the action functional (1) (written with the replacement $g_{\mu\nu} \to g^*_{\mu\nu}$) and where φ is a massless scalar field. [More generally, one can consider several massless scalar fields, with an action functional of the form of a general nonlinear σ model [13].] In other words, the action functional describing the dynamics of the spin 2 and spin 0 degrees of freedom contained in this generalized theory of gravitation reads

$$S_{\text{gravitational}}\,[g^*_{\mu\nu}, \varphi] = \frac{c^4}{16\pi\, G_*} \int \frac{\sqrt{g_*}}{c}\, \left[R(g_*) - 2 g_*^{\mu\nu}\, \partial_\mu\, \varphi \partial_\nu \varphi\right] d^4x. \tag{10}$$

Here, G_* denotes some bare gravitational coupling constant. This class of theories contains an arbitrary function, the "coupling function" $A(\varphi)$. When $A(\varphi) = \text{const.}$, the scalar field is not coupled to matter and one falls back (with suitable boundary conditions) on Einstein's theory. The simple, 1-parameter subclass $A(\varphi) = \exp(\alpha_0\, \varphi)$ with $\alpha_0 \in R$ is the Jordan-Fierz-Brans-Dicke theory [11], [14], [15]. In the general case, one can define the (field-dependent) coupling strength of φ to matter by

$$\alpha(\varphi) \equiv \frac{\partial \ln A(\varphi)}{\partial \varphi}. \tag{11}$$

It is possible to work out in detail the observable consequences of tensor-scalar theories and to contrast them with the general relativistic case (see, for example, [13]).

Let us now consider the experimental tests of the dynamics of the gravitational field that can be performed in the solar system. Because the planets move with slow velocities ($v/c \sim 10^{-4}$) in a very weak gravitational potential ($U/c^2 \sim (v/c)^2 \sim 10^{-8}$), solar system tests allow us only to probe the quasi-static, weak-field regime of relativistic gravity (technically described by the so-called "post-Newtonian" expansion). In the limit where one keeps only the first relativistic corrections to Newton's gravity (first post-Newtonian approximation), all solar-system gravitational experiments, interpreted within tensor-scalar theories, differ from Einstein's predictions only through the appearance of two "post-Einstein" parameters $\overline{\gamma}$ and $\overline{\beta}$ (related to the usually considered Eddington parameters γ and β through $\overline{\gamma} \equiv \gamma - 1$, $\overline{\beta} \equiv \beta - 1$). The parameters $\overline{\gamma}$ and $\overline{\beta}$ vanish in general relativity, and are given in tensor-scalar theories by

$$\overline{\gamma} = -2\,\frac{\alpha_0^2}{1+\alpha_0^2}, \tag{12}$$

$$\overline{\beta} = +\frac{1}{2}\,\frac{\beta_0\,\alpha_0^2}{(1+\alpha_0^2)^2}, \tag{13}$$

where $\alpha_0 \equiv \alpha(\varphi_0)$, $\beta_0 \equiv \partial\alpha(\varphi_0)/\partial\varphi_0$; φ_0 denoting the cosmologically determined value of the scalar field far away from the solar system. Essentially, the parameter $\overline{\gamma}$ depends only on the linearized structure of the gravitational theory (and is a direct measure of its field content, i.e., whether it is pure spin 2 or contains an admixture of spin 0), while the parameter $\overline{\beta}$ parametrizes some of the quadratic nonlinearities in the field equations (cubic vertex of the gravitational field).

All currently performed gravitational experiments in the solar system, including perihelion advances of planetary orbits, the bending and delay of electromagnetic signals passing near the Sun, and very accurate range data to the Moon obtained by laser echoes, are compatible with the general relativistic predictions $\overline{\gamma} = 0 = \overline{\beta}$ and give upper bounds on both $|\overline{\gamma}|$ and $|\overline{\beta}|$ (i.e., on possible fractional deviations from general relativity) of order 10^{-3}. More precisely: (i) the Viking mission measurement of the gravitational time delay of radar signals passing near the Sun

("Shapiro effect" [16]) gave [17]

$$|\overline{\gamma}| < 2 \times 10^{-3}\,, \tag{14}$$

with similar limits coming from VLBI measurements of the deflection of radio waves by the Sun [18]; (ii) the Lunar Laser Ranging measurements of a possible polarization of the orbit of the Moon toward the Sun ("Nordtvedt effect" [19]) give [9]

$$4\overline{\beta} - \overline{\gamma} = -0.0007 \pm 0.0010\,, \tag{15}$$

which, combined with the above constraint on $\overline{\gamma}$, gives

$$|\overline{\beta}| < 6 \times 10^{-4}\,; \tag{16}$$

and (iii) measurement of Mercury's orbit through planetary radar ranging gave [20]

$$|\overline{\beta}| < 3 \times 10^{-3}\,, \tag{17}$$

when assuming the above Viking limit on $\overline{\gamma}$ and a value of the Sun's quadrupole moment $J_2 \sim 2 \times 10^{-7}$. Recently, the parametrization of the weak-field deviations between generic tensor-multi-scalar theories and Einstein's theory has been extended to the second post-Newtonian order [21]. Only two post-post-Einstein parameters, ε and ζ, representing a deeper layer of structure of the gravitational interaction, show up. These parameters have been shown to be already significantly constrained by binary-pulsar data: $|\varepsilon| < 7 \times 10^{-2}$, $|\zeta| < 6 \times 10^{-3}$. See [21] for a detailed discussion, including the consequences for the interpretation of future, higher precision solar system tests.

12.4 Tests of the Dynamics of the Gravitational Field in the Strong-Field Regime

In spite of the diversity, number, and often high precision of solar system tests, they have an important qualitative weakness: they probe neither the radiation properties nor the strong-field aspects of relativistic gravity. Fortunately, the discovery [22] and continuous observational study of pulsars in gravitationally bound binary orbits has opened up an entirely new testing ground for relativistic gravity, giving us an experimental handle on the regime of strong and/or radiative gravitational fields.

The fact that binary pulsar data allow one to probe the propagation properties of the gravitational field is well known. This comes directly from the fact that the finite velocity of propagation of the gravitational interaction between the pulsar and its companion generates damping-like terms in the equations of motion, i.e., terms which are directed against the velocities. [This can be understood heuristically by considering that the finite velocity of propagation must cause the gravitational force on the pulsar to make an angle with the instantaneous position of the companion

[23], and was verified by a careful derivation of the general relativistic equations of motion of binary systems of compact objects [24].] These damping forces cause the binary orbit to shrink and its orbital period P_b to decrease. The remarkable stability of the pulsar clock, together with the cleanliness of the binary pulsar system, has allowed Taylor and collaborators to measure the secular orbital period decay $\dot{P}_b \equiv dP_b/dt$ [25], thereby giving us a direct experimental probe of the damping terms present in the equations of motion. Note that, contrary to what is commonly stated, the link between the observed quantity $\dot{P}_b$ and the propagation properties of the gravitational interaction is quite direct. [It appears indirect only when one goes through the common but unnecessary detour of a heuristic reasoning based on the consideration of the energy lost into gravitational waves emitted at infinity.]

The fact that binary pulsar data allow one to probe strong-field aspects of relativistic gravity is less well known. The a priori reason for saying that they should is that the surface gravitational potential of a neutron star $Gm/c^2R \simeq 0.2$ is a mere factor 2.5 below the black hole limit (and a factor $\sim 10^8$ above the surface potential of the Earth). Due to the peculiar "effacement" properties of strong-field effects taking place in general relativity [24], the fact that pulsar data probe the strong-gravitational-field regime can only be seen when contrasting Einstein's theory with more general theories. In particular, it has been found in tensor-scalar theories [26] that a self-gravity as strong as that of a neutron star can naturally (i.e., without fine tuning of parameters) induce order-unity deviations from general relativistic predictions in the orbital dynamics of a binary pulsar thanks to the existence of nonperturbative strong-field effects. [The adjective "nonperturbative" refers here to the fact that this phenomenon is nonanalytic in the coupling strength of the scalar field, Eq. (11), which can be as small as wished in the weak-field limit]. As far as we know, this is the first example where large deviations from general relativity, induced by strong self-gravity effects, occur in a theory which contains only positive energy excitations and whose post-Newtonian limit can be arbitrarily close to that of general relativity. [The strong-field deviations considered in previous studies [2], [13] arose in theories containing negative energy excitations.]

A comprehensive account of the use of binary pulsars as laboratories for testing strong-field gravity will be found in [27]. Two complementary approaches can be pursued: a phenomenological one ("Parametrized Post-Keplerian" formalism), or a theory-dependent one [13], [27].

The phenomenological analysis of binary pulsar timing data consists of fitting the observed sequence of pulse arrival times to the generic DD timing formula [28], whose functional form has been shown to be common to the whole class of tensor-multi-scalar theories. The least-squares fit between the timing data and the parameter-dependent DD timing formula allows one to measure, besides some "Keplerian" parameters ("orbital period" P_b, "eccentricity" e,...), a maximum of eight "post-Keplerian" parameters: k, γ, $\dot{P}_b$, r, s, δ_θ, $\dot{e}$, and $\dot{x}$. Here, $k \equiv \dot{\omega}P_b/2\pi$ is the fractional periastron advance per orbit, γ a time dilation parameter (not to be confused with its post-Newtonian namesake), $\dot{P}_b$ the orbital period derivative mentioned above, and r and s the "range" and "shape" parameters of the gravitational time delay caused by the companion. The important point is that the post-Keplerian parameters can be measured without assuming any specific theory of gravity. Now, each specific relativistic theory of gravity predicts that, for instance, k, γ, $\dot{P}_b$, r, and s (to quote parameters that have been successfully measured from some binary pulsar data) are some theory-dependent functions of the (unknown) masses m_1, m_2 of the pulsar and its companion. Therefore, in our example, the five simultaneous phenomenological measurements of k, γ, $\dot{P}_b$, r, and s determine, for each given theory, five corresponding theory-dependent curves in the $m_1 - m_2$ plane (through the 5 equations $k^{\text{measured}} = k^{\text{theory}}(m_1, m_2)$, etc...). This yields three ($3 = 5 - 2$) tests of the specified theory, according to whether the five curves meet at one point in the mass plane, as they should. In the most general (and optimistic) case, discussed in [27], one can phenomenologically analyze both timing data and pulse-structure data (pulse shape and polarization) to extract up to 19 post-Keplerian parameters. Simultaneous measurement of these 19 parameters in one binary pulsar system would yield 15 tests of relativistic gravity (here one must subtract 4 because, besides the two unknown masses m_1, m_2, generic post-Keplerian parameters can depend upon the two unknown Euler angles determining the direction of the spin of the pulsar). The theoretical significance of these tests depends upon the physics lying behind the post-Keplerian parameters involved in the tests. For instance, as we said above, a test involving $\dot{P}_b$ probes the propagation (and helicity) properties of the gravitational interaction. But a test involving, say, k, γ, r, or s probes (as shown by combining the results of [13] and [26]) strong self-gravity effects independently of radiative effects.

Besides the phenomenological analysis of binary pulsar data, one can also adopt a theory-dependent methodology [13], [27]. The idea here is to work from the start within a certain finite-dimensional "space of theories," i.e., within a specific class of gravitational theories labeled by some theory parameters. Then by fitting the raw pulsar data to the predictions of the considered class of theories, one can determine which regions of theory-space are compatible (at say the 90% confidence level) with the available experimental data. This method can be viewed as a strong-field generalization of the parametrized post-Newtonian formalism [2] used to analyze solar-system experiments. In fact, under the assumption that strong-gravity effects in neutron stars can be expanded in powers of the "compactness" $c_A \equiv -2\partial \ln m_A / \partial \ln G \sim G m_A / c^2 R_A$, reference [13] has shown that the observable predictions of generic tensor-multi-scalar theories could be parametrized by a sequence of "theory parameters,"

$$\overline{\gamma}\,,\ \overline{\beta}\,,\ \beta_2\,,\ \beta'\,,\ \beta''\,,\ \beta_3\,,\ (\beta\beta'), \ldots \tag{18}$$

representing deeper and deeper layers of structure of the relativistic gravitational interaction beyond the first-order post-Newtonian level parametrized by $\overline{\gamma}$ and $\overline{\beta}$ (the second layer β_2, β' being equivalent to the parameters ζ, ε describing the second-order post-Newtonian level [21], etc...).

When nonperturbative strong-field effects develop, one cannot use the multi-parameter approach just mentioned, based on expansions in powers of the "compactnesses." A useful alternative approach is then to work within specific, low-dimensional "mini-spaces of theories." Of particular interest is the two-dimensional mini-space of tensor-scalar theories defined by the coupling function $A(\varphi) = \exp\left(\alpha_0\varphi + \frac{1}{2}\beta_0\varphi^2\right)$. The predictions of this family of theories (parametrized by α_0 and β_0) are analytically described, in weak-field contexts, by the post-Einstein parameter (12), and can be studied in strong-field contexts by combining analytical and numerical methods [29].

After having reviewed the theory of pulsar tests, let us briefly summarize the current experimental situation. Concerning the first discovered binary pulsar PSR1913 + 16 [22], it has been possible to measure with accuracy the three post-Keplerian parameters k, γ, and $\dot{P}_b$. From what was said above, these three simultaneous measurements yield *one* test of gravitation theories. After subtracting a small ($\sim\ 10^{-14}$ level in $\dot{P}_b$!), but significant, perturbing effect caused by the galaxy [30], one finds that general relativity passes this $(k - \gamma - \dot{P}_b)_{1913+16}$ test with complete success at the 10^{-3} level. More precisely, one finds [31], [25]

$$\begin{aligned}\left[\frac{\dot{P}_b^{\mathrm{obs}} - \dot{P}_b^{\mathrm{galactic}}}{\dot{P}_b^{\mathrm{GR}}[k^{\mathrm{obs}}, \gamma^{\mathrm{obs}}]}\right]_{1913+16} &= 1.0032 \pm 0.0023(\mathrm{obs}) \pm 0.0026(\mathrm{galactic}) \\ &= 1.0032 \pm 0.0035\,, \end{aligned} \tag{19}$$

where $\dot{P}_b^{\mathrm{GR}}[k^{\mathrm{obs}}, \gamma^{\mathrm{obs}}]$ is the GR prediction for the orbital period decay computed from the observed values of the other two post-Keplerian parameters k and γ. [More explicitly, this means that the two measurements k^{obs} and γ^{obs} are used, together with the corresponding general relativistic predictions $k^{\mathrm{obs}} = k^{\mathrm{GR}}(m_1, m_2)$, $\gamma^{\mathrm{obs}} = \gamma^{\mathrm{GR}}(m_1, m_2)$, to compute the two masses m_1 and m_2 that enter the theoretical prediction for $\dot{P}_b$.]

This beautiful confirmation of general relativity is an embarrassment of riches in that it probes, at the same time, the propagation *and* strong-field properties of relativistic gravity! If the timing accuracy of PSR1913 + 16 could improve by a significant factor, two more post-Keplerian parameters (r and s) would become measurable and would allow one to probe separately the propagation and strong-field aspects [31]. Fortunately, the discovery of the binary pulsar PSR1534 + 12 [32] (which is significantly stronger than PSR1913 + 16 and has a more favorably oriented orbit) has opened a new testing ground, in which it has been possible to probe strong-field gravity independently of radiative effects. A phenomenological analysis of the timing data of PSR1534 + 12 has allowed one to measure the four post-Keplerian parameters k, γ, r, and s [31]. From what was said above, these four simultaneous measurements yield *two* tests of strong-field gravity, without mixing of radiative effects. General relativity is found to pass these tests with complete success within the measurement accuracy [31], [25]. The most precise of these new, pure strong-field tests is the one obtained by combining the measurements of k, γ, and s. Using the data reported in [33] (with, following [21], doubled statistical uncertainties to take care of systematic errors) one finds agreement at the 1% level:

$$\left[\frac{s^{\text{obs}}}{s^{\text{GR}}[k^{\text{obs}},\gamma^{\text{obs}}]}\right]_{1534+12} = 1.010 \pm 0.008\,. \tag{20}$$

More recently, it has been possible to extract also the "radiative" parameter $\dot{P}_b$ from the timing data of PSR1534 + 12. Again, general relativity is found to be fully consistent (at the current $\sim$ 20% level) with the additional test provided by the $\dot{P}_b$ measurement [35], [33]. Note that this gives our second direct experimental confirmation that the gravitational interaction propagates as predicted by Einstein's theory. Moreover, an analysis of the pulse shape of PSR1534 + 12 has shown that the misalignment between the spin vector of the pulsar and the orbital angular momentum was greater than 8° [27]. This opens the possibility that this system will soon allow one to test the spin precession induced by gravitational spin-orbit coupling.

To end this brief summary, let us mention that several other binary pulsar systems (of a different class than that of 1913 + 16 and 1534 + 12) can also be used to test relativistic gravity. We have here in mind nearly circular systems made of a neutron star and a white dwarf. Such dissymmetric systems are useful probes of the possible existence of dipolar gravitational waves [36] and/or of a possible violation of the universality of free fall linked to the strong self-gravity of the neutron star [37]. A theory-dependent analysis of the published pulsar data on PSRs 1913+16, 1534 + 12 and 0655 + 64 (a dissymmetric system constraining the existence of dipolar radiation) has been recently performed within the (α_0, β_0)-space of tensor-scalar theories introduced above [29]. This analysis proves that binary-pulsar data exclude large regions of theory-space which are compatible with solar system experiments. This is illustrated in Fig. 12.1 below (reproduced from Fig. 9 of [29]), which shows that β_0 must be larger than about -5, while any value of β_0 is compatible with weak-field tests as long as α_0 is small enough. Note that Fig. 12.1 is drawn in the framework of tensor-scalar theories respecting the equivalence principle. In the more general (and more plausible; see below) framework of theories where the scalar couplings violate the equivalence principle, one gets much stronger constraints on the coupling parameter α_0, of order $\alpha_0^2 \lesssim 10^{-7}$ [34].

For a general review of the use of pulsars as physics laboratories the reader can consult [39].

12.5 Cosmological Tests

The tests considered above have examined the gravitational interaction on scales smaller than a few astronomical units. In principle the universe is providing us with plenty of data concerning the behavior of gravity on large scales. However, most of these data cannot be used as clean tests of the law of gravity because of our lack of a priori knowledge of the matter distribution, and/or the low accuracy of the

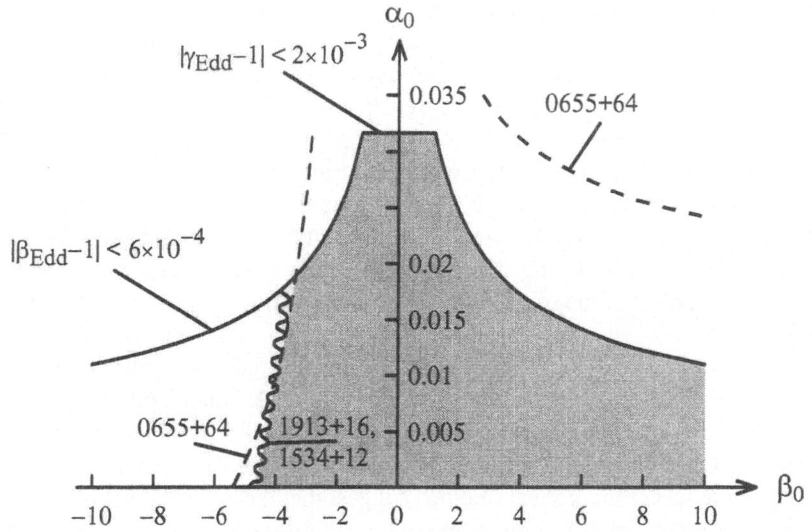

Figure 12.1 Regions of the (α_0, β_0)-plane allowed by (composition-independent) solar-system experiments and three binary pulsar experiments. The region simultaneously allowed by all the tests is shaded. Note that binary pulsar tests exclude a large portion of the region (below the solid line) allowed by solar system tests. (Figure taken from [29].)

data themselves. Fortunately, there are a few exceptions, notably when considering cosmological data.

Astrophysical and cosmological data have been used to put strong constraints on the possible space-time variability of the "constants" of physics. For instance, limits at the few $10^{-14}\,\text{yr}^{-1}$ level on the variability of the fine-structure constant α_{em} have been set by considering fine-structure splittings in cosmological spectra [40], [41]. [For more references on the variation of constants see [5].]

The general relativistic predictions for the action of gravity on light have been verified, within a precision of order 30%, on length scales $\sim$ 100 kiloparsec by considering gravitational lensing by clusters of galaxies. [See, for example, [42].]

Finally, by comparing the big bang computations of the abundances of light elements (helium, deuterium, lithium) with astronomical observations, one can set constraints on the gravitation theory ruling the evolution of the early universe; see, for example, [43], [44], [45], [46], [47], [48].

12.6 Was Einstein 100% Right ?

Summarizing the experimental evidence discussed above, we can say that Einstein's postulate of a pure metric coupling between matter and gravity ("equivalence principle") appears to be, at least, 99.999 999 999 9% right (because of universality-of-free-fall experiments), while Einstein's postulate (1) for the field content and dynamics of the gravitational field appears to be, at least, 99.9% correct both in the quasi-static weak-field limit appropriate to solar-system experiments, and in the radiative strong-field regime explored by binary pulsar experiments. Should one apply Occam's razor and decide that Einstein must have been 100% right, and then stop testing general relativity? My answer is definitely no!

First, one should continue testing a basic physical theory such as general relativity to the utmost precision available simply because it is one of the essential pillars of the framework of physics. This is the fundamental justification of an experiment such as Gravity Probe B (the Stanford gyroscope experiment), which will advance by two orders of magnitude our experimental knowledge of post-Newtonian gravity.

Second, some very crucial qualitative features of general relativity have not yet been verified: in particular the existence of black holes, and the direct detection on Earth of gravitational waves. Hopefully, the LIGO/VIRGO network of interferometric detectors will observe gravitational waves early in the next century.

Last, some recent theoretical findings suggest that the current level of precision of the experimental tests of gravity might be naturally (i.e., without fine tuning of parameters) compatible with Einstein being actually only 50% right! By this we mean that the correct theory of gravity could involve, on the same fundamental level as the Einsteinian tensor field $g^*_{\mu\nu}$, a massless scalar field φ.

Let us first question the traditional paradigm (initiated by Fierz [11] and enshrined by Dicke [15], Nordtvedt and Will [2]) according to which special attention should be given to tensor-scalar theories respecting the equivalence principle. This class of theories was, in fact, introduced in a purely *ad hoc* way so as to prevent too violent a contradiction with experiment. However, it is important to notice that the scalar couplings which arise naturally in theories unifying gravity with the other interactions systematically violate the equivalence principle. This is true both in Kaluza-Klein theories (which were the starting point of Jordan's theory) and in string theories. In particular, it is striking that (as first noted by Scherk and Schwarz [49]) the dilaton field Φ, which plays an essential role in string theory, appears as a necessary partner of the graviton field $g_{\mu\nu}$ in all string models. Let us recall that $g_s = e^{\Phi}$ is the basic string coupling constant (measuring the weight of successive string loop contributions) which determines, together with other scalar fields (the moduli), the values of all the coupling constants of the low-energy world. This means, for instance, that the fine-structure constant $\alpha_{\rm em}$ is a function of Φ (and possibly of other moduli fields). This spatiotemporal variability of coupling constants entails a clear violation of the equivalence principle. In particular, $\alpha_{\rm em}$ would be expected to vary on the Hubble time scale (in contradiction with the limit (6) above), and materials of different compositions would be expected to fall with different accelerations (in contradiction with the limits (7), (8) above).

The most popular idea for reconciling gravitational experiments with the existence, at a fundamental level, of scalar partners of $g_{\mu\nu}$ is to assume that all these scalar fields (which are massless before supersymmetry breaking) will acquire a mass after supersymmetry breaking. Typically one expects this mass m to be in the TeV range [50]. This would ensure that scalar exchange brings only negligible, exponentially small corrections $\propto \exp(-mr/\hbar c)$ to the general relativistic predictions concerning low-energy gravitational effects.

However, this idea is fraught with many cosmological difficulties. A first difficulty is that, the dilaton being protected from getting a mass to all orders of perturbation theory, any putative nonperturbative potential $V(\Phi)$ will be extremely shallow, which makes it difficult to fix the vacuum expectation value (VEV) of Φ without fine-tuning the initial conditions [51]. A second difficulty is that additional fine-tuning (or some new mechanism) is needed to ensure that the value of the potential $V(\Phi)$ at its minimum is zero, or at least 120 orders of magnitude smaller than the Planck density ("cosmological constant problem"). A third problem is that one generically expects a lot of potential energy to be stored initially in $V(\Phi)$. The cosmological decay of this energy is either too slow or leads to an overproduction of entropy ("Polonyi problem" [52]). Moreover, if cosmological strings exist, they tend to radiate a lot of dilatons, thereby causing a problem similar to the usual Polonyi problem [53].

Though these cosmological difficulties might be solved by a combination of ad hoc solutions (e.g., introducing a secondary stage of inflation to dilute previously produced dilatons [54], [55]), a more radical solution to the problem of reconciling the existence of the dilaton (or any moduli field) with experimental tests and cosmological data has been proposed [56] (see also [57], which considered an equivalence-principle-respecting scalar field). The main idea of [56] is that string-loop effects (i.e., corrections depending upon $g_s = e^{\Phi}$ induced by worldsheets of arbitrary genus in intermediate string states) may modify the low-energy, Kaluza-Klein type matter couplings ($\propto e^{-2\Phi} F_{\mu\nu} F^{\mu\nu}$) of the dilaton (or moduli) in such a manner that the VEV of Φ be cosmologically driven toward a finite value Φ_m where it decouples from matter. For such a "least coupling principle" to hold, the loop-modified coupling functions of the dilaton, $B_i(\Phi) = e^{-2\Phi} + c_0 + c_1 e^{2\Phi} + \cdots +$ (nonperturbative terms), must exhibit extrema for finite values of Φ, and these extrema must have certain universality properties. More precisely, the most general low-energy couplings induced by string-loop effects will be such that the various terms on the right-hand side of Eq. (3) will be multiplied by several different functions of the scalar field(s): say a factor $B_F(\varphi)$ in factor of the kinetic terms of the gauge fields, a factor $B_\psi(\varphi)$ in factor of the Dirac kinetic terms, etc. We work here in the Einstein frame, and with a canonically normalized scalar field φ, i.e., the Lagrangian density has the form

$$\mathcal{L} = \frac{1}{16\pi G_*} [R(g_*) - 2g_*^{\mu\nu} \partial_\mu \varphi \, \partial_\nu \varphi] - \frac{1}{4} B_F(\varphi) F_{\mu\nu} F^{\mu\nu} + \cdots \tag{21}$$

It has been shown in [56] that if the various coupling functions $B_i(\varphi), i = F, \psi, \ldots$, all admit an extremum (which must be a maximum for the "leading" B_i) at some common value φ_m of φ, the cosmological evolution of the coupled tensor-scalar-matter system will drive φ towards the value φ_m, at which φ decouples from matter. As suggested in [56], a natural way in which the required conditions could be satisfied is through the existence of a discrete symmetry in scalar space. [For instance, a symmetry under $\varphi \to -\varphi$ would guarantee that all the scalar coupling functions reach an extremum at the self-dual point $\varphi_m = 0$.] The existence of such symmetries have been proven for some of the scalar fields appearing in string theory (target-space duality for the moduli fields) and conjectured for others (S-duality for the dilaton). This gives us some hope that the mechanism of [56] could apply and thereby naturally reconcile the existence of massless scalar fields with experiment.

A study of the efficiency of attraction of φ towards φ_m estimates that the present vacuum expectation value φ_0 of the scalar field would differ (in an rms sense) from φ_m by

$$\varphi_0 - \varphi_m \sim 2.75 \times 10^{-9} \times \kappa^{-3}\, \Omega^{-3/4}\, \Delta\varphi, \tag{22}$$

where κ denotes the curvature of $\ln B_F(\varphi)$ around the maximum φ_m and $\Delta\varphi$ the deviation $\varphi - \varphi_m$ at the beginning of the (classical) radiation era. Equation (22) predicts (when $\Delta\varphi$ is of order unity[2]) the existence, at the present cosmological epoch, of many small, but not unmeasurably small, deviations from general relativity proportional to the *square* of $\varphi_0 - \varphi_m$. This provides a new incentive for trying to improve by several orders of magnitude the various experimental tests of Einstein's equivalence principle, i.e., of the consequences $C_1 - C_4$ recalled above. The most sensitive way to look for a small residual violation of the equivalence principle is to perform improved tests of the universality of free fall. The mechanism of [56] suggests a specific composition dependence of the residual differential acceleration of free fall and estimates that a nonzero signal could exist at a very small level as illustrated in Fig. 12.2 (taken from [56]). The dashed line in this figure is (as in Eq. (22) above) a rough analytical estimate (assuming random phases) which reads

$$\left(\frac{\Delta a}{a}\right)^{\max}_{\text{rms}} \sim 1.36 \times 10^{-18}\, \kappa^{-4}\, \Omega^{-3/2}\, (\Delta\varphi)^2, \tag{23}$$

where κ is expected to be of order unity (or smaller, leading to a larger signal, in the case where φ is a modulus rather than the dilaton).

Let us emphasize again that the strength of the cosmological scenario considered here as a counterargument to applying Occam's razor lies in the fact that the very small number on the right-hand side of Eq. (23) has been derived without any fine tuning or use of small parameters, and turns out to be naturally smaller than the 10^{-12} level presently tested by equivalence-principle experiments (see equations (7), (8)). The estimate (23) gives added significance to the project of a satellite test of the equivalence principle (nicknamed STEP, and currently considered by NASA and ESA) which aims at probing the universality of free fall of pairs of test masses orbiting the Earth at the 10^{-17} or 10^{-18} level [59].

---

[2]However, $\Delta\varphi$ could be $\ll 1$ if the attractor mechanism already applies during an early stage of potential-driven inflation [58].

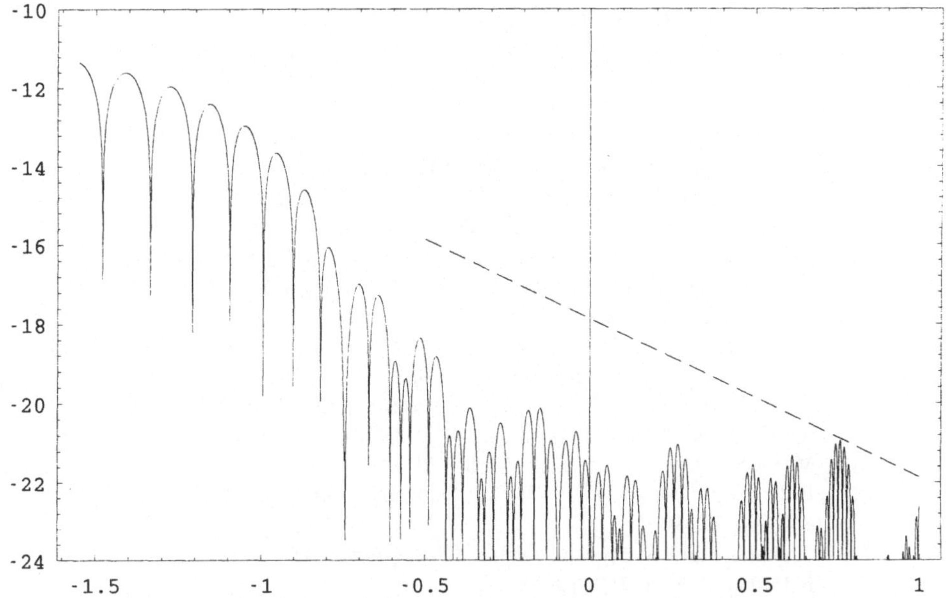

Figure 12.2 The solid line represents $\log_{10}(\Delta a/a)_{\max}$ as a function of $\log_{10}\kappa$, i.e., the expected present level of violation of the universality of free fall as a function of the curvature κ of the (string-loop induced) coupling function $\ln B_F^{-1}(\varphi)$ near a minimum φ. The dashed line is a rough analytical estimate (assuming random phases of oscillations). (Figure taken from [56].)

References

[1] S.N. Gupta, *Phys. Rev.* **96** (1954) 1683;
R.H. Kraichnan, *Phys. Rev.* **98** (1955) 1118;
R.P. Feynman, F.B. Morinigo and W.G. Wagner, *Feynman Lectures on Gravitation*, edited by Brian Hatfield (Addison-Wesley, Reading, 1995);
S. Weinberg, *Phys. Rev.* **138** (1965) B988;
V.I. Ogievetsky and I.V. Polubarinov, *Ann. Phys.* N.Y. **35** (1965) 167;
W. Wyss, *Helv. Phys. Acta* **38** (1965) 469;
S. Deser, *Gen. Rel. Grav.* **1** (1970) 9;
D.G. Boulware and S. Deser, *Ann. Phys.* N.Y. **89** (1975) 193;
J. Fang and C. Fronsdal, *J. Math. Phys.* **20** (1979) 2264;
R.M. Wald, *Phys. Rev. D* **33** (1986) 3613;
C. Cutler and R.M. Wald, *Class. Quantum Grav.* **4** (1987) 1267;
R.M. Wald, *Class. Quantum Grav.* **4** (1987) 1279.

[2] C.M. Will, *Theory and Experiment in Gravitational Physics*, 2nd edition (Cambridge University Press, Cambridge, 1993); and *Int. J. Mod. Phys. D* **1** (1992) 13.

[3] T. Damour, *Gravitation and Experiment* in *Gravitation and Quantizations*, eds B. Julia and J. Zinn-Justin, Les Houches, Session LVII (Elsevier, Amsterdam, 1995), pp 1–61.

[4] E. Fermi, *Atti Accad. Naz. Lincei Cl. Sci. Fis. Mat. & Nat.* **31** (1922) 184, 306; E. Cartan, *Lecons sur la Géométrie des Espaces de Riemann* (Gauthier-Villars, Paris, 1963).
[5] T. Damour and F. Dyson, *Nucl. Phys. B* **480** (1996) 37.
[6] J.D. Prestage, R.L. Tjoelker, and L. Maleki, *Phys. Rev.* Lett. **74** (1995) 3511.
[7] J.D. Prestage et al., *Phys. Rev. Lett.* **54** (1985) 2387;
S.K. Lamoreaux et al., *Phys. Rev. Lett.* **57** (1986) 3125;
T.E. Chupp et al., *Phys. Rev. Lett.* **63** (1989) 1541.
[8] P.G. Roll, R. Krotkov and R.H. Dicke, *Ann. Phys.* (N.Y.) **26** (1964) 442;
V.B. Braginsky and V.I. Panov, *Sov. Phys. JETP* **34** (1972) 463;
Y. Su et al., *Phys. Rev. D* **50** (1994) 3614.
[9] J.O. Dickey et al., *Science* **265** (1994) 482; J.G. Williams, X.X. Newhall, and J.O. Dickey, *Phys. Rev. D* **53** (1996) 6730.
[10] R.F.C. Vessot and M.W. Levine, *Gen. Rel. Grav.* **10** (1978) 181;
R.F.C. Vessot et al., *Phys. Rev. Lett.* **45** (1980) 2081.
[11] M. Fierz, *Helv. Phys. Acta* **29** (1956) 128.
[12] P. Jordan, *Nature* **164** (1949) 637; *Schwerkraft und Weltall* (Vieweg, Braunschweig, 1955).
[13] T. Damour and G. Esposito-Farèse, *Class. Quant. Grav.* **9** (1992) 2093.
[14] P. Jordan, *Z. Phys.* **157** (1959) 112.
[15] C. Brans and R.H. Dicke, *Phys. Rev.* **124** (1961) 925.
[16] I.I. Shapiro, *Phys. Rev. Lett.* **13** (1964) 789.
[17] R.D. Reasenberg et al., *Astrophys. J.* **234** (1979) L219.
[18] D.S. Robertson, W.E. Carter, and W.H. Dillinger, *Nature* **349** (1991) 768;
D.E. Lebach et al., *Phys. Rev. Lett.* **75** (1995) 1439.
[19] K. Nordtvedt, Phys. Rev. **170** (1968) 1186.
[20] I.I. Shapiro, in *general relativity and Gravitation 1989*, ed. N. Ashby, D.F. Bartlett and W. Wyss (Cambridge University Press, Cambridge, 1990), 313–330.
[21] T. Damour and G. Esposito-Farèse, *Phys. Rev. D* **53** (1996) 5541.
[22] R.A. Hulse and J.H. Taylor, *Astrophys. J. Lett.* **195** (1975) L51; see also the 1993 Nobel lectures in physics of Hulse (pp. 699–710) and Taylor (pp. 711–719) in *Rev. Mod. Phys.* **66,** n 0.3 (1994).
[23] P.S. Laplace, *Traité de Mécanique Céleste*, (Courcier, Paris, 1798–1825), second part: book 10, chapter 7.
[24] T. Damour and N. Deruelle, *Phys. Lett. A* **87** (1981) 81;
T. Damour, *C.R. Acad. Sci.* Paris **294** (1982) 1335;
T. Damour, in *Gravitational Radiation*, eds N. Deruelle and T. Piran (North-Holland, Amsterdam, 1983) pp 59–144.
[25] J.H. Taylor, *Class. Quant. Grav.* **10** (1993) S167 (Supplement 1993) and references therein; see also J.H. Taylor's Nobel lecture quoted in [22].
[26] T. Damour and G. Esposito-Farèse, *Phys. Rev. Lett.* **70** (1993) 2220.
[27] T. Damour and J.H. Taylor, *Phys. Rev. D.* **45** (1992) 1840.
[28] T. Damour and N. Deruelle, *Ann. Inst. H. Poincaré* **43** (1985) 107; **44** (1986) 263.
[29] T. Damour and G. Esposito-Farèse, *Phys. Rev. D* **54** (1996) 1474.
[30] T. Damour and J.H. Taylor, *Astrophys. J.* **366** (1991) 501.
[31] J.H. Taylor, A. Wolszczan, T. Damour and J.M. Weisberg, *Nature* **355** (1992) 132.
[32] A. Wolszczan, *Nature* **350** (1991) 688.

[33] Z. Arzoumanian, Ph.D. thesis, Princeton University, 1995.
[34] T. Damour and D. Vokrouhlicky, *Phys. Rev. D* **53** (1996) 4177.
[35] A. Wolszczan and J.H. Taylor, to be published (quoted in Taylor's Nobel lecture [22]).
[36] C.M. Will and H.W. Zaglauer, *Astrophys. J.* **346** (1989) 366.
[37] T. Damour and G. Schäfer, *Phys. Rev. Lett.* **66** (1991) 2549.
[38] T. Damour and D. Vokrouhlicky, *Phys. Rev. D* **53** (1996) 4177.
[39] R.D. Blandford et al. editors, *Pulsars as physics laboratories*, *Phil. Trans. R. Soc.* London **A341** (1992) pp 1–192; see notably the contributions by J.H. Taylor (pp. 117–134) and by T. Damour (pp. 135–149).
[40] J.N. Bahcall and M. Schmidt, *Phys. Rev. Lett.* **19** (1967) 1294.
[41] D.A. Varshalovich, V.E. Panchuk and A.V. Ivanchick, *Astronomy Letters* **22** (1996) 6.
[42] A. Dar, *Nucl. Phys. B* (Proc. Suppl.) **28** (1992) 321.
[43] G. Steigman, *Nature* (London) **261** (1976) 479; Essay on Gravitation, 1976, unpublished.
[44] J.D. Barrow, *Mon. Not. R. Astr. Soc.* **184** (1978) 677.
[45] J. Yang, D.N. Schramm, G. Steigman, and R.T. Rood, *Astrophys. J.* **227** (1979) 697; and **281** (1984) 493.
[46] T. Rothman and R. Matzner, *Astrophys. J.* **257** (1982) 450.
[47] T. Damour, G.W. Gibbons, and C. Gundlach, *Phys. Rev. Lett.* **64** (1990) 123;
T. Damour and C. Gundlach, *Phys. Rev. D* **43** (1991) 3873.
[48] D. Kalligas, K. Nordtvedt, and R.V. Wagoner, in *Proceedings of the Seventh Marcel Grossmann Meeting on general relativity* (World Scientific, Singapore, 1996);
D.I. Santiago, D. Kalligas, and R.V. Wagoner, gr-qc/9706017.
[49] J. Scherk and J.H. Schwarz, *Nucl. Phys. B* **81** (1974) 118; *Phys. Lett. B* **52** (1974) 347.
[50] B. de Carlos, J.A. Casas, F. Quevedo, and E. Roulet, *Phys. Lett. B* **318** (1993) 447.
[51] R. Brustein and P.J. Steinhardt, *Phys. Lett. B* **302** (1993) 196.
[52] G.D. Coughlan et al., *Phys. Lett. B* **131** (1983) 59;
J. Ellis, D.V. Nanopoulos and M. Quiros, *Phys. Lett. B* **174** (1986) 176;
T. Banks, D.B. Kaplan and A.E. Nelson, *Phys. Rev. D* **49** (1994) 779.
[53] T. Damour and A. Vilenkin, *Phys. Rev. Lett.* **78** (1997) 2288.
[54] L. Randall and S. Thomas, *Nucl. Phys. B* **449** (1995) 229.
[55] D.H. Lyth and E.D. Stewart, *Phys. Rev. Lett.* **75** (1995) 201; *Phys. Rev. D* **53** (1996) 1784.
[56] T. Damour and A.M. Polyakov, *Nucl. Phys. B* **423** (1994) 532; *Gen. Rel. Grav.* **26** (1994), 1171.
[57] T. Damour and K. Nordtvedt, *Phys. Rev. Lett.* **70** (1993) 2217; *Phys. Rev. D* **48** (1993) 3436.
[58] T. Damour and A. Vilenkin, *Phys. Rev. D* **53** (1996) 2981.
[59] P.W. Worden, in *Near Zero: New Frontiers of Physics*, eds J.D. Fairbank et al. (Freeman, San Francisco, 1988) p. 766;
J.P. Blaser et al., *STEP, Report on the Phase A Study*, ESA document SCI (96)5, March 1996;
Fundamental Physics in Space, special issue of *Class. Quant. Grav.* **13** (1996).
MiniSTEP, NASA-ESA report, December 1996 (second issue).

13

Some Developments in Newtonian Cosmology

Jürgen Ehlers

13.1 Introduction: The Curious History of Newtonian Cosmology

Although Newtonian cosmology, founded in 1934 by Edward Arthur Milne [19] and William McCrea [17], is approaching its retirement age, it is still of some interest—not only historically and for didactic reasons, but even for present research. Since its conceptual and formal basis is similar to, but simpler than, general relativistic cosmology, some problems of cosmology may first be solved within a Newtonian framework and then hopefully generalized to the Einsteinian theory. Pertinent examples to be outlined below include the problem of understanding the relation between an anisotropic, inhomogeneous world model and "its" isotropic, homogeneous background, and the task of setting up approximation methods capable of dealing with the evolution of large density contrasts and peculiar velocity fields from a nearly structureless, uniformly expanding matter distribution. Comparing the properties of Newtonian models with those of their Einsteinian counterparts may also help to elucidate when and why relativity is genuinely needed.

As mentioned already, Newtonian cosmology (in the present usage of the term) began with the work of Milne and McCrea in 1934, seventeen years after the advent of Einsteinian cosmology. Its original formulation, confined to homogeneous and isotropic world models, suffered from conceptual and mathematical problems originating in the notion "inertial frame" and in the divergence of the standard form of the Newtonian potential for mass distributions of infinite spatial support. These difficulties have in some sense been overcome, in response to criticisms by David Layzer [16] and William McCrea [18], by Otto Heckmann and Engelbert Schucking in 1955/1956 [13], [14], as I shall review in Section 13.2.1 below.

Strangely enough it seems to have been overlooked by all cosmologists that, in fact, a satisfactory "Newtonian" cosmology could have been put together straightforwardly on the basis of a generally covariant, local, spacetime reformulation (and generalization to spatially unbounded mass distribution) of Galilean-relativistic gravitational dynamics, all essential ingredients of which had been developed by Elie Cartan in 1923/24 [4] and Kurt Otto Friedrichs in 1927 [10]. Such an approach to Newtonian cosmology, which is free of the difficulties mentioned above, seems to have been realized first by Andrzej Trautman in 1964 [24], almost forty years after it could have been done. I shall return to this in Section 13.2.2 below.

13.2 Two Formulations of Newtonian Cosmology and its Relationship to Relativistic Cosmology

13.2.1 The Heckmann–Schucking Formulation

As is by now well recognized, Galileo's law of the universality of free fall and the fact that freely falling test particles exhibit relative accelerations show that the notion of an exact inertial frame lacks empirical support. Instead it is reasonable to assume, within classical kinematics, that there are "dynamically nonrotating" rigid frames of reference represented by coordinate systems (t, x^a) in spacetime, relative to which free fall is given by

$$\ddot{x}^a = g^a(t, x^b), \tag{1}$$

where g^a is a function depending on that coordinate system and related to the distribution of matter. Any two such coordinate systems are related to each other by a transformation of the type

$$\begin{aligned} t' &= t + \tau, \\ x' &= D \cdot x + b(t), \end{aligned} \tag{2}$$

where τ is a real number, b a time-dependent translation matrix, D, a constant rotation matrix, and the matrices x, x' represent the coordinates (x^a), $(x^{a'})$ of the two frames of reference in question. Obviously, the law (1) is maintained under a transformation (2), provided the gravitational field g^a is transformed as

$$g'(t', x') = D \cdot g(t, x) + \ddot{b}(t). \tag{3}$$

(Thus, g^a is not a frame-independent vector field.)

The transformations (2) form the Heckmann-Schucking group G_{HS}, and if a transformation (2) is denoted as (τ, D, b), the multiplication law reads

$$(\bar{\tau}, \bar{D}, \bar{b}) \cdot (\tau, D, b) = (\bar{\tau} + \tau, \bar{D} \cdot D, \bar{D} \cdot b + \bar{b}_\tau), \tag{4}$$

with the abbreviation

$$\bar{b}_\tau(t) := \bar{b}(\tau + t). \tag{4'}$$

(Thus, G_{HS} is the semidirect product of the direct product $\mathbb{R}^+ \times SO_3$ and the additive group $\mathbb{T} = \{(b(t))\}$ of t-dependent translations, via the homomorphism

h defined by (4') : $G_{HS} = (\mathbb{R}^+ \times SO_3) \overset{h}{\times} \mathbb{T}$.) Not only the free fall law (1), but also the Euler-Poisson laws of gravitational dynamics, slightly generalized by a cosmological Λ term,

$$\begin{aligned} &\rho_{,t} + (\rho V^a)_{,a} = 0 \,, \\ &\dot{V}^a := V^a_{,t} + V^b V^a_{,b} = g^a - \frac{p_{,a}}{\rho}, \qquad p = f(\rho)\,, \\ &g_a = -\Phi_{,a}\,, \qquad \Delta\Phi = 4\pi G\rho - \Lambda\,, \end{aligned} \tag{5}$$

are invariant under (2), provided the gravitational potential Φ transforms, in line with (3), not as a spacetime scalar but according to

$$\Phi'(t', x') = \Phi(t, x) - x' \cdot \ddot{b}(t)\,. \tag{6}$$

While Φ is not a scalar and g^a is not a vector, the second derivative $\Phi_{,ab}$ of Φ is a tensor. According to (1) and $(5)_3$, it determines the relative accelerations of freely falling test particles. It turns out to be useful to split $\Phi_{,ab}$ into its trace $\Delta\Phi$, which is determined pointwise by the density ρ, and its "nonlocally determined" trace-free part

$$E_{ab} := \Phi,_{\langle ab \rangle} := \Phi,_{ab} - \frac{1}{3}\delta_{ab}\Delta\Phi. \tag{7}$$

In analogy to general relativity, a nonrotating frame of reference whose origin is freely falling, $g^a(t, 0) = 0$, may be said to be "locally inertial at the origin." Such a reference frame approximates an inertial frame near the origin in a spatial region whose size depends on the accuracy required and on the scale on which g^a varies.

The foregoing concepts provide a *formal generalization* of Newtonian gravitational dynamics, formal in so far as no consequences of the laws, let alone empirically successful applications, have been stated. As it stands, it cannot be considered as a testable physical theory, since the local laws (5) do not determine the potential Φ in terms of ρ, and hence the future values of ρ and V^a are not determined by their initial values at one time. In other words, equations (5) do not define a dynamical system.

Newton's theory proper was designed for the solar system and more generally for isolated systems, and only for them was it successful in explaining facts. For such systems, the mass density ρ is spatially compactly supported or falls off at spatial infinity faster than r^{-2}, and the potential Φ can be and is required to vanish at spatial infinity. It is then possible and natural to restrict the class of nonrotating frames of reference to those for which g^a vanishes at infinity—frames which are "asymptotically inertial'—and correspondingly to restrict the Heckmann-Schucking group G_{HS} to the Galilean group ($\ddot{b} = 0$ in (2). According to (6), Φ then becomes a scalar, and Φ is given by the standard Newtonian potential of ρ and can be eliminated from (5). Equations (5) then reduce to an integro-differential system for ρ and V^a, the obvious generalization of Newton's laws to continuously distributed matter, particularly in the case of dust, $p = 0$.

Though the initial value problems for this system have not yet been shown to the well posed—due to "technical" mathematical difficulties associated either with the boundaries of compact bodies or with the fall-off of the density at spatial infinity—recent work indicates that these difficulties may be overcome and that, *for isolated systems*, the laws (5), augmented by the boundary condition $\Phi(t, \infty) = 0$, provide a valid extension of Newton's theory of mass points or rigid bodies to deformable bodies and to extended, bound gas clouds, as has been assumed as a matter of course in astrophysics all along.

In *cosmology*, however, the situation is essentially different, as has been emphasized by Heckmann and Schucking. The density cannot reasonably be taken to fall off at infinity, and therefore Φ as a solution of Poissons's equation cannot be required to vanish at infinity—a condition which is also incompatible with the transformation law (6)—and thus one has to face the difficulty mentioned above. Heckmann and Schucking proposed to impose the boundary condition

$$\lim_{|x|\to\infty} E_{ab}(t, x) = A_{ab}(t) \tag{8}$$

on the field E_{ab} defined in (7) and claimed without proof that, provided the density ρ has a limit at spatial infinity, the system (5) (at least for $p = 0$) "obeys the postulates of causality," i.e., it has unique solutions for given initial data for ρ and V^a. While I have verified uniqueness (assuming only $\rho > 0$), I have been unable to show existence. Apart from that, I fail to see a physical or mathematical motivation for postulating (8). If one is concerned with an inhomogenous model universe which is homogeneous on average only, then E_{ab} cannot be expected to have a direction-independent limit at infinity. If, on the other hand, one wants to construct strictly homogeneous models, it is easier to choose the rate of shear $\sigma_{ab}(t)$ arbitrarily instead of requiring (8), as pointed out and done by Heckmann and Schucking.

Thus, my proposal is to drop (8) and to admit that the formal, G_{HS}-covariant generalization of Newtonian dynamics based only on the local laws (5) does not, in general, determine a unique time evolution. Without a boundary condition, forces between parts of matter are no longer defined, and the scheme is not really a field theory either, since nothing specifies the evolution of the potential Φ. That the formalism is nevertheless useful for some purposes will be shown in sections 13.3–13.5.

13.2.2 The Cartan–Friedrichs Formulation

As remarked in the introduction, Cartan and Friedrichs independently conceived of the idea of formulating Newtonian gravitational theory in a generally covariant spacetime language and made first steps to realize this programme, and Friedrichs indicated how one may thus obtain Newton's theory as a degenerate limit of Einstein's. These ideas have been elaborated by several authors, notably by A. Trautman [23], P. Havas [11], and H. P. Künzle [15]; for additional references see my review article [7]. Based on these researches I formulated in 1981 a "frame theory" which comprises both GR and Newton's theory [6]. If a parameter λ which occurs in one of the laws of that theory is positive, the theory reduces to GR, if it is zero, it gives essentially Newton's theory. More precisely, for isolated, i.e., asymptotically flat systems, it does then give Newton's theory, while in general the case $\lambda = 0$ reduces to a slight generalization of it, "Newton-Cartan theory," which reduces to the Heckmann-Schucking theory if one adds a curvature condition found by Trautman [24].

The frame theory, which will not be reviewed here in detail, uses a pair of metrics $t_{\alpha\beta}$, $S^{\alpha\beta}$ which degenerate in the $\lambda = 0$ case, a symmetric linear connection $\Gamma^{\alpha}_{\beta\gamma}$ compatible with both metrics, and a mass-momentum-stress tensor $T^{\alpha\beta}$. The contracted curvature tensor of the connection is coupled to $T^{\alpha\beta}$; in addition some conditions imposed on the full curvature tensor and on $T^{\alpha\beta}$ ensure that in the degenerate case $\lambda = 0$, one indeed recovers the Newton-Cartan or Newton theory.

The merits of this frame theory are that it demonstrates in a precise way the limit relation

$$\text{General relativity} \rightarrow \text{Newton's theory;}$$

that it provides a rigorous definition of the Newtonian limit of a sequence of GR solutions (for examples, see [6] and [8]) and of post-Newtonian approximations [20]; and that it even enabled U. Heilig [12] to prove an existence theorem in GR concerning solutions for rigidly rotating fluid balls by means of corresponding Newtonian solutions. In particular, the limit relation of the frame theory permits one to derive (legitimately) Newtonian cosmological solutions directly as limits of relativistic ones. This works not only for all Friedmann solutions, but as well for Gödel's universe. Moreover, the intrinsic spacetime formulation allows one to construct "Newtonian" cosmological models with compact, evolving time slices; see Sections 13.4 and 13.5.

From the spacetime formulation of Newtonian theory, which allows the laws to be stated without reference to special coordinate systems, one recovers the Heckmann-Schucking version by reintroducing nonrotating, orthonormal coordinates with respect to which the metrics have components $t_{\alpha\beta} = \text{diag}(1, 0, 0, 0)$, $S^{\alpha\beta} = \text{diag}(0, 1, 1, 1)$. It then turns out that the inhomogeneous transformation law (6) for Φ emerges from the fact that, in those coordinates, the connection components are determined by Φ via $\Gamma^{\alpha}_{\beta\gamma} = t_{,\beta} t_{,\gamma} S^{\alpha\delta} \Phi_{,\delta}$. Φ thus appears as a connection potential. Also, the tensor E_{ab} defined in (7) turns out to be the Newtonian limit of the "electric" part of the Weyl conformal curvature tensor of GR.

13.2.3 Electrodynamics and Optics in Newtonian Cosmology

In order to relate a cosmological model to observations, one needs to describe light propagation in it. In GR that presents no problem, since that theory inherits from its birth an electrodynamics and (derivable from it) a geometrical optics. But in Newtonian theory it is not immediately clear how to proceed.

Using the Cartan-Friedrichs formulation, A. Trautman [24] found an elegant way out. Let $S^{\alpha\beta}$ be the degenerate (inverse) spatial metric and let V^{α} be the four-velocity of the cosmic substratum. Then

$$g^{\alpha\beta} := S^{\alpha\beta} - c^{-2} V^{\alpha} V^{\beta} \tag{9}$$

defines an (inverse) Lorentzian conformal structure, c denoting the speed of light. One can use that conformal structure and a 2-form $F_{\alpha\beta}$ to formulate Maxwell's equations, just as in GR. This amounts to an ether, dragged along by matter. Light rays are then null geodesics of that "optical" metric, which also provides an eikonal equation.

Applied to the Newtonian analogues of the Friedmann models one thus obtains, following H. Bondi [2], observable relations very similar to those of the genuine Friedmann models. Trautman's formulation, however, has the advantage of being applicable not just to homogenous-isotropic models, but to all models. Note that in this context, equation (9) is used for the purpose of electrodynamics and optics; the spacetime metric is given by the absolute time t (or $t_{\alpha\beta} = t_{\alpha} t_{,\beta}$) and $S^{\alpha\beta}$.

13.2.4 Relations Between Relativistic and Newtonian Cosmological Models

As stated in subsection 2, various Newtonian cosmological models can be obtained as limits of suitably parametrized relativistic ones. The limit refers to the metrics, connections, and matter variables. The several fields will in general not converge uniformly to their limits, but in subdomains where they do, the Newtonian limit fields approximate their relativistic progenitors. Thus, for example, a spherical Friedmann model cannot be globally approximated by one "elliptic" Newtonian model, but one can cover that model by open domains which are approximated by pieces of such a Newtonian model, as one would expect.

The limit relation does not include optics; that has to be added at the end, via (9). In general, the optical metric associated to a Newtonian model does not obey Einstein's field equation. There is one exception, though: The Newtonian limit of the Einstein–de Sitter model is the "parabolic" Newtonian model, and the optical metric of that equals the Einstein–de Sitter spacetime metric. In this case—and I guess *only* in this case—the original model and its limit are identical as far as cosmic time, cosmic 3-space with its Euclidean metric, the motion of the cosmic fluid, and light propagation are concerned. But even in this case the linear perturbations of the background models and the associated perturbed light deflections, though similar, are different.

An elegant account of "Newton-Cartan cosmology," which covers linear perturbation theory, has recently been given by Ch. Rüede and N. Straumann [22].

13.3 Observer-Homogenous, Bianchi-Type Models

From now on I shall use mostly the Heckmann-Schucking formulation of Section 13.2.1 and put the pressure p equal to zero.

How should one define homogeneity of a cosmological model? The most adequate definition, in my opinion, is *observer homogeneity*, which requires that all observers who participate in the mean motion of the cosmic matter experience the same history of the universe. More precisely, this means the following. Consider any local inertial frame (t, x^a) (as defined below equation (7), attached to a particle of the cosmic fluid; relative to it, the motion of that fluid is given by functions $\rho(t, x^b)$, $V^a(t, x^b)$. If, for a suitable choice of t and x^a for each fundamental observer, the functions ρ and V^a are the same for all observers, then the model is observer-homogeneous. Two such comoving local inertial systems are related by a transformation (2), and the set of all these transformations relating pairs of fundamental observers forms a subgroup H of the Heckmann-Schucking group G_{HS}. Note that the τ values occuring in H need not vanish, and the D-values need not be the unit matrix. Also, it is permitted that transformations of H relate two preferred coordinate systems attached to the same observer. In these three respects the concept of observer-homogeneity used here is more general than the "homogeneity postulate" employed in the classic papers by Heckmann and Schucking.

The task of determining observer-homogeneous models naturally splits into a kinematical and a dynamical part. The kinematical problem consists of finding those fields ρ, V, and subgroups H of G_{HS} which obey the functional equations

$$\begin{aligned} \rho(D \cdot x + b(t), t + \tau) &= \rho(x, t), \\ V(D \cdot x + b(t), t + \tau) &= D \cdot V(x, t) + \dot{b}(t), \end{aligned} \tag{10}$$

and where H is sufficiently large in the sense that, at least for some interval $t_1 < t < t_2$, $b(t)$ takes all values in $\mathbb{R}^3$. (This property says that all observers can be reached from one of them by applying a suitable transformation of H, assuming that the set of observers has the structure of the manifold $\mathbb{R}^3$. Another choice of topology is the 3-torus, $\mathbb{T}^3$; see section 13.4). The dynamical problem consists of selecting those triplets ρ, v, H where ρ, v obey equations (5) with $p = 0$.

In any kinematical solution, the second of equations (10) shows-put $x = 0$ and use $v(0, t) = 0$-that b satisfies the ordinary differential equation

$$V(b(t), t + \tau) = \dot{b}(t) \; ; \tag{11}$$

hence $b(t)$ is determined by V, τ, and any "initial" value $b(t_0)$. Thus, in contrast to G_{HS}, H is a Lie group whose dimension is at least 3 and at most 7. If H or a subgroup of H acts simply transitively in the set of fundamental observers, its dimension is 3. The corresponding solutions are Newtonian analogues of the well-known relativistic Bianchi models. Other cases correspond to the Friedmann, Gödel, or Kantowski-Sachs models.

The problem of finding the observer-homogeneous Newtonian dust models was treated and partly solved in the Ph.D. thesis of David L. Hibler, who worked with me at the University of Texas at Austin in 1970–71. Unfortunately, his work was neither submitted nor published although the results are quite interesting; I here report some of them without proofs:

1. Of the 9 Bianchi types, all but types VIII and IX are possible in Newtonian cosmology.
2. There are homogeneous Newtonian models with velocity fields which are nonlinear in the spatial coordinates.
3. There are models in which the density varies spatially at constant absolute time.

The results are also of some interest because they provide a variety of examples of self-gravitating hydrodynamic flows.

I presume that Hibler's work could be simplified, completed and related rigorously to the GR-Bianchi models by using the Cartan-Friedrichs formulation and the frame theory mentioned in Section 13.2.2, as a continuation of the Rüede-Straumann work quoted at the end of Section 13.2.4.

13.4 Averaging in Cosmology

The universe around us is very inhomogeneous, with density contrasts exceeding unity on all scales below a few megaparsecs. On the other hand, the microwave background radiation indicates (or, at least, is compatible with the assumption) that the universe has been very nearly homogeneous and isotropic at an early stage, and the present matter distribution on a scale of a few hundred megaparsecs seems to be compatible with statistical homogeneity and isotropy now. This raises problems, emphasized repeatedly in particular by George Ellis [9]: How can one model, in general relativity, a universe with the properties just indicated? Which averaging or smoothing method should be applied to an inhomogeneous model to produce "its" homogeneous—say, Robertson-Walker background? How do the equations governing the averaged variables (if any) differ from Einstein's equations, assumed to hold on a small (stellar) scale where they have been tested? Do these modified laws appreciably affect the comparison of theory with observations?

The purpose of this section is to illustrate these problems which, in my view, are not yet well understood. First, I shall review a few results concerning the average expansion rate of inhomogeneous dust model universes in Newtonian cosmology, and then mention corresponding results in GR.

Following Ehlers and Buchert 1997 [5], assume an arbitrary solution of equations (5) (for $p = 0$) on Euclidean space $\mathbb{R}^3$ and some interval of time. Split the velocity gradient as usual,

$$V_{a,b} = \omega_{ab} + \sigma_{ab} + \frac{1}{3}\Theta\delta_{ab}, \tag{12}$$

into the rates of rotation, shear, and expansion. Equations (5) then imply Raychaudhuri's equation

$$\dot{\Theta} = \Lambda - 4\pi G\rho - \frac{1}{3}\Theta^2 + 2(\omega^2 - \sigma^2), \tag{13}$$

where here and in the sequel the dot indicates the Lagrangian, comoving time derivative as in the second line of equation (5). Consider an arbitrary part of the fluid, occupying a compact domain $D(t)$ at time t, and write $V(t) = a^3(t)$ for its comoving volume and M for its (constant) mass. Introduce spatial averages

$$\langle A\rangle(t) := (V(t))^{-1} \cdot \int_{D(t)} A\, d^3x \tag{14}$$

of tensor fields $A(t, x^a)$. It is easily verified that

$$\langle\Theta\rangle = 3\frac{\dot{a}}{a}, \qquad \langle\rho\rangle = \frac{M}{a^3}, \tag{15}$$

and, for arbitrary tensor fields A,

$$\langle A\rangle^{\cdot} = \langle\dot{A}\rangle + \langle\Theta A\rangle - \langle\Theta\rangle\langle A\rangle. \tag{16}$$

Applying these results to (13) gives

$$3\frac{\ddot{a}}{a} + 4\pi G\frac{M}{a^3} - \Lambda = \frac{2}{3}(\langle\Theta^2\rangle - \langle\Theta\rangle^2) + 2\langle\omega^2 - \sigma^2\rangle. \tag{17}$$

Next, define a "background velocity field" W^a by

$$\begin{aligned} W_{a,b} &= \Omega_{ab} + \Sigma_{ab} + \frac{1}{3}\,\ominus\,\delta_{ab}, \\ \Omega_{ab} &: = \langle\omega_{ab}\rangle,\ \Sigma_{ab} := \langle\sigma_{ab}\rangle, \ominus := \langle\Theta\rangle, \end{aligned} \tag{18}$$

and write $\omega_{ab} = \Omega_{ab} + \hat{\omega}_{ab}$, etc. Then (17) can be rewritten as

$$3\frac{\ddot{a}}{a} + 4\pi G\frac{M}{a^3} - \Lambda = 2(\Omega^2 - \Sigma^2) + \frac{2}{3}\langle\hat{\Theta}^2\rangle + 2\langle\hat{\omega}^2 - \hat{\sigma}^2\rangle. \tag{19}$$

Finally, introduce the peculiar velocity field $U^a := V^a - W^a$. It turns out that the last two terms in (19) form a divergence; so one obtains a third version of the averaged Raychaudhuri equation:

$$3\frac{\ddot{a}}{a} + 4\pi G\frac{M}{a^3} - \Lambda = 2(\Omega^2 - \Sigma^2) + \Omega^a{}_{,a}, \tag{20}$$

where $\Omega^a := \hat{\Omega}U^a - U^bU^a{}_{,b}$.

Before drawing conclusions, I wish to point out that the foregoing reasoning applies as well to parts of, or to the whole of, a Newtonian "small universe," i.e., one whose 3-space has the topology of the torus $\mathbb{T}^3$ rather than that of $\mathbb{R}^3$.

Newtonian toroidal models have been treated by U. Brauer et al. in [3] by the Cartan-Friedrichs method, but they may as well be handled by the more elementary Heckmann-Schucking version. There is one point deserving special attention, however. Spatial velocities refer to a nonrotating coordinate system—they are not frame-independent vector fields, as the transformations (2) show. The torus cannot be covered by one such coordinate system. However, the objects on the right hand side of (12) as well as the difference U^a of two velocity fields *are* globally well-defined tensors on the torus, hence equations (13)–(20) are globally valid also in that case.

What lessons can we draw from this? First, equation (17) shows how the average expansion of some—small or large—part of a model universe is affected by inhomogeneities; it shows in particular that if there are typical regions representative of the whole, then the global expansion will be that of an underlying averaged Friedmann model only if the right hand side vanishes. Moreover, and somewhat surprisingly, equation (20), applied to the total space of a spatially compact model, shows that the global expansion *is* that of a homogeneous, in general anisotropic model of Bianchi type I. In fact, it has been shown in [3] that any Newtonian model with compact time slices determines, by averaging, a unique such background model with spatially constant expansion, shear, and rotation, so that also the peculiar velocity field U^a and the density contrast $(\rho - \langle\rho\rangle)/\langle\rho\rangle$ are uniquely determined by the original, inhomogeneous solution. (The proof depends essentially on the fact that in Newtonian theory spatial averages of tensors are well defined.)

How about averaging in GR? A *useful* general answer is not known, as far as I am aware. The analogue of (17) which, in contrast to (20), does not need tensor averaging or a peculiar velocity field, is readily obtained, though. Given an arbitrary 4-velocity field V^α on a spacetime M, one can introduce comoving coordinates x^a and use proper time t along the integral curves of V^α as the fourth coordinate. The metric then takes the form

$$ds^2 = h_{ab}(t, x^c)\, dx^a\, dx^b - (dt - u_a\, dx^a)^2 \,. \tag{21}$$

Moreover, with $h := \det(h_{ab})$,

$$\Theta := V^\alpha{}_{;\alpha} = (\log\sqrt{h})^{\cdot} \,. \tag{22}$$

Although the coordinates are not uniquely determined by the above prescription, the two parts on the right hand side of (21) are separately invariant under the remaining transformations. The leaves of the foliation given by $t =$ const. change under the "gauge transformations"

$$t \to t + f(x^a)\,;$$

nevertheless one may use, on the leaves of any such foliation, the metric h_{ab} instead of the induced metric. h_{ab} is positive definite and measures proper distances between the particles of the flow determined by V^α. For scalars, one can define spatial averages in strict analogy to (14),

$$\langle A\rangle(t) := (V(t))^{-1} \cdot \int_{D(t)} A\sqrt{h}\, d^3x, \tag{23}$$

where $D(t)$ denotes that part of the leaf $t =$ const. which is intercepted by the chosen subset of the "fluid," and $V(t) = \int_{D(t)} \sqrt{h}\, d^3x$.

Let us now assume that V^α is the 4-velocity of a dust model. Then one can use the decomposition analogous to (12), Raychaudhuri's equation (13) holds as it stands, and it is almost obvious that equations (15)–(18) again hold. If the flow is irrotational, one may use the unique foliation provided by the hypersurfaces orthogonal to V^α; this had been treated in [1] and, for compact orthogonal hypersurfaces, already in [25].

Are the perturbation terms in (17) quantitatively significant for realistic cosmological models? A partial answer has been given recently [22]. These authors used 2nd order, relatively increasing perturbations of a "small," i.e., toroidal, Einstein–de Sitter model with $\Lambda = 0$ and showed that, in that approximation, the right hand side of (17) vanishes, as it does exactly in the analogous Newtonian model. In addition, they averaged the Hamiltonian constraint, obtaining a small perturbation of the corresponding background equation containing the non-Newtonian term $\langle {}^3R \rangle$, the average of the spatial Ricci scalar. They then computed the influence of this term on the age of the universe, using various dark matter models and COBE data, obtaining that this influence is negligible, amounting to a fractional change of less than $2 \cdot 10^{-3}$. Considering that they chose the model closest to a Newtonian one, this result is perhaps not surprising. In view of the smallness of the number which they got, one will perhaps not expect significant deviations for, say, low density, open models or for higher approximations or approximations including gravitational waves, but such cases do not seem to have been investigated. (For the computation of Russ et al. [21] the assumption of spatial periodicity, equivalent to spatial compactness, is not just "technical," but essential.)

13.5 Lagrangian Perturbation Theory

Almost 30 years ago Y. B. Zel'dovich [26] suggested that the Lagrangian version of hydrodynamics may in some respects be superior to the traditionally preferred Eulerian form for studying structure formation in cosmology. The elaboration of this proposal confirmed this expectation, as shown, in particular, in papers of T. Buchert. The following brief account is based on [5], where more references can be found.

We start again with equations (5), setting $p = 0$. Let t_0 be any "initial" time, and denote the integral curve of V^a which passes through X^b at t_0, by $f^a(t, X^b)$. In other words, let the mapping $X^a \mapsto f^a(t, X^b) = x^a$ take the initial positions X^a of the dust particles to their positions x^a at t—all this with respect to a fixed, nonrotating coordinate system (t, x^a). Then, of course, $f^a(t_0, X^b) = X^a = x^a$; at t_0, the Lagrangian coordinates X^a which henceforth serve to label particles, coincide with their Eulerian coordinates. Further, let $J = \det(f^a{}_{,b})$ denote the Jacobian determinant of the above mapping. As before, a comma indicates Eulerian partial differentiation, while a prime indicates Lagrangian partial differentiation. As earlier, the dot signifies the Lagrangian time derivative.

If $\rho[t, X^a] := \rho(t, f^a(t, X^b))$ denotes the density in Lagrangian variables, local mass conservation is equivalent to

$$\rho[t, X^a] = J^{-1}(t, X^a)\rho_0(X^a), \tag{24}$$

where ρ_0 denotes the initial density. Because of this relation, the density need no longer be considered as a dynamical variable to be determined by evolution equations; only $f^a(t, X^b)$ is to be subjected to such equations, and ρ is then found from (24). Thus mass conservation is exactly and trivially taken care of, also if f^a is found from approximations. The remaining equations (5) say that the acceleration $\ddot{x}^a = \ddot{f}^a(t, X^b)$ is curl-free in Eulerian coordinates and that its Eulerian divergence equals $4\pi G\rho - \Lambda$. To express these laws conveniently, take d and δ to denote exterior differentiation and co-differentiation, respectively, on time slices, using the Hodge-star with respect to the Euclidean metric $\delta_{ab}dx^a dx^b$ on these slices. Then the required laws read, with summation over a,

$$d(\ddot{f}^a df^a) = 0, \qquad \delta(\ddot{f}^a df^a) = \Lambda - 4\pi G\rho, \tag{25}$$

where ρ has to be taken from (25). In components, these are four partial differential equations for the three functions $f^a(t, X^b)$.

A little thought shows that the locally isotropic, Friedmann-like solutions are given by

$$f^a(t, X^b) = a(t)X^a, \qquad a(t_0) = 1, \tag{26}$$

with a obeying Friedmann's equation. Finite deviations from these solutions can be obtained by making the *Ansatz*

$$f^a(t, X^b) = a(t)(X^a + P^a(t, X^b)). \tag{27}$$

Inserting this expression into (25) and using vector notation in "Lagrangian space," with $\vec{\nabla}_0$ denoting $\frac{\partial}{\partial X^a}$, leads to equations of the form

$$\begin{aligned} \frac{d}{dt}(\vec{\nabla}_0 \times \vec{P}) &= \vec{F}(\vec{\nabla}_0\vec{P}, \vec{\nabla}_0\dot{\vec{P}}) + a^{-2}(...), \\ \left(\frac{d^2}{dt^2} + 2\frac{\dot{a}}{a}\frac{d}{dt} + 3\frac{\ddot{a}}{a} - \Lambda\right)\left(\vec{\nabla}_0 \cdot \vec{P}\right) &= \vec{G}(\vec{P}, \vec{\nabla}_0\vec{P}, \dot{\vec{P}}, \ddot{\vec{P}}) + a^{-3}(...), \end{aligned} \tag{28}$$

where $\vec{F}$ and $\vec{G}$ do not contain terms linear in $\vec{P}$ and the expressions (...) contain only initial data. Therefore, these equations can be solved by iteration. Put $\vec{P} = 0$ on the right hand side to obtain the general linear approximation. Put that into the right hand side, keep only quadratic terms, and obtain the general 2nd approximation, etc. As shown in [5], this procedure can be carried out to any order. (Convergence is not known.)

I close with the following remarks.

1. If one takes as the 3-space underlying the model the space $\mathbb{R}^3$, the iteration provides, at any stage, $\vec{\nabla}_0 \times \vec{P}$ and $\vec{\nabla}_0 \cdot \vec{P}$, but no more. The general solution at any order therefore allows the addition of an arbitrary harmonic vector field.

Thus, there are infinitely many, quite different iterative solutions having the same initial data, in accordance with [3] and remarks in 2.1.

2. If one takes as the 3-space a torus, one may without loss of generality impose the gauge condition $\langle \vec{P} \rangle = 0$. Then the iteration provides unique solutions for given data, again in accordance with [3].
3. Since in the Lagrangian method one follows the evolution along the particle orbits, the convective term $v^b v^a{}_{,b}$ in the Eulerian description is not neglected compared to $v^a{}_{,t}$; this, besides mass conservation, may be the reason why Lagrangian approximations agree better with numerical computations (based on the full equations) than Eulerian ones.
4. The functions f^a carry information not only about the density and the particle orbits, but also about the physical Euclidean metric, since

$$\delta_{ab} dx^a dx^b = \delta_{ab} f^a{}_{,c} f^b{}_{,d} dX^c dX^d. \tag{29}$$

In GR one cannot expect functions of comoving coordinates to do the corresponding job. The closest analogues to the f^a are presumably orthonormal co-triads $e^{(a)} = e^{(a)}_\beta dX^\beta$ which generalize the differentials df^a. Perhaps this analogy is useful to relate the Newton-Lagrange perturbation theory to some version of the relativistic perturbation theory, and to pinpoint their differences.

REFERENCES

[1] Th. Buchert and J. Ehlers, *Astron. Astrophys.* **320**, 1 (1997).
[2] H. Bondi, *Cosmology*, 2nd ed. (Cambridge: University Press, 1960).
[3] U. Brauer, A. Rendall, and O. Reula, *Class. Quantum Grav.* **11**, 2283 (1994).
[4] E. Cartan, *A. Scient. de l' École Normale Supérieure* **40**, 325 (1923); **41**, 1 (1924).
[5] J. Ehlers and Th. Buchert, *Gen. Rel. Grav.* **29**, 733 (1997)
[6] J. Ehlers, *Grundlagenprobleme der modernen Physik*, eds. J. Nitsch et al, (Mannheim: Bibliogr. Institut).
[7] J. Ehlers, *Classical Mechanics and Relativity: Relationship and Consistency*, ed. G. Ferrarese (Naples: Bibliopolis, 95, 1991).
[8] J. Ehlers, *Class. Quantum Grav.* **14**, A 19 (1997).
[9] G. F. R. Ellis, *General Relativity and Gravitation*, eds. B. Bertotti, F. de Felice, and A. Pascolini (Dordrecht: Reidel, 1983), 215.
[10] K. O. Friedrichs, *Mathem. Annalen* **98**, 566 (1927).
[11] P. Havas, *Rev. Math. Phys.* **36**, 938 (1964).
[12] U. Heilig, *Commun. Math. Phys.* **166**, 457 (1995).
[13] O. Heckmann and E. Schücking, *Z. Astrophysik* **38**, 95 (1955); **40**, 81 (1956).
[14] *Encycl. of Physics* vol. L III, 489–519, S. Flügge (ed.), (Springer, Berlin 1959).
[15] H. P. Künzle, *Gen. Rel. Grav.* **7**, 495 (1976).
[16] D. Layzer, *Astron. J.* **59**, 268 (1954).
[17] W. McCrea and E. Milne, *Quart. J. Math.*, Oxford Ser. 5, 73–80 (1934).
[18] W. McCrea, *Math. Gazette* **39**, No. 330 (1955).
[19] E. A. Milne, *Quart. J. Math.*, Oxford, Ser. 5, 64–72 (1934).
[20] A. Rendall, *Proc. R. Soc. Lond.* A **438**, 341 (1992).
[21] H. Russ, M. H. Soffel, M. Kasai, and G. Börner, *Phys. Rev. D* **56**, in press.

[22] Ch. Rüede and N. Straumann, *Helv. Phys. Acta* **70**, 318 (1997).
[23] A. Trautman, *C. R. Acad. Sci.* Paris, **257**, 617 (1963).
[24] A. Trautman, *Perspectives in Geometry and Relativity*, ed. B. Hoffmann (Bloomington: Indiana Univ. Press), 413 (1966).
[25] P. Yodzis, *Proc. R. I. A.* **74**, Sect. A, 61 (1974).
[26] Ya. B. Zel'dovich, *Astron. Astrophys.* **5**, 84 (1970).

14

Deviation of Geodesics in FLRW Spacetime Geometries

George F.R. Ellis
Henk van Elst

ABSTRACT The geodesic deviation equation ("GDE") provides an elegant tool to investigate the timelike, null and spacelike structure of spacetime geometries. Here we employ the GDE to review these structures within the Friedmann–Lemaître–Robertson–Walker ("FLRW") models, where we assume the sources to be given by a noninteracting mixture of incoherent matter and radiation, and we also take a non-zero cosmological constant into account. For each causal case we present examples of solutions to the GDE and we discuss the interpretation of the related first integrals. The de Sitter spacetime geometry is treated separately.

14.1 Introduction

It has been known for a long while that the geodesic deviation equation ("GDE"), first obtained by J. L. Synge [23, 24], provides a very elegant way of understanding features of curved spaces, and, as pointed out by Pirani [14, 15], gives an invariant way of characterizing the nature of gravitational forces in spacetime. As such, it is a useful tool to use in examining specific exact solutions of the Einstein field equations ("EFE"). Indeed, it may be claimed that the GDE is one of the most important equations in relativity, as this is *how* one measures spacetime curvature.[1] This latter aspect has been discussed in some depth by Szekeres [26].

The GDE determines the second rate of change of the deviation vectors for a congruence of geodesics of arbitrary causal character, i.e., their relative acceleration. Consider the normalized tangent vector field V^a for such a congruence, parametrized by an affine parameter v. Then

$$V^a := \frac{dx^a(v)}{dv}\,, \qquad V_a V^a := \epsilon\,, \qquad 0 = \frac{\delta V^a}{\delta v} = V^b \nabla_b V^a\,, \tag{1}$$

[1] And so is analogous to the Lorentz force law in electrodynamics; cf. Misner, Thorne and Wheeler, Ch.3, Box 3.1 [12].

where $\epsilon = +1,\ 0,\ -1$ if the geodesics are spacelike, null, or timelike, respectively, and we define covariant derivativion *along* the geodesics by $\delta T^{a..}{}_{b..}/\delta v := V^c \nabla_c T^{a..}{}_{b..}$ for any tensor $T^{a..}{}_{b..}$. A deviation vector $\eta^a := dx^a(w)/dw$ for the congruence, which can be thought of as linking pairs of neighboring geodesics in the congruence, commutes[2] with V^a, so

$$\frac{\delta \eta^a}{\delta v} = \eta^b \nabla_b V^a \, . \tag{2}$$

It follows that their scalar product is constant along the geodesics:

$$\frac{\delta(\eta_a V^a)}{\delta v} = 0 \qquad \Leftrightarrow \qquad (\eta_a V^a) = \text{const.} \tag{3}$$

To simplify the relevant equations, we always choose them orthogonal:

$$\eta_a V^a = 0 \, . \tag{4}$$

The general GDE takes the form

$$\frac{\delta^2 \eta^a}{\delta v^2} = - R^a{}_{bcd} \, V^b \, \eta^c \, V^d \, , \tag{5}$$

(see, for example, Synge and Schild [25], Schouten [21], or Wald [29]). The general solution to this second-order differential equation along any geodesic γ will have two arbitrary constants (corresponding to the different congruences of geodesics that might have γ as a member). There is a *first integral* along any geodesic that relates the connecting vectors for two *different* congruences which have one central geodesic curve (with affine parameter v) in common. This is

$$\eta_{1a} \frac{\delta \eta_2{}^a}{\delta v} - \eta_{2a} \frac{\delta \eta_1{}^a}{\delta v} = \text{const.}, \tag{6}$$

and is completely independent of the curvature of the spacetime manifold.

The aim of this paper is to systematically use the GDE to explore the geometry of the standard Friedmann–Lemaître–Robertson–Walker ("FLRW") models of relativistic cosmology (see, for example, [18, 30, 5]), solving the GDE for timelike, null, and spacelike geodesic congruences in these geometries; hence, obtaining the Raychaudhuri equation [16], determining the time evolution of these models [2, 4], the Mattig observational relations [11] underlying the interpretation of cosmological data [19], and determining the nature of their spatial 3-geometry [18, 5]. Also, we identify in each case the first integral for the GDE and comment on its meaning, in the null case leading to the usual reciprocity theorem [4], and in the timelike case obtaining generic solutions of the GDE via this integral. Thus, our purpose is to characterize the major geometrical and physical features of these spacetimes by use of the GDE, hence showing the utility of this equation in obtaining all the essential geometrical and dynamical results of standard cosmology in

[2]The Lie derivative of η^a along the integral curves of V^a is zero; see, for example, Schouten [21].

a unified way. It is a pleasure to dedicate this paper to Engelbert Schücking, who has made a major contribution to obtaining clarity and elegance in understanding many features of relativistic cosmology.

14.1.1 The Cosmological Context

In the cosmological situation we consider, we assume the sources of the gravitational field to be a noninteracting mixture of incoherent matter and radiation, to each of which the phenomenological fluid description applies (see, for example, [2] and [4]). For completeness we also include a cosmological constant Λ.

Notation used is as follows: u^a is the normalized timelike tangent vector field ($u_a u^a = -1$) to the fundamental matter fluid flow, which is geodesic: $0 = u^b \nabla_b u^a := \dot{u}^a$. The integral curves of u^a are parameterized by the proper time t of comoving fundamental observers. We use standard FLRW comoving coordinates:

$$ds^2 = -dt^2 + a^2(t) f_{\mu\nu}(x^\rho)\, dx^\mu\, dx^\nu, \tag{7}$$

$$f_{\mu\nu}\, dx^\mu\, dx^\nu = dr^2 + f^2(r)(d\theta^2 + \sin^2\theta\, d\phi^2),$$

$$u^a = (\partial_t)^a = \delta^a{}_0\,, \tag{8}$$

where $a(t)$ denotes the time-dependent scale factor, and the function[3] $f(r)$ relates to the intrinsic curvature of the spacelike 3-surfaces $\{t = \text{const.}\}$ orthogonal to u^a. By spatial homogeneity and isotropy, the covariant derivative of u^a [2] reduces to

$$\nabla_a u_b = \frac{1}{3}\,\Theta\, h_{ab}\,, \qquad \Theta := \mathrm{D}_a u^a = 3\,\frac{\dot{a}}{a}\,. \tag{9}$$

Here, h_{ab} is the standard orthogonal projection tensor

$$h_{ab} = g_{ab} + u_a u_b \qquad \Rightarrow \qquad h_{\alpha\beta} = g_{\alpha\beta}\,, \tag{10}$$

Θ is the fluid rate of expansion, and the spatial derivative operator (projected orthogonal to u^a on all indices) is denoted by D_a (see [9]). It is a well-known consequence of Eq. (9) that FLRW spacetime geometries have vanishing Weyl curvature (see [2] and [4]),

$$C_{abcd} = 0\,; \tag{11}$$

the fluid matter flow neither generates tidal gravitational fields nor causes propagation of gravitational waves.

14.2 The Riemann Curvature Tensor

In order to determine the explicit form of the GDE (5), we need the Riemann curvature tensor R_{abcd}. Because of Eq. (11), R_{abcd} can be expressed purely in

[3] Determined later by use of the 3-D spatial GDE.

terms of the Ricci curvature tensor R_{ab}, its trace R, and the metric:

$$R_{abcd} = \frac{1}{2}\,(\,R_{ac}g_{bd} - R_{ad}g_{bc} + R_{bd}g_{ac} - R_{bc}g_{ad}\,) - \frac{1}{6}\,R\,(\,g_{ac}g_{bd} - g_{ad}g_{bc}\,)\,. \tag{12}$$

The EFE algebraically determine R_{ab} from the matter tensor T_{ab}:[4]

$$R_{ab} = T_{ab} - \frac{1}{2}\,T\,g_{ab} + \Lambda g_{ab} \quad \Rightarrow \quad R = -\,T + 4\Lambda\,. \tag{13}$$

When the matter takes a "perfect fluid" form:

$$T_{ab} = (\mu + p)u_a u_b + p g_{ab} \quad \Rightarrow \quad T = -\,(\mu - 3p)\,, \tag{14}$$

(μ is the total energy density and p the isotropic pressure), the Ricci tensor expression is

$$R_{ab} = (\mu + p)u_a u_b + \frac{1}{2}\,(\mu - p + 2\Lambda)g_{ab} \quad \Rightarrow \quad R = (\mu - 3p) + 4\Lambda\,. \tag{15}$$

Thus, from Eq. (12), the curvature tensor takes the form

$$R_{abcd} = \frac{1}{3}\,(\mu + \Lambda)(\,g_{ac}g_{bd} - g_{ad}g_{bc}\,) + \frac{1}{2}\,(\mu + p)(\,g_{ac}u_b u_d - g_{ad}u_b u_c + g_{bd}u_a u_c - g_{bc}u_a u_d\,)\,. \tag{16}$$

Then, for any normalized vector field V^a: $V_a V^a = \epsilon$, by a straightforward contraction one obtains from Eq. (16) the source term in the GDE:

$$R_{abcd}\,V^b V^d = \frac{1}{3}\,(\mu + \Lambda)(\,\epsilon g_{ac} - V_a V_c\,) + \frac{1}{2}\,(\mu + p)[\,(V_b u^b)^2 g_{ac} - 2\,(V_b u^b)\,u_{(a}\,V_{c)} + \epsilon\,u_a\,u_c\,]\,. \tag{17}$$

We will also want the GDE in the spacelike 3-surfaces $\{t = \text{const.}\}$ orthogonal to u^a, which are 3-spaces of maximal symmetry. In the FLRW case, the Gauss embedding equation provides the relation

$$^3R_{abcd} = (R_{abcd})_\perp - \frac{1}{9}\,\Theta^2(\,h_{ac}h_{bd} - h_{ad}h_{bc}\,) \tag{18}$$

for the 3-D Riemann curvature. From Eq. (16), which made use of the EFE, one has

$$(R_{abcd})_\perp = \frac{1}{3}\,(\mu + \Lambda)(\,h_{ac}h_{bd} - h_{ad}h_{bc}\,)\,, \tag{19}$$

[4]Geometrized units, characterized by $c = 1 = 8\pi G/c^2$, are used throughout.

so that Eq. (18) becomes

$$^3R_{abcd} = K(t)(h_{ac}h_{bd} - h_{ad}h_{bc})\,, \tag{20}$$

where the spatial curvature scalar $K(t)$ is given by

$$K(t) := \frac{1}{6}\,{}^3R = \frac{1}{3}\left(\mu - \frac{1}{3}\,\Theta^2 + \Lambda\right). \tag{21}$$

This factor will determine the 3-D spatial GDE[5] source term:

$$^3R_{abcd}\,V^b V^d = K\,(h_{ac} - V_a V_c)\,, \tag{22}$$

where $V_a\,V^a = 1$ and $V_a\,u^a = 0$.

14.3 The Geodesics

Before turning to address the GDE, we need to solve for the geodesic curves along which the GDE will be integrated. Now the fundamental 4-velocity $u^a = \delta^a{}_0$ is a geodesic vector field. Any other geodesic can be transformed to have a purely *radial* spatial part by suitable choice of local coordinates (because the FLRW geometry is isotropic about every point). Hence, without loss of generality, radial geodesics are considered, with the origin of the local coordinates $r = 0$ at the starting point $v = 0$, so that in all cases we will have $x^2 = \theta = \text{const.}$, $x^3 = \phi = \text{const.}$ $\Rightarrow 0 = V^2 = V^3$.

It is convenient to decompose a general geodesic tangent vector field V^a into parts parallel and orthogonal to u^a:

$$V^a := Eu^a + Pe^a\,, \tag{23}$$

where $e^a = a^{-1}\,(\partial_r)^a = a^{-1}\delta^a{}_1$, $e_a e^a = 1$, $e_a u^a = 0$, such that

$$V_a V^a = \epsilon = -\,E^2 + P^2\,, \qquad -(V_a u^a) = E\,, \qquad P = (\epsilon + E^2)^{1/2}\,. \tag{24}$$

As e^a spans a radial direction, $P \geq 0$. By spatial homogeneity and isotropy,[6] for a congruence of radial normalized geodesics, starting off isotropically from $r = 0 \Leftrightarrow v = 0$ (so $E\,|_{v=0} = \text{const.}$ for all of them),

$$0 = \mathrm{D}_a E = \mathrm{D}_a P\,. \tag{25}$$

To determine E, note that

$$\begin{aligned} -\,\frac{\delta(V_a u^a)}{\delta v} &= -\,V^b\nabla_b(V_a u^a) \;=\; -\,V^a(\nabla_b u_a)V^b \\ &= -\,\frac{1}{3}\,\Theta h_{ab}V^a V^b \;=\; -\,\frac{1}{3}\,\Theta[\,\epsilon + (V_a u^a)^2\,]\,. \end{aligned} \tag{26}$$

---

[5] See Section 14.4.4 below.

[6] Even though e^a is *not* invariantly defined, the $1+3$ covariant discussion of LRS perfect fluid spacetime geometries given in [7] still applies. As such, e^a is the Fermi-transported (along u^a) unit tangent of a geodesic and shearfree spacelike congruence. Furthermore, in the given context its spatial rotation also vanishes.

Thus, we need to solve

$$\frac{dt}{dv} = V^0 = E = -(V_a u^a)\,, \qquad \frac{1}{(\epsilon + E^2)}\,\frac{dE}{dv} = -\,\frac{1}{a}\,\frac{da}{dt}\,; \tag{27}$$

so

$$\frac{E}{(\epsilon + E^2)}\,\frac{dE}{dv} = -\,\frac{1}{a}\,\frac{da}{dt}\,\frac{dt}{dv}$$
$$\Leftrightarrow \frac{1}{2}\,\frac{d}{dv}\,\ln\left[\frac{(\epsilon + E^2)}{(\epsilon + E_0^2)}\right] = \frac{d}{dv}\,\ln\left[\left(\frac{a}{a_0}\right)\right]^{-1}. \tag{28}$$

Integrating, we obtain

$$\frac{(\epsilon + E^2)}{(\epsilon + E_0^2)} = \left(\frac{a_0}{a}\right)^2. \tag{29}$$

Now solving for E,

$$E^2 = (\epsilon + E_0^2)\,\left(\frac{a_0}{a}\right)^2 - \epsilon\,, \tag{30}$$

which implies

$$\frac{dt}{dv} = V^0 = E(a) = \pm\left[\,(\epsilon + E_0^2)\,\left(\frac{a_0}{a}\right)^2 - \epsilon\,\right]^{1/2}, \tag{31}$$

with a "+" for *future*-directed vectors V^a and a "−" for *past*-directed ones. Also, with Eq. (7),

$$\begin{aligned} V^a g_{ab} V^b &= -(V^0)^2 + a^2(V^1)^2 = \epsilon \\ \Leftrightarrow V^a h_{ab} V^b &= \epsilon + E^2 = a^2(V^1)^2\,, \end{aligned} \tag{32}$$

so

$$\frac{dr}{dv} = V^1 = \frac{P(a)}{a} = \left[\frac{\epsilon + E^2(a)}{a^2}\right]^{1/2}, \tag{33}$$

which, for later reference, can also be cast into the form

$$d\ell := a\,dr = (\epsilon + E_0^2)^{1/2}\left(\frac{a_0}{a}\right)dv\,, \tag{34}$$

the definition coming from Eq. (7). Hence,

$$\frac{dt}{dr} = \frac{dt/dv}{dr/dv} = \pm\frac{E(a)}{[a^{-2}\,(\epsilon + E^2(a))\,]^{1/2}} = \pm\frac{a^2\,E(a)}{a_0\,(\epsilon + E_0^2)^{1/2}}\,, \tag{35}$$

and so

$$\frac{dt}{dr} = \pm\,a(t)\left[\,1 - \epsilon\,\left(\frac{a(t)}{\alpha_0}\right)^2\,\right]^{1/2}, \qquad \alpha_0 := \pm\,a_0\,(\epsilon + E_0^2)^{1/2}\,. \tag{36}$$

14.3.1 Timelike

For timelike vector fields, $\epsilon = -1$. If we have V^a *initially* parallel to u^a, then $E_0^2 = 1$, and so $dt/dv = 1$ and $dr/dv = 0$, confirming that V^a then *remains* parallel to u^a (which is geodesic). Otherwise, for future-directed timelike geodesics V^a that have a *nonzero* initial hyperbolic angle of tilt with u^a (such that $E_0^2 > 1$), the following relations apply:

$$\frac{dt}{dv} = \left[1 + (E_0^2 - 1) \left(\frac{a_0}{a}\right)^2 \right]^{1/2}, \qquad \frac{dr}{dv} = (E_0^2 - 1)^{1/2} \left(\frac{a_0}{a^2}\right). \tag{37}$$

14.3.2 Spacelike

For spacelike vector fields, $\epsilon = +1$. Setting $E_0 = 0$ means starting off *orthogonally*, but these geodesics do *not* remain orthogonal to the flow lines, and so do not remain within the spacelike 3-surfaces $\{t = \text{const.}\}$. Indeed, from Eqs. (26) and (31)

$$\begin{aligned} &-(V_a u^a)\,|_P = 0\,, \qquad \Theta\,|_P > 0 \\ &\Rightarrow -(V_a u^a) = E < 0 \quad \text{nearby} \quad \Rightarrow \frac{dt}{dv} < 0\,, \end{aligned} \tag{38}$$

showing that the *geodesic*, nearby spacelike 3-surfaces bend *down* (into the past) relative to the spacelike 3-surfaces $\{t = \text{const.}\}$. In this case $\alpha_0 = \pm a_0$, and

$$\frac{dt}{dv} = -\left[\left(\frac{a_0}{a}\right)^2 - 1 \right]^{1/2}.$$

So, with $dr/dv = (a_0/a^2)$, we find

$$\frac{dt}{dr} = -a \left[1 - \left(\frac{a}{a_0}\right)^2 \right]^{1/2}. \tag{39}$$

The geodesic 3-surfaces give the best slicing of a spacetime in order to approximate Newtonian theory in a general spacetime—see the discussion by Ehlers [3]—and have been studied in the FLRW context by Rindler [17], Page [13], and Ellis and Matravers [6].

The simplest dynamical case is the spatially flat Einstein–de Sitter model, which has (pressure-free) incoherent matter as a source, and $\Lambda = 0$. Here, the length scale factor takes the functional form $a(t) = a_0\,[\,\frac{3}{2} H_0\, t\,]^{2/3}$, where H_0 is the value of the Hubble parameter $H := (1/a)\,(da/dt)$ at time t_0. Hence, we obtain (note that $t \le t_0$)

$$r(t, t_0) = -\frac{1}{a_0\,(3H_0/2)^{2/3}} \int_{t_0}^{t} \frac{dy}{y^{2/3}\,[\,1 - (3H_0/2)^{4/3}\, y^{4/3}\,]^{1/2}}\,, \tag{40}$$

leading to an elliptic integral which gives the value of r at time t, starting off orthogonally at $r = 0$ and time $t_0 = \frac{2}{3}\, H_0^{-1}$.

14.3.3 Null

In the case of null congruences, $\epsilon = 0$. Then it follows for the *past*-directed case that

$$\frac{dt}{dv} = -\left[(E_0^2)\left(\frac{a_0}{a}\right)^2\right]^{1/2} = -|E_0|\left(\frac{a_0}{a}\right),$$

and, as $dr/dv = |E_0|\,(a_0/a^2)$,

$$\frac{dt}{dr} = -\,a(t)\,. \tag{41}$$

Alternatively, we can use the fact that $\xi_a := a(t)\,u_a$ is a conformal Killing vector field: $\nabla_a\xi_b = \dot{a}(t)\,g_{ab}$. Thus, for any geodesic vector field k^a,

$$k^b\nabla_b(\xi_a k^a) = \dot{a}(t)\,k^a\,g_{ab}\,k^b = \dot{a}(t)\,k_a\,k^a\,, \tag{42}$$

and in the particular case that k^a is null:

$$\begin{aligned} k_a\,k^a = 0 \quad &\Rightarrow \quad \xi_a\,k^a = a(t)\,u_a\,k^a = \text{const.} \\ \Rightarrow (k_a u^a)(t) &= \frac{\text{const.}}{a(t)}\,. \end{aligned} \tag{43}$$

Relating this to the redshift, z, defined by

$$(1+z) := \frac{(k_a u^a)_e}{(k_b u^b)_0} = \frac{E_e}{E_0} = \left(\frac{a_0}{a(t_e)}\right), \tag{44}$$

for past-directed radial null geodesics it follows that

$$k^a = \frac{a_0}{a(t)}\left(-1,\,\frac{1}{a(t)},\,0,\,0\right)^T, \qquad \frac{dt}{dv} = k^0 = E = -\,(1+z)\,, \tag{45}$$

where we have set $E_0 = -\,1$ by choice of the affine parameter v.

14.4 The Geodesic Deviation Equation

14.4.1 The Deviation Vectors

The basic equations relating the geodesic vector V^a and orthogonal deviation vector η^a have been given above; see Eqs. (1)–(5). We now restrict the deviation vector further.

The screen space. When V^a is *not* parallel to u^a, the vector η^a lies in the *screen space* of u^a (i.e., the spacelike 2-surface orthogonal to both, u^a and V^a) iff, additionally to $(\eta_a V^a) = 0$, η^a also lies in the rest 3-space of u^a, i.e., $(\eta_a u^a) = 0$. We can choose this to be true *initially*; will it be maintained along the integral curves of any geodesic vector field V^a? With Eq. (9), we have

$$\frac{\delta(\eta_a u^a)}{\delta v} = \frac{1}{3}\,\Theta h_{ab}\eta^a V^b + u_a\eta^b\nabla_b V^a$$

$$= \frac{2}{3}\,\Theta h_{ab}\eta^{[a}V^{b]} + \eta^b\nabla_b(V_a u^a)\,, \tag{46}$$

so

$$\frac{\delta(\eta_a u^a)}{\delta v} = \eta^b\nabla_b(V_a u^a)\,, \tag{47}$$

which will be zero, if $\eta^b\nabla_b(V_a u^a) = 0$, and this will be true for the congruences we consider (cf. Eq. (25)). Propagation of condition (47) along the integral curves of u^a then confirms its preservation. This can be seen as follows. The fact that V^a and η^a commute, Eq. (2), gives rise to the relation

$$\begin{aligned} 0 = u_a\,[\,& u^c(\nabla_c V^b)(\nabla_b\eta^a) + V^b u^c\nabla_c\nabla_b\eta^a \\ & - u^c(\nabla_c\eta^b)(\nabla_b V^a) - \eta^b u^c\nabla_c\nabla_b V^a\,]\,, \end{aligned} \tag{48}$$

which is used to eliminate the respective terms in the "dot"-derivative of condition (47). Hence, with Eq. (9),

$$[\,V^b\nabla_b(\eta_a u^a) - \eta^b\nabla_b(V_a u^a)\,]^{\cdot} = \frac{2}{3}\,h_{ab}\,[\,\Theta\,\eta^{[a}\,V^{b]}\,]^{\cdot} = 0\,, \tag{49}$$

which vanishes because h_{ab} is symmetric in its indices. So, the consistent solution to these equations is

$$(\eta_a u^a) = 0\,, \qquad (\eta_a V^a) = 0\,, \qquad \mathrm{D}_a(V_b u^b) = 0\,; \tag{50}$$

i.e., η^a starts and remains within the rest 3-spaces of u^a, and it also remains orthogonal to V^a, which has a constant scalar product with u^a in these rest 3-spaces. From now on we will assume these relations hold.

The force term. The "force term" (see, for example, Pirani [14]) for the general GDE (5) for geodesic congruences of either timelike, null, or spacelike causal character, specialized to the FLRW case, can now be evaluated from Eqs. (17) and (50) to yield

$$R^a{}_{bcd}V^b\eta^c V^d = \left[\,\epsilon\,\frac{1}{3}\,(\mu+\Lambda) + \frac{1}{2}\,(\mu+p)E^2\,\right]\eta^a, \tag{51}$$

where, as before, $-(V_a u^a) = E$. Note that this force term is proportional to η^a itself, i.e., according to the GDE (5) only the magnitude η will change along a geodesic, while its spatial orientation will remain fixed.[7] Consequently, the GDE (5) reduces to give just a *single* differential relation for the scalar quantity η. This reflects the spatial isotropy of the Riemann curvature tensor about every point in the present situation; anisotropic effects as induced, e.g., by nonzero electric Weyl curvature, E_{ab}, or shear viscosity, π_{ab}, are not involved.

We deal, now, with three cases: the GDE for a fundamental observer, for past-directed geodesic null congruences, and for other families of geodesics.

[7] Also, Eq. (51) has *no* component proportional to u^a, confirming the consistency of the above screen space analysis.

14.4.2 Geodesic Deviation for a Fundamental Observer

Case 1: $V^a = u^a$ for the *central* geodesic. In this case the affine parameter coincides with the proper time of the central fundamental observer, i.e., $v = t$. From Eq. (51), with $\epsilon = -1$ and $E = 1$,

$$R_{abcd}u^b\eta^c u^d = \frac{1}{6}(\mu + 3p)\eta_a - \frac{1}{3}\Lambda\eta_a\,. \tag{52}$$

Let the deviation vector be $\eta^a = \ell\, e^a$, $e_a\, e^a = 1$, $e_a\, u^a = 0$, such that it connects neighboring flow lines in the radial direction. Then $\delta e^a/\delta t = u^b\nabla_b e^a = 0$ (as there is no shear or vorticity!), i.e., a basis is used which is parallelly propagated along u^a, and Eq. (5) gives

$$\frac{d^2\ell}{dt^2} = -\frac{1}{6}(\mu + 3p)\ell + \frac{1}{3}\Lambda\ell\,, \tag{53}$$

which is the Raychaudhuri equation [16]. However, this equation applies to *both* comoving matter of active gravitational mass density $(\mu + 3p)$, and to test matter that is not comoving. On the basis of this relation, it is clear that for positive active gravitational mass density and nonnegative cosmological constant[8] all families of past- and future-directed timelike geodesics will experience focusing, provided $(\mu + 3p) > 2\Lambda$, and so gives rise to the standard singularity theorems (see, for example, references [16, 2, 8, 4]).

Comoving matter. For comoving matter, $V^a = u^a \Rightarrow |E_0| = 1 \Rightarrow |E| = 1$ for the *whole* family of geodesics. Then, set $\ell = a$ and multiply by da/dt to get

$$0 = \frac{da}{dt}\frac{d^2a}{dt^2} + \frac{1}{6}(\mu + 3p)\,a\,\frac{da}{dt} - \frac{1}{3}\Lambda\,a\,\frac{da}{dt}\,. \tag{54}$$

Using the conservation equation for comoving matter,

$$\frac{d\mu}{dt} = -\frac{3}{a}\frac{da}{dt}(\mu + p) \quad\Rightarrow\quad \frac{d(\mu\, a^2)}{dt} = -(\mu + 3p)\,a\,\frac{da}{dt}\,, \tag{55}$$

one finds the familiar Friedmann equation

$$\left(\frac{da}{dt}\right)^2 - \frac{1}{3}\mu a^2 - \frac{1}{3}\Lambda a^2 = -k\,, \qquad k = \text{const.}\,, \tag{56}$$

giving the usual time evolution of $a(t)$ for a given equation of state. In terms of invariants,

$$\left(\frac{1}{a}\frac{da}{dt}\right)^2 - \frac{1}{3}\mu - \frac{1}{3}\Lambda = -\frac{k}{a^2}\,, \tag{57}$$

which is just the trace of the Gauss equation, Eq. (21), if we identify

$$K = \frac{k}{a^2} \tag{58}$$

[8]If $\Lambda < 0$, for $(\mu + 3p) > 0$ there will be focusing anyway.

as the constant curvature of the spacelike 3-surfaces $\{t = \text{const.}\}$. Hence, we recover the standard dynamical equations for the FLRW models from the GDE. As usual, whenever K is nonzero, by rescaling $a(t)$ by a constant the dimensionless quantity k can be normalized to ± 1, which is then the curvature of the 3-spaces of maximal symmetry with metric $f_{\mu\nu}\, dx^\mu\, dx^\nu$ (cf. Eq. (7)).

If one considers a noninteracting mixture of both incoherent matter and radiation, one has

$$\mu = 3H_0^2\Omega_{m_0}\left(\frac{a_0}{a}\right)^3 + 3H_0^2\Omega_{r_0}\left(\frac{a_0}{a}\right)^4, \qquad p = H_0^2\Omega_{r_0}\left(\frac{a_0}{a}\right)^4. \tag{59}$$

Then, evaluating Eq. (57) at $t = t_0$ shows that

$$H_0^2 - \frac{1}{3}(\mu_{m_0} + \mu_{r_0}) - \frac{1}{3}\Lambda = -\frac{k}{a_0^2} \quad \Leftrightarrow \quad H_0^2(\Omega_{m_0} + \Omega_{r_0} + \Omega_{\Lambda_0} - 1) = K_0\,, \tag{60}$$

where

$$K_0 := \frac{k}{a_0^2}\,, \tag{61}$$

and, as familiar, Ω_{i_0} denotes dimensionless cosmological density parameters $\Omega_i := \mu_i/(3H^2)$ at $t = t_0$; $\Omega_\Lambda := \Lambda/(3H^2)$ defines an analogous quantity for the cosmological constant. Similarly, evaluating the Raychaudhuri equation (53) at $t = t_0$ gives

$$q_0 = -\frac{1}{3H_0^2}\left(\frac{\ddot{a}}{a}\right)\bigg|_{t_0} = \frac{1}{2}(\Omega_{m_0} + 2\Omega_{r_0} - 2\Omega_{\Lambda_0}) \simeq \frac{1}{2}\Omega_{m_0} - \Omega_{\Lambda_0}\,, \tag{62}$$

the t_0 value of the dimensionless cosmological deceleration parameter $q := -(a\, d^2a/dt^2)/(da/dt)^2$. These results will be useful in deriving the observational relations for null data (see section 14.4.3).

Non-comoving matter. For isotropically distributed test matter moving with *other* 4-velocities about the fundamental observers, i.e., $V^a = v^a \Rightarrow |\, E_0 \,| > 1$, except for the *central* curve of the congruence v^a which coincides with u^a, we need to obtain *other* solutions to the GDE for timelike curves, evaluated along this central fundamental world line (where again proper time t is the same as the preferred affine parameter v, and also here the deviation vectors have radial orientation). There are two ways to do this.

One way is to fully specify the matter source in the equations of the previous discussion on the comoving matter case, solve these equations to obtain the source term in the GDE (53), and then solve the GDE to obtain its general solution (with two arbitrary constants). In the case of the de Sitter universe, we have $0 = \mu = p$,

$\Lambda \neq 0$, so Eq. (53) becomes[9]

$$0 = \frac{d^2\ell}{dt^2} - \frac{1}{3}\,\Lambda\,\ell\,, \tag{63}$$

and the solution is

$$\ell(t) = \begin{cases} C_1\,\cosh(\alpha\,t) + C_2\,\sinh(\alpha\,t) & \Lambda > 0, \\ C_1\,\cos(\alpha\,t) + C_2\,\sin(\alpha\,t) & \Lambda < 0, \end{cases} \tag{64}$$

with $\alpha := (\frac{1}{3} \mid \Lambda \mid)^{1/2}$ and C_1 and C_2 integration constants carrying the dimension of $\ell(t)$. This shows the deviation for *arbitrary* (i.e., independent of $\mid E_0 \mid > 1$) timelike geodesics in the de Sitter ($\Lambda > 0$) and anti-de Sitter ($\Lambda < 0$) cases.

When dynamical matter is present, life is more complex. Defining a dimensionless conformal time variable τ by $dt/d\tau := a \Rightarrow d^2t/d\tau^2 = da/d\tau$, for a matter source according to Eq. (59) the Friedmann equation (57) yields

$$\frac{da}{d\tau} = \left[\,\frac{1}{3}\,\Lambda\,a^4 - k\,a^2 + a_0^3\,H_0^2\,\Omega_{m_0}a + a_0^4\,H_0^2\,\Omega_{r_0}\,\right]^{1/2}. \tag{65}$$

This can easily be solved when $\Lambda = 0$, for a given value of the spatial curvature parameter k. It follows that the GDE for timelike congruences, Eq. (53), can be rewritten as

$$\begin{aligned} 0 = {} & \frac{d^2\ell}{d\tau^2} - \frac{1}{a}\,\frac{da}{d\tau}\,\frac{d\ell}{d\tau} \\ & + \frac{1}{2}\,a_0^2\,H_0^2\left[\,\Omega_{m_0}\left(\frac{a_0}{a}\right) + 2\,\Omega_{r_0}\left(\frac{a_0}{a}\right)^2\,\right]\ell - \frac{1}{3}\,\Lambda\,a^2\,\ell\,, \end{aligned} \tag{66}$$

where $a = a(\tau)$, and $da/d\tau$ is determined through Eq. (65). Unfortunately, this linear homogeneous second-order ordinary differential equation is very complicated, except for the de Sitter universe (where $0 = \Omega_{m_0} = \Omega_{r_0}$, $\Lambda \neq 0$), which we already considered.

To provide a simple example with dynamical matter, we fall back onto the Einstein–de Sitter model, where $\Lambda = 0$, $k = 0$, $\Omega_{r_0} = 0 \Rightarrow \Omega_{m_0} = 1$. In dimensionless conformal time, the length scale factor is $a(\tau) = \frac{1}{4}\,a_0^3\,H_0^2\,\tau^2$, and so the solution to Eq. (66), which then reduces to

$$0 = \frac{d^2\ell}{d\tau^2} - \frac{2}{\tau}\,\frac{d\ell}{d\tau} + \frac{2}{\tau^2}\,\ell\,, \tag{67}$$

is given by

$$\ell(\tau) = C_1\,\tau + C_2\,\tau^2\,; \tag{68}$$

again, the integration constants C_1 and C_2 carrying the dimension of $\ell(\tau)$. Fixing initial conditions so as to describe a set of test particles isotropically emanating from the central reference geodesic at $\eta = \eta_0$, one has $C_2 = -\,C_1/\eta_0$.

---

[9]Or, equivalently, $(\mu + p) = 0 \Rightarrow (\mu + 3p) = -\,2\,\mu =$ const., giving an effective cosmological constant.

Another way to obtain solutions to the timelike GDE (53) is to use the first integral which relates *different* solutions to the GDE along a central reference geodesic γ_0 (which is common to both congruences, and on which the affine parameters coincide and are equal to the preferred time coordinate, i.e., $v\,|_{\gamma_0} = t$). Let η_1 relate to the fundamental family of world lines and η_2 to another family. Then $\eta_1 = a(t)$, and as $dt/dv = -(v_a u^a) = E$ takes the value $E = 1$ on the central reference geodesic, $(1/\eta_1)(d\eta_1/dv) = H = \frac{1}{3}\Theta$. Considering parallel (radial) deviation vectors for the two families, we obtain for their magnitudes

$$\eta_1 \frac{d\eta_2}{dt} - \eta_2 \frac{d\eta_1}{dt} = \text{const.} \quad \Rightarrow \quad \frac{d\eta_2}{dt} - \eta_2\, H(t) = \frac{\text{const.}}{a(t)} . \tag{69}$$

In terms of initial data at time $t = t_0$,

$$\left.\frac{d\eta_2}{dt}\right|_{t_0} - \eta_2\,|_{t_0}\, H_0 = \frac{\text{const.}}{a_0} , \tag{70}$$

which leads to

$$\frac{d\eta_2}{dt} - \eta_2 \frac{1}{a(t)} \frac{da(t)}{dt} = \frac{a_0}{a(t)} \left[\left.\frac{d\eta_2}{dt}\right|_{t_0} - \eta_2\,|_{t_0}\, H_0 \right] , \tag{71}$$

and so

$$\frac{d}{dt}\left[\frac{\eta_2}{a(t)}\right] = \frac{a_0}{a^2(t)} \left[\left.\frac{d\eta_2}{dt}\right|_{t_0} - \eta_2\,|_{t_0}\, H_0 \right] . \tag{72}$$

Then, integration yields

$$\eta_2(t) = \eta_2\,|_{t_0} \left(\frac{a(t)}{a_0}\right) + \left[\left.\frac{d\eta_2}{dt}\right|_{t_0} - \eta_2\,|_{t_0}\, H_0 \right] a(t) \int_{t_0}^{t} \frac{a_0}{a^2(y)}\, dy . \tag{73}$$

For the Einstein–de Sitter example, which we referred to before, $a(t) = a_0\,(t/t_0)^{2/3}$ (as $H_0 = \frac{2}{3}\, t_0^{-1}$), and we find

$$\begin{aligned} \eta_2(t) &= \eta_2\,|_{t_0} \left(\frac{t}{t_0}\right)^{2/3} \\ &\quad + 3 \left[\left.\frac{d\eta_2}{dt}\right|_{t_0} - \frac{2}{3}\, \eta_2\,|_{t_0}\, t_0^{-1} \right] t_0^{2/3}\, t^{2/3} \left(t_0^{-1/3} - t^{-1/3}\right) . \end{aligned} \tag{74}$$

Special cases:
A: Suppose $\eta_2 = 0$ at $t = t_0$ (matter flowing out isotropically from the central line at that instant); then

$$\eta_2(t) = 3 \left.\frac{d\eta_2}{dt}\right|_{t_0} t_0^{2/3} t^{2/3} (t_0^{-1/3} - t^{-1/3}) , \tag{75}$$

giving the radial motion of free particles relative to the fundamental observers, that start off by diverging from them. The graph of Eq. (75) was plotted in Fig 14.1.

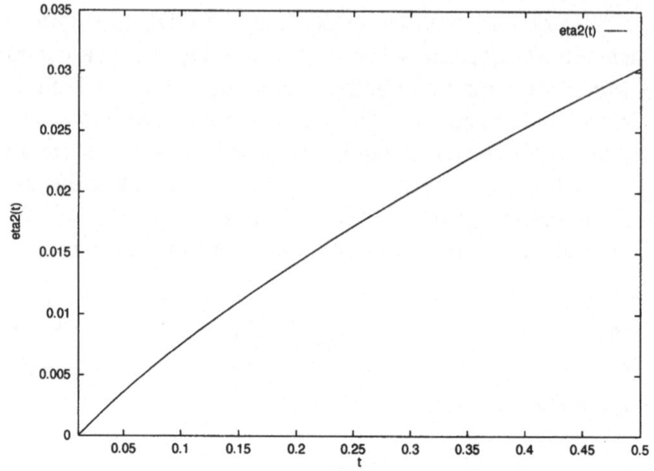

Figure 14.1 Plot of the deviation vector magnitude $\eta_2(t)$ according to Eq. (75). The parameter values chosen are $H_0 = 60$ km/s/Mpc, i.e., $t_0 = 0.01$ (Mpc/km)s, and $d\eta_2/dt\,|_{t_0} = 0.1$.

B: Suppose $d\eta_2/dt = 0$ at $t = t_0$ (matter released from rest at that instant, hence, not comoving with the expanding fundamental matter); then

$$\eta_2(t) = \eta_2\,|_{t_0} \left(\frac{t}{t_0}\right)^{1/3} \left[2 - \left(\frac{t}{t_0}\right)^{1/3}\right] , \tag{76}$$

gives their radial motion relative to the fundamental observers. The graph of Eq. (76) was plotted in Fig. 14.2.

C: Suppose $d\eta_2/dt = \eta_2\,|_{t_0} H_0$ at $t = t_0$ (matter initially comoving with the expanding fundamental matter); then the matter continues to move as the fundamental observers, i.e., $\eta_2(t) = \eta_2\,|_{t_0} (t/t_0)^{2/3}$.

Generically, the first integral (6), applied to this timelike case, relates the *out*-going and *in*-coming geodesics that link two (timelike separated) points O and P, on fixing boundary conditions for the first integral: namely it relates the positions and velocities of each congruence at O to those at P. Apart from the cases just considered, the other one that arises naturally is if particles 1 are at rest at O and coincide at P, whereas for particles 2 the situation is the converse: they are at rest at P but coincide at O. Then

$$\eta_1\,|_O \left.\frac{d\eta_2}{dt}\right|_O = \eta_2\,|_P \left.\frac{d\eta_1}{dt}\right|_P . \tag{77}$$

This relates the positions and velocities at O and P, showing that if both distances are the same (in *absolute*, not comoving terms), then the velocities will be the same.

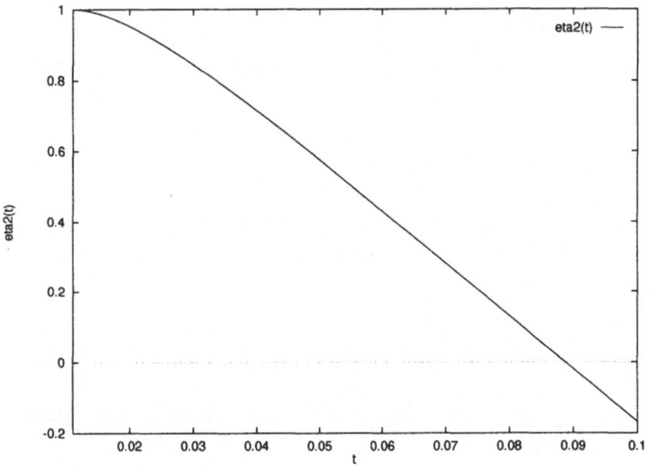

Figure 14.2 Plot of the deviation vector magnitude $\eta_2(t)$ according to Eq. (76). The parameter values chosen are $H_0 = 60\,\text{km/s/Mpc}$, i.e., $t_0 = 0.01\,(\text{Mpc/km})\text{s}$, and $\eta_2\,|_{t_0} = 1$ unit length.

14.4.3 Past Directed Null Vector Fields

Case 2: $V^a = k^a$, $k_a k^a = 0$, $k^0 < 0$. Equation (51) now gives

$$R_{abcd}\, k^b\, \eta^c\, k^d = \frac{1}{2}\,(\mu + p)\, E^2\, \eta_a\;, \tag{78}$$

so writing $\eta^a = \eta\, e^a$, $e_a\, e^a = 1$, $0 = e_a\, u^a = e_a\, k^a$, and using a parallelly propagated and aligned basis, $\delta e^a/\delta v = k^b \nabla_b e^a = 0$, we find from (5),

$$\frac{d^2\eta}{dv^2} = -\frac{1}{2}\,(\mu + p)\, E^2\, \eta\;. \tag{79}$$

Again, in line with the timelike case of Eq. (53), all families of past-directed (and future-directed) null geodesics experience focusing, provided $(\mu + p) > 0$ (while the sign of Λ has no influence). Equation (79) is easily solved in the case of the de Sitter universe, where $(\mu + p) = 0$, and the solution is $\eta(v) = C_1\, v + C_2$, equivalent to the (flat) Minkowski spacetime case. For null rays diverging from the origin, $C_2 = 0$, and we have the same angular size-distance relation as in flat space (provided we measure distance in terms of the affine parameter v).

When dynamical matter is present, we need to express the quantities contained in Eq. (79) in terms of the (*non*-affine parameter) redshift z, defined in Eq. (44). A standard collection of mathematical formulae [1] gives for the derivative operator of Eq. (79) the expression

$$\frac{d^2}{dv^2} = \left(\frac{dv}{dz}\right)^{-2}\left[\frac{d^2}{dz^2} - \left(\frac{dv}{dz}\right)^{-1}\frac{d^2v}{dz^2}\,\frac{d}{dz}\right]. \tag{80}$$

From Eq. (44) we know that

$$(1+z) = \frac{a_0}{a} = \frac{E}{E_0} \quad \Rightarrow \quad \frac{dz}{(1+z)} = -\frac{da}{a} = \frac{dE}{E} \, , \tag{81}$$

hence, (in the past-directed case),

$$dz = (1+z)\,\frac{1}{a}\,\frac{da}{dv}\,dv = (1+z)\,\frac{1}{a}\,\frac{da}{dt}\,E\,dv = E_0\,H\,(1+z)^2\,dv \, , \tag{82}$$

which leads to

$$\frac{dv}{dz} = \frac{1}{E_0\,H\,(1+z)^2} \, . \tag{83}$$

The Hubble parameter is to be determined via the Friedmann equation, Eq. (57), from which one obtains

$$H^2 = \frac{1}{3}\,\mu + \frac{1}{3}\,\Lambda + H_0^2\,(1 - \Omega_0 - \Omega_{\Lambda_0})(1+z)^2 \, . \tag{84}$$

By use of the Raychaudhuri equation, Eq. (53), one finds, furthermore,

$$\frac{d^2v}{dz^2} = -\frac{3}{E_0\,H\,(1+z)^3}\left[\,1 + \frac{1}{18H^2}\,(\mu + 3p) - \frac{1}{9H^2}\,\Lambda\,\right] . \tag{85}$$

So, altogether, the null GDE, Eq. (79), can be expressed in the new form

$$\begin{aligned} 0 = \frac{d^2\eta}{dz^2} &+ \frac{3}{(1+z)}\left[\,1 + \frac{1}{18H^2}\,(\mu + 3p) - \frac{1}{9H^2}\,\Lambda\,\right]\frac{d\eta}{dz} \\ &+ \frac{1}{2(1+z)^2}\,\frac{1}{H^2}\,(\mu + p)\,\eta \, . \end{aligned} \tag{86}$$

If we consider again the noninteracting mixture of incoherent matter and radiation, we have

$$\mu = 3H_0^2\Omega_{m_0}(1+z)^3 + 3H_0^2\Omega_{r_0}(1+z)^4 \, , \qquad p = H_0^2\Omega_{r_0}(1+z)^4 \, . \tag{87}$$

Then, from Eq. (84), for $\Lambda = 0$ the Hubble parameter evaluates to

$$H^2 = H_0^2\,(\,1 + \Omega_{m_0}\,z + \Omega_{r_0}\,z\,(2+z)\,)\,(1+z)^2 \, , \tag{88}$$

and Eq. (86) assumes the form

$$\begin{aligned} 0 = \frac{d^2\eta}{dz^2} &+ \frac{6 + \Omega_{m_0}(1+7z) + 2\Omega_{r_0}(1+8z+4z^2)}{2\,(1+z)\,(\,1 + \Omega_{m_0}z + \Omega_{r_0}z(2+z)\,)}\,\frac{d\eta}{dz} \\ &+ \frac{3\Omega_{m_0} + 4\Omega_{r_0}(1+z)}{2\,(1+z)\,(\,1 + \Omega_{m_0}z + \Omega_{r_0}z(2+z)\,)}\,\eta \, . \end{aligned} \tag{89}$$

When only incoherent matter is present (the dust case), then $\Omega_{r_0} = 0$, while a sole incoherent radiation matter source has $\Omega_{m_0} = 0$. The popular spatially flat FLRW case is contained for $\Omega_0 = \Omega_{m_0} + \Omega_{r_0} = 1$.

The general solution to this linear homogeneous second-order ordinary differential equation is given by

$$\eta(z) = \frac{1}{(1+z)^2}\,[\,C_1\,(2 - \Omega_{m_0} - 2\Omega_{r_0} + \Omega_{m_0} z\,) + C_2\,(\,1 + \Omega_{m_0} z + \Omega_{r_0} z(2+z))^{1/2}\,]\,, \tag{90}$$

which we obtained with support from some computer algebra packages. The integration constants C_1 and C_2 carry the dimension of $\eta(z)$. With this explicit form for the deviation vector of a (past-directed) geodesic null congruence at our hands, we are now in a position to easily infer an expression for the observer area distance, $r_0(z)$, originally derived by Mattig [11] for the dust case ($\Omega_{r_0} = 0$), which is of considerable astronomical importance (see, for example, [19] and [4]). Using $d/d\ell = E_0^{-1}(1+z)^{-1}\,d/dv = H\,(1+z)\,d/dz$ (see Eqs. (34) and (83)) and choosing the integration constants in Eq. (90) such that $\eta(z=0) = 0$, its definition,[10]

$$r_0(z) := \sqrt{\left|\frac{dA_0(z)}{d\Omega_0}\right|} = \left|\frac{\eta(z')\,|_{z'=z}}{d\eta(z')/d\ell\,|_{z'=0}}\right|\,,$$

yields

$$\begin{aligned} r_0(z) = H_0^{-1}\,[\,2\Omega_{m_0} - (\,2 - \Omega_{m_0} - 2\Omega_{r_0}\,)(\,\Omega_{m_0} + 2\Omega_{r_0}\,)\,]^{-1} \\ \times \frac{2}{(1+z)^2}\,[\,(\,2 - \Omega_{m_0} - 2\Omega_{r_0} + \Omega_{m_0} z\,) \\ - (\,2 - \Omega_{m_0} - 2\Omega_{r_0}\,) \\ \times (\,1 + \Omega_{m_0} z + \Omega_{r_0} z(2+z))^{1/2}\,]\,, \end{aligned} \tag{91}$$

giving the observer area distance as a function of the redshift z in units of the present-day Hubble radius H_0^{-1} for an arbitrary noninteracting mixture of matter and radiation (and containing as a special case the Mattig formula when $\Omega_{r_0} = 0$). The graph of Eq. (91) is plotted in Fig. 14.3.

The formula (91) is equivalent to the one stated earlier by Matravers and Aziz [10], but—unlike the usual calculations—is obtained in a uniform way from the null GDE (irrespective of the intrinsic curvature of the spacelike 3-surfaces $\{t = \text{const.}\}$). In the usual approach, three separate calculations are needed (one for each value of k), and it is a matter of some amazement that they all fit the same formula in the end. In the present approach, *one* integration is needed, leading to one formula—a considerable increase in clarity.

The first integral relation can be investigated analogously to the timelike case above. Consider null rays diverging from the observer at O and arriving at the source S, with deviation vector η_1, and null rays diverging from the source S and arriving at the observer O, with deviation vector η_2. The first integral is the same as before, but now we need to convert (for past-directed null rays) from the affine

---

[10] $d\Omega_0$ here denotes an infinitesimal solid angle rather than a change in density parameter.

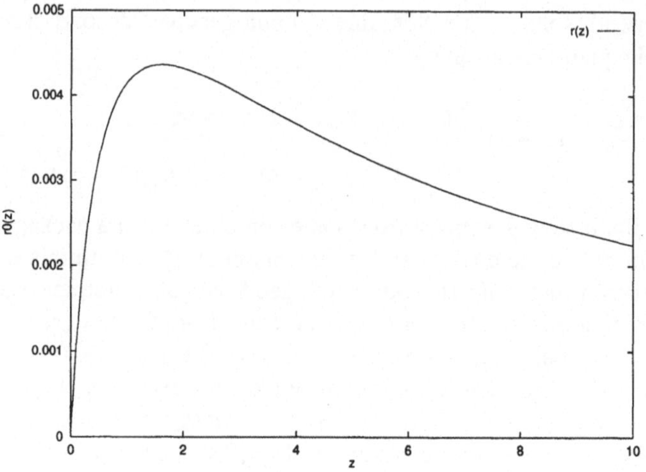

Figure 14.3 Plot of the observer area distance $r_0(z)$ according to Eq. (91), in units of H_0^{-1}. The parameter values chosen are $H_0 = 80\,\text{km/s/Mpc}$, $\Omega_{m_0} = 0.2$ and $\Omega_{r_0} = 0.1$.

parameter v to ℓ according to Eq. (34), with $a_0/a = (1+z)$. One obtains

$$\eta_2\,|_O\,\left.\frac{d\eta_1}{d\ell}\right|_O = \eta_1\,|_S\,\left.\frac{d\eta_2}{d\ell}\right|_S (1+z)\,, \tag{92}$$

where the terms $d\eta/d\ell$ are the angles subtended by the pairs of null rays corresponding to the deviation vectors. Expressed in terms of angular diameter distances, r_O and r_S, defined by

$$\eta_1\,|_S := r_O\,\left.\frac{d\eta_1}{d\ell}\right|_O\,, \qquad \eta_2\,|_O := r_S\,\left.\frac{d\eta_2}{d\ell}\right|_S\,, \tag{93}$$

(which, for FLRW geometry, are the *same* as area distances), we find the familiar null reciprocity theorem for FLRW models [30, 5]:

$$r_S = r_O\,(1+z)\,. \tag{94}$$

This underlies the equivalence (up to redshift factors) of area distance and luminosity distance, and the fact that measured radiation intensity is independent of area distance, depending only on redshift (see [4] for a more detailed discussion). These features are fundamental in analyzing observations of distant sources and measurements of the cosmic microwave background radiation.

14.4.4 Generic Geodesic Vector Fields

Case 3: $V_a V^a = \epsilon$, *not* parallel to u^a, *nor* null. The force term in the generic case is provided by Eq. (51). Writing $\eta^a = \ell e^a$, $e_a e^a = 1$, $0 = e_a V^a = e_a u^a$, and employing a parallelly propagated and aligned basis, $\delta e^a/\delta v = V^b \nabla_b e^a = 0$, we find from Eq. (5),

$$\frac{d^2\ell}{dv^2} = -\epsilon\,\frac{1}{3}\,(\mu+\Lambda)\,\ell - \frac{1}{2}\,(\mu+p)\,E^2\,\ell\,, \tag{95}$$

giving the spatial orthogonal separation of these geodesics within the 2-D screen space as they spread out in spacetime.

Orthogonal spacelike geodesics. A particular case is the spatial geodesics that start off *orthogonal* to u^a (so $E_0 = 0$, which implies that the corresponding geodesics are indeed spacelike), but then bend down towards the past thereafter (see the discussion in section 14.3.2). The above equation applies with $\epsilon = 1$. The simplest case is a de Sitter universe where $0 = \mu = p$, and then the solution for *all* $\mid E_0 \mid \geq 0$, i.e., *all* spacelike geodesics is

$$\ell(v) = \begin{cases} C_1\,\cos(\alpha\,v) + C_2\,\sin(\alpha\,v) & \Lambda > 0, \\ C_1\,\cosh(\alpha\,v) + C_2\,\sinh(\alpha\,v) & \Lambda < 0, \end{cases} \tag{96}$$

with $\alpha := (\frac{1}{3} \mid \Lambda \mid)^{1/2}$ (note this is just the exact converse to the timelike case of Eq. (64) above).

In the case of nonzero dynamical matter, however, μ and p are *not* constants along the initially orthogonal geodesics, as these geodesics do *not* remain within a spacelike 3-surface $\{t = \text{const.}\}$; we have to find $\mu[\,t(v)\,]$ or $\mu[\,t(r)\,]$ from the geodesic equation. However, near the starting point $v = 0$ at t_0 we have (for $\Lambda = 0$)

$$0 = \left.\frac{d^2\ell}{dv^2}\right|_{t_0} + \frac{1}{3}\,\mu_0\,\ell\,, \tag{97}$$

giving the solution near this origin, on carrying out a first-order expansion, by

$$\ell(v) = \ell_0\,\cos(\omega_0\,v) + \left.\frac{d\ell}{dv}\right|_{t_0}\,\sin(\omega_0\,v)\,, \qquad \omega_0 := \left(\frac{1}{3}\,\mu_0\right)^{1/2}. \tag{98}$$

This is always convergent for normal matter, irrespective of the intrinsic curvature of the particular spacelike 3-surface $\{t_0 = \text{const.}\}$ considered. However, as soon as the distance is appreciable, the geodesics will have bent down and lie below the initial 3-surface $\{t_0 = \text{const.}\}$, where the density of matter will be higher and the curvature greater. Thus, the geodesics will tend to converge even more strongly.

Geodesics in the orthogonal spacelike 3-surfaces. This is to be contrasted with geodesic congruences *within* the spacelike 3-surfaces $\{t = \text{const.}\}$ orthogonal to u^a, which are 3-spaces of maximal symmetry. In contrast to Eq. (1), these geodesics satisfy the 3-D equations

$$V^a := \frac{dx^a(v)}{dv}\,, \qquad V_a\,V^a = 1\,, \qquad V_a\,u^a = 0, \qquad 0 = V^b \mathrm{D}_b V^a\,. \tag{99}$$

From Eq. (22), the force term for the resulting 3-D spatial GDE[11] takes the form

$$ {}^3R_{abcd}\, V^b\, \eta^c\, V^d = \frac{1}{3}\left(\mu - \frac{1}{3}\,\Theta^2 + \Lambda\right)\eta_a = K\,\eta^a\,, \tag{100} $$

where $K(t)$ is the curvature of these 3-spaces (see Eq. (21)). Consequently, whether geodesics in these spacelike sections converge or diverge depends on the sign of K. Setting $\eta^a = \eta\, e^a$ where $e_a\, e^a = 1$ and $e_a\, u^a = 0$, as before we choose a congruence of vectors such that $\delta e^a/\delta v = V^b \nabla_b e^a = 0$, and the 3-D spatial GDE[12] becomes

$$ \frac{d^2\eta}{dv^2} = -\,K\,\eta\,. \tag{101} $$

$K = K(t)$ is indeed *constant* along these spatial geodesics (because they lie within the 3-surfaces $\{t = \text{const.}\}$). If $K > 0$, one deals again with the familiar oscillator equation, i.e., two neighboring spatial geodesics will harmonically converge to and diverge from each other as v increases. If $K < 0$, they will exponentially diverge, and if $K = 0$, they diverge linearly.

Focusing on radial spatial geodesics, the local FLRW coordinates of the spacelike 3-surfaces $\{t = \text{const.}\}$ arise as follows. We consider a 3-space with metric $f_{\mu\nu}\, dx^\mu\, dx^\nu$, and constant dimensionless scalar curvature, if nonzero, normalized to $k = \pm 1$, (cf. Eqs. (7) and (58)). Note that the full 3-space metric $h_{\mu\nu}(t)$ at arbitrary time t is just given by $h_{\mu\nu}(t) = a^2(t)\, f_{\mu\nu}$.[13] Choosing an affine parameter $v = r$, $V^a = (\partial_r)^a = \delta^a{}_1$ is the geodesic unit normal to the 2-surfaces $\{r = \text{const.}\}$, which are 2-spheres of area $4\pi\, f^2(r)$. Thus, it is tangent to the orthogonal coordinate curves $x^2 = \text{const.}$, $x^3 = \text{const.}$. A basis of deviation vectors in the 2-D screen space is given by ${\eta_1}^a = \delta^a{}_2$ and ${\eta_2}^a = \delta^a{}_3$ (these commute with the geodesic vector $V^a = \delta^a{}_1$, because each of these is a coordinate basis vector). Employing an *orthonormal* basis with components $(e_1)^a = \delta^a{}_1, (e_2)^a = f^{-1}(r)\,\delta^a{}_2$, $(e_3)^a = f^{-1}(r)\,(\sin\theta)^{-1}\,\delta^a{}_3$, parallelly propagated along the radial geodesics V^a, Eq. (101) yields

$$ 0 = \frac{d^2\eta}{dv^2} + k\,\eta \quad\Rightarrow\quad 0 = \frac{d^2 f}{dr^2} + k\,f\,; \tag{102} $$

the second relation following because relative to the orthonormal basis, ${\eta_1}^a = f(r)\,\delta^a{}_2$ and ${\eta_2}^a = f(r)\,\sin\theta\,\delta^a{}_3$ (apply the first equation to either vector to get the second). Then the solution we want corresponds to *that* solution for which

[11] Determined by Eqs. (2), (4), and (99).

[12] That is, the 3-D version of Eq. (5) that applies in these 3-spaces.

[13] When $a(t)$ is of unit magnitude, say at time $t = \tilde{t}$, then $f_{\mu\nu}$ is equal to the metric $h_{\mu\nu}(\tilde{t})$ on the 3-surface $\{\tilde{t} = \text{const.}\}$, except for a dimensional unit factor, and similarly for k and $K(\tilde{t})$.

$\eta(r = 0) = 0$; we find

$$f(r) = \begin{cases} \sin r & k = +1, \\ r & k = 0, \\ \sinh r & k = -1, \end{cases} \tag{103}$$

showing how the GDE within the spacelike 3-surfaces $\{t = \text{const.}\}$ determines the function $f(r)$ in Eq. (7). The corresponding solutions with $d\eta/dr = 0$ at $r = 0$ exhibit precisely how Euclid's parallel postulate breaks down for these curved 3-space sections, according to the spatial curvature.

In this context it is of interest to remark that the Lorentz-invariant[14] de Sitter spacetime geometry, which is the case $0 = \mu = p$, $\Lambda > 0$, can be sliced by spacelike 3-surfaces $\{t = \text{const.}\}$ of either constant positive, zero, or negative intrinsic curvature (see [22]), depending on the sign of the sum $3\,K = -\frac{1}{3}\,\Theta^2 + \Lambda$ (see Eq. (21)). For anti-de Sitter ($\Lambda < 0$) geometry only the negative curvature case applies. The different FLRW forms of the de Sitter spacetime metric follow from arguments essentially identical to that just given for the 3-space metric, because it is a 4-space of constant curvature, i.e., maximal symmetry (and the argument applies also to the 2-sphere, leading to the form of the terms in the last bracket in Eq. (7)). In each case, the GDE, together with the constant curvature condition (20), leads to the harmonic equation (102).

Similarly to the null case, the 3-D geometrical reciprocity theorem can be stated as

$$\eta_1\,|_O\,\left.\frac{d\eta_2}{dr}\right|_O = \eta_2\,|_P\,\left.\frac{d\eta_1}{dr}\right|_P , \tag{104}$$

showing how geodesics diverging about a central geodesic from P to O at an angle α_0 reach a separation d at O, and geodesics diverging from O at the same angle will reach the same distance apart at P (irrespective of the spatial curvature, which is constant). Corresponding statements hold for the families of geodesics that diverge from P and O, and end up parallel at O and P, respectively.

14.5 Conclusion

One way of solving the EFE is to treat them as *algebraic* equations relating R_{abcd} to R_{ab} and C_{abcd}, and then solve the GDE (which characterizes relative acceleration due to spacetime curvature) to determine both the spacetime geometry and its properties. In the case of a FLRW model, this can be carried out explicitly, as shown above: integrating the GDE (Eqs. (53), (86), and (95)) allows complete characterization of all interesting geometrical features of the exact FLRW geometry in an elegant manner—determining the timelike evolution, spacelike geometry, and

[14]That is, u^a is *not* uniquely defined.

null ray properties, which in turn determine the basic observational properties. The Newtonian analogue of some of this has been given by Tipler [27, 28].

An interesting project is to extend this calculation to perturbed FLRW models in order to work out the effects of linear anisotropies on the present results as regards all three causal cases (timelike, spacelike, null). This would allow investigation of both dynamical and observational features of such models, for example examining aspects of gravitational lensing theory [20].

Acknowledgments

This work has been supported by the South African Foundation for Research and Development (FRD). The integration of the GDEs was facilitated by application of the computer algebra packages MAPLE and REDUCE.

REFERENCES

[1] Bronstein, I. N., and K. A. Semendjajew: *Taschenbuch der Mathematik*, 23rd Edn. (Frankfurt/Main: Harri Deutsch, Thun, 1987).

[2] Ehlers, J.: Beiträge zur relativistischen Mechanik kontinuierlicher Medien, *Akad. Wiss. Lit. Mainz, Abhandl. Math.-Nat. Kl.* **11** (1961). Reprinted in *Gen. Rel. Grav.* **25** (1993), 1225.

[3] Ehlers, J.: Survey of General Relativity, in *Relativity, Cosmology, and Astrophysics*, Ed. W Israel (Dordrecht: Reidel, 1973).

[4] Ellis, G. F. R.: Relativistic Cosmology, in *General Relativity and Cosmology*, Proceedings of the XLVII Enrico Fermi Summer School, Ed. R K Sachs (New York: Academic Press, 1971).

[5] Ellis, G. F. R.: Standard Cosmology, in *Vth Brazilian School on Cosmology and Gravitation*, Ed. M Novello (Singapore: World Scientific, 1987).

[6] Ellis, G. F. R., and D. R. Matravers: Spatial Homogeneity and the Size of the Universe, in *A Random Walk in Relativity and Cosmology* (Raychaudhuri Festschrift), Eds. N Dadhich, J K Rao, J V Narlikar, and C V Vishveshswara (New Dehli: Wiley Eastern, 1985).

[7] van Elst, H., and G. F. R. Ellis: The Covariant Approach to LRS Perfect Fluid Spacetime Geometries, *Class. Quantum Grav.* **13** (1996), 1099.

[8] Hawking, S. W., and R. Penrose: The Singularities of Gravitational Collapse and Cosmology, *Proc. R. Soc. London A* **314** (1970), 529.

[9] Maartens, R.: Linearisation Instability of Gravity Waves? *Phys. Rev.* D **55** (1997), 463.

[10] Matravers, D. R., and A. M. Aziz: A Note on the Observer Area-Distance Formula, *Mon. Not. Astr. Soc. S.A.* **47** (1988), 124.

[11] Mattig, W.: Über den Zusammenhang zwischen Rotverschiebung und scheinbarer Helligkeit, *Astr. Nach.* **284** (1958), 109.

[12] Misner, C. W., K. S. Thorne, and J. A. Wheeler: *Gravitation* (San Francisco: Freeman and Co., 1973).

[13] Page, D. N.: How Big is the Universe Today?, *Gen. Rel. Grav.* **15** (1983), 181.

[14] Pirani, F. A. E.: On the Physical Significance of the Riemann Tensor, *Acta Phys. Polon.* **15** (1956), 389.
[15] Pirani, F. A. E.: Invariant Formulation of Gravitational Radiation Theory, *Phys. Rev.* **105** (1957), 1089.
[16] Raychaudhuri, A. K.: Relativistic Cosmology, *Phys. Rev.* **98** (1955), 1123.
[17] Rindler, W.: Public and Private Space Curvature in Robertson–Walker Universes, *Gen. Rel. Grav.* **13** (1981), 457.
[18] Robertson, H. P.: Relativistic Cosmology, *Rev. Mod. Phys.* **5** (1933), 62.
[19] Sandage, A. R.: The Ability of the 200-Inch Telescope to Discriminate between Selected World-Models, *Astrophys. J.* **133** (1961), 355.
[20] Schneider, P., J. Ehlers, and E. E. Falco: *Gravitational Lenses* (New York: Springer, 1992).
[21] Schouten, J. A.: *Ricci Calculus* (Berlin: Springer, 1954).
[22] Schrödinger, E.: *Expanding Universes* (Cambridge: Cambridge University Press, 1956).
[23] Synge, J. L.: On the Deviation of Geodesics and Null Geodesics, Particularly in Relation to the Properties of Spaces of Constant Curvature and Indefinite Line Element, *Ann. Math.* **35** (1934), 705.
[24] Synge, J. L.: *General Relativity* (Amsterdam: North Holland, 1960).
[25] Synge, J. L., and A. Schild: *Tensor Calculus*, (Toronto: University of Toronto Press, 1949). Reprinted: (New York: Dover Publ., 1978).
[26] Szekeres, P.: The Gravitational Compass, *J. Maths. Phys.* **6** (1965), 1387.
[27] Tipler, F. J.: Newtonian Cosmology Revisited, *Mon. Not. Roy. Astr. Soc.* **282** (1996), 206.
[28] Tipler, F. J.: Rigorous Newtonian Cosmology, *Am. J. Phys.* **64** (1996), 1311.
[29] Wald, R. M.: *General Relativity* (Chicago: University of Chicago Press, 1984).
[30] Weinberg, S.: *Gravitation and Cosmology* (New York: John Wiley and Sons, 1972).

15

Poincaré Pseudosymmetries in Asymptotically Flat Spacetimes

Simonetta Frittelli
Ezra T. Newman

ABSTRACT It is the purpose of this note to point out (or perhaps, more accurately, to argue) that for asymptotically flat spacetimes that are sufficiently close to flat space, there are global vector fields and associated global transformations (arising from the existence of the asymptotic symmetries) that can be identified as nonlinear counterparts of the Poincaré transformations. They are however not symmetries in any obvious sense, and hence the reference to them as "pseudosymmetries." They are obtained by rigidly pulling the asymptotic symmetries of null infinity into the interior of the spacetime, via the rigid light-cone structure of the spacetime. In the limiting flat-space case, when the radiation vanishes, these pseudosymmetries become the exact flat-space symmetries. It thus seems reasonable to think of the pseudosymmetries as being *approximate* global symmetries for these weak gravitational fields. Presumably this idea can be extended to the associated concept of *approximate* conservation laws.

Dedicated to Engelbert Schucking, who knows more mathematics and physics than most of us put together.

15.1 Introduction

Herman Bondi [1] in his pioneering work on gravitational radiation in asymptotically flat spacetimes (clarified and extended by R. K. Sachs [2] and Roger Penrose), discovered that even though the spacetime itself had no symmetries, there was an exact symmetry, now known as the Bondi-Metzner-Sachs (BMS) group, that acted as a transformation, not on the full spacetime but instead on the 3-space, future null infinity, known as scri ($\mathcal{I}^+$). Scri is defined by adding to the spacetime the future endpoints of all null geodesics; it forms part of the null boundary of the spacetime. For large classes of spacetimes, the BMS group can be restricted to the Poincaré group by an appropriate choice of gauge [3]. (More specifically, if the free characteristic data, the Bondi shear, has a limit in the distant future, a gauge can be chosen so that the so-called "electric part" of the shear vanishes in the future limit. It is this gauge that we use.) This (asymptotic) Poincaré group, which acts as a symmetry transformation only on $\mathcal{I}^+$, can be thought of, or imagined, as being generated by solutions of some (not even uniquely defined) asymptotic version of Killing's equation in the spacetime (referred to as the "Maiming" equation by A. Komar)—there being, of course, no solutions in general to Killing's equation in the radiative spacetimes. In this weak sense, the BMS group (or the Poincaré group) could be thought of as yielding (poorly defined) approximate symmetries in the "almost" flat region of the spacetime in the neighborhood of $\mathcal{I}^+$. It appeared that there was no means of extending the BMS or the asymptotic Poincaré group globally into the interior of the spacetime.

The purpose of this work is to argue that for asymptotically flat spacetimes that are sufficiently close to flat space, *there are* natural global vector fields and associated global transformations (that arise from the existence of the asymptotic Poincaré group) that can be identified as nonlinear realizations of global spacetime Poincaré transformations. They are however not symmetries in the sense of preserving the metric; hence, we refer to them as *pseudosymmetries*. They are obtained by pulling the asymptotic symmetries into the interior rigidly, via the rigid lightcone structure of the spacetime. In the limiting flat-space case, when the radiation vanishes, these pseudosymmetries become the exact flat-space symmetries. We thus think of these pseudosymmetries as being *approximate* global symmetries for the weak gravitational fields. It appears likely that this idea can be extended to the related concept of *approximate* conservation laws.

In Section 16.2, we first describe how in Minkowski space the standard symmetry action of the Poincaré group on the spacetime extends to its action on the future null boundary $\mathcal{I}^+$. We then show how this process can be inverted; i.e., how the (asymptotic) Poincaré action on $\mathcal{I}^+$ can, via the Minkowski light cones, be pulled back into the spacetime.

In Section 16.3, beginning with an asymptotically flat spacetime and its asymptotic Poincaré action, we generalize the argument of Section 16.2, using the lightcones of the asymptotically flat spacetime, to obtain the Poincaré action on the spacetime—obtaining a nonlinear realization of the Poincaré group. There is no implication here of a spacetime symmetry; nevertheless in the flat-space limit, the action does become the metric symmetries. The arguments used here depend heavily on technical results from the theory of linear (in general, infinite-dimensional reducible) representations of the homogeneous Lorentz group [4, 5].

Section 16.4 is devoted to a discussion of possible meanings to the preceding results as well as a discussion of other Lorentzian structures that arise due to the asymptotic symmetries.

Since the results of Gelfand [4], concerning the infinite-dimensional representations of the Lorentz group, are not widely know, we give a brief summary of them in the Appendix.

15.2 Minkowski Space

We begin with Minkowski space with standard Cartesian coordinates x^a and metric $\eta_{ab} = \mathrm{diag}(1, -1, -1, -1)$ and introduce new coordinates (null polar; a special case of Bondi coordinates) given by

$$x^a \to (u, r, \zeta, \bar{\zeta}), \qquad x^a = ut^a + r\ell^a(\zeta, \bar{\zeta}), \tag{1}$$

where u is a retarded time, r is an affine parameter along null geodesics, ℓ^a is a null vector for all values of $(\zeta, \bar{\zeta})$, which are the complex stereographic coordinates on the sphere of null directions at the spatial origin, and $t^a = \sqrt{2}(1, 0, 0, 0)$ is a timelike vector defining a boost frame (normalized such that $t^a \ell_a = 1$).

Explicitly we have that

$$\ell^a(\zeta, \bar{\zeta}) = \frac{1}{\sqrt{2}(1 + \zeta\bar{\zeta})}\Big(1 + \zeta\bar{\zeta}, \zeta + \bar{\zeta}, i(\bar{\zeta} - \zeta), -1 + \zeta\bar{\zeta}\Big). \tag{2}$$

It is also useful to have

$$\eth\ell^a(\zeta, \bar{\zeta}) \equiv m^a = \frac{1}{\sqrt{2}(1 + \zeta\bar{\zeta})}\Big(0, (1 - \bar{\zeta}^2), -i(1 + \bar{\zeta}^2), 2\bar{\zeta}\Big). \tag{3}$$

and

$$\bar{m}^a = \frac{1}{\sqrt{2}(1 + \zeta\bar{\zeta})}\Big(0, (1 - \zeta^2), i(1 + \zeta^2), 2\zeta\Big). \tag{4}$$

The two complex null vectors m^a and $\bar{m}^a$ defined in this way are orthogonal to and parallelly propagated along ℓ^a, and are normalized such that $m^a\bar{m}_a = -1$. (It is customary to define a fourth real null vector by $n^a \equiv \eth\bar{\eth}\ell^a + \ell^a$ which is orthogonal to ℓ^a and is normalized such that $n^a \ell_a = 1$. Then n^a is given explicitly by $n^a = 1/(\sqrt{2}(1 + \zeta\bar{\zeta}))((1 + \zeta\bar{\zeta}), -(\zeta + \bar{\zeta}), -i(\bar{\zeta} - \zeta), 1 - \zeta\bar{\zeta})$ and the timelike vector t^a can be obtained as $t^a = \ell^a + n^a$.)

From the differential of (1) we have that

$$dx^a = du\, t^a + dr\, \ell^a(\zeta, \bar{\zeta}) + \frac{r}{P} d\zeta\, m^a(\zeta, \bar{\zeta}) + \frac{r}{P} d\bar{\zeta}\, \bar{m}^a(\zeta, \bar{\zeta}), \tag{5}$$

where

$$P = 1 + \zeta\bar{\zeta}, \tag{6}$$

leading to the flat metric in the Bondi coordinates[1]

$$ds^2 = (dx^a)^2 = 2\,du^2 + 2\,du\,dr - 2r^2\,d\zeta\,d\bar{\zeta}/P^2. \tag{7}$$

Null infinity $\mathcal{I}^+$, defined as the future endpoint of all null geodesics, becomes with these coordinates the 3-surface given by $r = \text{constant} \to \infty$ and naturally

[1]The flat metric in Bondi coordinates is actually given by $ds^2 = du_B^2 + 2\,du_B\,dr_B - 4(r_B/P)^2\,d\zeta\,d\bar{\zeta}$. Our retarded time and affine parameter are related to the Bondi coordinates by $u = u_B/\sqrt{2}$ and $r = r_B\sqrt{2}$.

inherits the coordinate system $(u, \zeta, \bar{\zeta})$. As the metric (7) is not defined in this limit, Penrose [6] considered the conformally related metric (rescaled by r^{-2})

$$ds'^2 = r^{-2}\, ds^2 = r^{-2} 2\, du^2 + r^{-2} 2\, du\, dr - 2\, d\zeta\, d\bar{\zeta}/P^2 \tag{8}$$

which at null infinity, with Bondi coordinates $(u, \zeta, \bar{\zeta})$, becomes the degenerate S^2 metric

$$dl^2 = -2\, d\zeta\, d\bar{\zeta}/P^2. \tag{9}$$

We can now ask the following question: What would be the effect on $\mathcal{I}^+$ of a Lorentz or Poincaré transformation in the interior spacetime? More specifically, since, from (1), a Poincaré transformation on the x^a , i.e.,

$$x^a \rightarrow x'^a = \Lambda^a_b(x^b + d^b), \tag{10}$$

induces a transformation on the coordinates $(u, r, \zeta, \bar{\zeta})$; i.e.,

$$(u, r, \zeta, \bar{\zeta}) \rightarrow (u', r', \zeta', \bar{\zeta}'), \tag{11}$$

we want to know what specifically that transformation is when restricted to $\mathcal{I}^+$. (The transformed coordinates $(u', r', \zeta', \bar{\zeta}')$ are defined accordingly with respect to the second frame, namely $x'^a = u'T'^a + r'\ell^a(\zeta', \bar{\zeta}')$ where $T'^a = \sqrt{2}(1, 0, 0, 0)$ and $\ell^a(\zeta', \bar{\zeta}')$ has the same dependence on ζ' as ℓ^a has on ζ.) The answer, which is a well-known standard result (see, for instance, [7]), is remarkably simple: we have that

$$\zeta' = \frac{(a\zeta + b)}{(c\zeta + d)} \tag{12}$$

and

$$u' = K(u + \alpha_T(\zeta, \bar{\zeta})). \tag{13}$$

The homogeneous part of (10), equivalent to the $SL(2C)$ transformation

$$\begin{pmatrix} a & b \\ c & d \end{pmatrix}, \tag{14}$$

with $ad - bc = 1$, is given by the fractional linear transformation (12).

$K = K(a, b, c, d, \zeta, \bar{\zeta})$ is constructed in the following fashion: the new Lorentz frame defines a new timelike vector (i.e., the velocity of the boost) by

$$T^a = \Lambda^a_b(a, b, c, d)t^b \tag{15}$$

which has non-unit scalar product with the null vector ℓ^a. K is given by

$$K \equiv \frac{1}{T^a \ell_a(\zeta, \bar{\zeta})}. \tag{16}$$

Alternatively, K can be defined (via Eq. (12)) by

$$dl'^2 \equiv -\frac{2}{P'^2} d\zeta'\, d\bar{\zeta}' = -K^2 \frac{2}{P^2} d\zeta\, d\bar{\zeta} = -K^2\, dl^2. \tag{17}$$

The inhomogeneous part of (13) contains the Poincaré translations via

$$\alpha_T(\zeta,\bar{\zeta}) = d^a \ell_a(\zeta,\bar{\zeta}). \tag{18}$$

The transformations (12) and (13), with (16) and (18), which act on $\mathcal{I}^+$, are isomorphic to the Poincaré group; they give a realization of the Poincaré group on $\mathcal{I}^+$. (Note that these transformations preserve the conformal metric (9) as well as another object—Penrose's null angle [7]—for which we will have no use here.) These transformations play a fundamental role in later sections.

We have gone from the interior spacetime action of the Poincaré group to its action on $\mathcal{I}^+$; our immediate task is to reverse this and show how, beginning with $\mathcal{I}^+$ and (12)–(18), we can recover the ordinary Poincaré action in the spacetime interior. We must first give some relation between the interior and $\mathcal{I}^+$; in particular we will specify the light cone cuts of $\mathcal{I}^+$ from interior spacetime points x^a. By this we mean the following: Consider a spacetime point x^a and its future light cone, C_x. The light-cone cut is the intersection of C_x with $\mathcal{I}^+$, which is a 2-surface (topologically a sphere) that can be described by a function $u = Z_0(x^a,\zeta,\bar{\zeta})$. After a lengthy but straightforward exercise one can show [8] that Z_0 has the very simple form

$$u = Z_0(x^a,\zeta,\bar{\zeta}) = x^a \ell_a(\zeta,\bar{\zeta}). \tag{19}$$

The interior spacetime points are identified on $\mathcal{I}^+$ by their light-cone cuts; the internal transformation can be obtained from the asymptotic transformations (12)–(18) and (19).

In a new Lorentz frame (19) has the form

$$u' = x'^a \ell_a(\zeta',\bar{\zeta}'), \tag{20}$$

where $(u',\zeta',\bar{\zeta}')$ are given in terms of $(u,\zeta,\bar{\zeta})$ by (12) and (13), i.e., by

$$x'^a \ell_a(\zeta',\bar{\zeta}') = K\big(x^a \ell_a(\zeta,\bar{\zeta}) + \alpha_T(\zeta,\bar{\zeta})\big) \tag{21}$$

and

$$\zeta' = \frac{(a\zeta+b)}{(c\zeta+d)}. \tag{22}$$

By comparing powers of $(\zeta,\bar{\zeta})$ on the left and right sides of (21), one reconstructs that

$$x'^a = \Lambda^a_b(x^b + d^b), \tag{23}$$

the Poincaré transformation acting on the spacetime. It is the rigidity of the null cones from the interior spacetime points that allowed us to pull back uniquely the transformations of $\mathcal{I}^+$ to the interior. We will see in the next section that since the asymptotic symmetries of the asymptotically flat spacetimes are (essentially) the same as those for Minkowski space and the null cones are as rigid, the same idea is applicable. The difficulties, which are what lead to new and interesting results, are that light-cone cuts are far more complicated than those (19) of flat space.

Before beginning the nonflat space discussion, we want to point out a very important technical point that has been used but not emphasized. Equation (21), without the translation part $\alpha_T(\zeta,\bar{\zeta})$, namely,

$$x'^a \ell_a(\zeta',\bar{\zeta}') = K x^a \ell_a(\zeta,\bar{\zeta}) \tag{24}$$

is a special case of a more general class of transformations of functions on the sphere. We consider regular functions $\eta(\zeta,\bar{\zeta})$ that transform under (22) as

$$\eta(\zeta,\bar{\zeta}) \to \eta'(\zeta',\bar{\zeta}') = K(a,b,c,d,\zeta,\bar{\zeta})\eta(\zeta,\bar{\zeta}), \tag{25}$$

namely, objects of spin weight $s = 0$ and conformal weight $w = 1$. These functions can be thought of as infinite-dimensional vectors expandable in terms of the basis of spherical harmonics $Y_{lm}(\zeta,\bar{\zeta})$. The spin and conformal weights label different representations of the Lorentz group [5]. Transformations of the type (24) are within the special class of infinite-dimensional (reducible) representations of the Lorentz group labeled by the spin weight $s = 0$ and conformal weight $w = 1$, which contains an invariant subspace spanned by Y_{00}, Y_{1m}, $m = 0, \pm 1$. (See appendix for more details of the notation.) The transformation (24) is equivalent to the finite-dimensional representation known as the *vector representation;* it is the transformation on the invariant subspace of the more general transformation (25). In the next section, (25) becomes important.

15.3 Asymptotically Flat Spacetimes

We assume that the reader has some familiarity with the theory of asymptotically flat spaces and hence simply give a brief review of some of the relevant ideas. We confine ourselves to asymptotically flat spaces that are sufficiently close to Minkowski space and have a smooth $\mathcal{I}^+$. Again $\mathcal{I}^+$ is a 3-surface, with the same $(u,\zeta,\bar{\zeta})$ and the same conformal metric (9) as for flat space. The asymptotic symmetries (the BMS group) are similar but not the same; explicitly they are

$$\zeta' = \frac{(a\zeta + b)}{(c\zeta + d)} \tag{26}$$

and

$$u' = K(u + \alpha(\zeta,\bar{\zeta})), \tag{27}$$

the difference from the flat case being that $\alpha(\zeta,\bar{\zeta})$ is an arbitrary function of the angles $(\zeta,\bar{\zeta})$, the so-called supertranslations.

In addition, $\mathcal{I}^+$ possesses a further structure, namely the free radiation data which we can choose as the (complex) Bondi shear, written as $\sigma(u,\zeta,\bar{\zeta})$ ($\sigma = 0$ for flat space).

For the physically natural class of data such that the shear has a limit at $u \to \infty$, i.e., such that [7, 3]

$$u \to \infty, \qquad \sigma(u,\zeta,\bar{\zeta}) \to \Sigma(\zeta,\bar{\zeta}) \equiv \eth^2 S(\zeta,\bar{\zeta}), \tag{28}$$

one can use the $\alpha(\zeta, \bar{\zeta})$ of (27) to make the real part of S vanish. The remaining freedom in $\alpha(\zeta, \bar{\zeta})$ is precisely

$$\alpha(\zeta, \bar{\zeta}) = \alpha_T(\zeta, \bar{\zeta}) \tag{29}$$

of (13). In this case, to which we confine ourselves, the BMS group is restricted to the Poincaré group, and we have a situation very similar to the flat case.

We now investigate how the asymptotic symmetries (26), (27), and (29), can be pulled back to the spacetime interior. Again the basic idea is to connect interior spacetime points to $\mathcal{I}^+$ by their lightcone cuts—the intersection of the future light cones from interior points with $\mathcal{I}^+$.

Given a local coordinate chart y^a, we can describe the light-cone cut of the light cone from a point y^a by

$$u = Z(y^a, \zeta, \bar{\zeta}). \tag{30}$$

(Note that due to the formation of caustics on the light cones from the points y^a, $Z(y^a, \zeta, \bar{\zeta})$ will often not be a single-valued function on $\mathcal{I}^+$. The complications arising from this do not pose a serious difficulty. In the interest of simplicity we will avoid discussing this issue here [9].)

For a large class of spacetimes (its size is not yet fully understood but it contains all of the hyperboloidal Cauchy data spaces of Friedrich [10, 11, 12] and probably all of the asymptotically simple spacetimes of Penrose [6]) Eq. (30) can be expressed as an expansion in spherical harmonics,

$$u = Z(y^a, \zeta, \bar{\zeta}) = \sum_{l=0}^{\infty} \sum_{m=-l}^{l} \eta^{lm}(y^a) Y_{lm}(\zeta, \bar{\zeta}). \tag{31}$$

Furthermore, when these spaces are sufficiently close to Minkowski space there exists a canonical class of global coordinate systems, x^a, (pseudo-Minkowskian coordinates) obtained in the following manner: If we use the notation $x^a \longleftrightarrow x^{lm}$ (for $l = 0, 1$) via

$$\begin{aligned} x^0 &= \frac{1}{\sqrt{2\pi}} x^{00}, \\ x^1 &= \sqrt{\frac{3}{4\pi}} (x^{11} - x^{1-1}), \\ x^2 &= i\sqrt{\frac{3}{4\pi}} (x^{11} + x^{1-1}), \\ x^3 &= -\sqrt{\frac{3}{2\pi}} x^{10}, \end{aligned} \tag{32}$$

then

$$x^{lm} = \eta^{lm}(y^a) \quad \text{for } l = 0, 1. \tag{33}$$

It has been shown elsewhere [9] that (33) does define this special class of pseudo-Minkowskian coordinates.

Using these pseudo-Minkowskian coordinates, (31) takes the form

$$u = Z(x^a, \zeta, \bar{\zeta}) = x^a \ell_a(\zeta, \bar{\zeta}) + \sum_{l=2}^{\infty} \sum_{m=-l}^{l} \eta^{lm}(x^a) Y_{lm}(\zeta, \bar{\zeta}). \tag{34}$$

The first four terms in the spherical harmonic decomposition are exactly the same as for the flat space light-cone cut function; the nonflatness is built into the remaining terms, i.e., from $l = 2$ up. As a function on the sphere, Z can be represented as an infinite-dimensional vector with components (x^a, η^{lm}). This break-up of the components of Z into a 4-dimensional subset and the infinite-dimensional remainder acquires meaning from the following discussion, in which we study the transformation of the 2-surface given by $u - Z(x^a, \zeta, \bar{\zeta}) = 0$ under a Poincaré change of coordinates $(u, \zeta, \bar{\zeta}) \rightarrow (u', \zeta', \bar{\zeta}')$ at scri.

Under our asymptotic Poincaré transformation (26), (27) and (29), Z transforms as an $s = 0$, $w = 1$ infinite-dimensional vector (with a translation added on). We thus have that

$$Z'(x'^a, \zeta', \bar{\zeta}') = K(Z(x^a, \zeta, \bar{\zeta}) + \alpha_T(\zeta, \bar{\zeta})), \tag{35}$$

with

$$Z'(x'^a, \zeta', \bar{\zeta}') = x'^a \ell_a(\zeta', \bar{\zeta}') + \sum_{l=2}^{\infty} \sum_{m=-l}^{l} \eta'^{lm}(x'^a) Y_{lm}(\zeta', \bar{\zeta}') \tag{36}$$

and $\zeta' = (a\zeta + b)/(c\zeta + d)$. Therefore, Z belongs to the $s = 0$, $w = 1$ representation space of the Lorentz group, and its first four terms form the invariant subspace (the invariant subspace is characterized by the property that under a Lorentz transformation, it maps into itself). This singles out the difference with flat spacetime: the light-cone cut function of a flat spacetime constitutes a 4-dimensional vector representation of the Lorentz group, whereas in the general asymptotically flat case, the light-cone cut function is an *infinite-dimensional* representation. An essential point of this work is, however, that it is the existence of the invariant subspace which allows asymptotically flat spacetimes to inherit Lorentzian properties.

The next observation is that the rigidity of the conformal structure, embodied by the light-cone cut function Z, can be used to bring into the interior of the spacetime a coordinate transformation associated with the asymptotic Poincaré transformation at scri. By comparing spherical harmonic coefficients (in either variable ζ or ζ'), Eq. (35) leads to the transformation on the invariant subspace (the $l = 0, 1$ harmonics) or equivalently between pseudo-Minkowskian coordinate systems, $x'^a = x'^a(x^b)$, of the form

$$x'^a = \Lambda^a_b(x^b + d^b) + \Lambda^a_{lm}(a, b, c, d)\eta^{lm}(x^a), \qquad l \geq 2 \tag{37}$$

which is a nonlinear realization of the Poincaré group acting on the physical spacetime. Equation (37), along with the existence of the global pseudo-Minkowskian coordinates, is the main result of this work.

For the preceding discussion, leading to Eq. (37), we used the Poincaré transformation simply as a *passive* coordinate transformation: the 2-surface $u = Z(x^a, \zeta, \bar{\zeta})$ is simply described in terms of two different coordinate systems. In this passive sense, Eq. (37) describes the new label x'^a of the point originally described as x^a, but univocally determined as the apex of the light cone which meets scri at $u = Z(x^a, \zeta, \bar{\zeta})$.

Of more interest to us is the alternative view of Eq. (37) in which x'^a is thought of as a new spacetime point. In this view, the Poincaré transformation (37) is considered as an active diffeomorphism of the interior spacetime, and its asymptotic counterpart, (26), (27), and (29), actively transforms scri into itself. From this point of view, the original light-cone cut $u - Z(x^a, \zeta, \bar{\zeta}) = 0$ is mapped into a different cut $u - F(x'^a, \zeta, \bar{\zeta}) = 0$, which in general does not coincide with the light-cone cut of the point x'^a (if it did, the diffeomorphism would be preserving the conformal structure and would represent a conformal isometry). Thus the asymptotic Poincaré transformations (26), (27), and (29) do not send the set of light-cone cuts of a spacetime into itself (see Fig. 15.1).

In the following section we discuss the possible physical significance of the active interpretation. In addition, we point out several other examples of Lorentzian structures, inherited from the asymptotic structure, that also exist in the spacetime interior.

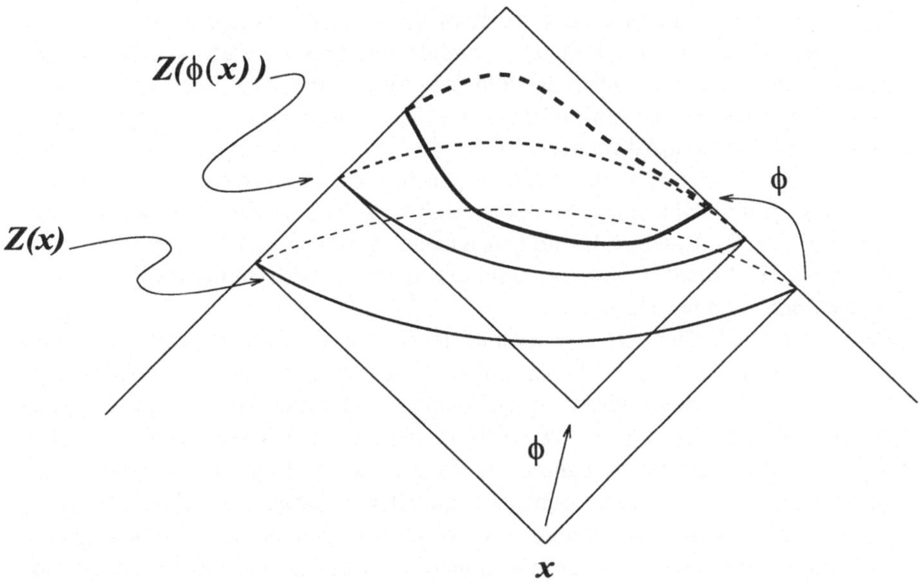

Figure 15.1 The Poincaré transformation considered as an active diffeomorphism consisting of $\phi_M : M \rightarrow M$ and $\phi_{\mathcal{I}} : \mathcal{I}^+ \rightarrow \mathcal{I}^+$ does not map light-cone cuts into light-cone cuts, namely $Z(\phi_M(x)) \neq \phi_{\mathcal{I}}(Z(x))$.

15.4 Discussion

Although, in a certain sense, asymptotically flat spacetimes do become "flat" at null infinity, there is no unique Minkowski spacetime that can be associated with a given asymptotically flat spacetime which approximates the spacetime in the neighborhood of null infinity. Essentially every fixed choice of Bondi coordinates at scri assigns an associated flat metric to the spacetime, in addition to its actual metric. Our interpretation of the pseudo-Minkowskian coordinates x^a is that for every fixed choice of Bondi frame at scri on a given spacetime, there is an associated *flat* spacetime with the same scri, in which spacetime points are labeled by the coordinates x^a.

In the active sense, and for a fixed choice of Bondi frame, the Poincaré transformations (37) generate a 10-parameter set of vector fields ξ_A^a, $A = 1, \ldots, 10$ on the spacetime, the ten parameters being the six contained in the homogeneous part, namely a, b, c, d with $ad - bc = 1$, plus the four translations contained in α_T. The 10 vector fields are obtained essentially by differentiating (37) with respect to each one of the parameters; i.e., from $x'^a = \phi(x^a, \lambda^A)$ one obtains

$$\xi_A^a = \frac{dx'^a}{d\lambda^A} = \frac{d\phi}{d\lambda^A}(x^a, \lambda^A) \quad \text{evaluated at } x^a = \phi^{-1}(x'^a, \lambda^A). \tag{38}$$

In the flat-space case, the ten vector fields obtained in this fashion are the standard Killing symmetries, since they satisfy $\xi_{(a,b)} = 0$. In the generic asymptotically flat case these ten vector fields cannot be expected to be symmetries of the spacetime, but can be regarded as the couterparts of the Killing symmetries of flat space, or *pseudosymmetries*. The Poincaré pseudosymmetries are thus ten global vector fields arising from the action of the Poincaré group on the spacetime which become exact symmetries in the limit of flat space. There are thus 10 pseudosymmetries for each choice of Bondi frame.

Presumably, since by assumption our spacetimes are close to flat space in a global sense (not just in the neighborhood of scri), for each ξ_A^a we can define a geometric object $K_{ab} \equiv \nabla_{(a}\xi_{b)}$ which would give a measure of the lack of a symmetry, since K_{ab} should be "small." Thus one could expect the pseudosymmetries to represent *approximate* symmetries as well.

In view of the fact that conservation laws arise from exact symmetries, and adopting this point of view of approximate symmetries, one could hope to be able to formalize the related idea of *approximate* conservation laws; e.g., divergence laws that, though not vanishing, would be expressed in terms of the small geometric K_{ab}. With this interpretation comes the caveat that the 10 pseudosymmetries are not unique; for each Lorentz frame (or equivalently for each Bondi frame at $\mathcal{I}^+$) there is an associated 10-parameter set of pseudosymmetries, as argued above. Presumably, the two sets of pseudosymmetries associated with two Bondi frames are "approximately" the same—in some sense yet to be made specific—since in the case of flat spacetime all Bondi frames yield *the same* 10 symmetries.

We have argued here that there is a surprising Lorentzian structure of global pseudosymmetries in asymptotically flat spacetimes that arise from the asymptotic symmetries. Though we will not explore it any further here, we want to point out that there are other Lorentzian structures that exist in the spacetime interior that result from the asymptotic symmetries. They are consequences of other (different from the $s = 0$, $w = 1$) representations of the Lorentz group. In particular, the Bondi mass aspect is an $s = 0$, $w = -3$ quantity, and via the intertwining operator (see Appendix) it yields a Lorentzian vector field at each spacetime point that resembles an energy-momentum vector. This structure, as well as others, is being investigated.

Acknowledgments

We thank Roger Penrose, Abhay Ashtekar, Helmut Friedrich, and Bernd Schmidt for many valuable conversations. We are pleased to gratefully acknowledge the hospitality and support of the Erwin Schrödinger Institute in Vienna, where part of this research was carried out. This work was also supported by the NSF grant No. PHY 92-05109.

Appendix: Representations of the Homogeneous Lorentz Group

We give a brief review of some of the ideas associated with certain finite- and infinite-dimensional representations of the homogeneous Lorentz group. These are discussed in full detail in [4, 5], and are further reviewed in [13] and [14].

In general, representations of the Lorentz group are labeled by two numbers, k_0 and c, where k_0 is integer or half integer and c is any complex number [4]. Alternatively, in the case in which the vector spaces associated with them are chosen as sets of functions on the sphere, the representations can also be labeled by the spin weight s and conformal weight w, with s being either integer or half integer and w being complex. However, for the sake of simplicity and because we do not have an immediate interest in the entire set of representations, here we confine ourselves to a particular class of these representations, containing the representations with $s = 0$ and integer $w \neq -1$. The vector space associated with each of these representations has a dual space which also lies in this same class. The representations in this class can thus be organized in dual pairs, the pairs being

$$(w, w') = (-2, 0), (-3, 1), (-4, 2), \ldots, (-n, n-2), \qquad n \geq 2. \tag{39}$$

The vectors, in any one of these $s = 0$ representations, can be expressed as regular functions on the sphere; i.e., by

$$\eta_{(w)}(\zeta, \bar{\zeta}) = \sum_{l=0}^{\infty} \sum_{m=-l}^{l} \eta_{(w)}^{lm} Y_{lm}(\zeta, \bar{\zeta}), \tag{40}$$

with the constants $\eta^{lm}_{(w)}$ being the components of the vector in the $Y_{lm}(\zeta, \bar{\zeta})$ basis. The (w) labels the representation and also describes how the vectors of the representation transform under a Lorentz transformation. A Lorentz transformation of the sphere takes $(\zeta, \bar{\zeta})$ into $(\zeta', \bar{\zeta}')$ in the form of the fractional linear transformation (a Mobius transformation, or (almost) equivalently an $SL(2, C)$ transformation). Specifically we have

$$\zeta' = \frac{a\zeta + b}{c\zeta + d}, \qquad ad - bc = 1 \tag{41}$$

for (a, b, c, d) complex. The vectors $\eta_{(w)}$ transform according to

$$\eta'_{(w)}(\zeta', \bar{\zeta}') = K^{w}\eta_{(w)}(\zeta, \bar{\zeta}), \tag{42}$$

where $\eta'_{(w)}$ is represented as

$$\eta'_{(w)}(\zeta', \bar{\zeta}') = \sum_{l=0}^{\infty}\sum_{m=-l}^{l} \eta'^{lm}_{(w)} Y_{lm}(\zeta', \bar{\zeta}') \tag{43}$$

and K is an explicit object given by

$$\begin{aligned} K &= v^{-1}, \\ v &= v^{I}\ell_{I}(\zeta, \bar{\zeta}) \\ &= \sum_{l=0}^{1}\sum_{m=-l}^{l} v^{lm} Y_{lm}(\zeta, \bar{\zeta}) \\ &= (1 + \zeta\bar{\zeta})^{-1}\Big((a\zeta + b)(\bar{a}\bar{\zeta} + \bar{b}) + (c\zeta + d)(\bar{c}\bar{\zeta} + \bar{d})\Big). \end{aligned} \tag{44}$$

The four components v^{I} represent a Lorentz vector, namely the velocity of the "boost." For a rotation, for example, $v^{I} = \sqrt{2}(1, 0, 0, 0)$ and we have that $K = 1$.

Equations (40)–(44) contain the full description of the $s = 0$ and integer $w \neq -1$ representations. These representations are all infinite-dimensional but not totally reducible; they do contain *invariant subspaces*, namely, subsets which under a Lorentz transformation map into themselves. For the cases of $w \geq 0$, the invariant subspaces are finite-dimensional and yield the finite-dimensional representations; for $w < 0$ the invariant subspaces are all infinite dimensional. Specifically, for fixed $w \geq 0$, the invariant subspace contains the vectors of the form

$$\eta^{inv}_{(w)} = \sum_{l=0}^{w}\sum_{m=-l}^{l} \eta^{lm}_{(w)} Y_{lm}(\zeta, \bar{\zeta}). \tag{45}$$

The invariant subspace of the scalar represntation ($w = 0$) is the set of constant functions on the sphere, since

$$\eta^{inv}_{(0)} = \eta^{00}_{(0)} Y_{00}(\zeta, \bar{\zeta}) = \frac{1}{\sqrt{4\pi}}\eta^{00}_{(0)}. \tag{46}$$

The invariant subspace of the ordinary vector representation ($w = 1$) is the set of functions on the sphere with no spherical harmonics of order 2 nor higher, namely

$$\eta_{(1)}^{inv} = \sum_{l=0}^{1} \sum_{m=-l}^{l} \eta_{(1)}^{lm} Y_{lm}(\zeta, \bar{\zeta}); \tag{47}$$

and the invariant subspace of the symmetric trace-free representation ($w = 2$) is the set of functions with spherical harmonics of order 2 or lower; i.e,

$$\eta_{(2)} = \sum_{l=0}^{2} \sum_{m=-l}^{l} \eta_{(2)}^{lm} Y_{lm}(\zeta, \bar{\zeta}). \tag{48}$$

There are "intertwining" operators [4, 5] that map vectors from one representation to another. We have an interest in the special case where the map is from an infinite-dimensional negative w representation to *the invariant subspace of its dual*; i.e., from $w < -1$ to the finite dimensional representation $w' = -w - 2$; e.g., from $w = -2$ to $w' = 0$ or $w = -3$ to $w' = 1$. Explicitly, the mappings are given by

$$\eta'_{(w')}(\zeta, \bar{\zeta}) = \oint G_{(w)}(\zeta, \bar{\zeta}; \lambda, \bar{\lambda}) \eta_{(w)}(\lambda, \bar{\lambda}) \frac{d\lambda\, d\bar{\lambda}}{(1 + \lambda\bar{\lambda})^2}, \tag{49}$$

where $G_{(w)}(\zeta, \bar{\zeta}; \lambda, \bar{\lambda})$ represents a Green function for every w. For $w = -2$, $G_{(-2)} = 1$, while for $w = -3$, $G_{(-3)} = \ell^a(\lambda)\ell_a(\zeta)$. These intertwining operators allow us to obtain finite-dimensional Lorentz covariant objects (scalars and four-vectors in our example) from infinite-dimensional ones. This idea is exploited in the main text to call attention to the existence of Lorentz structures arising from conformally weighted objects at scri.

References

[1] H. Bondi, M. G. J. van der Burg, and A. W. K. Metzner, *Proc. R. Soc. Lond.* **A269**, 21 (1962).
[2] R. K. Sachs, *Proc. R. Soc. Lond.* **A270**, 103 (1962).
[3] E. T. Newman and R. Penrose, *J. Math. Phys.* **5**, 863 (1966).
[4] I. M. Gelfand, M. I. Graev, and N. Velenkin, *Generalized Functions* (Academic Press, New York, 1966), Vol. V.
[5] E. T. Newman and R. Posadas, *J. Math. Phys.* **11**, 3145 (1970).
[6] R. Penrose and W. Rindler, *Spinors and Space-time* (Cambridge University Press, Cambridge, 1984), Vol. II.
[7] R. Penrose, in *Group Theory in Nonlinear Problems*, edited by A. O. Barut (D. Reidel, Boston, 1974).
[8] S. L. Kent, C. N. Kozameh, and E. T. Newman, *J. Math. Phys.* **26**, 300 (1985).
[9] S. Frittelli and E. T. Newman, *Phys. Rev. D* **55**, 1971 (1997).
[10] H. Friedrich, *Comm. Math. Phys.* **107**, 587 (1986).
[11] L. Andersson, P. T. Chruściel, and H. Friedrich, *Comm. Math. Phys.* **149**, 587 (1992).
[12] L. Andersson and P. T. Chruściel, *Comm. Math. Phys.* **161**, 533 (1994).
[13] R. W. Lind, J. Messmer, and E. T. Newman, *J. Math. Phys.* **13**, 1879 (1972).
[14] S. Frittelli, C. N. Kozameh, and E. T. Newman, "On the dynamics of lightcone cuts of null infinity", preprint.

16

Taub Numbers and Asymptotic Invariants

Edward N. Glass

ABSTRACT This paper reviews Taub numbers and their role as asymptotic invariants. For asymptotically flat spacetimes, they are studied on backgrounds with Killing symmetries. For each independent Killing vector of the background, Taub showed that there is a conservation law for the associated gravitational perturbations. Taub's results are obtained from functional derivatives of the Hilbert action on a function space of Lorentz metrics.

This article is dedicated to Engelbert Schucking on the occasion of his 70th year.

16.1 Introduction

In an early paper about asymptotic invariants and radiation fields [1], Goldberg discussed the fundamental structure of tensorial integrals over 2-surfaces which count the mass-energy contained inside the 2-surfaces. One can find an anticipation of the Penrose-Goldberg (PG) superpotential [2], [3] in his discussion of the Riemann tensor. The final part of the paper calculated the Bondi mass as a two-surface integral over a spherical cut of $\mathcal{I}^+$ using the Einstein pseudotensor and an associated superpotential, and a transformation from asymptotically rectangular coordinates to Bondi coordinates. Now, thirty-five years later, we have a number of tensorial methods for computing asymptotic invariants but no single one which fits all gravitating systems and derives from a variational principle.

In stationary asymptotically flat spacetimes, the symmetry generated by the timelike Killing vector k_t has an associated conserved integral for the mass of the source

$$M = \int_\sigma (-g)^{1/2} T^\alpha_\beta k_t^\beta \, dS_\alpha,$$

where dS_α is the 3-volume tensor density of the spacelike or null hypersurface σ. One of the goals of gravitational theory has been to provide a tensorial link to integrals over the 2-surfaces cutting σ. The first truly tensorial structure which provided a 2-surface integral was the Komar [4] superpotential, yielding the Komar mass

$$M_{(K)} = -\frac{1}{16\pi} \oint_{\partial\sigma} U^{\alpha\beta}_{(K)}(k_t)\, dS_{\alpha\beta}, \tag{1}$$

where $U^{\alpha\beta}_{(K)} = (-g)^{1/2} 2\nabla^{[\alpha} k^{\beta]}$. Since $\nabla_\beta U^{\alpha\beta}_{(K)} = -2(-g)^{1/2} R^{\alpha}{}_{\beta} k^{\beta}$, Komar's superpotential is associated with the Riemann and Ricci tensors rather than the Einstein tensor, and that difference gives rise to the well known "factor of 2" mass anomaly [5]. Systems with axial symmetry have a similar integral for the axial component of angular momentum

$$J_{(K)} = -\frac{1}{16\pi} \oint_{\partial\sigma} U^{\alpha\beta}_{(K)}(k_\varphi)\, dS_{\alpha\beta}. \tag{2}$$

A more recently developed superpotential, the PG superpotential, is

$$U^{\alpha\beta}_{(PG)} = (-g)^{1/2} \frac{1}{2} G^{\alpha\beta}{}_{\mu\nu} Q^{\mu\nu}, \tag{3}$$

where $G^{\alpha\beta}{}_{\mu\nu}$ is the negative right and left dual of the Riemann tensor and antisymmetric $Q^{\mu\nu}$ is a Killing potential. In order for

$$\nabla_\beta U^{\alpha\beta}_{(PG)} = (-g)^{1/2} G^{\alpha\nu} k_\nu \tag{4}$$

it is necessary that the Killing potential, with $3k^\mu = \nabla_\nu Q^{\mu\nu}$, satisfy Penrose's equation [2]

$$\nabla^{(\alpha} Q^{\mu)\nu} - \nabla^{(\alpha} Q^{\nu)\mu} + g^{\alpha[\mu} Q^{\nu]\beta}{}_{;\beta} = 0, \tag{5}$$

whose spinor form is $\nabla_{A'}{}^{(A} W^{BC)} = 0$ for symmetric spinor W^{BC}. When $Q^{\mu\nu}$ exists (global existence seems to be restricted to Petrov type D systems), isolated systems can have their mass and angular momentum computed by integrating the PG superpotential over a 2-surface containing the source. The PG mass is

$$M_{(PG)} = -\frac{1}{8\pi} \oint_{\partial\sigma} U^{\alpha\beta}_{(PG)}\, dS_{\alpha\beta}. \tag{6}$$

Global Killing potentials exist for the timelike Killing vectors of the Schwarzschild and Kerr solutions [5], and so for both solutions one can obtain the quasi-local value $M_{(PG)} = m$, however a global axial Killing potential does not exist [6].

Working at $\mathcal{I}^+$ in a particular Bondi frame, Goldberg [3] solved Eq. (5) and obtained the $Q^{\mu\nu}$ for the asymptotic Killing vectors of the BMS group. $U^{\alpha\beta}_{(PG)}$ can be used with Goldberg's solutions to obtain asymptotic invariants, and since the PG superpotential is associated with the Einstein tensor by Eq. (4) there is no "factor of 2" anomaly.

For asymptotically flat spacetimes defined by the properties of null infinity, another type of conserved integral, arising from a generalization of Green's identity, has been shown [7] to yield asymptotic invariants which are the Newman-Penrose constants [8][9]. The generalization of Green's identity is derived from the field equations and the linearized field equations for a metric perturbation.

Additional asymptotic invariants are obtained from a modification of the Komar superpotential, the Winicour-Tamburino [10] superpotential (whose integral over a 2-sphere at $\mathcal{I}^+$ is called a “inkage”), which uses the generators of the BMS group of asymptotic symmetries and yields the Bondi mass [11] when integrated over a cross section of null infinity. Unfortunately, the Winicour-Tamburino linkages do not derive from a variational principle.

This work reviews the Taub method of computing asymptotic invariants. The method is tensorial and follows from a variational principle. One obtains Taub numbers by integrating the Taub superpotential over a 2-surface containing the sources. Asymptotic values are found by taking the 2-surface out to $\mathcal{I}^+$ if the 2-surface foliates an outgoing null 3-surface, or to $\mathcal{I}^0$ if the 3-surface is spacelike. Our focus will be restricted to future null infinity, since that is where the flux due to radiative motions of the sources ends up. The main theory of Taub numbers has been developed in [12],[13],[14], which will be referred to in the following as Taub I, Taub II, and Taub III respectively. The theory has been extended to include Einstein-Maxwell spacetimes [15].

Each of the Taub numbers in the sequence $\tau_1, \tau_2, \ldots, \tau_n$ derives from a variational derivative of $\int \sqrt{-g}\, g^{\alpha\beta} R_{\alpha\beta}\, d^4x$, the nth number from the $n+1^{th}$ variation. If the system is radiating, we find that the Bondi mass at $\mathcal{I}^+$ (with respect to a flat background) is the sum of τ_1 and τ_2, where τ_1 gives the curvature part of the mass and τ_2 contains the news function.

In this review, Greek indices range over $0, 1, 2, 3$ and uppercase Latin indices range over $2, 3$. Our sign conventions are $2A_{\nu;[\alpha\beta]} = A_\mu R^\mu{}_{\nu\alpha\beta}$, and $R_{\mu\nu} = R^\alpha{}_{\mu\nu\alpha}$. We use $\eth$ to symbolize the differential operator edth acting on 2-spheres in Minkowski space.

16.2 Taub Numbers and Superpotential

We examine curves of solutions in a function space of Lorentzian metrics. A Taub number τ_n is defined with respect to tensors $h^{(n)}_{\mu\nu}$ on a curve of asymptotically flat Einstein solutions $\hat{g}_{\mu\nu}$ where

$$\begin{aligned}\hat{g}_{\mu\nu}(\lambda) &= g_{\mu\nu} + h_{\mu\nu}(\lambda), \\ h_{\mu\nu}(\lambda) &= \lambda h^{(1)}_{\mu\nu} + \lambda^2 h^{(2)}_{\mu\nu} + \cdots,\end{aligned} \tag{7}$$

and a linearized Einstein operator $D_g G_{\alpha\beta} \cdot h^{(n)}$ (a directional derivative on the space of Lorentz metrics in the direction $h^{(n)}_{\mu\nu}$ evaluated at $g_{\mu\nu}$). The $h^{(n)}_{\mu\nu}$ are given by

$$h^{(n)}_{\mu\nu} := \frac{1}{n!}\left[\frac{d^n \hat{g}_{\mu\nu}(\lambda)}{d\lambda^n}\right]_{\lambda=0}.$$

Taub's theorem [16][17], namely $\nabla^\alpha(D_g G_{\alpha\beta} \cdot h^{(n)}) = 0$ for any symmetric $h^{(n)}_{\mu\nu}$, must hold in order for all Taub numbers to be well defined. Taub's theorem is true when $G_{\alpha\beta}(g) = 0$. All curves $\hat{g}_{\mu\nu}(\lambda)$ pass through the background $g_{\mu\nu}$ and the background metric is required to satisfy $G_{\alpha\beta}(g) = 0$. A sequence of field equations is determined by the coefficients of $G_{\alpha\beta}(\hat{g})$ expanded as a series in λ along the curve of solutions $\hat{g}_{\mu\nu}(\lambda)$.

$$G_{\alpha\beta}(g) = 0, \tag{8}$$

$$D_g G_{\alpha\beta} \cdot h^{(1)} = 0, \tag{9}$$

$$D_g^2 G_{\alpha\beta} \cdot (h^{(1)}, h^{(1)}) + D_g G_{\alpha\beta} \cdot h^{(2)} = 0, \tag{10}$$

etc.

The first equation of the sequence is the vacuum Einstein equation for the background metric. The next equation is the linearized Einstein equation for $h^{(1)}_{\mu\nu}$, and the subsequent equations each determine an $h^{(n)}_{\mu\nu}$. Since we consider solutions of operator equations on noncompact Cauchy surfaces, we must find the Cauchy sequences which approach those solutions in "weighted" Sobolev spaces [18].

Taub numbers result from integrating a vector density $t^{\alpha}_{(n)} = (-g)^{1/2}(D_g G^{\alpha}{}_{\beta} \cdot h^{(n)})k^{\beta}$, conserved by virtue of Taub's theorem and Killing's equation, over a 3-surface.

$$\tau_n := \int_{\Sigma\to\mathcal{N}} t^{\alpha}_{(n)}\, dS_{\alpha}. \tag{11}$$

$t^{\alpha}_{(n)}$ is integrated over a 4-dimensional region, D, given in Taub I. D is bounded by two 3-surfaces Σ_1 and Σ_2, which meet in the same S^2 cut of $\mathcal{I}^+$. Σ_1 is a null surface in the vacuum region (becoming smoothly spacelike in the interior source region), and Σ_2 lies to the future of Σ_1; it is spacelike in the source and vacuum regions and becomes null asymptotically where both Σ_1 and $\Sigma_2 \to \mathcal{N}$ and intersect $\mathcal{I}^+$ in the same cut.

Taub numbers are invariant with respect to infinitesimal diffeomorphisms [19]. If $h_{\mu\nu} = \mathcal{L}_y(g_{\mu\nu})$ for vector field y^{α}, then $\tau_n(h_{\mu\nu}) = 0$.

A superpotential $U^{\alpha\beta}_{Taub}(h^{(n)})$ for all $\tau_n, n \geq 1$, has been found. The superpotential has the same functional form for all the $h^{(n)}$. When $h_{\mu\nu}$ is known, one can compute the entire sum of Taub numbers by using $h_{\mu\nu}$ in the superpotential:

$$U^{\alpha\beta}_{\text{Taub}} = (-g)^{1/2}\left(k^{[\alpha}h^{\beta]}{}_{\mu}{}^{;\mu} - k^{[\alpha}h^{;\beta]} + \frac{1}{2}hk^{[\alpha;\beta]} + k^{\mu}h_{\mu}{}^{[\alpha;\beta]} + k^{\mu;[\alpha}h^{\beta]}{}_{\mu}\right),$$

where

$$\nabla_{\beta} U^{\alpha\beta}_{\text{Taub}} = (-g)^{1/2}(D_g G^{\alpha}{}_{\beta} \cdot h)k^{\beta}. \tag{12}$$

For Einstein solutions which admit Kerr-Schild form, such as the Kerr metric given below in Eq. (20), $\hat{g}^{\text{Kerr}}_{\mu\nu} = \eta_{\mu\nu} - (2m/r_a)N_{\mu}N_{\nu}$, one knows $h_{\mu\nu}$ since $(2m/r_a)N_{\mu}N_{\nu}$ is a solution of the linearized Einstein equations. The sequence of solutions $h^{(n)}_{\mu\nu}$ terminates with $h^{(1)}_{\mu\nu} = h_{\mu\nu}$.

Any point on the curve of solutions can be chosen as a background metric provided it has a Killing symmetry for the physical quantity one wishes to compute. The superpotential is a function of the backgound metric, a Killing vector k^{μ} on the background (*timelike* for mass), and $h_{\mu\nu}$. The covariant derivatives in $U^{\alpha\beta}_{\text{Taub}}$ are with respect to the background $g_{\mu\nu}$ and $h := g^{\mu\nu}h_{\mu\nu}$. The sum of all Taub numbers is

$$\tau_\Sigma = \sum_{n=1}^{\infty} \tau_n. \tag{13}$$

Global Taub numbers (the asymptotic invariants reviewed here) are evaluated on an S^2 cut of $\mathcal{I}^+$ as one goes out on a null surface $\mathcal{N}$ to future null infinity. The global Taub mass for k_t and axial angular momentum component for k_φ are

$$\tau_\Sigma(k, h) = -\frac{1}{8\pi} \oint_{\partial\mathcal{N}} U^{\alpha\beta}_{\text{Taub}}(k, h)\, dS_{\alpha\beta}. \tag{14}$$

Calculation is done in the coordinate system of the background metric. The mass of solution $\hat{g}$ is

$$\text{Mass}\,(\hat{g}) = M_0 + \tau_\Sigma(k_t, h), \tag{15}$$

and the angular momentum is

$$J(\hat{g}) = J_0 + \tau_\Sigma(k_\varphi, h). \tag{16}$$

Background $\eta_{\mu\nu}$ is flat with $M_0 = J_0 = 0$. This is an obvious background for Kerr-Schild solutions and is also the background used with the Bondi-Sachs metric to find the Bondi mass. The vacuum Schwarzschild solution can also be an appropriate background with $g^{(Sch)}_{\mu\nu}$ having background mass $M_0 = m$, the Schwarzschild mass.

There are an infinite number of solution curves that go between the background solution $g_{\mu\nu}$ at $\lambda = 0$ and the asymptotically flat solution $\hat{g}_{\mu\nu}$ at $\lambda = 1$. One can select a family of curves (a one-jet) by requiring the curve tangent, $[\frac{d\hat{g}}{d\lambda}]_{\lambda=0}$, to be the linearized solution which gives the monopole moment of the source. (Asymptotically flat systems are linearization stable [20].) Janis and Newman [21] have defined the multipole structure of gravitational sources in terms of initial data for asymptotic solutions of the Newman-Penrose equations. We can use their data for the monopole to fix $h^{(1)}$ when $\eta_{\mu\nu}$ is the background:

$$\Psi_0 = \Psi_1 = \Psi_3 = \Psi_4 = 0, \qquad \sigma^0 = 0,$$
$$\Psi_2 = a_0/r^3, \qquad a_0 \text{ real, const.}$$

Of course a unique curve will be known only when all the $h^{(n)}$ and their respective initial data are given.

16.3 Null Infinity

A spacetime $(M, g_{\mu\nu})$ is asymptotically simple if there exists a space $(\tilde{M}, \tilde{g}_{\mu\nu})$, $\tilde{M} = M \cup \mathcal{I}$, where M is imbedded in $\tilde{M}$ with boundary $\mathcal{I}$ such that

1. There is a smooth scalar field Ω on $\tilde{M}$ with conformal map $\tilde{g}_{\mu\nu} = \Omega^2 g_{\mu\nu}$;

2. On boundary $\mathcal{I}$, $\Omega = 0$, $\tilde{n}_\mu = \tilde{\nabla}_\mu \Omega \neq 0$, and $\tilde{n}_\mu$ is null;
3. Every maximally extended null geodesic in M has two endpoints on $\mathcal{I}$.

The boundary $\mathcal{I}$ consists of two disjoint parts, $\mathcal{I}^+$ (future null infinity) and $\mathcal{I}^-$ (past null infinity).

Not all perturbations preserve $\mathcal{I}^+$. Here we state the Geroch-Xanthopoulos [22] result concerning perturbations and null infinity.

Theorem: *In an asymptotically simple spacetime, when $g_{\mu\nu}$ solves $G_{\mu\nu} = 0$ and $h^{(1)}_{\mu\nu}$ solves $G^{(1)}_{\mu\nu} = 0$, asymptotic simplicity is preserved and $h^{(1)}_{\mu\nu}$ satisfies three conditions on $\mathcal{I}$ (here we only need $\mathcal{I}^+$). For $\tilde{h}^{(1)}_{\mu\nu} = \Omega^2 h^{(1)}_{\mu\nu}$ smoothly extending from M to $\tilde{M}$, on $\mathcal{I}^+$:*

1. $\tilde{h}^{(1)}_{\mu\nu}|_{\mathcal{I}^+} = 0$;
2. $\Omega^{-1}\tilde{h}^{(1)}_{\mu\nu}\tilde{n}^\nu|_{\mathcal{I}^+} = 0$;
3. $\Omega^{-2}\tilde{h}^{(1)}_{\mu\nu}\tilde{n}^\mu\tilde{n}^\nu|_{\mathcal{I}^+} = 0$.

16.4 Kerr–Schild Solutions

The Schwarzschild metric in outgoing Minkowski null coordinates (u, r, θ, φ) is given by

$$ds^2 = (1 - 2m/r)\,du^2 + 2\,du\,dr - r^2(d\theta^2 + \sin^2\theta\, d\varphi^2). \tag{17}$$

With null geodesics labeled by $l_\mu dx^\mu = du = dt - dr$, null spherical coordinates admit the Kerr-Schild form

$$\hat{g}^{\mathrm{Sch}}_{\mu\nu} = \eta_{\mu\nu} - (2m/r) l_\mu l_\nu. \tag{18}$$

The Kerr solution in Minkowski (t, x, y, z) coordinates is

$$\hat{g}^{\mathrm{Kerr}}_{\mu\nu} dx^\mu\, dx^\nu = dt^2 - dx^2 - dy^2 - dz^2 - \frac{2mr_a^3}{r_a^4 + a^2z^2}\, dw^2, \tag{19}$$

$$dw = dt - \frac{r_a(x\,dx + y\,dy) + a(y\,dx - x\,dy)}{r_a^2 + a^2} - \frac{z\,dz}{r_a},$$

where $r_a^4 - (x^2 + y^2 + z^2 - a^2)r_a^2 - a^2z^2 = 0$ determines r_a in terms of x, y, z up to a sign. With $z = r\cos\theta$ and $r^2 = x^2 + y^2 + z^2$,

$$r_a = r - a^2\sin^2\theta/(2r) + O(1/r^3).$$

In standard Kerr-Schild form

$$\hat{g}^{\mathrm{Kerr}}_{\mu\nu} = \eta_{\mu\nu} - (2m/r_a) N_\mu N_\nu, \tag{20}$$

where $g^{\mu\nu}N_\mu N_\nu = \eta^{\mu\nu}N_\mu N_\nu = 0$. In the (t, x, y, z) coordinate frame, null geodesic N_μ has components $N_\mu = A(1, N_1, N_2, N_3)$, $A^2 = 1/(1 + a^2z^2r_a^{-4})$,

$$N_1 = -(r_a x + ay)/(r_a^2 + a^2),$$

$$N_2 = -(r_a y - ax)/(r_a^2 + a^2),$$
$$N_3 = -z/r_a.$$

If we calculate Taub numbers with background $\eta_{\mu\nu}$, then metrics (18) and (20) result in

$$M(\hat{g}^{\text{Sch}}) = \tau_1(k_t) = m, \qquad J(\hat{g}^{\text{Sch}}) = \tau_1(k_\varphi) = 0,$$

$$M(\hat{g}^{\text{Kerr}}) = \tau_1(k_t) = m, \qquad J(\hat{g}^{\text{Kerr}}) = \tau_1(k_\varphi) = -ma.$$

Gürses and Gürsey [23] have shown that the Einstein and Landau-Lifshitz pseudotensors coincide for Kerr-Schild metrics and are proportional to the Einstein tensor. The pseudotensors gave the earliest correct values for conserved quantities in Einstein's theory. Nahmad-Achar and Schutz [24] have extended the domain of both pseudotensors from asymptotically Cartesian coordinates to asymptotically curvilinear. Taub numbers on a Minkowski background give the same mass and angular momentum values as the pseudotensors for Kerr-Schild metrics, and so agree with accepted values.

16.5 Bondi–Sachs Solutions

We consider radiative systems described by

$$\hat{g}^{\text{B-S}}_{\mu\nu} dx^\mu\, dx^\nu = \frac{V e^{2b}}{r} du^2 + 2e^{2b} du\, dr - r^2 H_{AB}(dx^A - U^A\, du)(dx^B - U^B\, du), \tag{21}$$

where outgoing null hypersurfaces are labeled by $x^0 = u = \text{const}$. The Bondi-Sachs metric [25] extends Bondi's original metric [11] to include φ dependence and has six independent functions $V,\ b, U^A,\ y$, and q, of u, r, ϑ, φ. The rays of each $u = \text{const}$. null surface are null geodesics $x^\alpha(r)$ with tangent dx^α/dr, where $x^1 = r$ is a luminosity distance. Coordinates $x^2 = \vartheta$ and $x^3 = \varphi$ are constant along each ray. The luminosity distance is defined by $r^4 \sin^2\vartheta = \det(g_{AB}) = \det(r^2 H_{AB})$, where

$$H_{AB} = \begin{bmatrix} e^{2y}\cosh(2q) & \sinh(2q)\sin\vartheta \\ \sinh(2q)\sin\vartheta & e^{-2y}\cosh(2q)\sin^2\vartheta \end{bmatrix}.$$

We use the asymptotic solution given in Glass and Goldberg [7], where notation was chosen to avoid confusion between Sachs metric functions and Newman-Penrose spin coefficients:

$$2y = \gamma + \delta \text{ (Sachs)}, \quad 2q = \gamma - \delta \text{ (Sachs)}, \quad b = \beta \text{ (Sachs)}.$$

Metric function V is found in Eq. (B16) of [7]:

$$V = r - 2M + O(1/r),$$

where $M(u, \vartheta, \varphi)$ is the Bondi mass aspect given by

$$-2M = \Psi_2^0 + \bar{\Psi}_2^0 + \partial_u(\sigma^0\bar{\sigma}^0).$$

The Bondi mass is the 2-surface integral of the mass aspect over a topological 2-sphere at $\mathcal{I}^+$

$$M_{\text{Bondi}} = -\frac{1}{8\pi}\oint_{S^2}[\Psi_2^0 + \bar{\Psi}_2^0 + \partial_u(\sigma^0\bar{\sigma}^0)]\,d\Omega. \tag{22}$$

To construct the background, we use characteristic initial data for the Newman-Penrose equations

$$\Psi_0(u_0, r, \vartheta, \varphi) = \Psi_1^0(u_0, \vartheta, \varphi) = 0,$$
$$\Psi_2^0(u_0, \vartheta, \varphi) + \bar{\Psi}_2^0(u_0, \vartheta, \varphi) = \sigma^0(u, \vartheta, \varphi) = 0,$$

which yields Bondi-Sachs metric functions $U^A = b = y = q = 0,\ \ V = r$. The Bondi-Sachs solution is then the flat Minkowski metric

$$\eta_{\mu\nu}\,dx^\mu\,dx^\nu = du^2 + 2\,du\,dr - r^2 H_{AB}\,dx^A\,dx^B. \tag{23}$$

This is the background metric for perturbation calculations. The Bondi-Sachs metric covers the vacuum region outside the sources, and in that region the coordinates of the Minkowski metric (23) and the Bondi-Sachs metric (21) coincide.

The linearized contribution to $h_{\mu\nu}$ is given by [14]

$$\begin{aligned} h_{\mu\nu}^{(1)} = {} & [(\Psi_2^0 + \bar{\Psi}_2^0)/r + O_3]l_\mu l_\nu - (2\bar{\sigma}^0/r + O_3)m_\mu m_\nu \\ & - (2\sigma^0/r + O_3)\bar{m}_\mu\bar{m}_\nu - (\eth\bar{\sigma}^0/r + O_3)(l_\mu m_\nu + m_\mu l_\nu) \\ & - (\bar{\eth}\sigma^0/r + O_3)(l_\mu\bar{m}_\nu + \bar{m}_\mu l_\nu). \end{aligned}$$

Here the components of $h_{\mu\nu}^{(1)}$ are given in the coordinates and null tetrad of the background Minkowski frame. $U_{\text{Taub}}^{\alpha\beta}$ is integrated over a $u =$ const., $r =$ const. 2-surface at $\mathcal{I}^+$ with $dS_{\alpha\beta} = l_{[\alpha}n_{\beta]}d\vartheta d\varphi$. Taking the limit to $\mathcal{I}^+$ yields the first-order Taub mass

$$\begin{aligned} \tau_1(k_t, h^{(1)}) &= -\frac{1}{8\pi}\oint_{S^2}(\Psi_2^0 + \bar{\Psi}_2^0 + \frac{1}{2}\eth^2\bar{\sigma}^0 + \frac{1}{2}\bar{\eth}^2\sigma^0)\,d\Omega \\ &= -\frac{1}{8\pi}\oint_{S^2}(\Psi_2^0 + \bar{\Psi}_2^0)\,d\Omega, \end{aligned}$$

which is the monopole mass moment, the Bondi mass when the news $\partial_u\sigma^0$ is zero. The Komar superpotential also gives this mass (up to a factor of 2). For the Schwarzschild solution, where $\sigma = 0$ and $\Psi_2^0 = -m$, we note that $\eta_{\mu\nu} + h_{\mu\nu}^{(1)}$ is a Kerr-Schild representation of the Schwarzschild metric with Bondi mass m.

The second-order contribution to $h_{\mu\nu}$ is given by

$$\begin{aligned} h_{\mu\nu}^{(2)} = {} & [\partial_u(\sigma^0\bar{\sigma}^0)/r + O_2]l_\mu l_\nu + [-\sigma^0\bar{\sigma}^0/2r^2 + O_4](l_\mu n_\nu + n_\mu l_\nu) \\ & + [O_2](l_\mu m_\nu + m_\mu l_\nu) + [O_2](l_\mu\bar{m}_\nu + \bar{m}_\mu l_\nu) + [O_3]m_\mu m_\nu \\ & + [O_3]\bar{m}_\mu\bar{m}_\nu + [-2\sigma^0\bar{\sigma}^0/r^2 + O_3](m_\mu\bar{m}_\nu + \bar{m}_\mu m_\nu). \end{aligned}$$

Again $U^{\alpha\beta}_{\text{Taub}}$ is integrated over a 2-surface at $\mathcal{I}^+$ with $dS_{\alpha\beta} = l_{[\alpha}n_{\beta]}d\vartheta d\varphi$. This provides

$$\tau_2(k_t, h^{(2)}) = -\frac{1}{8\pi}\int\limits_{S^2} [\partial_u(\sigma^0\bar{\sigma}^0)]\, d\Omega.$$

$h^{(3)}_{\mu\nu}$ and higher orders yield terms which fall off too rapidly in r to contribute at $\mathcal{I}^+$. Adding the first- and second-order Taub masses yields the Bondi mass

$$\begin{aligned} M_{\text{Bondi}} &= \tau_1 + \tau_2 \\ &= -\frac{1}{8\pi}\int\limits_{S^2} [\Psi_2^0 + \bar{\Psi}_2^0 + \partial_u(\sigma^0\bar{\sigma}^0)]\, d\Omega. \end{aligned} \tag{24}$$

16.6 Summary

The early history of Taub numbers showed their usefulness in the straightforward establishment of linearization instability for a class of cosmologies. It was shown that closed cosmologies with isometries and compact Cauchy surfaces had constant Taub numbers with zero values and these constants over-constrained a well-posed initial value problem leading directly to linearization instability [26]. On the other hand, asymptotically flat systems have been shown to be linearization stable, and for those systems Taub numbers organize a set of physical parameters in a logical tensorial manner through Noether's theorem. In particular, the global mass and angular momentum of nonradiative systems have been computed as conserved Taub numbers generated by time translations and axial symmetries respectively.

The Bondi mass of radiative systems has previously been computed [1], [10] by using time translations of the asymptotic BMS group. The Taub method uses the exact time symmetry of a Minkowski background and first- and second-order perturbations from that background manifold. Using the Taub superpotential, the Bondi mass then appeared directly as the 2-surface integral of the mass aspect over a spherical cut of $\mathcal{I}^+$. The second and third variational derivatives of the Hilbert action provide Taub numbers τ_1 and τ_2 which comprise the Bondi mass.

Acknowledgments

I thank the Physics Department of the University of Michigan and Professor Jean Krisch for their hospitality. This work has been partially supported by an NSERC of Canada grant.

References

[1] J. N. Goldberg, *Phys. Rev.* **131**, 1367 (1963).

[2] R. Penrose, *Proc. Roy. Soc. Lond.* **A381**, 53 (1982).
[3] J.N. Goldberg, *Phys. Rev. D* **41**, 410 (1990).
[4] A. Komar, *Phys. Rev.* **113**, 934 (1959).
[5] E. N. Glass and M. G. Naber, *J. Math. Phys.* **35**, 4178 (1994).
[6] E. N. Glass, *J. Math. Phys.* **37**, 421 (1996).
[7] E. Glass and J. Goldberg, *J. Math. Phys.* **11**, 3400 (1970).
[8] E. T. Newman and R. Penrose, *Phys. Rev. Let.* **15**, 231 (1965).
[9] E. T. Newman and R. Penrose, *Proc. R. Soc.* London, **A305**, 175 (1968).
[10] L. Tamburino and J. Winicour, *Phys. Rev.* **150**, 1039 (1966).
[11] H. Bondi, M. van der Berg, and A. Metzner, *Proc. R. Soc. London* **A269**, 21 (1962).
[12] E. N. Glass, *Phys. Rev.* **D 47**, 474 (1993).
[13] M. G. Naber and E. N. Glass, *J. Math. Phys.* **35**, 5969 (1994).
[14] E. N. Glass and M. G. Naber, *Class. Quantum Grav.* **14**, 1889 (1997).
[15] E. N. Glass and Mark Naber, *J. Math. Phys.* **35**, 1834 (1994).
[16] A. Taub, *J. Math. Phys.* **2**, 787 (1961).
[17] A. Taub, in *Relativistic Fluid Dynamics*, ed. C. Cattaneo (Lectures at the Centro Internazionale Matematico Estivo, Bressanone, 1970), Edizioni Cremonese, Rome (1971).
[18] M. Cantor, *Compositio Math.* **38**, 3 (1979).
[19] A. E. Fischer, J. E. Marsden, and V. Moncrief, *Ann. Inst. Henri Poincaré* **33**, 147 (1980).
[20] D. Brill, O. Reula, and B. Schmidt, *J. Math. Phys.* **28**, 1844 (1987).
[21] A. I. Janis and E. T. Newman, *J. Math. Phys.* **6**, 902 (1965).
[22] R. Geroch and B. C. Xanthopoulos, *J. Math. Phys.* **19**, 714 (1978).
[23] M. Gürses and F. Gürsey, *J. Math. Phys.* **16**,
[24] E. Nahmad-Achar and B. Schutz, *Gen. Rel. Grav.* **19**, 655 (1987).
[25] R. K. Sachs, *Proc. Roy. Soc. London* **A270**, 103 (1962).
[26] V. Moncrief, *J. Math. Phys.* **17**, 1893 (1976).

17

Second-Class Constraints

Joshua N. Goldberg

ABSTRACT The properties of second-class constraints and their Dirac brackets are reviewed using a finite-dimensional mechanical model. Following the general formulation, the simplification which results when the constraints can be solved for canonical pairs of variables is obtained. Finally, the case where all the constrained momenta vanish is treated. This case corresponds to the second-class constraints which arise in the canonical formulation of a field theory on a null surface. In this case we show that not only are the constrained variables given by the explicit solution, but the Dirac brackets of the unconstrained variables are equal to their original Poisson brackets.

17.1 Introduction

In the Dirac theory of constrained systems [1–6], new constraints arise from the propagation of previously known constraints until the system closes. One then separates the constraints into two classes: the first-class constraints are those combinations whose Poisson brackets with all the constraints vanish on the constraint surface; second-class constraints are those whose Poisson brackets do not vanish on the constraint surface. First-class constraints may be even or odd in number, while second-class constraints are always even. They represent redundant pairs of canonical variables on the phase space.

The first-class constraints generate the invariant transformations of the theory while the second-class constraints do not. Therefore, the second-class constraints should be removed as dynamical variables before making the transition to a quantum theory. As a result, first-class constraints occur in all theories whose invariant mappings form a function group. Second-class constraints rarely occur. However, they do occur when a field theory is cast into Hamiltonian form based on a null surface [7–10]. Most of this work is concerned with the Hamiltonian structure of general relativity on a null surface. The second-class constraints have been identified in [9,11], but nowhere has one considered the elimination of the second-class constraints explicitly. Here we show how that may be accomplished when the constraints have a particularly simple form.

In this brief paper, I want to consider the general treatment of the second-class constraints in a finite-dimensional model. The treatment will include the particular algebraic properties which arise in the canonical formalism on a null surface. This will be carried out in the next section. Following that, in the third section, the method is demonstrated by the trivial example of the electromagnetic field on a null surface. The treatment of the gravitational field will appear subsequently.

17.2 The Mechanical System

Consider a Hamiltonian, $H(q, p)$, for a mechanical system defined on a $2n$-dimensional phase space $\Gamma(q^\alpha, p_\alpha)$, $\alpha = 1, \cdots, 2n$. We assume that, in addition to any first-class constraints which may exist, there are $2m$ second-class constraints $\phi_I(q, p) = 0, I, J, \cdots = 1, \cdots, 2m$. Here we will only treat the second-class constraints and omit consideration of the first-class constraints.

The Poisson brackets among the ϕ_I form a nonsingular matrix

$$C_{IJ} = \{\phi_I, \phi_J\}$$

with an inverse $\check{C}^{JK}$,

$$C_{IJ}\check{C}^{JK} = \delta_I{}^K. \tag{1}$$

In order to eliminate the second-class constraints, Dirac defined a new bracket algebra on the phase space such that the bracket of any phase space function with a constraint vanishes. The Dirac bracket of two functions on the phase space $F(q, p)$ and $G(q, p)$ is defined to be

$$\{F, G\}_D = \{F, G\} - \sum_{J,K}\{F, C_J\}\check{C}^{JK}\{C_K, G\}. \tag{2}$$

This bracket satisfies three conditions: i) it vanishes if either F or G is a constraint; ii) it is antisymmetric; iii) it satisties the Jacobi identities. Therefore, if one replaces the Poisson bracket by this Dirac bracket, the constraints may be treated as strong conditions and eliminated from the theory.

Bergmann and Komar took an alternative, but equivalent, view by defining new phase space variables $q^{*\alpha}$, $p^*{}_\alpha$:

$$q^{*\alpha} = q^\alpha + \sum_I A^I C_I,$$
$$p^*_\alpha = p_\alpha + \sum_I B^I C_I. \tag{3}$$

Choose the coefficients A^I and B^I such that the new variables have vanishing Poisson brackets with the constraints

$$A^I = -\sum_J \{q^a, C_J\}\check{C}^{JI},$$
$$B^I = -\sum_J \{p_a, C_J\}\check{C}^{JI}. \tag{4}$$

One finds that the Poisson bracket of any functions of the new variables will then be equal to the Dirac bracket. Whether one should use the Dirac brackets directly or define the starred variables depends on the individual case. In field theories it is usually simpler to use the starred variables.

In general, neither the use of Dirac brackets directly or the use of the starred variables will select m canonical pairs of variables for elimination. However, in some cases it may be possible to solve the constraints for $(q^i, p_i), i = 1, \cdots, m$ so that the $2m$ constraints may be written

$$\begin{aligned} C_i &= q^i - Q^i(q^a, p_a) = 0, \\ C_{m+i} &= p_i - P_i(q^a, p_a) = 0 \end{aligned} \tag{5}$$

$a, b, \cdots = m + 1, \cdots, n$. The Poisson bracket matrix then has the form

$$C = \begin{pmatrix} Q & R \\ -\tilde{R} & P \end{pmatrix}, \tag{6}$$

where

$$\begin{aligned} C_{ij} = Q_{ij} &= \{Q^i, Q^j\} \\ C_{i,m+j} = R_{ij} &= \delta^i{}_j + \{Q^i, P_j\}, \\ C_{m+i,m+j} = P_{ij} &= \{P_i, P_j\}, \end{aligned} \tag{7}$$

and the tilde indicates the transpose matrix. The inverse matrix can be written

$$\check{C} = \begin{pmatrix} X & Z \\ -\tilde{Z} & Y \end{pmatrix}. \tag{8}$$

Since C is not singular, the entries can easily be found in terms of Q, P, and R:

$$\begin{aligned} X &= PR^{-1}(PR^{-1}Q + \tilde{R})^{-1}, \\ Y &= Q\tilde{R}^{-1}(R + Q\tilde{R}^{-1}P)^{-1}, \\ Z &= -\tilde{R}^{-1}(I - PY) \equiv -I + z. \end{aligned} \tag{9}$$

From the above, we then find

$$\begin{aligned} q^{*i} &= Q^i - \tilde{z}^{ij}C_j - Y^{ij}C_{m+j}, \\ p_i^* &= P_i + X^{ij}C_j + z^{ij}C_{m+j}, \\ q^{*a} &= q^a - \{q^a, C_j\}[X^{jk}C_k + Z^{jk}C_{m+k}] + \{q^a, C_{m+j}\}[\tilde{Z}^{jk}C_k - Y^{jk}C_{m+k}], \\ p_a^* &= p_a - \{p_a, C_j\}[X^{jk}C_k + Z^{jk}C_{m+k}] + \{p_a, C_{m+j}\}[\tilde{Z}^{jk}C_k - Y^{jk}C_{m+k}]. \end{aligned} \tag{10}$$

From this structure, it is easy to show that on the constraint surface q^{*i} and $p*_i$ are just the solutions given in equations (4). But, the Poisson bracket algebra of the starred variables is not that of the original variables. It is, of course, the Dirac bracket algebra. On the other hand, it is clear that any unconstrained variable whose canonical conjugate does not occur in Q^i or P_i will have Dirac brackets equal to the original Poisson brackets. The situation simplifies considerably if $P_a = 0$, as is true for the canonical formalism on a null surface. In that case, $P = X = 0$ and $R = \tilde{Z} = I$, so that the Poisson bracket matrix becomes

$$C = \begin{pmatrix} Q & I \\ -I & 0 \end{pmatrix} \tag{11}$$

and the inverse matrix is

$$\check{C} = \begin{pmatrix} 0 & -I \\ I & Q \end{pmatrix}. \tag{11a}$$

In this case, the unconstrained variables (q^a, p_a) continue to satisfy the original Poisson bracket algebra.

17.3 The Self-Dual Maxwell Field

As a simple example of the above, we shall treat the self-dual Maxwell field on a null plane. We introduce the Minkowski space metric in terms of bi-null real and spacelike complex coordinates $u = t - z$, $v = \frac{1}{2}(t + z)$, $\zeta = \frac{1}{\sqrt{2}}(x - iy)$, $\bar{\zeta} = \frac{1}{\sqrt{2}}(x + iy)$ so that the metric takes the form

$$ds^2 = 2\,du\,dv - 2\,d\zeta\,d\bar{\zeta}. \tag{12}$$

In the following, Greek letters will take the values 0–3. Lowercase Latin indices from the middle of the alphabet, $i, j, \cdots$, and uppercase indices will take the values 1–3, while lowercase Latin indices from the beginning of the alphabet, $a, b, \cdots$, will take the values 2, 3. Thus, $x^\alpha = (u, v, \zeta, \bar{\zeta})$. However, we will let x without an index represent $(v, \zeta, \bar{\zeta})$ as the argument of a function.

The Maxwell field consists of the vector potential A_α and the electromagnetic field 2-form $F_{\mu\nu}$. We introduce the self-dual field by

$$\mathcal{F} := \frac{1}{2}(F - iF^*), \tag{13}$$

where

$$F^{*\rho\sigma} := \frac{1}{2}\epsilon^{\rho\sigma\mu\nu}F_{\mu\nu}.$$

The Levi-Civita tensor has the value $\epsilon^{0123} = \epsilon_{0123} = -i$, so that $\mathcal{F}^* = i\mathcal{F}$. We find that

$$\begin{aligned} \mathcal{F}^{01} = \mathcal{F}^{23} &= \frac{1}{2}\left(F^{01} + F^{23}\right), \\ \mathcal{F}^{21} = F^{21}, \qquad \mathcal{F}^{03} &= F^{03}. \end{aligned} \tag{14}$$

The action can be written

$$S = -\frac{1}{2}\int d^4x\left[dA \wedge \mathcal{F} - \frac{1}{4}\mathcal{F} \wedge \mathcal{F}\right]. \tag{15}$$

Decompose this with respect to the null foliation $u =$ constant to obtain

$$\begin{aligned} S = \int du \int d^3x\Big\{ & A_{A,0}\mathcal{F}^A + A_0\mathcal{F}^A{}_{,A} \\ & - \mathcal{F}^1\left[2A_{[2,3]} + \mathcal{F}^1\right] + B\left[\mathcal{F}^3 + 2A_{[1,2]}\right] + \lambda\mathcal{F}^2\Big\}. \end{aligned} \tag{16}$$

In the above we have defined

$$\mathcal{F}^A = \mathcal{F}^{0A}, \quad A = 1,3; \qquad \mathcal{F}^2 = 0; \qquad B = \mathcal{F}^{21}, \tag{17}$$

and have used $du\, d^3x = i du\, dv\, d\zeta\, d\bar{\zeta}$. The configuration space for this action is 6-dimensional, $(A_A,\ \mathcal{F}^A)$. A_0, B, and λ are Lagrange multipliers. The latter enforces the constraint $\mathcal{F}^2 = 0$. The action above is already in canonical form. That is, in the form of a Legendre transformation. The canonical pairs defining the phase space are $(A_A, \mathcal{F}^A)$, $A = 1, 2, 3$ with the Poisson brackets

$$\{A_A(x),\ \mathcal{F}^B(x'')\} = \delta_A{}^B \delta(x - x'), \tag{18}$$

$x = (v, \zeta, \bar{\zeta})$. By varying the three Lagrange multipliers, we find three constraints on the phase space variables:

$$\begin{aligned} G &:= \mathcal{F}^A{}_{,A} = 0, \\ C_2 &:= A_{1,2} - A_{2,1} + \mathcal{F}^3 = 0, \\ C_3 &:= \mathcal{F}^2 = 0. \end{aligned} \tag{19}$$

It is easy to show that G is a first-class constraint. It has a vanishing Poisson bracket with all the constraints, whereas

$$\{C_2(x),\ C_3(x')\} = \delta(x - x') \equiv' C_{23}(x, x'). \tag{20}$$

C_3 already gives us $\mathcal{F}^2 = 0$, and we can solve C_2 for A_2 and rewrite it as

$$C_2 = A_2 - \int_{-\infty}^{v} [\mathcal{F}^3 + A_{1,2}]\, dv' + A(-\infty) = 0. \tag{21}$$

Now the Poisson bracket matrix is

$$C(x, x') = \begin{pmatrix} 0 & \delta(x - x') \\ -\delta(x - x') & 0 \end{pmatrix}. \tag{22}$$

As a result, A_2^* and $\mathcal{F}^{*2}$ are defined by (21) and (20) respectively. Furthemore, A_1 and $\mathcal{F}^3$ are unchanged while A_3 and $\mathcal{F}^1$ add some $\mathcal{F}^2$. Therefore, the Dirac bracket algebra of $(A_1, A_3, \mathcal{F}^1, \mathcal{F}^3)$ is just the original Poisson bracket algebra, and $(A_2, \mathcal{F}^2)$ may be eliminated.

17.4 Conclusion

We have shown that second-class constraints are easily eliminated from a theory when they can be cast so that the constrained canonical momenta can be set equal to zero. In that case the Poisson bracket algebra of the unconstrained variables is unchanged by elimination of the constraints. This situation is what occurs when casting the canonical formalism of a field theory on a null surface. Our particular interest is with general relativity. While the general algebraic structure of the constraints is as discussed in this paper, other problems arise because the geometry itself is dynamical. Therefore, the integration which is needed in order to solve the constraints requires the introduction of a parameter along the null generators of the base null surface. This parameter itself depends on the dynamical variables. The difficulties presented by this complication are currently under study.

It is a pleasure to dedicate this paper to Engelbert Schucking, whose friendship and collegial relationship has been a source of enjoyment for most of my scientific life.

This work was supported in part by a NATO collaborative research grant, CGR Programme SA.5-2-05 (CRG.960083).

References

[1] P. A. M. Dirac, *Proc. Roy. Soc.* London **A246**, 333 (1958).
[2] P. A. M. Dirac, *Phys. Rev.* **114**, 924 (1959).
[3] P. G. Bergmann and A. Komar, *Phys. Rev. Lett.* **4**, 432 (1960).
[4] K. Sundermeyer, *Constrained Dynamics*, Springer-Verlag, Berlin (1982).
[5] E. C. G. Sudarshan and N. Mukunda, *Classical Mechanics: A Modern Perspective,* John Wiley and Sons, New York (1974).
[6] M. Henneaux and C. Teitelboim, *Quantization of Gauge Systems*, Princeton University Press, Princeton, NJ (1992).
[7] J. N. Goldberg, *Found. Phys.* **15**, 439 (1985).
[8] C. Torre, *Class. Quant. Grav.* **3**, 773 (1986).
[9] J. N. Goldberg, D.C. Robinson, and C Soteriou, *Class. Quant. Grav.* **9**, 1309 (1992).
[10] R. d'Inverno and J. Vickers, *Class. Quant. Grav.* **12**, 753 (1995).
[11] J. N. Goldberg and C Soteriou, *Class. Quant. Grav.* **12**, 2779 (1995).

18

On the Structure of the Energy-Momentum and the Spin Currents in Dirac's Electron Theory

Friedrich W. Hehl
Alfredo Macías
Eckehard W. Mielke
Yuri N. Obukhov

ABSTRACT We consider a classical Dirac field in flat Minkowski spacetime. We perform a Gordon decomposition of its canonical energy-momentum and spin currents, respectively. Thereby we find for each of these currents a convective and a polarization piece. The polarization pieces can be expressed as exterior covariant derivatives of the 2-forms $\check{M}_\alpha$ and $M_{\alpha\beta} = -M_{\beta\alpha}$, respectively. In analogy to the magnetic moment in electrodynamics, we identify these 2-forms as *gravitational moments* connected with the translation group and the Lorentz group, respectively. We point out the relation between the Gordon decomposition of the energy-momentum current and its Belinfante-Rosenfeld symmetrization. In the nonrelativistic limit, the translational gravitational moment of the Dirac field is found to be proportional to the spin covector of the electron.[1]

18.1 Introduction

Fermi systems do not possess classical analogs. Nevertheless, for such systems, we can define a limit for $\hbar \to 0$, even if, in this limit, we do not have ordinary classical theory. We feel that a description of such a limit would be interesting in view of the possibility to develop a better intuition and to build up new models [1]. Conventionally Fermi systems can be quantized by the path integral method. By introducing semiclassical Grassmann variables, for instance, we get a deeper understanding of this method [2].

In some dual theories with Fermi variables, a detailed knowledge of the classical limit is also interesting, because it allows a reinterpretation in terms of superstrings

---

[1] We would like to dedicate this article to Engelbert Schucking on the occasion of his 70th birthday.

[4, 3]. This unified treatment of commuting and anticommuting variables is obviously of particular interest for supersymmetric theories, because in this case one is obliged to treat Fermi and Bose fields in a symmetrical way [3, 4, 5].

Knowing the usefulness of understanding the classical limit, we will investigate, in this article, the structure of the energy-momentum and the spin current of the *classical* Dirac theory. Of course, the classical Dirac theory can only be made consistent by second quantization [6]. However, for low energy phenomena, when particle creation and annihilation don't play a role, the classical Dirac theory and its nonrelativistic limit convey an appropriate picture of the underlying physics.

An electron carries a spin of $s_z = \hbar/2$. It is legitimate to visualize this spin as some sort of intrinsic circular *motion*. This seems to be clear from the relation of spin to the Lorentz or the 3-dimensional rotation group [7], but it can also be made explicit by studying the quantum mechanics of a Dirac particle in a Coulomb potential. Even in an $s_{1/2}$ state, an electric current is present, i.e., "a polarized Dirac electron is a *rotating* particle"[8].

If we take this for granted, it is obvious that the electric charge, which is housed in the electron, induces an Ampère-type ring current which, in turn, according to the Oersted-Ampère law, acts as a magnetic moment μ_{e}. The specific nature of spin angular momentum yields a gyromagnetic ratio $\mu_{\mathrm{e}}/(s_z/\hbar) = -2\mu_{\mathrm{B}}$, where $\mu_{\mathrm{B}} := e\hbar/2m_{\mathrm{e}}$ is Bohr's magneton (e is the charge of the electron, m_{e} its mass). This gyromagnetic ratio turns out to be twice as large as that of ordinary *orbital* angular momentum known from Newtonian mechanics.

The electron, besides the electric charge, carries also *gravitational charges*. In the framework of general relativity, the electron's mass is the only gravitational charge, and we expect a mass-energy ring current inducing a gravitational moment μ_{GR}. In the Einstein-Cartan theory, which is the appropriate gravitational theory for a Dirac field in much the same way as general relativity, i.e., Einstein's theory, is appropriate for classical point particles (without spin), there feature *two* types of gravitational charges: mass-energy and spin. In such a theoretical framework, we would search, already in the special-relativistic realm, for a translational gravitational moment M_{T}, linked to the mass, and a Lorentz (or rotational) gravitational moment M_{L}, linked to the spin aspect. The latter one is to be understood as the spin charge which is carried around by itself.

The expected units of the translational and Lorentz gravitomoments $\mu_{\mathrm{G}_1} = \hbar/2$ and $\mu_{\mathrm{G}_2} = \hbar^2/4m_{\mathrm{e}}$ are found by substituting in Bohr's magneton μ_{B} the electric charge by the mass or the spin of the electron, respectively. Then, in analogy to electrodynamics, we should expect *gyrogravitational* ratios for the different gravitational moments, namely $\mu_{\mathrm{GR}}/(s_z/\hbar) = g_0\,\hbar/2$, $M_{\mathrm{T}}/(s_z/\hbar) = g_1\,\hbar/2$, and $M_L/(s_z/\hbar) = g_2\,\hbar^2/4m_{\mathrm{e}}$, with g_0, g_1, g_2 as the dimensionless gravitational g-factors. For the gravitational g-factors we wouldn't be able to predict the values before going into the corresponding computations. But our guess, for reasons of analogy with electrodynamics, would be $g_0 = g_1 = g_2 = 2$.

How could we thoroughly check these ideas? Again appealing to analogies to electrodynamics, the following procedure is near at hand: By using a Gordon-type decomposition of the energy-momentum and the spin currents in *special* relativity, we can provisionally define the gravitational moments. Then we couple the Dirac Lagrangian minimally to the gravitational field, reshuffle the Lagrangian suitably, and identify the factor(s) in front of the gravitational field strength(s) as gravitational moment(s). In general relativity, the gravitational field strength is the Riemannian curvature, in the Einstein-Cartan theory, both the torsion and the (Riemann-Cartan) curvature represent the field strengths, respectively. In this article we make the first step in identifying the gravitational moments of the classical Dirac field within the framework of special relativity.

We find it surprising that, apart form Kobzarev and Okun [9], nobody seems to care about the gravitational moments of the Dirac field. It is true that the effects will be so small that there is no hope for measuring these moments in the near future. In any case, as a matter of principle, the gravitational moments of a fundamental particle belong to its basic properties and will enter any unification attempt.

18.2 Dirac–Yang-Mills Theory

The Dirac Lagrangian is given by the *Hermitian* 4-form

$$L_{\mathrm{D}} = L(\vartheta^\alpha, \Psi, D\Psi) = \frac{i}{2}\hbar\left\{\overline{\Psi}\,{}^*\gamma \wedge D\Psi + \overline{D\Psi} \wedge {}^*\gamma\,\Psi\right\} + {}^*mc\,\overline{\Psi}\Psi\,. \tag{1}$$

The coframe ϑ^α necessarily occurs in the Dirac Lagrangian, even in special relativity. We are using the formalism of Clifford-valued exterior forms; see the Appendix for the basic definitions. For the mass term, we use the short-hand notation ${}^*m := m\eta = m\,{}^*1$. The hermiticity of the Lagrangian (1) leads to a charge current which admits the usual probabilistic interpretation.

Here we assume a Minkowski spacetime geometry (no gravity). However, the fermions may carry various *internal* charges. This is reflected in the gauge covariant derivatives, $D = d + \mathcal{A}$, where $\mathcal{A} = \frac{e}{\hbar}A^K\tau_K$ is a Lie algebra-valued 1-form, the gauge potential, with a set of matrices τ_K specifying a representation of the generators of the gauge symmetry. The charge e stands for the coupling constant of a fermion–gauge field interaction. In the commutator $[\tau_K, \tau_L] = f^M{}_{KL}\tau_M$, the structure constants $f^M{}_{KL}$ are totally antisymmetric for semisimple groups. The Abelian case is also trivially included, when the internal symmetry reduces to the one-parameter group with a single generator $\tau_1 = i$ (then $f^M{}_{KL} \equiv 0$). This corresponds to electrodynamics with e as the usual electric charge.

The Lagrangian (1) yields the Dirac equation[2] and its adjoint, which are obtained by independent variation with respect to the spinor fields $\overline{\Psi}$, Ψ:

$$i\hbar{}^*\gamma \wedge D\,\Psi + {}^*mc\,\Psi = 0, \tag{2}$$

$$i\hbar\overline{D\Psi} \wedge {}^*\gamma + {}^*mc\,\overline{\Psi} = 0\,. \tag{3}$$

[2]For a comprehensive study of the Dirac equation, see Thaller [10].

The 3-form

$$J_K := \frac{\delta L_{\mathrm{D}}}{\delta A^K} = -ie\,\overline{\Psi}\tau_K{}^*\gamma\Psi \tag{4}$$

is the canonical Noether current, the "isospin" 3-form, which turns out to be covariantly conserved, provided the equations of motion (2)–(3) are substituted:

$$DJ_K = dJ_K + \frac{e}{\hbar} f^M{}_{KL} A^L \wedge J_M = 0. \tag{5}$$

Using the Dirac equation, we can, following Gordon [11], decompose the Noether current into two pieces, the *convective* and the *polarization* currents. The derivation is rather short: *assuming the mass to be nonzero*, one can resolve (2)–(3) with respect to the spinor fields:

$$\Psi = \frac{i\hbar}{mc}{}^*({}^*\gamma \wedge D\Psi), \qquad \overline{\Psi} = \frac{i\hbar}{mc}{}^*(\overline{D\Psi} \wedge {}^*\gamma). \tag{6}$$

Now substitute this into (4) and find

$$J_K = J^{(c)}_K + J^{(p)}_K, \tag{7}$$

where the convective 3-form current reads

$$J^{(c)}_K := \frac{e\hbar}{2mc}\,{}^*(\overline{\Psi}\tau_K D\Psi - \overline{D\Psi}\tau_K\Psi), \tag{8}$$

and the polarization current is constructed as a covariant exterior differential of the *polarization* 2-form P_K:

$$J^{(p)}_K := DP_K, \qquad P_K := -i\,\frac{e\hbar}{2mc}\,\overline{\Psi}\tau_K{}^*\widehat{\sigma}\Psi. \tag{9}$$

The covariant exterior differential D is defined as in (5).

Unlike the total current, its convective and polarization parts are *not* conserved separately in non-Abelian gauge theory. For example, we can immediately check that

$$DJ^{(p)}_K = DDP_K = \frac{e}{\hbar} f^N{}_{KL} F^L \wedge P_N, \tag{10}$$

where

$$F^K := dA^K + \frac{e}{2\hbar} f^K{}_{MN} A^M \wedge A^N \tag{11}$$

is the 2-form of the gauge field strength. [We can also write the gauge field strength in the form $\mathcal{F} := d\mathcal{A} + \mathcal{A} \wedge \mathcal{A} = \frac{e}{\hbar} F^K \tau_K$. Note that $DD\Psi = \mathcal{F}\Psi$.] Only in the *Abelian* case one finds Gordon's separate conservation of the two currents.

In order to understand the physical meaning of these currents, it is useful to study the nonrelativistic approximation for fermions in a weak external gauge field. In the nonrelativistic approximation, a Dirac 4-spinor is represented by a pair of 2-component spinors, $\Psi = \begin{pmatrix} \psi \\ \chi \end{pmatrix}$, with $\chi = \mathcal{O}(\frac{v}{c})\,\psi$ (for positive energy solutions). The polarization 2-form is then

$$P_\kappa = \frac{e\hbar}{m}\, dt \wedge S_\kappa + \mathcal{O}\left(\frac{v}{c}\right), \qquad S_\kappa := \psi^\dagger(-i\tau_\kappa)\frac{1}{2}\boldsymbol{\sigma}\psi\, d\boldsymbol{x}. \tag{12}$$

Consider the interaction term in the Lagrangian, $A^\kappa \wedge J_\kappa$. It is clear that the convective contribution describes a (non-Abelian generalization of the) usual Schrödinger current,

$$J_\kappa^{(c)} \approx \frac{e\hbar}{2mc}\, {}^*[\psi^\dagger\tau_\kappa d\psi - (d\psi^\dagger)\tau_\kappa\psi], \tag{13}$$

whereas the polarization contribution reads

$$A^\kappa \wedge J_\kappa^{(p)} \approx F^\kappa \wedge P_\kappa \approx \frac{e\hbar}{m}\, dt \wedge S_\kappa \wedge F^\kappa. \tag{14}$$

In the Abelian case (for $\tau_\kappa = i$), the standard result of the Dirac electron theory is recovered, $A \wedge J^{(p)} \approx \boldsymbol{\mu}\boldsymbol{B}\, dt \wedge {}^{(3)}\eta$. Here ${}^{(3)}\eta$ is the volume form of a spatial hypersurface, $\boldsymbol{B}$ is the magnetic field strength, and $\boldsymbol{\mu} = \frac{e\hbar}{2m}\psi^\dagger\boldsymbol{\sigma}\psi$.

Technically, we can derive the two currents as follows. Substituting (6) back into the Dirac equation (2)–(3), we find the *squared* equation:

$$\left[D^*D - i\,{}^*\widehat{\sigma} \wedge \mathcal{F} + {}^*(mc/\hbar)^2\right]\Psi = 0. \tag{15}$$

Equation (15) can be derived from the Lagrange form $L_{D^2} = L^{(c)} + L^{(p)}$, with

$$L^{(c)} := \frac{1}{2}\left(\frac{\hbar^2}{mc}\, {}^*\overline{D\Psi} \wedge D\Psi + {}^*mc\,\overline{\Psi}\,\Psi\right), \tag{16}$$

$$L^{(p)} := P_\kappa \wedge F^\kappa. \tag{17}$$

Since both pieces of the Lagrangian are separately gauge invariant, we straightforwardly recover the convective and polarization currents, (8) and (9), as the respective Noether currents:

$$J_\kappa^{(c)} = \frac{\delta L^{(c)}}{\delta A^\kappa}, \qquad J_\kappa^{(p)} = \frac{\delta L^{(p)}}{\delta A^\kappa}. \tag{18}$$

18.3 Gordon Decomposition of Energy-Momentum and Spin Currents

The *canonical* energy-momentum and spin 3-forms are defined in the standard way:

$$\Sigma_\alpha := e_\alpha \rfloor L - \frac{\partial L}{\partial D\Psi}(e_\alpha \rfloor D\Psi) - (e_\alpha \rfloor \overline{D\Psi})\frac{\partial L}{\partial \overline{D\Psi}}, \tag{19}$$

$$\tau_{\alpha\beta} := \frac{\partial L}{\partial D\Psi}\,\ell_{\alpha\beta}\Psi + \overline{\Psi}\ell_{\alpha\beta}\frac{\partial L}{\partial \overline{D\Psi}}. \tag{20}$$

Recalling that the spinor generators of the Lorentz group are $\ell_{\alpha\beta} = \frac{i}{4}\widehat{\sigma}_{\alpha\beta}$, we find for the Dirac Lagrangian (1), with $L = L_{\rm D}$ inserted into (19)–(20),

$$\Sigma_\alpha = \frac{i\hbar}{2}\left(\overline{\Psi}\,{}^*\gamma D_\alpha \Psi - D_\alpha \overline{\Psi}\,{}^*\gamma \Psi\right), \tag{21}$$

$$\tau_{\alpha\beta} = \frac{\hbar}{4}\,\vartheta_\alpha \wedge \vartheta_\beta \wedge \overline{\Psi}\gamma\gamma_5\Psi. \tag{22}$$

Here we denoted $D_\alpha := e_\alpha \rfloor D$ and took into account that $L_{\rm D} \cong 0$ (upon substitution of the field equations).

Translational and Lorentz invariance of the Dirac Lagrangian yield the well-known conservation laws for energy-momentum and angular momentum:

$$D\Sigma_\alpha = 0, \qquad D\tau_{\alpha\beta} + \vartheta_{[\alpha} \wedge \Sigma_{\beta]} = 0. \tag{23}$$

Here D denotes the Lorentz covariant exterior derivative containing the Levi-Civita connection $\Gamma_\alpha{}^\beta$ of flat Minkowski space. Thus we have, e.g., $D\Sigma_\alpha = d\Sigma_\alpha - \Gamma_\alpha{}^\beta \wedge \Sigma_\beta$.

Let us perform the Gordon-type decomposition of the energy-momentum current [12, 13, 14, 15, 16]. At first, we substitute (6) into (21) and obtain

$$\begin{aligned}\Sigma_\alpha = \frac{\hbar^2}{2mc}\Big[& {}^*\,(\overline{D\Psi})D_\alpha\Psi + D_\alpha\overline{\Psi}{}^*D\Psi \\ & + i\left(\overline{D\Psi} \wedge {}^*\widehat{\sigma} D_\alpha\Psi - D_\alpha\overline{\Psi}{}^*\widehat{\sigma} \wedge D\Psi\right)\Big].\end{aligned} \tag{24}$$

The Dirac equation (2)–(3) yields

$$i\hbar^*\gamma D_\alpha \Psi = i\hbar(e_\alpha\rfloor{}^*\gamma) \wedge D\Psi + \eta_\alpha mc\,\Psi, \tag{25}$$

$$i\hbar D_\alpha\overline{\Psi}{}^*\gamma = i\hbar\overline{D\Psi} \wedge (e_\alpha\rfloor{}^*\gamma) - \eta_\alpha mc\,\overline{\Psi}. \tag{26}$$

Substituting this into (21), we find another representation for the energy-momentum, provided we use (6) again:

$$\begin{aligned}\Sigma_\alpha &= \frac{i\hbar}{2}\left((e_\alpha\rfloor\overline{\Psi}{}^*\gamma) \wedge D\Psi - \overline{D\Psi} \wedge (e_\alpha\rfloor{}^*\gamma\Psi)\right) + \eta_\alpha mc\,\overline{\Psi}\Psi \\ &= \frac{\hbar^2}{2mc}\Big[(e_\alpha\rfloor{}^*\overline{D\Psi}) \wedge D\Psi + \overline{D\Psi} \wedge (e_\alpha\rfloor{}^*D\Psi) \\ &\quad + i\left(D_\alpha\overline{\Psi}{}^*\widehat{\sigma} \wedge D\Psi - \overline{D\Psi} \wedge {}^*\widehat{\sigma} D_\alpha\Psi\right) \\ &\quad - 2i\,\overline{D\Psi} \wedge (e_\alpha\rfloor{}^*\widehat{\sigma}) \wedge D\Psi\Big] + \eta_\alpha mc\,\overline{\Psi}\Psi.\end{aligned} \tag{27}$$

Then, the sum of (24) and (27) yields

$$\Sigma_\alpha = \Sigma_\alpha^{(c)} + \Sigma_\alpha^{(p)}, \tag{28}$$

where

$$\begin{aligned}\Sigma_\alpha^{(c)} := \frac{mc}{2}\,\overline{\Psi}\Psi\,\eta_\alpha + \frac{\hbar^2}{4mc}\Big[& {}^*(\overline{D\Psi})D_\alpha\Psi + D_\alpha\overline{\Psi}{}^*D\Psi \\ & + (e_\alpha\rfloor{}^*\overline{D\Psi}) \wedge D\Psi + \overline{D\Psi} \wedge (e_\alpha\rfloor{}^*D\Psi)\Big],\end{aligned} \tag{29}$$

$$\Sigma_\alpha^{(p)} := D\check{M}_\alpha, \tag{30}$$

$$\check{M}_\alpha := -\frac{i\hbar^2}{4mc}\,[\overline{\Psi}\,(e_\alpha\rfloor^*\hat{\sigma}) \wedge D\Psi + \overline{D\Psi} \wedge (e_\alpha\rfloor^*\hat{\sigma})\,\Psi]. \tag{31}$$

Since we are in flat Minkowski spacetime, the last term in (28), the *polarization* part of the energy-momentum current, is identically conserved:

$$D\Sigma_\alpha^{(p)} = DD\check{M}_\alpha = 0. \tag{32}$$

Recalling (23), we immediately obtain a separate conservation of the first term on the right-hand side of (28), which we naturally call the *convective energy-momentum* current:

$$D\Sigma_\alpha^{(c)} = 0. \tag{33}$$

Moreover, one can immediately check that the convective current is symmetric:

$$\vartheta_{[\alpha} \wedge \Sigma_{\beta]}^{(c)} = 0. \tag{34}$$

It is remarkable to observe that, similar to how the convective current (8) represents a Noether current for the convective Lagrangian (16), the convective energy-momentum (29) precisely turns out to be the canonical Noether current (19) for $L = L^{(c)}$. The proof is straightforward: substitute (16) into (19) and compare with (29).

As we saw, the ordinary polarization current (9) also emerges as a Noether current for the "Pauli-type" polarization Lagrangian (17) which describes the interaction of the *moment 2-form* with the background gauge field strength F^κ. A guess would be that the polarization energy-momentum should arise in a similar way as a canonical current from the respective "Pauli-type" polarization Lagrangian when the gravitational field is "switched on." The discussion of the precise form of this Lagrangian (as well as of the nature of the gravitational field represented by the coframe) will be considered in a separate publication; see also [17, 18, 19, 20, 21, 22].

Here, however, we have to complete our derivations and to consider the Gordon type decomposition of the spin current. We have good reasons to expect a similar structure of the decomposed 3-form $\tau_{\alpha\beta}$. At first, we note that the canonical spin current (20) for the *convective* Lagrangian (16) reads

$$\tau_{\alpha\beta}^{(c)} = \frac{\partial L^{(c)}}{\partial D\Psi}\,\ell_{\alpha\beta}\Psi + \overline{\Psi}\ell_{\alpha\beta}\frac{\partial L^{(c)}}{\partial \overline{D\Psi}} = -\frac{i\hbar^2}{8mc}\left({}^*\overline{D\Psi}\widehat{\sigma}_{\alpha\beta}\Psi - \overline{\Psi}\widehat{\sigma}_{\alpha\beta}{}^*D\Psi\right). \tag{35}$$

Now we use the standard trick by substituting (6) into the spin current (22):

$$\begin{aligned}\tau_{\alpha\beta} &= \frac{\hbar}{4}\,\vartheta_\alpha \wedge \vartheta_\beta \wedge \overline{\Psi}\gamma\gamma_5\Psi = \frac{\hbar}{8}\,\overline{\Psi}({}^*\gamma\widehat{\sigma}_{\alpha\beta} + \widehat{\sigma}_{\alpha\beta}{}^*\gamma)\Psi \\ &= \frac{i\hbar^2}{8mc}\Big(-{}^*\overline{D\Psi}\widehat{\sigma}_{\alpha\beta}\Psi + \overline{\Psi}\widehat{\sigma}_{\alpha\beta}{}^*D\Psi - i\overline{D\Psi} \wedge {}^*\widehat{\sigma}\widehat{\sigma}_{\alpha\beta}\Psi \\ &\qquad - i\overline{\Psi}\widehat{\sigma}_{\alpha\beta}{}^*\widehat{\sigma} \wedge D\Psi\Big).\end{aligned} \tag{36}$$

We immediately recognize the second line as the *convective spin* (35). The last line should thus be related to the polarization spin current. The identities (97)–(98) (see the Appendix) play the crucial role here. Substituting them into (36), we obtain straightforwardly

$$\tau_{\alpha\beta} = \tau^{(c)}_{\alpha\beta} + DM_{\alpha\beta} + \vartheta_{[\alpha} \wedge \check{M}_{\beta]}, \tag{37}$$

where we define the *Lorentz gravitational moment* 2-form by

$$M_{\alpha\beta} := \frac{\hbar^2}{16mc}\overline{\Psi}({}^*\widehat{\sigma}\widehat{\sigma}_{\alpha\beta} + \widehat{\sigma}_{\alpha\beta}{}^*\widehat{\sigma})\Psi \tag{38}$$

$$= \frac{\hbar^2}{8mc}\left(\overline{\Psi}\Psi\,\eta_{\alpha\beta} - i\,\overline{\Psi}\gamma_5\Psi\,\vartheta_\alpha \wedge \vartheta_\beta\right), \tag{39}$$

whereas $\check{M}_\alpha$ is given in (31). The Lorentz moment $M_{\alpha\beta}$ is very simple in structure. It is additively built up from a scalar and a pseudoscalar piece, i.e., from its 36 components only 2 are independent.

Substituting the Gordon decompositions (28) and (37) into the conservation law of angular momentum (23), we find the *separate* conservation of the convective spin,

$$D\tau^{(c)}_{\alpha\beta} = 0. \tag{40}$$

18.4 Relocalization of Energy-Momentum and Spin

Like for internal symmetries, also in the case of the Poincaré group, the Noether currents are only determined up to an exact 2-form. This non-uniqueness has troubled physicists already for quite some time. Within gravitational theory, the question of the "correct" energy-momentum current of matter is as old as general relativity itself [23, 24]; see also the review [25]. But only Belinfante [26], in the framework of special relativity, and Rosenfeld [27], within general relativity, gave a general prescription of how one can find the metric or *Hilbert* energy-momentum current from the canonical or *Noether* energy-momentum current of an arbitrary matter field Ψ. The Hilbert current acts as source on the right-hand side of the Einstein field equation, whereas the Noether current is of central importance in special-relativistic canonical field theory. We will now turn our attention to this interrelationship between the different energy-momentum currents within the framework of special relativity.

The Noether law for energy-momentum in (23) also holds for an energy-momentum current which is supplemented by a D-exact form:

$$\hat{\Sigma}_\alpha(X) := \Sigma_\alpha - DX_\alpha. \tag{41}$$

In special relativity, $DD = 0$. This is the only property of D that is needed in this context. The X_α does not interfere with the Noether law:

$$D\Sigma_\alpha = D\hat{\Sigma}_\alpha + DDX_\alpha = D\hat{\Sigma}_\alpha = 0\,. \tag{42}$$

If we insert $\Sigma_\alpha = \hat{\Sigma}_\alpha(X) + DX_\alpha$ into the left-hand side of the Noether law for angular momentum in (23), we find

$$D\tau_{\alpha\beta} + \vartheta_{[\alpha} \wedge \Sigma_{\beta]} = D(\tau_{\alpha\beta} - \vartheta_{[\alpha} \wedge X_{\beta]}) + \vartheta_{[\alpha} \wedge \hat{\Sigma}_{\beta]} . \tag{43}$$

If a relocalized spin $\hat{\tau}_{\alpha\beta}$ is required to fulfill again a law of the type given in (23), i.e., $D\hat{\tau}_{\alpha\beta} + \vartheta_{[\alpha} \wedge \hat{\Sigma}_{\beta]} = 0$, then

$$\hat{\tau}_{\alpha\beta}(X, Y) := \tau_{\alpha\beta} - \vartheta_{[\alpha} \wedge X_{\beta]} - DY_{\alpha\beta} , \tag{44}$$

where $DY_{\alpha\beta}$ is an additional D-exact form with $Y_{\alpha\beta} = -Y_{\beta\alpha}$. Thus a relocalization of the energy-momentum is, up to a D-exact form, accompanied by an induced transformation of the canonical spin. Therefore we have the following result:

The canonical currents $(\Sigma_\alpha, \tau_{\alpha\beta})$ *fulfill the Noether laws (23). Take arbitrary 2-forms* X_α *and* $Y_{\alpha\beta} = -Y_{\beta\alpha}$ *as superpotentials. Then the relocalized currents*

$$\Sigma_\alpha \rightarrow \hat{\Sigma}_\alpha(X) = \Sigma_\alpha - DX_\alpha , \tag{45}$$

$$\tau_{\alpha\beta} \rightarrow \hat{\tau}_{\alpha\beta}(X, Y) = \tau_{\alpha\beta} - \vartheta_{[\alpha} \wedge X_{\beta]} - DY_{\alpha\beta} , \tag{46}$$

satisfy the same relations

$$D\hat{\Sigma}_\alpha = 0 , \qquad D\hat{\tau}_{\alpha\beta} + \vartheta_{[\alpha} \wedge \hat{\Sigma}_{\beta]} = 0 . \tag{47}$$

Accordingly, the Noether identities turn out to be invariant under the *relocalization transformation* (45)–(46). As a consequence, the total energy-momentum P_α and the total angular momentum $J_{\alpha\beta}$, up to boundary terms, remain invariant under (45)–(46),

$$\hat{P}_\alpha \stackrel{*}{=} P_\alpha - \int_{\partial H_t} X_\alpha , \qquad \hat{J}_{\alpha\beta} \stackrel{*}{=} J_{\alpha\beta} - \int_{\partial H_t} (x_{[\alpha} \wedge X_{\beta]} + Y_{\alpha\beta}) , \tag{48}$$

where H_t denotes a timelike hypersurface in Minkowski space and ∂H_t its 2-dimensional boundary. Provided the superpotentials X_α and $Y_{\alpha\beta}$ approach zero at spacelike asymptotic infinity sufficiently fast, the total quantities are not affected by the relocalization procedure.

Let us put our results of the *Gordon decomposition* in the last section into this general framework. If we choose as superpotentials the respective gravitational moments,

$$X_\alpha = \check{M}_\alpha , \qquad Y_{\alpha\beta} = M_{\alpha\beta} , \tag{49}$$

then the relocalized currents turn out to be the *convective* pieces

$$\Sigma_\alpha^{(c)} = \Sigma_\alpha - D\check{M}_\alpha = \Sigma_\alpha - \Sigma_\alpha^{(p)} , \tag{50}$$

$$\tau_{\alpha\beta}^{(c)} = \tau_{\alpha\beta} - \vartheta_{[\alpha} \wedge \check{M}_{\beta]} - DM_{\alpha\beta} = \tau_{\alpha\beta} - \tau_{\alpha\beta}^{(p)} . \tag{51}$$

From explicit calculations we know (see (34)) that the convective current (50) is symmetric, as one would expect for a Schrödinger-type energy-momentum current. Consequently, in special relativity, the decomposed currents have the following properties:

$$D\Sigma_\alpha^{(c)} = 0 , \qquad D\Sigma_\alpha^{(p)} = 0 , \qquad \vartheta_{[\alpha} \wedge \Sigma_{\beta]}^{(c)} = 0 , \tag{52}$$

$$D\tau^{(c)}_{\alpha\beta} = 0\,, \qquad D\tau^{(p)}_{\alpha\beta} + \vartheta_{[\alpha} \wedge \Sigma^{(p)}_{\beta]} = 0\,. \tag{53}$$

Thus a Gordon decomposition in Minkowski spacetime is nothing else but a *specific relocalization* of the currents. It yields a symmetric energy-momentum current $\Sigma^{(c)}_{\alpha}$ with a nonvanishing *conserved* spin current $\tau^{(c)}_{\alpha\beta}$. The spin tensor of Hilgevoord et al. [28, 29], which was constructed outside of the framework of Lagrangian formalism, coincides with our convective spin current[3] $\tau^{(c)}_{\alpha\beta}$.

18.5 Trivial Lagrangians and Relocalization

For the purpose of understanding relocalization from a Lagrangian point of view, let us consider an arbitrary *three*–form $U(\Psi, D\Psi)$ constructed from a matter field Ψ and its derivatives. Here we discard other possible arguments of U (like the coframe, e.g.). In general, Ψ can be any matter field, not necessarily the Dirac 4-spinor.

If the form U is invariant under spacetime translations (coordinate transformations) and Lorentz transformations, then, using the standard Lagrange-Noether machinery (see [30]) one derives the first and the second Noether *identities*:

$$D\overset{\circ}{X}_{\alpha} \equiv -e_{\alpha}\rfloor d\,U + (e_{\alpha}\rfloor D\Psi)\,\frac{\delta U}{\delta\Psi}, \tag{54}$$

$$D\overset{\circ}{Y}_{\alpha\beta} + \vartheta_{[\alpha} \wedge \overset{\circ}{X}_{\beta]} \equiv -\,\ell_{\alpha\beta}\Psi\,\frac{\delta U}{\delta\Psi}. \tag{55}$$

Here the *two*-forms

$$\overset{\circ}{X}_{\alpha} := e_{\alpha}\rfloor U - (e_{\alpha}\rfloor D\Psi)\,\frac{\partial U}{\partial D\Psi}, \tag{56}$$

$$\overset{\circ}{Y}_{\alpha\beta} := \ell_{\alpha\beta}\Psi\,\frac{\partial U}{\partial D\Psi}, \tag{57}$$

are analogs of the 3-forms of canonical energy-momentum and spin derived from the "Lagrangian" U. We use the standard notation

$$\frac{\delta U}{\delta\Psi} := \frac{\partial U}{\partial\Psi} - D\frac{\partial U}{\partial D\Psi}. \tag{58}$$

We will *not* assume, however, that the matter field Ψ satisfies the "equation of motion" $\frac{\delta U}{\delta\Psi} = 0$. The relations (54) and (55) are thus *strong identities* valid for all matter field configurations. The first term on the right-hand side of (54) is usually absent in the first Noether identity due to the fact that the Lagrangian is a form of maximal rank (i.e. 4 in standard spacetime). Here we have a 3-form U in 4-dimensional spacetime, and the 4-form $d\,U$ is, in general, nontrivial.

---

[3]... up to a factor of 2 due to different conventions.

Let us now treat the form $L = dU$ as a specific Lagrangian. It is straightforward to find for the variation

$$\delta U = \delta\Psi \frac{\partial U}{\partial \Psi} + \delta D\Psi \wedge \frac{\partial U}{\partial D\Psi} = \delta\Psi \frac{\delta U}{\delta \Psi} + d\left(\delta\Psi \frac{\partial U}{\partial D\Psi}\right). \tag{59}$$

Hence, for our specific Lagrangian, we have

$$\delta L = d(\delta U) = \delta\Psi\, D\left(\frac{\delta U}{\delta\Psi}\right) + \delta(D\Psi) \wedge \frac{\delta U}{\delta \Psi}. \tag{60}$$

For the partial derivatives, this yields

$$\frac{\partial L}{\partial \Psi} = D\left(\frac{\delta U}{\delta \Psi}\right), \tag{61}$$

$$\frac{\partial L}{\partial D\Psi} = \frac{\delta U}{\delta \Psi}. \tag{62}$$

Consequently, for the Lagrangian $L = dU$, the equation of motion

$$\frac{\delta L}{\delta \Psi} = \frac{\partial L}{\partial \Psi} - D\frac{\partial L}{\partial D\Psi} \equiv 0 \tag{63}$$

is identically satisfied. Actually, this is no surprise: it is well known that a Lagrangian, which is a total differential, has trivial dynamics.

Nevertheless, although the dynamics is trivial, the conserved currents are *not* trivial. In particular, the canonical energy-momentum and spin 3-forms are defined as usual by

$$\Sigma_\alpha = e_\alpha \rfloor L - \frac{\partial L}{\partial D\Psi}(e_\alpha \rfloor D\Psi), \tag{64}$$

$$\tau_{\alpha\beta} = \frac{\partial L}{\partial D\Psi}\, \ell_{\alpha\beta}\Psi. \tag{65}$$

And now we are approaching the crucial point. We recall that $L = dU$. Thus we can substitute (62) into these currents. Then, for the corresponding energy-momentum, we immediately find

$$\Sigma_\alpha = e_\alpha \rfloor d\, U - (e_\alpha \rfloor D\Psi)\frac{\delta U}{\delta\Psi} = -D\overset{\circ}{X}_\alpha, \tag{66}$$

where we made use of the first Noether identity (54). Analogously, for the spin current, we have

$$\tau_{\alpha\beta} = \ell_{\alpha\beta}\Psi \frac{\delta U}{\delta \Psi} = -D\overset{\circ}{Y}_{\alpha\beta} - \vartheta_{[\alpha} \wedge \overset{\circ}{X}_{\beta]}, \tag{67}$$

where we used the second Noether identity (55).

This observation underlies the relocalization described above of energy-momentum and spin. Indeed, our derivation shows that if one adds to any matter field Lagrangian L_Ψ a total divergence[4] $d\,U$, then, for the new Lagrangian $L_\Psi + d\,U$, the canonical energy-momentum and spin currents are relocalized according to (45)–(46), with the superpotentials

$$X_\alpha = \overset{\circ}{X}_\alpha, \qquad Y_{\alpha\beta} = \overset{\circ}{Y}_{\alpha\beta}. \tag{68}$$

In this sense, one can say that a relocalization is *generated* by the 3-form U via (56)–(57).

Again looking back to the Gordon decomposition of energy-momentum and spin as a special case of the relocalization procedure, we are now able to *generate* the corresponding results by means of the simple *three-form*

$$U = \frac{1}{2}\,\vartheta^\alpha \wedge \check{M}_\alpha = -\frac{i\hbar^2}{4mc}\left(\overline{\Psi}\,{}^*\widehat{\sigma} \wedge D\Psi - \overline{D\Psi} \wedge {}^*\widehat{\sigma}\Psi\right). \tag{69}$$

If we substitute it into (56)–(57), we find, indeed,

$$\overset{\circ}{X}_\alpha = \check{M}_\alpha \quad \text{and} \quad \overset{\circ}{Y}_{\alpha\beta} = M_{\alpha\beta}\,; \tag{70}$$

compare with (49) and (68). It is remarkable that the *translational* moment $\check{M}_\alpha$, via U, also generates the corresponding *Lorentz* moment $M_{\alpha\beta}$.

Moreover, the *same* 3-form U generates the relocalization of the isospin current J_K,

$$J_K \to J_K - D\overset{\circ}{Z}_K, \tag{71}$$

where

$$\overset{\circ}{Z}_K = \frac{e}{\hbar}\,\tau_K \Psi\,\frac{\partial U}{\partial D\Psi}. \tag{72}$$

The proof goes along the same lines as above. We just formulate the relevant Noether identity which arises from the invariance of the 3-form U with respect to the gauge transformations under consideration. Substituting (69) into (72), we recover the polarization moment 2-form

$$\overset{\circ}{Z}_K = P_K = -i\,\frac{e\hbar}{2mc}\,\overline{\Psi}\tau_K{}^*\check{\sigma}\Psi. \tag{73}$$

18.6 Belinfante Symmetrization of the Energy-Momentum Current

A simple way, within special relativity, to arrive at the "generic" symmetric energy-momentum current, i.e., at the Hilbert current, is to require that the *relocalized spin current vanishes*. This is what the Belinfante-Rosenfeld symmetrization

---

[4] In previous papers (see [31]) this Lagrangian prescription was used "on shell" to generate the transition to chiral fermions. However, in the massless limit these fields carry no spin but rather helicity.

amounts to. Therefore the Belinfante-Rosenfeld energy-momentum current t_α can be defined as

$$t_\alpha := \hat{\Sigma}_\alpha(X) \quad \text{with} \quad \hat{\tau}_{\alpha\beta}(X, Y) = 0\,. \tag{74}$$

The last equation, together with (46), yields $\tau_{\alpha\beta} = \vartheta_{[\alpha} \wedge X_{\beta]} + DY_{\alpha\beta}$, which can be resolved with respect to the superpotential X^β as follows:

$$X^\beta = \mu^\beta - 2e_\gamma \rfloor DY^{\gamma\beta} - \frac{1}{2}\vartheta^\beta \wedge \left(e_\gamma \rfloor e_\delta \rfloor DY^{\gamma\delta}\right). \tag{75}$$

Here

$$\mu^\beta := 2e_\gamma \rfloor \tau^{\gamma\beta} + \frac{1}{2}\vartheta^\beta \wedge \left(e_\gamma \rfloor e_\delta \rfloor \tau^{\gamma\delta}\right) \tag{76}$$

is the *spin energy potential.* Then the first Noether law in (47) reads alternatively $Dt_\alpha = 0$.

Let us collect the key formulae for our Belinfante-Rosenfeld current with the *specific* superpotential X_α of (75)–(76):

$$t_\alpha = \Sigma_\alpha - DX_\alpha\,, \qquad \vartheta_{[\alpha} \wedge t_{\beta]} = 0\,, \qquad Dt_\alpha = 0\,. \tag{77}$$

For $Y_{\alpha\beta} = 0$, these are the familiar Belinfante-Rosenfeld relations [26, 27]. For particles with spin zero, the improved energy-momentum current can be derived by a suitable choice of the superpotential $Y_{\alpha\beta}$; cf. [19].

It is remarkable that, for a matter field of any spin, we can find a relocalized Belinfante-Rosenfeld energy-momentum current t_α, with $Dt_\alpha = 0$. If we consider the motion of a "test" field in a Minkowski spacetime, then our procedure shows that we can always attach to this motion a geodesic line, irrespective of the spin.

As we saw in the previous section, a relocalization of energy-momentum and spin can be generated by a superpotential 3-form U. However, the finding of an explicit U for the Belinfante relocalization turns out to be a nontrivial problem. Although the general prescription (75) involves $Y_{\alpha\beta}$, a symmetrization of the energy-momentum is already achieved for $Y_{\alpha\beta} = 0$. Moreover, our discussion here was confined to flat Minkowski geometry; but in Riemannian spacetime (which is the arena of general relativity theory) the Belinfante relocalization necessarily demands $Y_{\alpha\beta} = 0$, as was shown in [32].

Accordingly, a puzzling feature of the Belinfante relocalization for the Dirac energy-momentum is the apparent impossibility of constructing a generating form U, with $Y_{\alpha\beta} = 0$. Indeed, for the Dirac field, the spin current is given by (22). Hence the spin energy potential (76) reads

$$\mu_\alpha = \frac{\hbar}{4}\vartheta_\alpha \wedge \overline{\Psi}\gamma\gamma_5\Psi. \tag{78}$$

From (57) it is clear that $Y_{\alpha\beta} = 0$ if and only if U does not depend on the differentials $D\Psi$, i.e., $\partial U/\partial D\Psi = 0$. Then (56) and (75) yield

$$\mu_\alpha = e_\alpha \rfloor U. \tag{79}$$

Contracting with the coframe ϑ^α, we find $\vartheta^\alpha \wedge \mu_\alpha = 3U$. However, for the Dirac spin energy potential (78) we get $\vartheta^\alpha \wedge \mu_\alpha \equiv 0$. Therefore we have to conclude that there is no such 3-form U which can generate the Belinfante relocalization in the Dirac theory—provided one starts with the canonical currents (21)–(22). Probably the latter requirement has to be given up. One could start with the convective currents (29) and (35) as well. But we will leave that for future consideration.

18.7 Properties of the Gravitational Moments and Nonrelativistic Limit

Let us find out the dimensions of the gravitational moments. Recall that the Dirac fields has dimension $[\Psi] = [\overline{\Psi}] = \text{length}^{-3/2}$, whereas $[\vartheta^\alpha] = \text{length}$, and $[e_\alpha] = \text{length}^{-1}$. Thus, we have for the 2-forms $[\widehat{\sigma}] = [{}^*\widehat{\sigma}] = \text{length}^2$, and we immediately get

$$[\check{M}_\alpha] = [mc], \qquad [M_{\alpha\beta}] = [\hbar]. \tag{80}$$

This is consistent with the analogous result for the polarization moment (9) in the Dirac–Yang-Mills theory, where one finds $[P_\kappa] = [e]$ (with e as the non-Abelian charge or the usual electric charge in the Abelian case). Dimensionwise, the "translational charge," which defines the translational moment, is thus a *momentum* and the "Lorentz charge" an *angular momentum*.

In a remarkable way, the gravitational moments are closely related to the spin of a Dirac particle. The relocalization superpotential U and the translational moment $\check{M}_\alpha$ both can be expressed in terms of the convective spin alone via the *identities*

$$U \equiv -\,\eta_{\mu\nu} \wedge {}^*\tau^{(c)\mu\nu}, \qquad \check{M}_\alpha \equiv -\,\eta_{\alpha\mu\nu} \wedge {}^*\tau^{(c)\mu\nu}. \tag{81}$$

Since $\check{M}_\alpha$ and $\tau^{(c)\mu\nu}$ have the same number of independent components (namely 24), the last algebraic identity may be inverted, giving the convective spin current in terms of the translational moment.

At first sight, it may be unclear that the Lorentz moment (38) is also related to spin. However, let us consider its square invariant:

$$\begin{aligned} M_{\alpha\beta} \wedge {}^*M^{\alpha\beta} &= \left(\frac{\hbar^2}{8mc}\right)^2 \left[-(\overline{\Psi}\Psi)^2 + (\overline{\Psi}\gamma_5\Psi)^2\right] \eta_{\alpha\beta} \wedge \vartheta^\alpha \wedge \vartheta^\beta \\ &= 3\left(\frac{\hbar^2}{4mc}\right)^2 (\overline{\Psi}\gamma_\alpha\gamma_5\Psi)(\overline{\Psi}\gamma^\alpha\gamma_5\Psi)\,\eta \\ &= \frac{1}{2}\tau_{\alpha\beta} \wedge {}^*\tau^{\alpha\beta}. \end{aligned} \tag{82}$$

Here, in the first line, we used the representation (39). Then we rearranged the products, which are bilinear in the spinor fields, by means of the Fierz identity. As we recognize, in contrast to the translational moment, the Lorentz moment is directly related to the *complete* Dirac spin $\tau_{\alpha\beta}$. Note that in applying the Gordon decomposition non-zero mass was assumed. In the limit of massless Dirac particles, the expression in (82) would vanish.

Some further insight can be obtained if we calculate the gravitational moments for specific spinor field configurations. The plane waves

$$\Psi = \Psi_0(p) \exp\left(-\frac{i}{\hbar} p_\alpha x^\alpha\right), \tag{83}$$

as general solution of the free Dirac equation, are very important in this context. Substituting them into (31) and (39), we find

$$\check{M}_\alpha = p_\alpha \frac{\hbar}{2mc} \,\overline{\Psi}_0{}^*\widehat{\sigma}\Psi_0, \tag{84}$$

$$M_{\alpha\beta} = \hbar \frac{\hbar}{8mc} \,\overline{\Psi}_0\Psi_0 \,\eta_{\alpha\beta}. \tag{85}$$

The (reduced) Compton wavelength $\hbar/mc$ in (84)–(85) evidently provides the correct dimension for these 2-forms. In deriving (84), we used the identity which holds for any solution of the Dirac equation (i.e., when equations (2)–(3) are satisfied):

$$i\left(\overline{\Psi}{}^*\widehat{\sigma}\wedge D\Psi - \overline{D\Psi}\wedge {}^*\widehat{\sigma}\Psi\right) = {}^*D(\overline{\Psi}\Psi). \tag{86}$$

Since for the plane waves (83) the scalar $\overline{\Psi}\Psi = \overline{\Psi}_0\Psi_0$ is constant (standard normalization is then $\overline{\Psi}_0\Psi_0 = (mc/\hbar)^3$), we immediately find that the generating 3-form (69) vanishes, $U = 0$.

In the nonrelativistic approximation, we get for the translational moment (84):

$$\check{M}_\alpha = \frac{p_\alpha \hbar}{m} dt \wedge S + \mathcal{O}\left(\frac{v}{c}\right), \qquad S := \psi_0^\dagger \frac{1}{2}\sigma\psi_0 \, dx. \tag{87}$$

This result is a complete analog of (12) for the nonrelativistic polarization moment in Dirac–Yang-Mills theory.

At the same time, no further clarification of the structure of the Lorentz moment (85) occurs in the nonrelativistic limit. In particular, it is *not* proportional to the spin 1-form S, unlike the translational moment $\check{M}_\alpha$ and the polarization current P_κ.

18.8 Discussion

Inherent in the structure of the "inertial currents" $(\Sigma_\alpha, \tau_{\alpha\beta})$ of the Dirac field, namely of energy-momentum and spin, is the existence of convective and polarization pieces, the latter ones being exterior covariant derivatives of the gravitational moments $(\check{M}_\alpha, M_{\alpha\beta})$ of translational and Lorentz type, respectively. This discovery is made on the level of special-relativistic field theory, i.e., in (flat) Minkowski spacetime. In other words, we were able to identify the gravitational moments of the Dirac field *without* any involvement of the gravitational field itself. Rather, we only assumed that the canonical *energy-momentum* and *spin* currents are the *sources* of gravity. With this assumption, which is in accord with the Einstein-Cartan theory of gravity, we can tell from our results that the field strengths of gravity have to be represented by two 2-forms with the following structure: $\left(T_\alpha,\, R_{\alpha\beta} = -R_{\beta\alpha}\right)$. In a future paper we will show, as a final proof of our conception, that the moments $\left(\check{M}_\alpha,\, M_{\alpha\beta}\right)$ couple, indeed, to the field strengths $\left(T_\alpha,\, R_{\alpha\beta}\right)$, provided the field strengths are interpreted as torsion and curvature of spacetime.

Acknowledgments

This work was partially supported by CONACyT, grant No. 3544–E9311, and by the joint German-Mexican project KFA-Conacyt E130-2924 and DLR-Conacyt 6.B0A.6A. Moreover, EWM acknowledges support by the short-term fellowship 9616160156 of the DAAD (Bonn) and YNO by the project He 528/17-2 of the DFG (Bonn).

Appendix

Our general notation is as follows: The spacetime is 4-dimensional with a metric g of signature $(+,-,-,-)$. A local frame is denoted by e_α ($\alpha = 0,1,2,3$; $\Xi = 1,2,3$) and the dual coframe by ϑ^α. They fulfill the relation $e_\alpha \rfloor \vartheta^\beta = \delta^\beta_\alpha$, with $\rfloor$ denoting the interior product. For a holonomic or coordinate basis, we have $d\vartheta^\alpha = 0$; then there exists a local coordinate system $\{x^i = (x^0, x^a)\}$ ($i = 0,1,2,3$; $a = 1,2,3$) such that $e_\alpha \stackrel{*}{=} \delta^i_\alpha \partial_i$ and $\vartheta^\alpha \stackrel{*}{=} \delta^\alpha_i dx^i$. Anholonomic indices are always taken from the Greek and holonomic indices from the Latin alphabet. The Hodge star operator is denoted by *. Let $\eta := {}^*1$ be the volume 4-form. The following forms span the exterior algebra at each point of spacetime:

$$\eta_\alpha := e_\alpha \rfloor \eta = {}^*\vartheta_\alpha, \tag{88}$$

$$\eta_{\alpha\beta} := e_\beta \rfloor \eta_\alpha = {}^*(\vartheta_\alpha \wedge \vartheta_\beta), \tag{89}$$

$$\eta_{\alpha\beta\gamma} := e_\gamma \rfloor \eta_{\alpha\beta} = {}^*(\vartheta_\alpha \wedge \vartheta_\beta \wedge \vartheta_\gamma), \tag{90}$$

$$\eta_{\alpha\beta\gamma\delta} := e_\delta \rfloor \eta_{\alpha\beta\gamma} = {}^*(\vartheta_\alpha \wedge \vartheta_\beta \wedge \vartheta_\gamma \wedge \vartheta_\delta). \tag{91}$$

For the flat metric of Minkowski spacetime, $o_{\alpha\beta} = \mathrm{diag}(+1,-1,-1,-1)$, we choose the Dirac matrices in the form

$$\gamma^{\hat{0}} = \begin{pmatrix} \mathbf{1} & \mathbf{0} \\ \mathbf{0} & -\mathbf{1} \end{pmatrix}, \quad \gamma^a = \begin{pmatrix} \mathbf{0} & \sigma^a \\ -\sigma^a & \mathbf{0} \end{pmatrix}, \quad a = 1,2,3. \tag{92}$$

Here σ^a are the standard 2×2 Pauli matrices. Two important elements of the Dirac algebra are

$$\widehat{\sigma}_{\alpha\beta} := \frac{i}{2}(\gamma_\alpha \gamma_\beta - \gamma_\beta \gamma_\alpha), \tag{93}$$

$$\gamma_5 := -\frac{i}{4!}\, \eta_{\alpha\beta\mu\nu}\, \gamma^\alpha \gamma^\beta \gamma^\mu \gamma^\nu = -i\, \gamma^{\hat{0}} \gamma^{\hat{1}} \gamma^{\hat{2}} \gamma^{\hat{3}} = \begin{pmatrix} \mathbf{0} & -\mathbf{1} \\ -\mathbf{1} & \mathbf{0} \end{pmatrix}. \tag{94}$$

It is convenient to convert the constant γ_α matrices into Clifford algebra-valued 1- or 3-forms, respectively:

$$\gamma := \gamma_\alpha\, \vartheta^\alpha\,, \qquad {}^*\gamma = \gamma^\alpha\, \eta_\alpha\,. \tag{95}$$

Correspondingly, we obtain a 2-form:

$$\widehat{\sigma} := \frac{1}{2}\,\widehat{\sigma}_{\alpha\beta}\, \vartheta^\alpha \wedge \vartheta^\beta = \frac{i}{2}\,\gamma \wedge \gamma\,. \tag{96}$$

Two important identities hold for these Clifford algebra-valued objects:

$$\widehat{\sigma}_{\alpha\beta}\, {}^*\widehat{\sigma} = \eta_{\alpha\beta} - i\,\gamma_5\, \vartheta_\alpha \wedge \vartheta_\beta - 2i\, \vartheta_{[\alpha} \wedge e_{\beta]} \rfloor {}^*\widehat{\sigma}\,, \tag{97}$$

$${}^*\widehat{\sigma}\, \widehat{\sigma}_{\alpha\beta} = \eta_{\alpha\beta} - i\,\gamma_5\, \vartheta_\alpha \wedge \vartheta_\beta + 2i\, \vartheta_{[\alpha} \wedge e_{\beta]} \rfloor {}^*\widehat{\sigma}\,. \tag{98}$$

REFERENCES

[1] R. Casalbuoni: "On the quantization of systems with anticommuting variables," *Nuovo Cimento* **A33** (1976) 115–125; "The classical mechanics for Bose-Fermi systems," *Nuovo Cimento* **A33** (1976) 389–431.

[2] J. L. Martin: "The Feynman principle for a Fermi system," *Proc. Roy. Soc. London* **A251** (1959) 543–549.

[3] P. Ramond: "Dual theory for free fermions," *Phys. Rev.* **D3** (1971) 2415–2418.

[4] M. B. Green, J.H. Schwarz, and E. Witten: *Superstring Theory*, 2 volumes (Cambridge University Press: Cambridge, 1987).

[5] L. Corwin, Y. Ne'eman, and S. Sternberg: "Graded Lie algebras in mathematics and physics (Bose-Fermi symmetry)," *Rev. Mod. Phys.* **47** (1975) 573–603.

[6] R. Jost: *The General Theory of Quantized Fields* (Amer. Math. Soc.: Providence, Rhode Island, 1965) p. 39.

[7] E. Wigner: "On unitary representations of the inhomogeneous Lorentz group," *Ann. of Math.* **40** (1939) 149–204.

[8] T. T. Chou, C. N. Yang: "Hadronic matter current distribution inside a polarized nucleus and a polarized hadron," *Nucl. Phys.* **B107** (1976) 1–20.

[9] I. Yu. Kobzarev and L. B. Okun: "Gravitational interaction of fermions," *Sov. Phys. JETP* **16** (1963) 1343–1346 [*ZhETF* **43** (1962) 1904–1909 (in Russian)].

[10] B. Thaller: *The Dirac Equation* (Springer: Berlin, 1992).

[11] W. Gordon: "Der Strom der Diracschen Elektronentheorie," *Z. Physik* **50** (1928) 630–632.

[12] P. von der Heyde: "Is gravitation mediated by the torsion of space-time?," *Z. Naturforsch.* **31a** (1976) 1725–1726.

[13] F. W. Hehl: "Four lectures on Poincaré gauge field theory," in *Proc. of the 6th Course of Internat. School on Cosmology and Gravitation: Spin, Torsion, Rotation and Supergravity (Erice, 1979)* P. G. Bergmann and V. De Sabbata, eds. (Plenum: New York, 1980) 5–61.

[14] J. Audretsch: "Dirac electron in space-times with torsion: Spinor propagation, spin precession, and nongeodesic orbits," *Phys. Rev.* **D24** (1981) 1470–1477; erratum **D25** (1982) 605.

[15] M. Seitz: "The gravitational moment densities of classical matter fields with spin," *Ann. Physik (Leipzig)* **41** (1984) 280–290.

[16] M. Seitz: "A quadratic Lagrangian for the Poincaré gauge field theory of gravity," *Class. Quantum Grav.* **3** (1986) 175–182.

[17] F. W. Hehl and W. -T. Ni: "Inertial effects of a Dirac particle," *Phys. Rev.* **D42** (1990) 2045–2048.

[18] F. W. Hehl, J. Lemke, and E. W. Mielke: "Two lectures on fermions and gravity," in *Geometry and Theoretical Physics* (Bad Honnef School, 1990) J. Debrus and A.C.Hirshfeld, eds. (Springer: Berlin, 1992) 56–140.

[19] F. W. Hehl and E. W. Mielke: "Improved expressions for the energy–momentum current of matter," Festschrift für E. Schmutzer, *Wiss. Zeitschr. Friedrich–Schiller–Universität Jena, Naturw. Reihe* **39** (1990) 58–65.

[20] J. Audretsch, F. W. Hehl, and C. Lämmerzahl: "Matter wave interferometry and why quantum objects are fundamental for establishing a gravitational theory," in *Relativistiv Gravity Research, Proc., Bad Honnef, Germany 1991*, J. Ehlers and G. Schäfer, eds., Lecture Notes in Physics (Springer) **410** (1992) 368–407.

[21] J. Lemke: "On the gravitational interaction of elementary particles" (in German). Ph.D. thesis, University of Cologne (1992).

[22] A. Macias, E. W. Mielke, and H. A. Morales–Técotl: "Gravitational–geometric phases and translations," in *New Frontiers in Gravitation*, G.A. Sardanashvily, ed. (Hadronic Press: Palm Harbor, Florida, 1996) 227–242.

[23] D. Hilbert: "Die Grundlagen der Physik (Erste Mitteilung)," *Königl. Gesellsch. d. Wiss. Göttingen*, Nachr. Math.-Phys. Kl. (1915) 395–407.

[24] A. Einstein: "Hamiltonsches Prinzip und allgemeine Relativitätstheorie," *Sitzber. Königl. Preuss. Akad. Wiss. Berlin* (1916) 1111–1116.

[25] F. W. Hehl: "On the energy tensor of spinning massive matter in classical field theory and general relativity," *Rep. on Math. Phys. (Toruń)* **9** (1976) 55–82.

[26] F. J. Belinfante: "On the spin angular momentum of mesons," *Physica* **6** (1939) 887-898; "On the current and the density of the electric charge, the energy, the linear momentum and the angular momentum of arbitrary fields", *Physica* **7** (1940) 449–474.

[27] L. Rosenfeld: "Sur le tenseur d'impulsion-energie" *Mém. Acad. Roy. Belgique, cl. sc.* **18**, fasc. 6 (1940).

[28] J. Hilgevoord and S. A. Wouthuysen: "On the spin angular momentum of the Dirac particle," *Nucl. Phys.* **40** (1963) 1–12.

[29] J. Hilgevoord and E. A. De Kerf: "The covariant definition of spin in relativistic quantum field theory," *Physica* **31** (1965) 1002–1016.

[30] F. W. Hehl, J. D. McCrea, E. W. Mielke, and Y. Ne'eman, "Metric-affine gauge theory of gravity: field equations, Noether identities, world spinors, and breaking of dilaton invariance," *Phys. Rep.* **258** (1995) 1–171.

[31] E. W. Mielke, A. Macias, and H.A. Morales–Técotl: "Chiral fermions coupled to chiral gravity," *Phys. Lett.* **A215** (1996) 14–20.

[32] E. W. Mielke, F. W. Hehl, and J. D. McCrea: "Belinfante type invariance of the Noether identities in a Riemannian and Weitzenböck space-time," *Phys. Lett.* **A140** (1989) 368–372.

19

The Physical Reality of the Quantum Wave Function

Arthur Komar

ABSTRACT By examining a particular thought experiment, we show that two observers, outside of "each other's light cone", when examining the same quantum system, will ascribe very different quantum wave functions to that system. Nevertheless, both observers will agree on the statistical distribution of all observations performed upon the system. Thus the quantum state of the system does not appear to possess an invariant objective status. We conclude from this example that the quantum wave function which describes the state of the system under observation is observer dependent. Therefore it is reasonable and consistent to regard the wave function as but a convenient algorithmic device enabling observers to predict the response of quantum systems to observations. We briefly apply this point of view to demystify the well-known problem of Schrödinger's Cat.

19.1 Introduction

The classical physics of Newton provided so close a model of our intuitive view of reality, the mathematical elements employed bearing so close an identification with the observed quantities in nature, that physicists had come to regard natural reality as, in some fundamental sense, "being" that model. Essentially, it was believed that reality was isomorphic to the mathematical model that the classical physicist constructed. Initially there was some concern, or at least puzzlement, with regard to the nature of Newton's required action-at-a-distance forces. However, that was resolved by Faraday's introduction of the force field. Thus by the end of the 19th century physicists had succeeded in constructing a rather satisfactory objective model of physical reality.

The first intimations of subjective difficulties arose at the dawn of the 20th century with Einstein's introduction of the relativity of simultaneity. In a classical thought experiment concerning lightning bolts striking both the front and back of a speeding train, Einstein demonstrated that, as a consequence of the invariance of the velocity of light, observers on the train would not agree with those on the ground concerning whether the bolts of lightning struck simultaneously. It is important to understand that the observers would never be in disagreement as to the temporal ordering of the sequence of observations each of them made, that is, the objective observed facts of the situation are not in dispute. What was at issue in Einstein's experiment was the interpretation of the facts, i.e., the model of reality implied by the facts.

Since a disagreement with regard to simultaneity translates immediately into a disagreement with regard to the temporal ordering of events, and since the "cause" of an event must always precede the "effect" of that event, it was initially feared that Einstein's apparent subjectivization of the temporal order of events would result in the demise of coherent reasoning in physical theories, thereby undermining any hope of obtaining a sensible model of physical reality. In order to recover an objective model of reality in the face of this dilemma, Einstein was compelled to abandon the Newtonian conception of space and time and employ Minkowski's model of a 4-dimensional space-time.

However, in order for the Minkowski space-time to provide a satisfactory arena for the description of causal phenomena, it became essential to restrict permissible physical theories exclusively to those which do not permit the propagation of energy and/or information with speeds greater than the velocity of light. With that caveat, objective models of macroscopic physical reality again became possible. Indeed the Minkowski model of space-time proved to be so successful that it serves as the standard surrogate for the arena of reality in almost all current physical theories.

The situation for microscopic physical reality has proven to be more intractable. In contrast to the fundamental predictability property of the classical physics of the macroworld, identically prepared microsystems do not in general have identical predictable behaviors when subject to subsequent observations. The appropriate physical theory for such phenomena, quantum mechanics, correctly predicts the observed dispersion in the results obtained when performing repeated measurements upon identically prepared replicas. The guise in which this information is encoded, a complex field called, for historical reasons, the wave function, is initially determined by the precise statement of the experimental arrangement. It then propagates in time in a deterministic fashion via Schrödinger's wave equation. However the relationship between the wave function's predictions and the results of physical measurements performed upon the system is necessarily statistical in character, as is required by the dispersion of the experimental results. By performing a broad range of repeated observations on identically prepared replicas, the wave function of the selected ensemble of replicas can, to a large extent, be effectively mapped out.

The question that we wish to examine is whether this wave function is an objective physical structure. That is, can the wave function be regarded as a physical field in a sense similar to Faraday's force fields? What calls this into question is the fact that once an observation is made upon a system, the associated wave function generally changes radically. This so-called "collapse of the wave function" for the measured system is quite understandable, since the new information obtained as a consequence of the measurement alters the ensemble of replicas of which our system is a member. However when combined with the restriction imposed upon permissible physical systems by Einstein's caveat, which limits propagation of information to speeds not greater than that of light, the instantaneous nature of the "collapse" raises doubts about the compatiblity of the wave function with the objective model of physical reality achieved by relativity theory.

In this paper we raise yet another, and in some ways more serious, challenge to the point of view which ascribes a physical reality to the wave function. Employing a generalization of a version of the EPR thought experiment, we shall demonstrate that spacelike separated observers who examine correlated portions of a composite system will disagree on the resulting "collapsed" wave function even though they will agree on all observed statistical distributions of actual data.

19.2 The Thought Experiment

We consider a composite system consisting of two independent subsystems which are initially completely correlated and whose combined state is describable by a single quantum wave function. The system then fissions and the two components fly off in opposite directions. When the components are well separated, local observers perform measurements on the respective components in their neighborhoods. We shall assume that identical replicas are available, so that the observers can repeat and even alter their observational methods as frequently as they wish in order to build up a body of statistical data concerning the subsystem in their immediate neighborhood. Each observer is then asked to determine the quantum states of both subsystems.

For simplicity let us consider a composite system whose two subparts are correlated in the following fashion. The quantum state of the combined system is given by the wave function

$$|\psi\rangle = \frac{1}{\sqrt{N}} \sum_{i=1}^{N} |a_i, 1\rangle \otimes |b_i, 2\rangle, \tag{1}$$

where a_i denotes the N eigenvalues of some Hermitian operator A which acts on the first subsystem, and b_i denotes the N eigenvalues of a correlated Hermitian operator B which acts solely on the second system. (As an example, envision the eigenvalues of angular momentum about parallel axes A_z and B_z respectively for each subsystem, the correlation established by the requirement that the overall angular momentum of the system vanishes.) We are assuming that the two sets of bras and kets comprise complete orthonormal sets of one-particle states in their respective spaces.

Strictly speaking the subsystems also have their respective position observables as well, but we choose to suppress reference to them in order to reduce the complexity of the notation. The positions will of course play an essential role, since we are most interested in the situation when the subsystems are far from each other. We shall therefore employ the notation 1 or 2 in the bras and kets to keep track of this positional information, which incidentally also serves to identify the applicable subsystem.

Suppose the first observer performs the measurement of observable A and obtains the eigenvalue a_0, a situation which has the probability of occurring on the average of one time out of N. He would then determine that the state of the system as described by the state vector in Eq. (1) has collapsed suddenly into

$$|\psi'\rangle = |a_0, 1\rangle \otimes |b_0, 2\rangle, \tag{2}$$

and more particularly, that the second system has now to be described by the state vector

$$|\psi_2\rangle = |b_0, 2\rangle. \tag{3}$$

Note that the location of the second subsystem can be taken to be sufficiently far away that the presumed "collapse" occurs faster than light can travel between the two subsystems. Clearly, if the theory is to be consistent, a measurement of B performed on the second system must yield with absolute certainty the eigenvalue b_0. Repeated tests of this correlation between the eigenvalues of A and those of B can experimentally verify this assertion, provided we employ as initial states replicas of the initial preparation.

Let us now suppose that the second observer did not select the observable B for measurements performed on the subsystem in his location, but rather C, which can be obtained by a unitary transformation of B. (For example, in the case of the z component of angular momentum, an arbitrarily rotated axis could have been chosen as the new z axis.) The eigenvalues of C of course will be numerically the same as those of B, however the numerical ordering might be permuted. If we denote the eigenvalues of C as c_i, the eigenvectors of C are related to those of B thus

$$|b_i, 2\rangle = \sum_{m=1}^{N} |c_m, 2\rangle\langle c_m, 2|b_i, 2\rangle. \tag{4}$$

It thereby follows that the fact that the second subsystem has collapsed into the eigenstate $|b_0, 2\rangle$ would be reflected in the observation that the probability that the second observer would find the eigenvalue c_i occurring in repeated trials is given by the expression

$$P_{oi}(2) = |\langle c_i, 2|b_0, 2\rangle|^2. \tag{5}$$

In this fashion both observers can establish that indeed whenever the first observer finds the eigenvalue a_0, it is consistent to account for the state of the second subsystem as having collapsed into the state $|b_0, 2\rangle$ despite the fact that the two subsystems may be separated outside each other's light cone.

If we now substitute Eq. (4) into Eq. (1), we find, after a small rearrangement of the terms,

$$|\psi\rangle = \frac{1}{\sqrt{N}} \sum_{i,m=1}^{N} \langle c_m, 2|b_i, 2\rangle |a_i, 1\rangle \otimes |c_m, 2\rangle \tag{6}$$

Defining a new orthonormal basis for the first subsystem by the unitary transformation

$$|d_n, 1\rangle = \sum_{i=1}^{N} \langle c_n, 2|b_i, 2\rangle |a_i, 1\rangle, \tag{7}$$

the initial correlated state can now be written

$$|\psi\rangle = \frac{1}{\sqrt{N}} \sum_{m=1}^{N} |d_m, 1\rangle \otimes |c_m, 2\rangle \tag{8}$$

Consider now the second observer, who is not in communication with the first observer, and who upon measuring C finds some eigenvalue c_i. He can legitimately conclude that his system is in state $|c_i, 2\rangle$. He would then conclude from Eq. (8) that the first subsystem is in the state $|d_i, 1\rangle$. The probability of obtaining the eigenvalue a_0 for the first system would therefore be

$$P_{oi}(1) = |\langle a_0, 1|d_i, 1\rangle|^2 \tag{9}$$

Indeed, examining the data for the subensemble defined by selecting a given c_i, one similarly finds the probability distribution for the eigenvalues a_j to be

$$P_{ij}(1) = |\langle d_i, 1|a_j, 1\rangle|^2 \tag{10}$$

The second observer then can correctly confirm that his determination of the state of the first system as $|d_i, 1\rangle$ is valid.

Returning to the consideration of the subensemble of data given by the eigenvalues a_0 and b_0, it is easy to see from Eq. (7) that

$$\langle a_0, 1|d_i, 1\rangle = \langle c_i, 2|b_0, 2\rangle \tag{11}$$

and consequently

$$P_{oi}(1) = P_{oi}(2). \tag{12}$$

It follows from the above considerations that although the two observers give radically different descriptions of their respective quantum states, and can exhibit the data to document the validity of their respective descriptions, they in fact will always agree on the probability distribution of observed, measured quantities which occur in any well-specified subensemble. Much as in the case of the relativistic observers who agree on events but not on the "explanation" of them, our quantum observers agree on the observed probability distributions of well-specified subensembles but do not agree on the quantum states which presumably brought them about. For example we have shown above that upon measuring A and finding the eigenvalue a_0, the first observer *correctly* concludes that states of the two subsystems have collapsed into $|a_0, 1\rangle$ and $|b_0, 2\rangle$ respectively; the second observer, however, upon measuring C and finding the eigenvalue c_0, *correctly* concludes that states of the two subsystems have collapsed into $|d_0, 1\rangle$ and $|c_0, 2\rangle$ respectively. As the measurements of A and C are arranged to be performed in regions which are at spacelike separation from each other, they cannot interfere or influence each other.

Both observers' conclusions are equally valid and can be confirmed by repeated testing.

We therefore find that quantum states are not objects of objective reality, but rather they are a convenient, subjective (i.e., observer-dependent) description of physical phenomena which provide an important and valid algorithm for computing the dispersion of results which occur for identically prepared experimental situations. From this point of view, the concept of a quantum state has very much the same standing as the concept of simultaneity in relativistic physics. We note however that, in the quantum case, even though both observers are in the same Lorentz frame, they have different descriptions of "reality".

I should add that had we taken a more general form for the initial correlated state than that given in Eq. (1), for example, the coefficients of the various eigenstates in the summation being unequal, a description of the resulting subensembles would not even have been available in terms of quantum states. We would have been forced to employ density matrices to describe them. Apart from the fact that the resulting formalism would be somewhat more cumbersome than that employing "collapsing" wave functions, more to the point, its statistical nature, that is, its distance from "reality," would be far more blatant. Thus, pursuing this generalization, while rather straightforward, would not significantly alter our conclusions. We need only note that if we define the density matrix

$$\rho = |\psi\rangle\langle\psi|, \tag{13}$$

the reader can easily confirm that the probability, P, for the subensemble of measurements which yield, say, the eigenvalues a_0 for the first subsystem and c_0 for the second system, is given by

$$P = \mathrm{Tr}(\rho M), \tag{14}$$

where

$$M = (|a_0, 1\rangle \otimes |c_0, 2\rangle)(\langle c_0, 2| \otimes \langle a_0 1|) \tag{15}$$

Curiously, after such an observation, the state of the measured system is naturally described by the density matrix M, a situation which neither observer would subscribe to, nor would they have the necessary information immediately available to do so. This discrepancy results from the fact that different subensembles are being considered in each case. The first observer's probability distribution is based solely on his observation of a_0; the second observer, solely on his observation of c_0. The density matrix M specifies the statistical properties of the subensemble of data defined by the results of the two independent simultaneous measurements, A and C. Clearly all three subensembles are different.

19.3 Schrödinger's Cats

In the infamous Schrödinger Cat experiment, a cat's life is dependent upon the occurrence of a quantum transition of an atomic system. That is, the state of the cat, idealized as quantum and having only two possibilties, is correlated with the state of another quantum system. As the second system evolves, the cat's state becomes entangled with it, resulting in the cat being described at any given time by a state vector which coherently entangles "living" and "dead" states. For those who seriously believe in the objective reality of quantum states, such a situation poses a dilemma: is the act of observing the cat running the risk of killing him in the process (admittedly at the appropriate probability level)?

However, since we have shown that the quantum state is not an objective element of reality, but merely an algorithmic device for encoding and entangling probability information, the Schrödinger Cat problem becomes a nonproblem. The statement that, after some interval of time, the cat has say a 10% probability of being dead, simply means that if the experiment is identically repeated one hundred times (clearly not always with the same cat!), on approximately ten occasions it will be found that the poor cat has died. Pace Humane Society.

Alas, as a consequence of prosaic operationalism, the mysticism associated with quantum theory has dissipated, at least to some extent.

Pace Engelbert.

20

The Ultimate Extension of the Bianchi Classification for Rotating Dust Models

Andrzej Krasiński

ABSTRACT For a rotating dust with a 3-dimensional symmetry group all possible metric forms can be classified and, within each class, explicitly written out. With respect to the structures of the groups, this is just the Bianchi classification, but with all possible orientations of the orbits taken into account. This result follows from the formalism introduced by Plebański. This paper is a brief overview of results that will be published elsewhere.

20.1 Introduction and Summary

The theorem of Darboux (see Section 20.2) allows one to introduce invariantly defined coordinates in which the velocity field of a fluid acquires a preferred form. It is assumed in addition that the fluid moves with zero acceleration and nonzero rotation. These assumptions result in a simplification of the metric tensor and in limitations imposed on the Killing vectors, if any exist. A Killing field k^α may be spanned on velocity u^α and rotation w^α or may be linearly independent of u^α and w^α. This gives rise to a classification of possible symmetries in the presence of rotating matter.

When there exist three linearly independent Killing fields, the classification described above gives rise to a complete classification of all possible metric forms. With respect to the algebras of the symmetry groups, this is just the Bianchi classification, but with all possible orientations of the orbits taken into account (i.e., they may be timelike, spacelike or null).

In every case that emerges, the commutation relations of the algebra have been solved, resulting in explicit formulae for the Killing fields, and then the Killing equations have been solved, resulting in the formulae for the metric tensors compatible with the symmetry group considered. The degree of success in solving the Einstein equations varied very strongly from case to case. In most cases, no progress was made. In some cases the Einstein equations have been integrated either to an autonomous set of first-order equations or to a single nonlinear differential equation of second or third order. Several solutions known earlier were identified in the present scheme (those by Lanczos [1], Gödel [2], Maitra [3], Ellis [4], Vishveshwara and Winocour [6], and a few solutions with rotating charged dust; see below).

The Darboux theorem was first applied as a tool for investigating the equations of motion and the Einstein equations by Plebański [7]. The approach of Plebański was used by this author [8–12] to find a collection of stationary, cylindrically symmetric solutions of Einstein's equations with berotropic perfect fluid sources.

In Section 20.2 the Darboux theorem and the associated classification of first-order differential forms are introduced. In Section 20.3 the classification is applied to geodesic vector fields with rotation. When the vector field is the velocity field of a fluid, a class of preferred coordinates results (which shall be termed "Plebański coordinates"). In Section 20.4 it is shown that each Killing vector field that might possibly exist is a spacetime with a geodesic and rotating fluid source is determined by two functions of two variables. If the Killing field is not spanned on velocity and rotation, then the Plebański coordinates may be adapted to it so that it acquires the unique form $k^\alpha = \delta^\alpha_1$.

In Section 20.5, the consideration of Section 20.4 is applied to the case of three Killing vector fields existing on a manifold. When all three of them are spanned on u^α and w^α, the group becomes 2-dimensional, and this case is not considered here. When two of them are spanned on u^α and w^α while the third one is not, two cases arise. In one of them (Bianchi type II), the Einstein equations reduce to a single ordinary differential equation of third order (of second order when $\Lambda = 0$). In the other case (Bianchi type I), the Einstein equations are reduced to an autonomous set of first-order equations. The solutions of Lanczos [1] and of Gödel [2] emerge as special cases of both these classes.

The remaining cases are sketched only briefly. Section 20.6 contains the description of the case when two of the Killing fields are linearly independent of u^α and w^α; Section 20.7 of the case when all three Killing fields are linearly independent of u^α and w^α. With the increasing number of Killing vectors that are linearly independent of u^α and w^α, the equations become progressively more complicated, the number of subcases requiring a separate treatment increases, while the progress in integrating the Einstein equations decreases. The full results of the investigation will be published elsewhere.

20.2 The Classification of Differential Forms of First Order and the Darboux Theorem

Definition. *Let q be a differential form of first order.*

If $Q_{2l} := dq \wedge \ldots \wedge dq$ [multiplied l times] $\neq 0$, but $q \wedge Q_{2l} = 0$, then q is said to be of class $2l$.

If $Q_{2l+1} := q \wedge Q_{2l} \neq 0$, but $dQ_{2l+1} \equiv dq \wedge Q_{2l} = 0$, then q is said to be of class $(2l+1)$. □

Then the following holds:

The Theorem of Darboux. *The form q is of class $2l$ if and only if there exists a set of $2l$ independent functions $(\xi_1, \dots, \xi_l, \eta_1, \dots, \eta_l)$ such that*

$$q = \eta_1 d\xi_1 + \eta_2 d\xi_2 + \cdots + \eta_l d\xi_l. \tag{1}$$

The form q is of class $(2l+1)$ if and only if there exists a set of $(2l+1)$ independent functions $(\tau, \xi_1, \dots, \xi_l, \eta_1, \dots, \eta_l)$ such that

$$q = d\tau + \eta_1 d\xi_1 + \eta_2 d\xi_2 + \cdots + \eta_l d\xi_l. \tag{2}$$

□

(See [13] for a proof).

The Darboux theorem implies that in a 4-dimensional spacetime V_4 the most general differential form of first order can be represented as

$$q = \sigma d\tau + \eta\, d\xi, \tag{3}$$

where σ, τ, η, and ξ are scalar functions on V_4.

Any vector field u^α on V_4 defines the following form of first order:

$$q_u := u_\alpha\, dx^\alpha. \tag{4}$$

According to (3), in the most general case there exist scalar functions σ, τ, η, and ξ such that

$$u_\alpha = \sigma\tau,_\alpha + \eta\xi,_\alpha . \tag{5}$$

In general, the four functions are independent, i.e.,

$$\frac{\partial(\sigma, \tau, \eta, \xi)}{\partial(x^0, x^1, x^2, x^3)} \neq 0. \tag{6}$$

In that case, they can be chosen as coordinates in the spacetime.

20.3 Geodesically Moving Fluids

To any timelike vector field u_α normalized to unity (so that $u_\alpha u^\alpha = 1$), the decomposition described in [14] and [15] may be applied:

$$u_{\alpha;\beta} = \dot{u}_\alpha u_\beta + \sigma_{\alpha\beta} + \omega_{\alpha\beta} + \frac{1}{3}\theta h_{\alpha\beta}, \tag{7}$$

where $\dot{u}^\alpha$, θ, $\sigma_{\alpha\beta}$ and $\omega_{\alpha\beta}$ are, respectively, acceleration, expansion, shear, and rotation. In the signature (+ - - -) used here, the projection tensor $h_{\alpha\beta}$ is

$$h_{\alpha\beta} = g_{\alpha\beta} - u_\alpha u_\beta. \tag{8}$$

The following equations hold:

$$\dot{u}_\alpha u^\alpha = 0, \qquad \sigma_{\alpha\beta} u^\beta = \omega_{\alpha\beta} u^\beta = 0. \tag{9}$$

We shall assume from now on that u_α is the velocity field of a fluid and that $\dot{u}_\alpha = 0$, i.e., that the particles of the fluid move on geodesics. Then, from (5) we have

$$\omega_{\alpha\beta} = \frac{1}{2}(\sigma,_\beta \tau,_\alpha - \sigma,_\alpha \tau,_\beta + \eta,_\beta \xi,_\alpha - \eta,_\alpha \xi,_\beta), \tag{10}$$

and from (9) we have

$$(u^\beta \sigma,_\beta)\tau,_\alpha - (u^\beta \tau,_\beta)\sigma,_\alpha + (u^\beta \eta,_\beta)\xi,_\alpha - (u^\beta \xi,_\beta)\eta,_\alpha = 0. \tag{11}$$

It is easy to see that, in virtue of (11), the form (4) cannot be of class 4. Hence, for a geodesically moving fluid the form (4) is of class at most 3, i.e., at most 3 independent functions τ, η, ξ exist such that

$$u_\alpha = \tau,_\alpha + \eta \xi,_\alpha . \tag{12}$$

The functions $\{\tau, \xi, \eta\}$ in (12) are determined up to the following transformations:

$$\xi = F(\xi', \eta'), \qquad \eta = G(\xi', \eta'), \tag{13}$$

$$\tau = \tau' - S(\xi', \eta'), \tag{14}$$

where the functions F and G must obey the equation

$$F,_{\xi'} G,_{\eta'} - F,_{\eta'} G,_{\xi'} = 1, \tag{15}$$

and then S is determined by

$$S,_{\xi'} = GF,_{\xi'} - \eta', \qquad S,_{\eta'} = GF,_{\eta'} . \tag{16}$$

Let us now make the additional assumption that the number of particles of the fluid is conserved, i.e.,

$$(\sqrt{-g} n u^\alpha)_{,\alpha} = 0, \tag{17}$$

where g is the determinant of the metric tensor and n is the particle number density. This equation is a necessary and sufficient condition for the existence of a function ζ such that

$$\sqrt{-g} n u^\alpha = \varepsilon^{\alpha\beta\gamma\delta} \xi,_\beta \eta,_\gamma \zeta,_\delta . \tag{18}$$

Note that (12) implies that

$$u^\alpha \tau,_\alpha = 1, \tag{19}$$

and then Eq. (18) implies that

$$\varepsilon^{\alpha\beta\gamma\delta} \tau,_\alpha \xi,_\beta \eta,_\gamma \zeta,_\delta \equiv \frac{\partial(\tau, \eta, \xi, \zeta)}{\partial(x^0, x^1, x^2, x^3)} = \sqrt{-g} n \neq 0. \tag{20}$$

Equation (18) also implies that

$$u^\alpha \zeta,_\alpha = 0. \tag{21}$$

The function ζ is determined by (18) up to the transformations

$$\zeta = \zeta' + T(\xi', \eta'), \tag{22}$$

where T is an arbitrary function. Equation (20) certifies that $\{\tau, \xi, \eta, \zeta\}$ can be used as coordinates in the spacetime. If they are chosen as the $\{x^0, x^1, x^2, x^3\} = \{t, x, y, z\}$ coordinates, respectively, then Eq. (12) implies

$$u_0 = 1, \qquad u_1 = y, \qquad u_2 = u_3 = 0. \tag{23}$$

We will use these coordinates throughout the remaining part of the paper and call them "Plebański coordinates." Equation (20) implies now

$$g = -n^{-2}, \tag{24}$$

and Eq. (18) implies

$$u^\alpha = \delta^\alpha_0, \tag{25}$$

i.e., the Plebański coordinates are comoving. The rotation vector defined by

$$w^\alpha = -(1/\sqrt{-g})\varepsilon^{\alpha\beta\gamma\delta}u_\beta u_{\gamma,\delta} \tag{26}$$

assumes the form

$$w^\alpha = n\delta^\alpha_3. \tag{27}$$

Equations (23) and (25) imply that

$$g_{00} = 1, \qquad g_{01} = y, \qquad g_{02} = g_{03} = 0, \tag{28}$$

and also that the only nonvanishing components of the rotation tensor are

$$\omega_{12} = -\omega_{21} = 1/2. \tag{29}$$

If we now assume that the fluid is perfect, then we conclude from the equations of motion $T^{\alpha\beta}_{;\beta} = 0$ that either $\omega = 0$ or $p =$ const. (see also [16]). This means that a geodesic perfect fluid can be rotating only if it is in fact dust; the constant p can be reinterpreted as the cosmological constant. In this case, the energy-density obeys the conservation equation $(\sqrt{-g}\epsilon u^\alpha)_{,\alpha} = 0$ and Eq. (17) need not be assumed separately. A more detailed exposition of the same material can be found in [8].

Note that the rotating dust is not the only example to which this approach may be applied. In several papers, rotating charged dust was considered under the additional assumptions that all charges are attached to the dust particles, that the only current is the one created by the flow of dust, and that the Lorentz force acting on the dust particles $F^\mu{}_\nu u^\nu$ is zero, (i.e., that the electric and magnetic fields are such that they cancel each other's influence on the charged dust particles). Under these assumptions the dust particles move on geodesics and the formalism of this section applies. Such solutions of the Einstein-Maxwell equations were considered in references [18–25]; they will be mentioned in Section 20.5.

20.4 The Killing Vector Fields Compatible with Rotation

We shall assume that the symmetries of the spacetime (if any exist) are inherited by the source, i.e., that if the Lie derivative of the metric tensor $g_{\alpha\beta}$ along the vector field k^α is zero, $\pounds_k g_{\alpha\beta} = 0$, then the velocity field and the particle number density are also invariant: $\pounds_k u^\alpha = 0 = \pounds_k n$. (For a pure perfect fluid source the inheritance is guaranteed.) It follows that the rotation tensor must also be invariant, $\pounds_k \omega_{\alpha\beta} = 0$. All these equations imply that

$$k^0 = C + \phi - y\phi,_y\,, \qquad k^1 = \phi,_y\,, \qquad k^2 = -\phi,_x\,, \qquad k^3 = \lambda, \tag{30}$$

where $\phi(x, y)$ and $\lambda(x, y)$ are arbitrary functions and C is an arbitrary constant. If there are no symmetries, then $\phi = \lambda = C = 0$. However, if any symmetries are present, then the Killing vector fields must have the form (30).

Suppose that ϕ is not a constant, i.e., that a Killing vector field k^α exists that has a nonzero component in the x- or y-direction (in invariant terms this means that the vector field k^α is not spanned on the vector fields of velocity, u^α, and rotation, w^α). We can then, within the Plebański class defined in Section 20.3, adapt the coordinates to k^α in such a way that $k^{\alpha'} = \delta^{\alpha'}_1$, i.e., so that the metric becomes independent of x'. From (13)–(16) and (22) the transformation functions are

$$t' = t - S(x, y), \qquad x' = F(x, y), \qquad y' = G(x, y), \qquad z' = z + T(x, y), \tag{31}$$

where T is arbitrary, while F, G, and S obey

$$F,_x G,_y - F,_y G,_x = 1, \qquad S,_x = GF,_x - y, \qquad S,_y = GF,_y\,. \tag{32}$$

The condition $k^{\alpha'} = \delta^{\alpha'}_1$ then implies

$$G = \phi + C, \tag{33}$$

$$T,_x \phi,_y - T,_y \phi,_x = -\lambda. \tag{34}$$

Equations (32) and (34) simply define the accompanying F, S, and T, which are seen to exist always. Since ϕ was assumed nonconstant, the transformation is nonsingular, and results in $\phi = y$ in the new coordinates; the metric becomes independent of x after the transformation. This property is preserved by the transformations (31), but with F, G, S, and T restricted now by

$$G = y, \qquad F = x + H(y), \qquad T = T(y), \qquad S = \int yH,_y\, dy + A, \tag{35}$$

where A is an arbitrary constant and H and T are arbitrary functions.

Suppose that three Killing vector fields exist and all three are spanned on u^α and w^α, so that ϕ = const. in (30) for each of them, i.e.,

$$k^\alpha_{(i)} = C_i \delta^\alpha_0 + \lambda_i(x, y)\delta^\alpha_3, \qquad i = 1, 2, 3. \tag{36}$$

From the Killing equations it follows then that constants α_1, α_2 and α_3 exist such that $\alpha_1 k_{(1)} + \alpha_2 k_{(2)} + \alpha_3 k_{(3)} = 0$, i.e., the symmetry group is in fact 2-dimensional. Hence, for a 3-dimensional group at least one of the generators must be linearly independent of u^α and w^α at every point of the spacetime region under consideration.

20.5 The Case of Two Generators Spanned on u^α and w^α

In this section we shall assume that exactly one generator, $k^\alpha_{(1)}$, is linearly independent of u^α and w^α, while the other two, $k^\alpha_{(2)}$ and $k^\alpha_{(3)}$, are of the form (36). In agreement with the result of Section 20.4, the Plebański coordinates can be adapted to $k^\alpha_{(1)}$ so that

$$k^\alpha_{(1)} = \delta^\alpha_1, \tag{37}$$

while

$$k^\alpha_{(2)} = C_2\delta^\alpha_0 + \lambda_2(x, y)\delta^\alpha_3, \qquad k^\alpha_{(3)} = C_3\delta^\alpha_0 + \lambda_3(x, y)\delta^\alpha_3, \tag{38}$$

and the coordinate transformations preserving (37) and (38) are (35). Note that C_2 and C_3 cannot vanish simultaneously because otherwise the Killing equations immediately imply that either $k^\alpha_{(3)}$ = const. $k^\alpha_{(2)}$ (in which case the symmetry group is 2-dimensional) or the metric is singular. However, with no loss of generality we can assume that

$$C_2 \neq 0 = C_3 \tag{39}$$

because the Killing vector fields are determined up to linear combinations among them. This implies $\lambda_3 \neq 0$. The commutators of the Killing vectors are then

$$[k_{(1)}, k_{(2)}]^\alpha = (\lambda_{2,x}/\lambda_3)k^\alpha_{(3)}, \qquad [k_{(2)}, k_{(3)}]^\alpha = 0, \tag{40}$$

$$[k_{(1)}, k_{(3)}]^\alpha = (\lambda_{3,x}/\lambda_3)k^\alpha_{(3)}. \tag{41}$$

The Killing vector fields will thus form a Lie algebra when

$$\lambda_{2,x} = b\lambda_3, \qquad \lambda_{3,x} = c\lambda_3, \tag{42}$$

where b and c are arbitrary constants. The cases $c \neq 0$ and $c = 0$ have to be considered separately. However, when $c \neq 0$, it follows that

$$\lambda_3 = \beta(y)e^{cx}, \qquad \lambda_2 = (b/c)\beta(y)e^{cx} + \alpha(y), \tag{43}$$

and so with no loss of generality we can assume $b = 0$. The Einstein equations then imply that either $c = 0$ or there is no rotation. Since we are interested in rotating solutions only, this case need not be followed. We thus consider

Case I: $c = 0, b \neq 0$. Then

$$\lambda_3 = \beta(y), \qquad \lambda_2 = b\beta(y)x + \alpha(y). \tag{44}$$

The algebra of the Killing vector fields is of Bianchi type II when $b \neq 0$ and of Bianchi type I when $b = 0$.

In order to simplify the Killing vectors, we now transform the coordinates as follows:

$$(t', x', y') = (t, x, y), \qquad z' = -(\alpha/C_2)t + z/\beta. \tag{45}$$

The transformation is not of the form (35), so the new coordinates do not belong to the Plebański class, and the forms of velocity, rotation, and the metric will no longer agree with (23) - (29). The Killing vector fields in the new coordinates become

$$k_{(1)}^{\alpha} = \delta_1^{\alpha}, \qquad k_{(2)}^{\alpha} = \delta_0^{\alpha} + bx\delta_3^{\alpha}, \qquad k_{(3)}^{\alpha} = \delta_3^{\alpha}, \tag{46}$$

while the velocity and rotation fields become

$$u^{\alpha} = \delta_0^{\alpha} - (\alpha/C_2)\delta_3^{\alpha}, \qquad w^{\alpha} = (n/\beta)\delta_3^{\alpha}. \tag{47}$$

The transformed metric is independent of x and z.

The orbits of the symmetry group are the $\{y = \text{const}\}$ hypersurfaces. In order to follow the standard technique of the Bianchi-type spaces, we should now carry out a coordinate transformation that preserves (46) and makes the y-coordinate curves orthogonal to the group orbits, so that $g'_{02} = g'_{12} = g'_{23} = 0$ after the transformation. This step is not in fact necessary for solving the Einstein equations (in general it only reshuffles the unknown functions without eliminating any of them), but in the case under consideration it leads to a simplification.

After the transformation, and with the Killing equations solved, the metric becomes (primes dropped; details of the derivation will be given in another paper):

$$ds^2 = (dt + Y\,dx)^2 - (F\,dx)^2 - dy^2 - G^2[A\,dt - (bt - k_{13})\,dx + dz]^2, \tag{48}$$

where $G(y)$, $A(y)$, $k_{13}(y)$, and $F(y)$ are new names for the unknown functions. The velocity field in the coordinates of (48) is

$$u^{\alpha} = \delta_0^{\alpha} - A\delta_3^{\alpha}. \tag{49}$$

The components of the Einstein tensor will be referred to the orthonormal tetrad of forms $e^i = e^i_{\alpha}\,dx^{\alpha}, i = 0, 1, 2, 3,$ uniquely implied by (48). Note that $e^0 = u_{\alpha}\,dx^{\alpha}$.

The equation $G_{12} = 0$ implies that $bA,_y = 0$. The case $b = 0$ will be considered separately below, so we take here

$$A = \text{const}. \tag{50}$$

Then other field equations, together with simplifying coordinate transformations, lead to

$$A = k_{13} = 0, \tag{51}$$

$$Y,_y\,G/F = B = \text{const}. \tag{52}$$

and we can assume $B \neq 0$ because rotation would be zero with $B = 0 = Y,_y$.

It is convenient to introduce $Y(y)$ as the new variable. The equation $G_{11}+G_{22} = 2\Lambda$ can then be written, with the help of (52), as

$$(F^2 G,_Y / G),_Y = \frac{2\Lambda G^2}{B^2} - \frac{1}{2}, \tag{53}$$

and so

$$F^2 = \left(C - \frac{1}{2}Y + 2\frac{\Lambda}{B^2}\int G^2\, dY\right) G/G,_Y\,, \tag{54}$$

where C is a new arbitrary constant (we can assume $G,_Y \neq 0$ because $G,_Y = 0$ immediately implies $b = 0$ from $G_{11} - G_{22} = 0$, and $b = 0$ will be considered separately). Using (54) in $G_{22} = \Lambda$, we obtain the following integro-differential equation that determines G:

$$-\frac{1}{4}b^2 G G,_Y + \frac{1}{2}(B/G)^2 \left(C - \frac{1}{2}Y + 2\frac{\Lambda}{B^2}\int G^2\, dY\right)^2 (G,_Y/G - G,_{YY}/G,_Y)$$
$$= 0. \tag{55}$$

In the special case $\Lambda = 0$, this becomes an ordinary second-order differential equation. It is easy to get rid of the integral by transforming (55) appropriately and differentiating the result by Y (in this way a third-order differential equation for $G(Y)$ is obtained) or by introducing the new variable $u(Y)$ by $dY/du = 1/G^2$ (this results in a second-order equation for $G(u)$). However, no progress toward solving (55) results in either case.

The formula for energy-density may be simplified to

$$(8\pi G/c^4)\epsilon = (B/G)^2 - (bG)^2 - 2\Lambda. \tag{56}$$

Note that the solutions considered here have a meaningful limit $b = 0$.

When $G =$ const., equations (54) and (55) no longer apply and one has to go back to the Einstein equations; the resulting metric is the Gödel solution.

When $G,_Y \neq 0 = b$, Eq. (55) implies $G = e^{DY+E}$, and this leads to the Lanczos solution (see Ref. [8]).

Case II: $b = c = 0$ in (40)–(42).

The reasoning up to Eq. (49) applies also here, but (50) no longer follows. Instead, the equation $G_{13} = 0$ can be integrated with the result

$$k_{13,y} = BF/G^3 - YA,_y\,, \tag{57}$$

where B is an arbitrary constant; the equation $G_{01} =$ can be integrated to

$$Y,_y = (C - BA)F/G, \tag{58}$$

where C is an arbitrary constant; and the equation $G_{03} = 0$ can be integrated to

$$A,_y = (BY - D)/(FG^3), \tag{59}$$

where D is one more arbitrary constant.

At this point, only the diagonal components of the Einstein tensor survive, and $G_{00} = (8\pi G/c^4)\epsilon - \Lambda$ just defines the energy-density. The equations $G_{11} = \Lambda =$

$G_{22} = G_{33}$ can be integrated to the first-order set[1]

$$FG,_y = \frac{1}{2}Bk_{13} + \frac{1}{2}BAY - \frac{1}{2}CY + 2\Lambda \int FG\,dy + H_0, \tag{60}$$

$$GF,_y = -\frac{1}{2}Bk_{13} - \frac{1}{2}DA - E + 2\Lambda \int FG\,dy + H_0. \tag{61}$$

where H_0 is an arbitrary constant. The integral can be calculated if the new variable $u(y)$ is introduced by

$$dy/du = 1/(FG). \tag{62}$$

In terms of the variable u from (61), equations (57)–(61) form an autonomous set of first-order equations that can be investigated further by qualitative methods (see for example [17]). This is left as a subject for a separate study.

The functions $A(y)$ and $k_{13}(y)$ have invariant meaning: they are proportional to the scalar products of the Killing vectors (see equations (46) and (48) with $b = 0$):

$$A = -g_{\alpha\beta}k^{\alpha}_{(2)}k^{\beta}_{(3)}/G^2, \qquad k_{13} = -g_{\alpha\beta}k^{\alpha}_{(1)}k^{\beta}_{(3)}/G^2 \tag{63}$$

(note that $G^2 = -g_{\alpha\beta}k^{\alpha}_{(3)}k^{\beta}_{(3)}$, i.e., it is a scalar, too). Hence, $A = 0$ and $k_{13} = 0$ are invariant properties. Note that $A = 0$ implies, through (59), that either $Y =$ const. (in which case there is no rotation) or $B = D = 0$. In the latter case, $k_{13} =$ const. and the coordinate transformation $z = z' - k_{13}x$ leads to $k_{13} = 0$ in the new coordinates. With $A = k_{13} = 0$, the Lanczos and Gödel models result from the Einstein equations as the only solutions.

In references [18–25], charged dust solutions with zero Lorentz force were considered. Apart from the one in [18], they are cylindrically symmetric and stationary, and so they would emerge in this section, had we allowed charged dust as a source and considered the Einstein-Maxwell equations. However, not all of these solutions allow nonempty limits of vanishing electromagnetic fields. The ones from [19] and [21] become a vacuum solution and the Minkowski spacetime, respectively, in that limit, the one from [20] does not allow the limit at all. The solution by Som and Raychaudhuri [22] is a generalization to charged dust of the $\Lambda = 0$ subcase of the Lanczos solution [1]; the first of the six solutions by Banerjee and Banerji [23] is a generalization of the Gödel solution. (The other solutions from Ref. [23] have the following properties: the second and the sixth become vacuum solutions in the limit $F_{\mu\nu} = 0$, the third does not allow this limit at all, the fourth has a 2-dimensional symmetry group, and the fifth reduces to the Minkowski metric in the limit.) The two solutions by Mitskiévič and Tsalakou [24] are, respectively, generalizations of the full Lanczos solution and of the Gödel solution to a charged dust source, and those by Upornikov [25] are coordinate transforms of those from [24].

This short overview of literature deliberately omits plain rediscoveries of solutions known earlier; they will be listed in the main paper.

---

[1] The derivation will be published separately.

20.6 The Case of One Generator Spanned on u^α and w^α

The number of cases that require separate treatment is larger here, and the equations are more complicated, so the results will be described only briefly.

One of the two generators that are not spanned on u^α and w^α may be given the simple form (37) by a coordinate transformation, the other will have the general form (30). The one that is spanned on u^α and w^α will have the form (38). Hence, the generators are

$$k^\alpha_{(1)} = \delta^\alpha_1,$$

$$k^\alpha_{(2)} = (C_2 + \phi - y\phi,_y)\delta^\alpha_0 + \phi,_y \delta^\alpha_1 - \phi,_x \delta^\alpha_2 + \lambda_2(x, y)\delta^\alpha_3,$$

$$k^\alpha_{(3)} = C_3\delta^\alpha_0 + \lambda_3(x, y)\delta^\alpha_3, \tag{64}$$

and the remaining freedom of coordinate transformations is given by equations (31) and (35). The Killing fields (64) will form a Lie algebra when

$$[k_{(1)}, k_{(2)}] = ak_{(1)} + bk_{(2)} + ck_{(3)},$$

$$[k_{(1)}, k_{(3)}] = dk_{(1)} + ek_{(2)} + fk_{(3)},$$

$$[k_{(2)}, k_{(3)}] = gk_{(1)} + hk_{(2)} + jk_{(3)}, \tag{65}$$

where $a, \ldots, j$ are constants to be determined from (64) and (65). The set (65) written out explicitly is

$$\begin{aligned}
&\phi,_x - y\phi,_{xy} = b(C_2 + \phi - y\phi,_y) + cC_3,\\
&\phi,_{xy} = a + b\phi,_y, \qquad \phi,_{xx} = b\phi,_x,\\
&\lambda_{2,x} = b\lambda_2 + c\lambda_3,\\
&e(C_2 + \phi - y\phi,_y) + fC_3 = 0,\\
&d + e\phi,_y = 0, \qquad e\phi,_x = 0,\\
&\lambda_{3,x} = e\lambda_2 + f\lambda_3,\\
&h(C_2 + \phi - y\phi,_y) + jC_3 = 0,\\
&g + h\phi,_y = 0, \qquad h\phi,_x = 0,\\
&\phi,_y \lambda_{3,x} - \phi,_x \lambda_{3,y} = h\lambda_2 + j\lambda_3.
\end{aligned} \tag{66}$$

Two equations in this set form an alternative: either $\phi,_x = 0$ or $e = h = 0$. Such alternatives also occur at later stages of integration, and they give rise to the large number of separate cases. In three of them simple explicit solutions, most probably new, were derived; they will be published in the main papers. The case that contains the solutions considered by Maitra [3], King [5], and Vishveshwara and Winicour [6] will be presented here.

$$\phi,_x = \lambda_{2,x} = \lambda_{3,x} = C_3 = 0 \neq \lambda_3, \qquad \text{all constants } a, \ldots, j = 0. \tag{67}$$

Now $\phi_{,y} \neq 0$ follows from the definition of the case; otherwise $k_{(2)}$ would be spanned on u and w. The resulting Killing fields are given by the appropriate subcase of (64)–(65) and all commute to zero.[2] Hence, the Bianchi type of the algebra is I, but the solution is different from (57)–(61) because of the orientation of the orbit of the symmetry group in the spacetime.

Through coordinate transformations the Killing vector fields can now be transformed to the simplest form:

$$k^{\alpha}_{(1)} = \delta^{\alpha}_{1}, \qquad k^{\alpha}_{(2)} = \delta^{\alpha}_{0}, \qquad k^{\alpha}_{(3)} = \delta^{\alpha}_{3}, \tag{68}$$

and the velocity, rotation, and the metric then acquire the forms

$$\begin{aligned} u^{\alpha} &= F^{-1}(\delta^{\alpha}_{0} - P\delta^{\alpha}_{1} - \lambda_2\delta^{\alpha}_{3}), \qquad w^{\alpha} = (n/\lambda_3)\delta^{\alpha}_{3}, \\ ds^2 &= [(C_2 + \phi)dt + Ydx]^2 - k^2_{11}(Pdt + dx)^2 - dy^2 \\ &\quad - k^2_{33}[(\lambda_2 + h_{13}P)dt + h_{13}dx + dz]^2, \end{aligned} \tag{69}$$

where all symbols are functions of y only, and

$$P = \phi_{,Y} \equiv \{\phi_{,y}\}/Y_{,y}, \qquad F = C_2 + \phi - YP. \tag{70}$$

King [5] considered rotating dust metrics with the same symmetry, but in addition assumed reflection symmetries that in the coordinates of (69) correspond to $z \to -z$ and $(t, x) \to (-t, -x)$. With these additional assumptions, $\lambda_2 = h_{13} = 0$. Even in this case, King found that the problem is underdetermined: one of the functions (in our notation it is k_{33}) may be chosen arbitrarily. King's paper contains a few examples of explicit solutions resulting from different choices of it (among them are the solutions of Lanczos [1] that goes by the name of Ehlers - van Stockum, and of Maitra [3]). Another specific example was found by Vishveshwara and Winicour [6].

King's metric ansatz can be derived from the following assumptions:

1. The algebra of the symmetry group is of Bianchi type I.
2. One Killing field ($k_{(3)}$) is collinear with rotation, the two others are linearly independent of u and w.
3. The velocity vector field is spanned on $k_{(1)}$ and $k_{(2)}$ only (i.e., $\lambda_2 = 0$).
4. The Killing fields $k_{(1)}$ and $k_{(2)}$ are both orthogonal to $k_{(3)}$ (i.e., $h_{13} = 0$).

Also in this class is the Maitra solution [3] which has the following invariant property in addition:

5. The timelike Killing field $k_{(2)}$ has unit length so that $(C_2 + \phi)^2 - k^2_{11}P^2 = 1$.

These conditions are still insufficient to reduce (68)–(69) to the Maitra solution; the following coordinate-dependent relations must hold in addition:

$$\begin{aligned} (C_2 + \phi)Y - k^2_{11}P &= m, \\ Y^2 + [1 - (C_2 + \phi)^2]/P^2 &= m^2 - r^2, \end{aligned} \tag{71}$$

[2] Some of the equations in (6.7) follow as necessary consequences of some others through (6.6), so the total number of different cases is not as large as (6.7) might suggest.

where $r(y)$ is a new coordinate defined by

$$\ln\left(\frac{dy}{dr}\right) = -\frac{1}{4u^2}\left\{(1+u^2)^{1/2} - 1 + \frac{1}{8} - \frac{1}{4}\ln[\frac{1}{2}(1+u^2)^{1/2} + \frac{1}{2}]\right\}$$

$$u := 2r/a, \qquad a = \text{const}, \tag{72}$$

and $m(r)$ is the function

$$m = -\frac{1}{2}a\left\{(1+u^2)^{1/2} - 1 - \ln\left[\frac{1}{2}(1+u^2)^{1/2} + \frac{1}{2}\right]\right\}. \tag{73}$$

This author was not able to interpret (71) in invariant terms.

The collection of models described in this section has a nonempty (in fact, quite large) common subset with those by Ellis [4]. However, the interrelations are somewhat complicated and require a more elaborate explanation. In short, the following classes of Ellis with nonzero rotation are not contained in this collection:

1. The generic case Ib; because it has a 4-dimensional symmetry group acting multiply transitively on 3-dimensional orbits, and the group has in general no 3-dimensional simply transitive subgroups. However, in special cases such subgroups do exist and the corresponding solutions of the Einstein equations are found in the present scheme.

2. The generic case Cii of the shearfree solutions; because it has only a 2-dimensional symmetry group.

All other rotating solutions of Ellis do belong to the present collection.

The first three of the six solutions by Ozsváth [26] also belong here, and they are subcases of the class considered by King. All of Ozsváth's solutions have 4-dimensional symmetry groups whose orbits are the whole 4-dimensional manifolds. In order to place specific Ozsváth's solutions in the classification considered here, one has to identify 3-dimensional subgroups of Ozsváth's groups. Examples can be spotted by inspection in Ref. 26 in which different non-isomorphic 3-dimensional subgroups are contained in the same 4-dimensional group. Hence, the same Ozsváth's solutions should come up as limits in different classes of the present investigation. For unique and complete identification, the formulae for group generators are necessary, and these are not given for most of Ozsváth's solutions. Among the cases that could not be identified are the other solutions of Ozsváth [27] and the "finite rotating Universe" of Ozsváth and Schücking [28-30]; this is where the present investigation makes contact with the legacy of the patron of this volume.

Other results in this class will be presented in a separate paper. Most of them do not lead to any explicit solution of the Einstein equations.

20.7 The Case of All Three Generators Being Linearly Independent of u^α and w^α

The commutator equations are still more complicated here, and the number of cases is still larger than in Section 20.6. The Killing fields $k_{(1)}$ and $k_{(2)}$ are the same as in (64), while $k_{(3)}$ is

$$k^\alpha_{(3)} = (C_3 + \psi - y\psi_{,y})\delta^\alpha_0 + \psi_{,y}\,\delta^\alpha_1 - \psi_{,x}\,\delta^\alpha_2 + \lambda_3(x, y)\delta^\alpha_3; \tag{74}$$

ϕ and ψ are functions of x and y. The commutator equations (65) can be partly integrated without going into separate cases. The components (t, x, y) of the first equation in (65) are integrated to

$$\phi_{,x} = ay + b\phi + c\psi + bC_2 + cC_3; \tag{75}$$

the components (t, x, y) of the second equation in (65) are integrated to

$$\psi_{,x} = dy + e\phi + f\psi + eC_2 + fC_3; \tag{76}$$

and the components (x, y) of the third equation in (65) are integrated to

$$\phi_{,y}\,\psi_{,x} - \phi_{,y}\,\psi_{,x} = gy + h\phi + j\psi + C, \tag{77}$$

where C is an arbitrary constant. The remaining equations are

$$\lambda_{2,x} = b\lambda_2 + c\lambda_3, \qquad \lambda_{3,x} = e\lambda_2 + f\lambda_3, \tag{78}$$

$$\phi_{,y}(\psi_{,x} - y\psi_{,xy}) + y\phi_{,x}\,\psi_{,yy} - \psi_{,y}(\phi_{,x} - y\phi_{,xy}) - y\psi_{,x}\,\phi_{,yy}$$

$$= h(C_2 + \phi - y\phi_{,y}) + j(C_3 + \psi - y\psi_{,y}),$$

$$\phi_{,y}\lambda_{3,x} - \phi_{,x}\lambda_{3,y} - \psi_{,y}\lambda_{2,x} + \psi_{,x}\lambda_{2,y} = h\lambda_2 + j\lambda_3. \tag{79}$$

The equations to solve are (75), (76), and (78); the remaining ones are consistency conditions to be imposed on the solutions. The set (75)–(76) and the set (78) are both (ordinary differential) linear vector equations of the same form:

$$U_{,x} = AU + W, \tag{80}$$

where, for (75)–(76), the constant matrix A and the vectors U and W are

$$A = \begin{pmatrix} b & c \\ e & f \end{pmatrix}, \quad U = \begin{pmatrix} \phi \\ \psi \end{pmatrix}, \quad W = y\begin{pmatrix} a \\ d \end{pmatrix} + A\begin{pmatrix} C_2 \\ C_3 \end{pmatrix}, \tag{81}$$

while for (78) the matrix A is the same, $W = 0$, and $U = \begin{pmatrix} \lambda_2 \\ \lambda_3 \end{pmatrix}$.

Solving Eq. (80) is a textbook exercise, but, since the constants $a, \ldots, j$ are all arbitrary, a multitude of separate cases arises: the matrix A may be nonsingular with two complex eigenvalues, with two real eigenvalues, one double eigenvalue or a single eigenvalue, it may be singular with two different eigenvalues, nilpotent, etc. Some of the subcases turn out to be equivalent in the end (in the sense that they generate the same algebra), some others turn out to be reducible (by changes of the basis of Killing vectors) to those considered in Section 20.6. Still, the number of cases to be considered is large. No explicit solutions of the Einstein equations were identified in this case, and so it will not be described here in any more detail.

20.8 Conclusion

The investigation should be useful as an intermediate step in looking for more general solutions: perfect fluid solutions with the same symmetries and any solutions with lower symmetries. The progress with respect to earlier knowledge on hypersurface-homogeneous geometries with a rotating dust source consists in the fact that such solutions have been looked for by trial and error, beginning from certain metric ansatzes. The collection of possible ansatzes was hereby reduced to a well-defined, not-too-large set.

*****

The algebraic calculation for this paper were done with use of the program Ortocartan [31, 32].

Note added in proof: The full results of the research reported here have already been published in *J. Math. Phys.* **39** (1998), 380–400, 401–422, 2148–2179.

References

[1] K. Lanczos, *Z. Physik* **21**, 73 (1924); English trans. in *Gen. Rel. Grav.* **29** 363 (1997).
[2] K. Gödel, *Rev. Mod. Phys.* **21**, 447 (1949).
[3] S. C. Maitra, *J. Math. Phys.* **7**, 1025 (1966).
[4] G. F. R. Ellis, *J. Math. Phys.* **8**, 1171 (1967).
[5] A. R. King, *Commun. Math. Phys.* **38**, 157 (1974).
[6] C. V. Vishveshwara & J. Winicour, *J. Math. Phys.* **18**, 1280 (1977).
[7] J. Plebański, *Lectures on non-linear electrodynamics*, Nordita, Copenhagen, 1970, pp. 107–115 and 130–141.
[8] A. Krasiński, *Acta Phys. Polon.* **B5**, 411 (1974).
[9] A. Krasiński, *Acta Phys. Polon.* **B6**, 223 (1975).
[10] A. Krasiński, *Acta Phys. Polon.* **B6**, 239 (1975) (available in extended form as a preprint).
[11] A. Krasiński, *J. Math. Phys.* **16**, 125 (1975).
[12] A. Krasiński, *Rep. Math. Phys.* **14**, 225 (1978).
[13] S. Sternberg, *Lectures on differential geometry*, Prentice Hall, Englewood Cliffs, N. J. 1964, p. 141.
[14] G. F. R. Ellis, in *General relativity and cosmology* (Proceedings of the International School of Physics "Enrico Fermi," Course 47). Edited by R. K. Sachs. Academic Press, New York and London, 1971, p. 104.
[15] J. Ehlers, *Abhandl. Math. Naturw., Kl. Akad. Wiss. Lit. Mainz* **11**, 791 (1961); English translation in *Gen. Rel. Grav.* **25**, 1225 (1993).
[16] C. B. Collins, *Canad. J. Phys.* **64**, 191 (1986).

[17] O. J. Bogoyavlenskii, *Methods in the qualitative theory of dynamical systems in astrophysics and gas dynamics*, Springer-Verlag, New York, 1980.
[18] J. N. Islam, *Proc. Roy. Soc.* London **A353**, 523 (1977).
[19] J. N. Islam, *Proc. Roy. Soc.* London **A385**, 189 (1983).
[20] P. Wils & N. van den Bergh, *Proc. Roy. Soc.* London **A394**, 437 (1984).
[21] A. Georgiou, *Nuovo Cimento,* **B108**, 69 (1993).
[22] M. M. Som & A. K. Raychaudhuri, *Proc. Roy. Soc.* London **A304**, 81 (1968).
[23] A. Banerjee & S. Banerji, *J. Phys.* **A1**, 188 (1968).
[24] N. V. Mitskevič & G. A. Tsalakou, *Class. Quant. Grav.* **8**, 209 (1991).
[25] A. M. Upornikov, *Class. Quant. Grav.* **11**, 2085 (1994).
[26] I. Ozsváth, *J. Math. Phys.* **6**, 590 (1965).
[27] I. Ozsváth, *J. Math. Phys.* **11**, 2871 (1970).
[28] I. Ozsváth, *Abhandl. Math. Naturw. Kl., Akad. Wiss. Lit. Mainz* No.13 (1962).
[29] I. Ozsváth & E. Schücking, *Nature* **193**, 1168 (1962).
[30] I. Ozsváth & E. Schücking, *Ann. Phys.* **55**, 166 (1969).
[31] A. Krasiński, *Gen. Rel. Grav.* **25**, 165 (1993).
[32] A. Krasiński & M. Perkowski, *Gen. Rel. Grav.* **13**, 67 (1981).

21

On the Classification of the Real Four-Dimensional Lie Algebras

M.A.H. MacCallum

ABSTRACT The distinct real 4-dimensional Lie algebras are enumerated. The method of enumeration uses a result due to Farnsworth and Kerr, and provides a basis for an algorithm for classifying an algebra given in terms of its structure constants. The different types are given labels referring to their invariant properties. Alternative schemes in the literature are reviewed, and a table cross-referencing all these schemes is provided. The 3-dimensional algebras are similarly treated in an appendix.

Dedication

It is a pleasure to dedicate this paper to Prof. Engelbert Schucking, in acknowledgment of his inspirational research and teaching, and in recognition of pleasant memories of social occasions. I feel, and I hope he will, that it is a particularly appropriate one to be able to offer, because it continues a line of work in which Schucking's own contributions directly influenced my Ph.D. supervisor, George Ellis, and hence my own first scientific paper (Ellis and MacCallum, 1969), in which Schucking's name appears on the second page.

A previous version of this work was circulated in 1979. I prepared it as a useful background for work in homogeneous space-times and related problems, but it had (and has) the weakness of being rather specific to that purpose and using methods whose adaptation to higher dimensions, while that may be possible, is not instantly evident. However, since its circulation in preprint form it has been referred to or used by by a number of authors (e.g. Bradfield and Kantowski (1982), Demaret and Hanquin (1985), Demianski *et al* (1987), Turkowski (1988), Schöbel (1992), Marklund (1997)), and I am therefore glad to have the extra stimulus of contributing to this volume for Prof. Schucking to prompt me to make it suitable for publication. The present revision has introduced a number of changes, but the classification has only been modified slightly, so the form given here is reasonably compatible with the citations of the preprint.

21.1 Introduction

The enumeration of the distinct real 4-dimensional Lie algebras may appear to a pure mathematician to be too trivial and special a problem to be of interest. However, it is of some use in applications, for instance in relation to solutions of Einstein's equations with high but not maximal symmetry. To make the paper self-contained and thus more accessible to readers with interests in applications, I have used only rather elementary methods from, and provided a summary of some well-known results in, Lie algebra theory. Lie (1893), in his classic work, enumerated all complex 4-dimensional algebras. Kruchkovich (1954) adapted these results to the real case, and Petrov (1969) and, independently, Ellis and Sciama (1966), completed this classification. A second enumeration was given by Bratzlavsky (1959). (As this work is not obtainable, I have relied on the accounts by Sengier-Diels (1974a, 1974b and private communication) and Fee (1979), who constructed all possible (pseudo-)Riemannian metrics on the corresponding connected Lie groups.) A third enumeration was given by Mubarakzyanov (1963a), in illustration of his general method for enumerating all solvable Lie algebras, and modified forms of that scheme have been used by Patera et al. in several papers (Patera *et al*, 1976; Patera and Winternitz, 1977; Patera and Zassenhaus, 1990)). These three classifications do not exhaust all such enumerations in the literature; for example Fee (1979) mentions an early enumeration by Dobrescu (1953) which I have not been able to obtain, and there may be others. There have also been some discussions of subclasses, in particular the nilpotent algebras, for example Sund (1979), Magnin (1986), Seeley (1993) and references therein.

Moreover, the authors of the three schemes for the real case noted above, and described in Section 21.3, seem to have been unaware of each other's work. Of those three, the Mubarakzyanov/Patera et al. scheme, which is complete and nonredundant, seems to be best. The scheme developed here could be considered superior only in the following respects: (i) it is based on a method which naturally provides the basis of an algorithm for classifying an algebra given the commutators of a basis, (ii) it follows the same pattern as the well-known schemes for algebras of dimensions 2 and 3, and (iii) the labeling of the cases is less arbitrary than those which just assign sequential integer numbers, in that it represents invariant properties. For use by relativists it has the extra advantage of being adapted to the treatment of homogeneous spacetimes developed by Ozsváth (see for example Ozsváth (1965), Hiromoto and Ozsváth (1978)) and by Farnsworth and Kerr (1966), and indeed depends on a result given in the latter paper.

The detailed enumeration is given in Section 21.2. Section 21.3 is in effect an extended caption to the accompanying table cross-referencing the various schemes. It includes a review of the other schemes. An appendix gives a similar table for 3-dimensional Lie algebras, since this is referred to in Mubarakzyanov's treatment of the 4-dimensional case. Section 21.4 discusses extensions, applications, and other related work.

The following notation is used. L denotes a Lie algebra, and L_n a Lie algebra of dimension n. If $\{X_a;\ i = 1, 2, \ldots, n\}$ is a basis of L_n,

$$[X_a, X_b] = C^d{}_{ab} X_d, \tag{1}$$

where $[\ ,\ ]$ is the Lie product (commutator) and summation over repeated indices is implied, defines the structure constants $C^d{}_{ab}$. These obey

$$C^d{}_{ab} = -C^d{}_{ba}, \tag{2}$$

$$C^e{}_{[bc} C^a{}_{d]e} = 0, \tag{3}$$

where indices contained within square (respectively, round) brackets are to be skewed over (respectively, symmetrized over). The left sides of (1) are the basis of a subalgebra called the first derived algebra, which will be denoted L' (or L'_n). Equation (3) is the Jacobi identity. When dealing with the L_4, Greek indices will run from 1 to 3 and Latin indices from 1 to 4.

We now recall some basic results from the theory of Lie algebras[1] (see for example Jacobson (1962)). The operation of constructing the derived algebra can be repeated. If after a finite number of repetitions of this operation the resulting derived algebra is the trivial one (consisting just of the vector 0), then the algebra is called *solvable* or *integrable*. The largest solvable ideal of L (an ideal being a subalgebra $M \subseteq L$ such that $ML \subseteq M$) is unique and is called the (*solvable*) *radical* R (or $R(L)$). If $R = 0$, the algebra is called *semisimple* and is necessarily a sum of simple algebras (where *simple* means there are no nontrivial subalgebras). $R = L$ is the solvable case.

Algebras which are neither semisimple nor solvable are covered by a theorem of Levi, proving that they have the structure $R + S$ where S is a semisimple algebra which is unique up to a conjugacy as shown by a theorem of Mal'cev and Harish-Chandra. Here the $+$ refers to the vector space structure: the Lie products are such that $N^2 \subset N$, $S^2 = S$, and $SN \subseteq N$. This is called a *semidirect* product (*direct*, denoted $\oplus$, if $SN = 0$), and the action of S on N by the Lie product gives a representation of S on N and, as a consequence of the Jacobi identities, (3) is also a derivation[2] in N. There are algorithmic ways to find this decomposition (Rand, 1986; Zassenhaus *et al*, 1987; Rand, 1987; Rand *et al*, 1988; de Graaf *et al*, 1997).

The simple (and hence the semisimple) algebras over $\mathbb{C}$ are completely known. However, there is no complete list of the solvable algebras (and hence, no such list of those which are neither semisimple nor solvable, or of all algebras) over arbitrary dimension n. For $n = 1$ to 4 the results are included here. Because the smallest real semisimple algebras have dimension 3, there are no 4-dimensional algebras with (nontrivial) Levi decompositions to consider: all the cases considered here are either solvable or are direct products of a semisimple algebra and $\mathbb{R}$.

[1]The underlying field here will be taken to have characteristic zero, since this paper actually deals only with $\mathbb{R}$ and $\mathbb{C}$.

[2]A derivation is a map D such that for any vectors $\mathbf{x}$ and $\mathbf{y}$, $D(\mathbf{xy}) = (D\mathbf{x})\mathbf{y} + \mathbf{x}(D\mathbf{y})$.

Mubarakzyanov (1963b) gave a complete list of Lie algebras for $n \leq 5$ (see also Patera *et al* (1976)). Morozov (1958) gave the nilpotent algebras of dimension 6, Mubarakzyanov (1963c) gave the solvable cases of dimension 6 containing a nilpotent ideal of dimension 5, and Turkowski (1990) completed the list of solvable algebras of dimension 6. (An algebra is nilpotent if for some n, $L^n = 0$. All nilpotent algebras are solvable, but not vice versa.) Magnin (1986) and Seeley (1993) gave the nilpotent algebras of dimension 7. Turkowski (1988, 1992) gave all algebras with a nontrivial Levi decomposition up to dimension 9 (since the smallest semisimple algebra has dimension 3, this requires the solvable algebras only up to dimension 6), and this has been extended to some cases of higher dimension of relevance to multidimensional cosmologies (Turkowski, 1987), (Demianski *et al*, 1987).

Following Vranceanu (1947) and various later authors (Ozsváth (1965), Estabrook *et al* (1968), Ellis and MacCallum (1969)) it is useful to define the structure constant vector $\boldsymbol{c}$ by

$$c_b = C^a{}_{ab}. \tag{4}$$

The basic division in the present scheme is to separate the unimodular algebras[3] which have $\boldsymbol{c} = 0$, from the non-unimodular algebras which have $\boldsymbol{c} \neq 0$. This completely classifies the L_2 (see e.g. Lie (1893), Bianchi (1918)) and divides the L_3 (see Estabrook *et al* (1968), Ellis and MacCallum (1969)) and L_4 into two main classes. The unimodular algebras L_4 are then subdivided into two groups using the result of Farnsworth and Kerr (1966).

One may note that there are only two distinct real semisimple (in fact, simple) algebras of dimension 4 or less, both of dimension 3 (and equivalent over $\mathbb{C}$). Only unimodular algebras can be simple, since it is easy to see that (3) implies

$$c_e C^e{}_{ab} = 0, \tag{5}$$

so that $\dim L'_n \leq (n-1)$ if $c_e \neq 0$.

21.2 An Enumeration of the 4-Dimensional Algebras

Let us first consider the non-unimodular algebras. I will call this class[4] N. Choosing a basis such that $c_a = (0, 0, 0, c)$, (5) shows $C^\alpha{}_{\beta\gamma}$ and $C^\alpha{}_{\beta 4} = \theta^\alpha{}_\beta$ remain to be

[3] A Lie group has left-invariant and right-invariant Haar measures. The ratio of these measures is in general nonconstant. It is constant, and so can be taken to be 1, if and only if the associated Lie algebra has $\mathbf{c} = 0$. This ratio is called the modulus, and this is one reason for the name unimodular. Another is that a unimodular group can be represented by unimodular matrices.

[4] In the preliminary version this was called C, to show $\mathbf{c} \neq 0$, but that now seems too dependent on choice of notation. N stands for "non-unimodular."

discussed. The former can be represented by

$$n^{ab} = \frac{1}{2c} C^a{}_{de} \epsilon^{debf} c_f, \tag{6}$$

where ϵ^{abcd} are the components of an arbitrary but fixed nonzero 4-form on L_4. From this definition it follows that n^{ab} is symmetric and $n^{a4} = 0$. One can choose a basis of the subspace V spanned by $\{X_1, X_2, X_3\}$ (which is, in the present case, a subalgebra) in such a way that $n^{\alpha\beta} = \text{diag}(n_1, n_2, n_3)$ where each of the n_i is 1, -1, or 0. (3) now gives

$$n_1(2\theta^1{}_1 - c) = n_2(2\theta^2{}_2 - c) = n_3(2\theta^3{}_3 - c) = 0, \tag{7}$$

$$n_2\theta^3{}_2 + n_3\theta^2{}_3 = n_3\theta^1{}_3 + n_1\theta^3{}_1 = n_1\theta^2{}_1 + n_2\theta^1{}_2 = 0. \tag{8}$$

If rank$(n^{ab}) = 3$, (7) easily leads to a contradiction with $c = \theta^i{}_i$. If rank$(n^{ab}) = 2$, there are two cases distinguished by the (modulus of the) signature of n^{ab}. If rank$(n^{ab}) = 1$, one can take $n_1 = 1$, $n_2 = n_3 = 0$: then $\theta^1{}_1 = c/2 \neq 0$ and by making the basis change $X'_2 = X_2 - (\theta^1{}_2/\theta^1{}_1)X_1$, $X'_3 = X_3 - (\theta^1{}_3/\theta^1{}_1)X_1$, $X'_4 = 4X_4/c$, one can set $c = 4$ and $\theta^1{}_2 = \theta^2{}_1 = \theta^1{}_3 = \theta^3{}_1 = 0$, leaving only $\theta^A{}_B$ $(A, B = 2, 3)$ to consider. If $n^{ab} = 0$, $\theta^\alpha{}_\beta$ remains to be classified.

These cases will be denoted as follows. The letter N (for class N) is followed by the rank of n^{ab}. In class N2 this is followed by the value of the modulus of the signature of n^{ab}. In classes N1 and N0 the Segre (1884) type of the remaining matrix ($\theta^A{}_B$ or $\theta^\alpha{}_\beta$) is attached. The Segre type consists of a list of numbers each of which is the dimension of a distinct[5] invariant subspace of the transformation represented by the matrix: round brackets are used to bracket together subspaces of equal eigenvalue. The Segre type thus gives the algebraic and geometric multiplicities of the eigenvalues. On occasion it is useful to subscript the Segre types with a list of the corresponding values of the eigenvalues. Finally one should note that the Segre type is defined over the complex field; this is overcome by showing pairs of complex conjugate eigenvalues as Z, $\bar{Z}$ instead of 1, 1, the corresponding real invariant subspace being 2-dimensional and containing no real eigenvector. The full list of resulting types follows the explanation of the method for the unimodular algebras.

For the unimodular algebras the following holds:

Theorem 21.2.1 (Farnsworth and Kerr, 1966) *For a unimodular algebra L_4, either there exists a p_a such that*

$$C^a{}_{bd} = \theta^a{}_{[b} p_{d]}, \qquad \theta^a{}_b p_a = 0\,, \tag{9}$$

or there is no such p_a and there exists a nonzero ℓ^c such that

$$C^a{}_{bd}\ell^d = 0. \tag{10}$$

[5]"Distinct" meaning each is the maximal invariant subspace containing exactly one eigendirection.

Proof. Take a basis of the L_4. Define $\boldsymbol{C}^a$ to be the 2-form on L_4 with components $C^a{}_{bc}$. (3) is equivalent to

$$\epsilon^{abcd} C^e{}_{bc} C^f{}_{de} = 0. \tag{11}$$

Since there are no nonzero 5-forms on L_4,

$$\epsilon^{[abcd} C^{e]}{}_{bc} C^f{}_{de} = 0. \tag{12}$$

Expanding this and using $c_a = 0$ and (11), one obtains

$$\epsilon^{bcde} C^a{}_{bc} C^f{}_{de} = 0 \Leftrightarrow \boldsymbol{C}^a \wedge \boldsymbol{C}^f = 0. \tag{13}$$

Taking $a = f$, (13) shows that $\boldsymbol{C}^a$ is simple and can thus be considered geometrically as a plane. Taking $a \neq f$ in (13) shows that the planes meet in common lines. Either all four planes meet in a common line, or there are three meeting in three distinct lines. In the latter case, take ℓ^a to be any annihilator of the space spanned by the 1-form factors of the $\boldsymbol{C}^a$, i.e., any vector whose contraction with each of these forms is zero; this immediately gives (10). In the former case, take p_a to be the common 1-form factor. $\boldsymbol{c} = 0$ implies $\theta^a{}_b p_a = \theta^a{}_a p_b$; since $\theta^a{}_b$ is so far only defined up to multiples of p_b, we can choose a new $\theta^a{}_b$ so that (9) holds (for example, in terms of a basis with $p_a = (0, 0, 0, 1)$, take $\theta'^a{}_b = \theta^a{}_b - \theta^c{}_c \delta^a{}_4 p_b$).

Corollary 21.2.1 *Every real L_4 contains an invariant L_3.*

Proof. From (5), (9), and (10) it is easy to see that each case has dim $L' \leq 3$. Any subalgebra L_3 containing L' is invariant, and any 3-dimensional vector subspace of L_4 containing L' is such an L_3.

The corollary is a theorem of which Kantowski (1966) gave a proof and which is a special case of Fubini's result (1904) that any isometry group G_r ($r \leq 7$) acting on a V_4 contains a subgroup G_{r-1}, since one can always impose a Riemannian metric on a G_4 (see also Egorov (1955), Petrov (1969), Collins (1977)). Its complex counterpart is given in Section 137 of Lie (1893). Farnsworth and Kerr (1966) call the algebras obeying (9) class A, and those obeying (10) Class B. Here I will call them U1 and U3, these names standing for unimodular algebras whose 2-forms $\boldsymbol{C}^a$ meet in one common or (at least) three lines, respectively. In class U1 only the matrix $\theta^\alpha{}_\beta$ (in the basis used in the proof of the theorem) remains to be classified; as in the N0 types, for which U1 can be regarded as a limit, this is done by Segre type. In class U3, take a basis such that $\ell^a = (0, 0, 0, 1)$. Then either $L' = V$ (V being as defined above), in which case V is semisimple (and hence simple), which is denoted class U3S, or $\dim(L' \cap V) = 2$ and the L_4 is integrable,[6] which I call class U3I. Each of these gives rise to two cases which can be distinguished by the signature of the quantity $n^{\alpha\beta} = \frac{1}{2}\epsilon^{\alpha\gamma\delta} C^\beta{}_{\gamma\delta}$, where $\epsilon^{\alpha\beta\gamma}$ are the components of a nonzero 3-form; the modulus of this signature is therefore appended to complete the symbols describing these cases. The U3I cases can be considered as the $c_a = 0$ limits of class N1[1,1].

[6]The name "solvable" here would lead to confusing nomenclature!

It can be seen from the above description that classifying a given algebra is a straightforward matter involving only commutators and elementary operations in linear algebra. The detailed list of possibilities, each in a canonical form, now follows.

Class N2:

Class N22: Choosing a basis in which $n_1 = 0$, $n_2 = n_3 = 1$, (7) and (8) imply that $\theta^1{}_1 = 0$, $\theta^2{}_2 = \theta^3{}_3 = c/2$, $\theta^1{}_2 = \theta^1{}_3 = 0$, and $\theta^2{}_3 = -\theta^3{}_2$. Then after the basis change $X_4' = 2(X_4 - \theta^2{}_3 X_1)/c$, $X_1' = X_1 - \theta^2{}_1 X_1 - \theta^3{}_1 X_1$, the canonical form is given by

$$\begin{aligned} &[X_2, X_3] = 0, \quad [X_3, X_1] = X_2, \quad [X_1, X_2] = X_3, \\ &[X_1, X_4] = 0, \quad [X_2, X_4] = X_2, \quad [X_3, X_4] = X_3. \end{aligned} \tag{14}$$

Class N20: By a method similar to that in Class N22, this case has a canonical form

$$\begin{aligned} &[X_2, X_3] = 0, \quad [X_3, X_1] = X_2, \quad [X_1, X_2] = -X_3, \\ &[X_1, X_4] = 0, \quad [X_2, X_4] = X_2, \quad [X_3, X_4] = X_3. \end{aligned} \tag{15}$$

An alternative basis $Y_1 = (-X_1 - X_4)/2$, $Y_2 = X_2 + X_3$, $Y_3 = (X_1 - X_4)/2$, and $Y_4 = X_2 - X_3$ gives commutators

$$\begin{aligned} &[Y_1, Y_2] = Y_2, \quad [Y_1, Y_3] = 0, \quad [Y_1, Y_4] = 0, \\ &[Y_2, Y_3] = 0, \quad [Y_2, Y_4] = 0, \quad [Y_3, Y_4] = Y_4. \end{aligned} \tag{16}$$

showing that this is a direct product of two 2-dimensional non-Abelian algebras.

Class N1:

In all cases in this class $c = 4$ and the canonical form has

$$[X_2, X_3] = X_1, \quad [X_3, X_1] = [X_1, X_2] = 0, \quad [X_1, X_4] = 2X_1. \tag{17}$$

Class N1[1,1]:

$$[X_2, X_4] = (1 + \lambda)X_2, \qquad [X_3, X_4] = (1 - \lambda)X_3, \qquad \lambda \geq 0. \tag{18}$$

The values $\lambda = 0$, class N1[(1,1)], and $\lambda = 1$, class N1$[1,1]_{2,0}$, give special subcases.

Class N1[Z, $\bar{Z}$]:

$$[X_2, X_4] = X_2 - \mu X_3, \qquad [X_3, X_4] = \mu X_2 + X_3, \qquad \mu \neq 0. \tag{19}$$

Class N1[2]:

$$[X_2, X_4] = X_2 + X_3, \qquad [X_3, X_4] = X_3. \tag{20}$$

Class N0:

Throughout this class the canonical forms have

$$[X_2, X_3] = [X_3, X_1] = [X_1, X_2] = 0. \tag{21}$$

Class N0[1,1,1]:

$$[X_1, X_4] = \lambda \mathbf{X}_1, \qquad [X_2, X_4] = \mu X_2, \qquad [X_3, X_4] = \nu X_3, \tag{22}$$

$$c = \lambda + \mu + \nu \neq 0. \tag{23}$$

One can order and scale the basis so that $\lambda = 1 \geq \mu \geq \nu \geq -1$. Special subcases with equalities of eigenvalues give the classes N0[(1,1)1], N0[(1,1,1)], and particular eigenvalues give the classes $\mathrm{N0}[1,1,1]_{1,\mu,0}$, $\mathrm{N0}[(1,1)1]_{1,1,0}$, and $\mathrm{N0}[(1,1)1]_{0,0,1}$.

Class N0[Z, $\bar{Z}$,1]:

$$[X_1, X_4] = \lambda X_1 + X_2, \qquad [X_2, X_4] = -X_1 + \lambda X_2,$$

$$[X_3, X_4] = \mu X_3 \tag{24}$$

$$c = 2\lambda + \mu \neq 0. \tag{25}$$

Special cases arise when one or both of λ and μ are either 1 or 0.

Class N0[2,1]:

$$[X_1, X_4] = \lambda X_1 + X_2, \qquad [X_2, X_4] = \lambda X_2,$$

$$[X_3, X_4] = \mu X_3 \tag{26}$$

$$c = 2\lambda + \mu \neq 0. \tag{27}$$

Here one can take μ to be 0 or 1. Special cases arise when $\lambda = \mu$, N0[(2,1)], $\lambda = 0$ and $\mu = 0$.

Class N0[3]:

$$[X_1, X_4] = \lambda X_1 + X_2, \qquad [X_2, X_4] = \lambda X_2 + X_3, \qquad [X_3, X_4] = \lambda X_3. \tag{28}$$

Here $\lambda = c/3$ can be set to 1 but is retained to make manifest the limit to Class U1[3].

Class U1:

These cases can be viewed as limits of Class N0 and (21) holds.

Class U1[1,1,1]:

(22) holds with $\lambda + \mu + \nu = 0$. One can still set $\lambda = 1$ except in the case $\lambda = \mu = \nu = 0$, Class U1[(1,1,1)]. Other special cases are U1[(1,1)1] and $\mathrm{U1}[1,1,1]_{1,-1,0}$.

Class U1[Z, $\bar{Z}$,1]:

(24) holds but $\lambda = -\mu/2$. $\lambda = 0$ is a special case.

Class U1[2,1]:

(26) holds and μ is 0 or 1 but $\lambda = -\mu/2$.

Class U1[3]:

(28) holds with $\lambda = 0$.

Class U3:

Throughout this class

$$[X_1, X_4] = [X_2, X_4] = [X_3, X_4] = 0. \tag{29}$$

Class U3I0:

$$[X_2, X_3] = X_4, \qquad [X_3, X_1] = X_2, \qquad [X_1, X_2] = -X_3. \tag{30}$$

Class U3I2:

$$[X_2, X_3] = -X_4, \qquad [X_3, X_1] = X_2, \qquad [X_1, X_2] = X_3. \tag{31}$$

Class U3S1:

$$[X_2, X_3] = X_1, \qquad [X_3, X_1] = X_2, \qquad [X_1, X_2] = -X_3. \tag{32}$$

Class U3S3:

$$[X_2, X_3] = X_1, \qquad [X_3, X_1] = X_2, \qquad [X_1, X_2] = X_3. \tag{33}$$

21.3 Comparison with Other Enumerations

As mentioned earlier, this section is really an extended caption to Table 21.1. The first column of this table lists the present types. When special subcases are given separately (as for instance in Class N1), the general form considered first is supposed not to include the special subcases. The second column shows L':

L_2I is the Abelian (unimodular) L_2 and L_2II the non-Abelian one, while the L_3 are classified by Bianchi type (for which, see the Appendix). The third column shows whether or not the L_4 contains an Abelian L_3. Columns 4–8 relate the present enumeration to those elsewhere in the literature. In each case the arbitrary constants employed by those other authors are defined in terms of the ones used in Section 21.2, and are superscripted or subscripted in accordance with the usage of the other authors concerned.

The first of these classifications (column 4) is Lie's. As this was constructed over $\mathbb{C}$, algebras distinct over $\mathbb{R}$ may have the same Lie class. When a complex basis transformation is required to bring the real algebra into Lie's form, I have followed Mubarakzyanov (1963a) in appending an asterisk to the Lie type. The Lie types have been labeled by their equation numbers in Lie (1893), section 137. Lie began with the nonsolvable cases, which he characterized as those L_4 containing a noninvariant L_3 itself not containing an invariant L_2 also contained in another L_3 invariant in L_4. He then listed the solvable L_4 not containing an Abelian L_3 in descending order of dim L', and lastly gave the L_4 containing an Abelian L_3. His list is complete over $\mathbb{C}$.

Kruchkovich (1954) does not give details of his method, but he appears to have taken Lie's list and subdivided those types not containing an Abelian L_3 into their distinct real forms; the L_4 containing an Abelian L_3 are treated as one class. The nonsolvable algebras are put at the end, not the start, of the list, and the types labeled by Roman numerals. Petrov (1969) subdivided Kruchkovich's class VI of L_4 containing Abelian L_3. Unfortunately not all his subtypes are in fact distinct, and this is responsible for some of the confusion between different editions of his work pointed out by Ray and Zimmerman (1977). The Kruchkovich-Petrov list is complete. Its disadvantages are (i) the labels are sometimes confused with the Bianchi types of L_3, to which they are not related, (ii) the necessary subdivisions of the classes (especially Class VI) are numerous and rather cumbersomely notated, and (iii) the scheme contains redundancies, details of which follow.

Petrov divided Class VI (in which (21) holds in a canonical basis) as follows

$$VI_1 : [X_1, X_4] = aX_1 + bX_4, \qquad [X_2, X_4] = cX_2 + dX_4, \qquad [X_3, X_4] = eX_3 + fX_4, \tag{34}$$

$$VI_2 : [X_1, X_4] = kX_1 + X_2, \qquad [X_2, X_4] = kX_2, \qquad [X_3, X_4] = \epsilon X_3, \tag{35}$$

$$VI_3 : [X_1, X_4] = kX_1 + X_2, \qquad [X_2, X_4] = kX_2 + X_3, \qquad [X_3, X_4] = \epsilon X_3, \tag{36}$$

$$VI_4 : [X_1, X_4] = kX_1 + X_2, \qquad [X_2, X_4] = -X_1 + kX_2, \qquad [X_3, X_4] = \ell X_3, \tag{37}$$

where $\epsilon = 0$ or 1. Petrov states that essentially the only cases of type VI_1 are (1) $a = b = c = d = e = f = 0$, (2) $a = c = e = f = 0$, $b = d = 1$, (3) $a = c = e = 0$, $b = d = f = 1$, (4) $c = d = e = f = 0$, $a = b = 1$, and (5) $b = d = f = 0$, $a = 1$. Of these, (2), (3), and (4) are in fact redundant.

Explicitly (2) can be transformed to (5) by the basis change $X_1' = X_4/2$, $X_2' = X_1 - X_2$, $X_3' = X_3$, $X_4' = X_1 + X_2$. (3) can be transformed to (5) by $X_1' = X_4/3$, $X_2' = X_2 - X_3$, $X_3' = X_3 - X_1$, $X_4' = X_1 + X_2 + X_3$. (4) can be transformed to (5) by $X_1' = X_1 + X_4$, $X_2' = X_2$, $X_3' = X_3$, $X_4' = X_4$. Thus case VI_1 may be characterized, in all cases, by (21) and

$$VI_1: \quad [X_1, X_4] = \epsilon X_1, \quad [X_2, X_4] = cX_2, \quad [X_3, X_4] = eX_3, \tag{38}$$

with $\epsilon = 1$ unless $\epsilon = c = e = 0$. The case VI_3 with $k \neq \epsilon$ can be transformed to VI_2 by $X_1' = X_1 - (X_3/(k-\epsilon)^2)$, $X_2' = X_2 + (X_3/(k-\epsilon))$, $X_3' = X_3$, $X_4' = X_4$. Thus there are only the following distinct cases: VI_2 and a restricted VI_3 given by (21) and

$$VI_3: \quad [X_1, X_4] = \epsilon X_1 + X_2, \quad [X_2, X_4] = \epsilon X_2 + X_3, \quad [X_3, X_4] = \epsilon X_3, \tag{39}$$

where $\epsilon = 0$ or 1. These defects of Petrov's list were independently noted by Fee (1979) (and are implicit in the list given by Ellis and Sciama (1966)).

Bratzlavsky's paper (1959) was unavailable to me, but a reworking is given by Fee (1979) and the results in Sengier-Diels (1974a). It gives the algebras in ascending order of dim L'. The coefficients used are allowed to be zero, and the Abelian algebra was included, although these points are not discussed explicitly by Sengier-Diels (1974a). The Class N0[2,1] with $\lambda = 0$, $\mu = 1$, is misprinted in Sengier-Diels (1974a), where it is case III.

Mubarakzyanov's classification (1963a) lists first the decomposable algebras (direct sums of algebras of lower dimension) and then the indecomposable algebras. The notation is that $g_{i,j}$ is the j-th type of i-dimensional algebra. g_1 is unique. The two dimensional algebras are $2g_1$ (or $g_1 \oplus g_1$) and the non-Abelian one, which is denoted g_2. Mubarakzyanov's technique for listing the higher-dimensional algebras is to take the (semi-)simple algebras from the usual lists and complete the enumeration of the solvable algebras by considerations on the maximal nilpotent ideal M and the possible structures involving non-nilpotent elements. His method leads to seven types of indecomposable g_3 and two decomposable g_3. These in turn lead to ten decomposable g_4, when $2g_2$ has been added, and the indecomposable types give ten more. Of these $g_{4,1}$ is nilpotent, i.e., has dim $M = 4$, $g_{4,2}$ to $g_{4,6}$ have dim $M = 3$ and M Abelian, $g_{4,7}$ to $g_{4,9}$ have $M = g_{3,1}$, and $g_{4,10}$ has dim $M = 2$. Since the decomposable algebras involve the $g_{3,j}$, a list of these, cross-referencing various authors' enumerations, and similar to that given in Section 22.2 and this section for the L_4, is given as an appendix.

Patera *et al* (1976) and Patera and Winternitz (1977) change Mubarakzyanov's notation from $g_{i,j}$ to $A_{i,k}$ and add the values of the arbitrary parameters as superscripts. Thus $g_{4,6}$ becomes $A_{4,6}^{a,b}$. Patera et al. also increase the number of types by subdividing Mubarakzyanov's. They divide $g_{3,4}$ into $A_{3,4}$ and $A_{3,5}$, and $g_{3,5}$ into $A_{3,6}$ and $A_{3,7}$, and similarly divide $g_{4,8}$ into $A_{4,8}$ and $A_{4,9}$, and $g_{4,9}$ into $A_{4,10}$ and $A_{4,11}$, with the effect that subsequent cases are renumbered so that $g_{i,j} \equiv A_{i,j+2}$. It is of some interest that when they considered the subalgebra structure of the L_4, Patera and Winternitz (1977) found it necessary to distinguish between $A_{4,2}^1$, $A_{4,2}^{-1}$

and other $A^a_{4,2}$ $(a \neq 0)$, between $A^{a,a}_{4,5}$ $(a \neq 1)$, $A^{a,b}_{4,5}$ $(a \neq b \neq 1)$, $A^{a,1}_{4,5}$ and $A^{1,1}_{4,5}$, and between $A^b_{4,9}$ $(0 < b < 1)$, $A^1_{4,9}$, $A^0_{4,9}$ and $A^{-1/2}_{4,9}$.

One may note that only the cases $\mathrm{U1}[(1,1,1)]_{0,0,0}$, U1[(2,1)], and U1[3] are nilpotent.

21.4 Extensions, Applications and Other Work

The various classifications described in Section 22.3 have been used as a basis for computing the possible metrics of manifolds on which the groups act simply-transitively and finding those satisfying Einstein's equations for various energy-momentum tensors (see for example Kruchkovich (1957), Petrov *et al* (1960), Ozsváth (1965), Farnsworth and Kerr (1966), Ozsváth and Schucking (1969), Petrov (1969), Sengier-Diels (1974a), Sengier-Diels (1974b), Hiromoto and Ozsváth (1978), Fee (1979)).

The classification introduced here was reorganized in work of F. Schöbel (1992, 1993) from which a REDUCE program was developed which carries out the classification (for dimensions up to 4) and computes the transformation of a given basis to the canonical one. She also gave some examples of occurrence of the algebras as symmetry groups of differential equations and in other contexts. The regrouping of the types was in order of the dimension of L' and L''. C. Schöbel (1992, 1993) has similarly studied algebras of any dimension for which $\dim L'$ is low. The resulting programs are now included as the LIE package in the REDUCE distribution.

S. T. C. Siklos (private communication) has remarked that the non-unimodular algebras correspond to homothety groups of identifiable surfaces and that they may generate subgroups of unimodular groups of higher dimension. Koutras (1992) found the possible generators corresponding to nontrivial homothetic scalings when the groups act as homothety groups.

Acknowledgments

I am grateful to Prof. G. F. R. Ellis and Drs. C. and F. Schöbel, J. Sengier-Diels, and P. Turkowski for correspondence, to them and Drs. W. de Graaf, G. Fee, A. Koutras and I. Ozsváth for copies of their works, and to the Schöbels for advance copies of their programs.

Appendix A: The Enumeration of the Real L_3

Here again the classification over $\mathbb{C}$ was done by Lie (1893) (see also Jacobson (1962), section 1.4). The standard enumeration of the real cases is that of Bianchi (1918), who worked in ascending order of $\dim L'$. A very similar enumeration

Table 21.1 Enumeration of the distinct L_4 (part 1: the non-unimodular cases)

Type	L'	Abelian $L_3 \subset L_4$?	Lie	Kruchkovich Petrov	Bratzlavsky	Mubarakzyanov	Patera et al.
N22	L_2I	No	(64)*	V	V	$g_{4,10}$	$A_{4,12}$
N20	L_2I	No	(64)	IV	IV	$2g_2$	$2A_2$
N1[1,1]	L_3II	No	(62), $c(1+\lambda)=2$	I_c	XI_q, $q=c-1$	$g_{4,8}^q$	$A_{4,9}^q$
$\mathrm{N1}[1,1]_{2,0}$	L_2I	No	(65)	$I, c=1$			
N1[Z, $\bar{Z}$]	L_3II	No	(62)*, $c(1+i\mu)=2$	III_q, $q^2(1+\mu^2)=4$	XII_p, $p\mu=1$	$g_{4,9}^p$	$A_{4,11}^p$
N1[2]	L_3II	No	(63)	II	X	$g_{4,7}$	$A_{4,7}$
$\mathrm{N0}[1,1,1]_{1,\mu,\nu}$	L_3I		(67), $a=\mu$	$VI_1, a=1$		$g_{4,5}^{\mu,\nu}$	$A_{4,5}^{\mu,\nu}$
$\mathrm{N0}[1,1,1]_{1,\mu,0}$	L_2I	Yes	$c=\nu$	$c=\mu$	$VIII, p=\mu$	$g_{3,4}^{\mu} \oplus g_1$	$A_{3,5}^{\mu} \oplus A_1$
$\mathrm{N0}[1,1,1]_{1,1,0}$	L_2I		$\lambda=1$	$e=\nu$	$q=\nu$	$g_{3,3} \oplus g_1$	$A_{3,3} \oplus A_1$
$\mathrm{N0}[1,1,1]_{1,0,0}$	L_1				$VIII_{0,0}$	$g_2 \oplus 2g_1$	$A_2 \oplus 2A_1$
N0[Z, $\bar{Z}$,1]	L_3I	Yes	(67)*, $\mu a=\lambda-i$, $\mu c=\lambda+i$	$VI_4, k=\lambda$	$IX, p=\lambda$	$g_{4,6}^{\mu\lambda}$	$A_{4,6}^{\mu\lambda}$
$\mathrm{N0}[Z,\bar{Z},1]_{\mu=0}$	L_2I	Yes	(67)*, $c=0$, $(\lambda+i)a=\lambda-i$	$\ell=\mu$	$q=\mu$	$g_{3,5}^{\lambda} \oplus g_1$	$A_{3,7}^{\lambda} \oplus A_1$
N0[2,1]	L_3I		(68)		$VII, \lambda q=\mu$	$g_{4,2}^q$	$A_{4,2}^q$
$\mathrm{N0}[2,1]_{\lambda=1,\mu=0}$	L_2I	Yes	$(\mu-\lambda)c=\lambda$	$VI_2, k=\lambda$		$g_{3,2} \oplus g_1$	$A_{3,2} \oplus A_1$
$\mathrm{N0}[2,1]_{\lambda=0,\mu=1}$	L_2I				III	$g_{4,3}$	$A_{4,3}$
N0[(2,1)]	L_3I		(72)		$VII, q=1$	$g_{4,2}^1$	$A_{4,2}^1$
N0[3]	L_3I	Yes	(70)	$VI_3, \lambda=1$	VI	$g_{4,4}$	$A_{4,4}$

Table 21.2 Enumeration of the distinct L_4 (part 2: the unimodular cases). For full description see the text of Section 3.

Type	L'	Abelian $L_3 \subset L_4$?	Lie	Kruchkovich Petrov	Bratzlavsky	Mubarakzyanov	Patera et al.
U1[1,1,1]	L_3I	Yes	(67), $a=\lambda$ $c=\mu$	$VI_1, \epsilon=1$ $c=\lambda,$ $1+c+e=0$	$VIII, p=\mu$ $q=\nu$	$g_{4,5}^{\mu,\nu}$	$A_{4,5}^{\mu,\nu}$
$\mathrm{U1}[1,1,1]_{\nu=0}$	L_2I	Yes				$g_{3,4}^{-1} \oplus g_1$	$A_{3,4} \oplus A_1$
U1[(1,1,1)]	0	Yes	(73)	$VI_1, \epsilon=0$	0	$4g_1$	$4A_1$
U1[$Z, \bar{Z}$,1]	L_3I	Yes	(67)*, $a=\bar{c}$ $=(-1+i/\lambda)/2$	VI_4 $k=-\ell/2=\lambda$	IX $p=-q/2=\lambda$	$g_{4,6}^{-2\lambda,\lambda}$	$A_{4,6}^{-2\lambda,\lambda}$
$\mathrm{U1}[Z, \bar{Z},1]_{\lambda=0}$	L_2I	Yes	(67)* $a=-1,\ c=0$			$g_{3,5}^{0} \oplus g_1$	$A_{3,6} \oplus A_1$
U1[2,1]	L_3I	Yes	(68), $c=-1/3$	$VI_2,$ $\epsilon=1=-2k$	$VII, q=-2$	$g_{4,2}^{-2}$	$A_{4,2}^{-2}$
U1[(2,1)]	L_1	Yes	(71)	$VI_2,$ $\epsilon=0=k$	I	$g_{3,1} \oplus g_1$	$A_{3,1} \oplus A_1$
U1[3]	L_2I	Yes	(69)	$VI_3, \epsilon=0$	II	$g_{4,1}$	$A_{4,1}$
U3I0	L_3II	No	(62)	$I, c=0$	$XI, q=-1$	$g_{4,8}^{-1}$	$A_{4,8}$
U3I2	L_3II	No	(62)*	$III, q=0$	$XII, p=0$	$g_{4,9}^{0}$	$A_{4,10}$
U3S1	L_3VIII	No	(58)	VII	XIV	$g_{3,6} \oplus g_1$	$A_{3,8} \oplus A_1$
U3S3	L_3IX	No	(58)*	VIII	XIII	$g_{3,7} \oplus g_1$	$A_{3,9} \oplus A_1$

Table 21.3 Classifications of the L_3. For explanation see the text of the Appendix.

Bianchi	modified Bianchi	Lie	Lee	Vranceanu	Mubarakzyanov	Patera et al.	New names
I	I	(46)	III	10	$3g_1$	$3A_1$	U1[(1,1)]$_{0,0}$
II	II	(47)	V	9	$g_{3,1}$	$A_{3,1}$	U1[2]
$VI_{q=-1}$	VI_0	(44)	$(B)_{p=0}$	8	$g^0_{3,4}$	$A_{3,4}$	U1[1,1]
VII_0	VII_0		$(A)_{p=0}$	7	$g^0_{3,5}$	$A_{3,6}$	U1[Z, Z]
VIII	VIII	(43)	II	5	$g_{3,6}$	$A_{3,8}$	U31
IX	IX		I	6	$g_{3,7}$	$A_{3,9}$	U33
V	V	(44)	VI	$1, a = 1$	$g_{3,3}$	$A_{3,3}$	N[(1,1)]$_{0,0}$
IV	IV	(45)	VII	3	$g_{3,2}$	$A_{3,2}$	N[2]
III	$VI_{-1} = III$		IV	4	$g_1 \oplus g_2$	$A_1 \oplus A_2$	N[1,1]$_{1,0}$
VI_q	VI_h $h = -((1+q)/(1-q))^2$	(44)	$(B)_{p^2=h}$	1, $(a-1)^2h = -1$	$g^k_{3,4}, k = -q$	$A^q_{3,5}$	N[1,1]
VII_q	$VII_h, (4-q^2)h = q^2$		$(A)_{p^2=h}$	$2, k^2h = 1$	$g^p_{3,5}$	$A^p_{3,5}$	N[Z, Z]

was given by Lee (1947), except that Lee took the simple algebras as types I and II rather than VIII and IX. Bianchi types VI and VII can be combined, as in Lee's class (C).

Vranceanu's method (1947) is similar to that used here for the L_4. He considers the c_a defined by (4) and the canonical forms of a bilinear form

$$c_{\alpha\beta} = C^{\gamma}{}_{\alpha\epsilon} C^{\epsilon}{}_{\beta\gamma}, \tag{40}$$

The resulting types have been numbered here in accordance with the order in which they appear in Vranceanu's work.

A different method was introduced by Behr and others (Estabrook *et al*, 1968), (Ellis and MacCallum, 1969). In this approach one takes $\boldsymbol{c} = (2c,\ 0,\ 0)$ and uses a nonzero 3-form $\epsilon^{\alpha\beta\gamma}$ on L_3 to define

$$n_{\alpha\beta} = \frac{1}{2} C^{(\alpha}{}_{\gamma\delta}\epsilon^{\beta)\gamma\delta}. \tag{41}$$

(Lee had used a similar but unsymmetrized quantity.) One can then, by choice of basis, set

$$n^{\alpha\beta} = \mathrm{diag}(N_1,\ N_2,\ N_3), \quad \text{each } N_i = 1,\ 0 \text{ or } -1, \quad cN_1 = 0. \tag{42}$$

A parameter h is defined, when $N_2N_3 \neq 0$, by $hN_2N_3 = c^2$. The resulting canonical forms differ slightly from Bianchi's. It has become common to refer to the unimodular cases as Class A and the rest as Class B (Ellis and MacCallum, 1969).

The method used by Mubarakzyanov (1963a), Patera *et al* (1976), and Patera and Winternitz (1977) has been described above.

Table 21.3 cross-references the above enumerations. (42) gives a canonical form for the commutators, namely

$$[\boldsymbol{X}_2,\ \boldsymbol{X}_3] = N_1\boldsymbol{X}_1,\quad [\boldsymbol{X}_3,\ \boldsymbol{X}_1] = N_2\boldsymbol{X}_2 - c\boldsymbol{X}_3,\quad [\boldsymbol{X}_1,\ \boldsymbol{X}_2] = c\boldsymbol{X}_2 + N_3\boldsymbol{X}_3. \tag{43}$$

The final column in Table 21.3 gives an enumeration similar to that used in this paper for the L_4. The non-unimodular L_3 leave a 2×2 matrix to classify. The $\boldsymbol{C}^a$ in the unimodular case again represent planes and hence must intersect (being in 3-dimensional space). If they have a common line, case U1 results, which has a residual 2×2 matrix to classify, and if there is no common line the class U3, which has two members, both simple, arises. The procedure and notation runs in close analogy with the L_4 cases.

REFERENCES

Bianchi, L. (1918). *Lezioni sulla teoria dei gruppi continui finiti di trasformazioni* (Pisa: Enrico Spoerri).

Bradfield, T. and Kantowski, R. (1982). Jordan-Kaluza-Klein type unified theories of gauge and gravity fields. *J. Math. Phys.* **23**, 128–131.

Bratzlavsky, F. (1959). Sur les algebres et les groupes de Lie résolubles de dimension trois et quatre. Memoire de Licence, Université Libre de Bruxelles.

Collins, C.B. (1977). Global structure of the 'Kantowski-Sachs' cosmological models. *J. Math. Phys.* **18**, 2116.

de Graaf, W.A., Ivanyos, G., Küronya, A., and Rónyai, L. (1997). Computing Levi decompositions in Lie algebras. *Appl. Alg. Eng. Comm. Comp.***8**, 291–303.
Demaret, J. and Hanquin, J.L. (1985). Anisotropic Kaluza-Klein cosmologies. *Phys. Rev. D***31**, 258–261.
Demianski, M., Golda, Z., Sokolowski, L.M., Szydlowski, M., and Turkowski, P. (1987). The group-theoretical classification of the 11-dimensional classical homogeneous Kaluza-Klein cosmologies. *J. Math. Phys.***28**, 171–173.
Dobrescu, A. (1953). La classification des groupes de Lie réels a quatre parametres. *Acad. Repub. Pop. Roumaine Stud. Cerc. Mat.***4**, 395.
Egorov, I.P. (1955). Motions in spaces with affine connections (in Russian). Ph.D. thesis, Moscow.
Ellis, G.F.R. and MacCallum, M.A.H. (1969). A class of homogeneous cosmological models. *Comm. math. phys.***12**, 108–141.
Ellis, G.F.R. and Sciama, D.W. (1966). On a class of model universes satisfying the perfect cosmological principle. In Hoffman, B., editor, *Perspectives in Geometry and Relativity (Essays in honour of V. Hlavaty)* (Bloomington: Indiana University Press).
Estabrook, F.B., Wahlquist, H.D., and Behr, C.G. (1968). Dyadic analysis of spatially homogeneous world models. *J. Math. Phys.***9**, 497.
Farnsworth, D.L. and Kerr, R.P. (1966). Homogeneous dust-filled cosmological solutions. *J. Math. Phys.***7**, 1625.
Fee, G.J. (1979). Homogeneous spacetimes. M. Math. thesis, University of Waterloo.
Fubini, G. (1904). Sugli spazii a quattro dimensioni che ammettono un gruppo continuo di movimenti. *Ann. di Mat.***9**, 33.
Hiromoto, R.E. and Ozsváth, I. (1978). On homogeneous solutions of Einstein's field equations. *Gen. Rel. Grav.***9**, 299–327.
Jacobson, N. (1962). *Lie algebras* (New York: Wiley Interscience). Reprinted by Dover, New York, 1979.
Kantowski, R. (1966). Some relativistic cosmological models. Ph.D. Thesis, Univ. of Texas at Austin.
Koutras, A. (1992). Mathematical properties of homothetic space-times. Ph.D. thesis, Queen Mary and Westfield College.
Kruchkovich, G.I. (1954). The classification of three-dimensional Riemannian spaces by groups of motions (in Russian). *Usp. Matem. Nauk SSSR***9, part 1 (59)**, 3.
Kruchkovich, G.I. (1957). On motions in Riemannian spaces V_4 (in Russian). *Mat. Sbornik***41 (83)**, 195.
Lee, H.C. (1947). Sur les groupes de Lie réels à trois paramètres. *Journ. de Math.***XXVI (3)**, 33.
Lie, S. (1893). *Theorie der Transformationsgruppen*. (Leipzig: Teubner Verlag). With F. Engel.
Magnin, L. (1986). Sur les algebres de Lie nilpotentes de dimension ≤ 7. *J. Geom. Phys.***3**, 119–144.
Marklund, M. (1997). Invariant construction of solutions to Einstein's field equations: LRS perfect fluids I. *Class. Quant. Grav.***14**, 1267–1284.
Morozov, V.V. (1958). Classification of nilpotent Lie algebras of order $n \leq 6$ (in Russian). *Isv. Vyss. Uch. Zav. Mat.***5**, 161.
Mubarakzyanov, G.M. (1963c). On solvable Lie algebras (in Russian). *Isv. Vyss. Uch. Zav. Mat.***1(32)**, 99–106.
Mubarakzyanov, G.M. (1963b). Classification of the structure constants of Lie algebras of five dimensions (in Russian). *Isv. Vyss. Uch. Zav. Mat.***3(34)**, 114–123.

Mubarakzyanov, G.M. (1963a). Classification of solvable Lie algebras of six dimensions with one nilpotent basis element (in Russian). *Isv. Vyss. Uch. Zav. Mat.***4(35)**, 104–116.

Ozsváth, I. (1965). New homogeneous solutions of Einstein's field equations with incoherent matter obtained by a spinor technique. *J. Math. Phys.***6**, 590.

Ozsváth, I. and Schucking, E. (1969). The finite rotating universe. *Ann. Phys.***55**, 166.

Patera, J., Sharp, R.T., Winternitz, P., and Zassenhaus, H. (1976). Invariants of low dimensional Lie algebras. *J. Math. Phys.***17**, 986–994.

Patera, J. and Winternitz, P. (1977). Subalgebras of real three and four dimensional Lie algebras. *J. Math. Phys.***18**, 1449.

Patera, J. and Zassenhaus, H. (1990). Solvable Lie algebras of dimension ≤ 4 over perfect fields. *Linear algebra and its applications***142**, 1–17.

Petrov, A.Z. (1969). *Einstein spaces* (Oxford: Pergamon Press). Translation by R.F. Kelleher of Russian edition published by Fitzmatlit, Moscow, 1961.

Petrov, A.Z., Kaigorodov, V.R., and Abdullin, V.N. (1960). Classification of general-relativistic gravitational fields by groups of motions III. *Isv. Vyss. Uch. Zav. Mat.***4(17)**, 158–169.

Rand, D.W. (1986). Pascal programs for identification of Lie algebras I. Radical: a program to calculate the radical and nil radical of parameter-free and parameter-dependent Lie algebras. *Computer Physics Communications***41**, 105–125 (Erratum: Computer Physics Communications, vol. 47 p. 369 (1988)).

Rand, D. (1987). Pascal programs for the identification of Lie algebras II: Levi decomposition and canonical basis. *Comp. Phys. Comm.***46**, 311.

Rand, D., Winternitz, P., and Zassenhaus, H. (1988). On the identification of a Lie algebra given by its structure constants: I. Direct decomposition, Levi decomposition, and nilradicals. *Linear algebra and its applications*, **109**, 197–246.

Ray, J.R. and Zimmerman, J.C. (1977). A systematic investigation of the Petrov G_4 types. *J. Math. Phys.***18**, 881.

Schöbel, C. (1992). A classification of real finite-dimensional Lie algebras with a low-dimensional derived algebra. Preprint, Leipzig.

Schöbel, C. (1993). On the classification of real finite-dimensional Lie algebras with a low-dimensional derived algebra. Ph.D. thesis, Leipzig.

Schöbel, F. (1992). The symbolic classification of real 4-dimensional Lie algebras. Preprint 27/92, Leipzig.

Schöbel, F. (1993). Constructive classification of real 4-dimensional Lie algebras. Ph.D. thesis, Leipzig.

Seeley, C. (1993). 7-dimensional nilpotent Lie algebras. *Trans. A.M.S.***335**, 479–496.

Segre, C. (1884). Sulla teoria e sulla classificazione delle omografie in uno spazio lineare ad un numero qualqunque di dimensioni. *Memorie dell R. Accad. Lincei, serie 3a***XIX**, 127–148.

Sengier-Diels, J. (1974a). Espaces pseudo-riemannians homogènes à quatre dimensions. *Bull. Cl. Sci., Acad. Roy. Belg.***60**, 1469–1485.

Sengier-Diels, J. (1974b). Sur les espaces homogènes de la relativité. Ph.D. thesis, Université Libre de Bruxelles.

Sund, T. (1979). On the structure of solvable Lie algebras. *Math. Scand.***44**, 235–242.

Turkowski, P. (1987). Classification of multidimensional spacetimes. *J. Geom. Phys.***119-132**.

Turkowski, P. (1988). Low-dimensional real Lie algebras. *J. Math. Phys.* **29**, 2139–2144.

Turkowski, P. (1990). Solvable Lie algebras of dimension 6. *J. Math. Phys.* **31**, 1344–1350.

Turkowski, P. (1992). Structure of real Lie algebras. *Linear algebra and its applications* **171**, 197–212.

Vranceanu, G. (1947). *Lecons de geometrie differentielle, vol 1.* (Bucharest: Editions de l'Academie de la Republique Populaire Roumaine). Second edition: Gauthier-Villars, Paris (1956).

Zassenhaus, H., Rand, D., and Winternitz, P. (1987). Pascal programs for the identification of Lie algebras II: SPLIT - a program to decompose parameter-free and parameter-dependent Lie algebras into direct sums. *Comp. Phys. Comm.* **46**, 297.

22

Spinning Universes in Newtonian Cosmology

Jayant V. Narlikar

ABSTRACT In this paper we shall review some of the interesting work on spinning universes in Newtonian cosmology, which was inspired by the early work of Heckmann and Schucking. We will show that there are two ways of interpreting the boundary conditions at infinity, which lead to opposite conclusions regarding the inevitability of the Newtonian singularity. A brief comparison with relativity is made.

22.1 Introduction

The first problem that my Ph.D. supervisor Fred Hoyle asked me to look at was that of rotating universes. This was back in 1960, and the pioneering papers in this field that I had to consult were those by K. Gödel [2], A. K. Raychaudhuri [7], and O. Heckmann and E. Schucking [3, 4]. I had the pleasure of meeting Engelbert Schucking much later at one of the early Texas Symposia. It took me some time to discover the sense of humour residing beneath his quiet and unassuming personality. This article carries my best wishes on the occasion of his 70th birthday.

The revival of Isaac Newton's efforts to describe cosmology within the framework of his theory of gravitation was carried out, ironically enough, *after* the general theory of relativity had settled down as the proper theory for describing cosmological models. Although Newton had found a model for a static homogeneous but infinite universe, he had soon realized that the model was highly unstable and would collapse into concentrations of matter.

With the advent of relativity, the inadequacies of the Newtonian framework for cosmology became clear: (i) the theory was based on instantaneous action at a distance, which therefore was unsuitable for a subject involving enormous distances; (ii) the dynamics of Newtonian theory was inconsistent with Lorentz invariance of special relativity.

Nevertheless, it was the work of E. A. Milne and W. H. McCrea [5] that showed that with suitable interpretation of Newtonian gravitation, it is possible to obtain cosmological models of the expanding universe. Moreover, these models turn out to be surprisingly similar to the homogeneous and isotropic models of general relativity, models discovered in the 1920s by Alexander Friedmann, Abbé Lemaitre, and H. P. Robertson.

However, in the 1950s there was an interest in general relativistic models of a universe that is homogeneous but *anisotropic*, for the following reason. Could the introduction of anisotropy in the form of shear and rotation lead to nonsingular models? If so, one could have models which oscillate between finite radii and do not have the "age problem" of the isotropic ones. No explicit models of this kind were available, and the work of A. K. Raychaudhuri [7] showed that while rotation did help oppose the singularity formation, shear went the other way. Heckmann and Schucking, as well as Raychaudhury, however, considered the anisotropic problem in the Newtonian framework, partly because its equations were easier to handle and partly because it is always interesting to see to what extent the conclusions of the Newtonian and Einsteinian approaches ran parallel and where they diverged from each other.

Eventually the work of Penrose and Hawking in the mid-1960s demonstrated that the spacetime singularity is inevitable in relativistic cosmology, unless the equation of state of the physical contents of the universe violated certain reasonable energy conditions. Nevertheless, the interest in Newtonian cosmology for anisotropic models continued, and we shall here concentrate on the results obtained in this field. The question we wish to answer is, Do we get nonsingular models in Newtonian cosmology if we admit anisotropy?

22.2 Homogeneous and Anisotropic Cosmologies

In the Milne-McCrea approach, the universe was assumed to be homogeneous and isotropic at any time t (the time of course is the absolute time of Newton), and to follow the standard Newtonian laws of motion and gravitation. The simplest models assumed the matter in the form of "dust," that is, with zero pressure. The postulate of homogeneity means that the density ρ of dust will be a function of time only. Also, if we assume that the Cartesian coordinates of a typical galaxy are x_μ, and its velocity, v_μ, μ=1, 2, 3, then the homogeneity part of the cosmological principle implies that

$$v_\mu = H_{\mu\nu} x_\nu. \tag{1}$$

In the case of isotropy, we get $H_{\mu\nu} = H(t)\delta_{\mu\nu}$, and we recover the velocity distance relation of Hubble:

$$\mathbf{v} = H(t)\mathbf{r}. \tag{2}$$

The above relation can be integrated to give

$$\mathbf{r} = S(t)\mathbf{r}_0, \tag{3}$$

where the constant vector $\mathbf{r}_0$ denotes the comoving coordinate of the galaxy. The "scale factor" $S(t)$ tells us how the universe expands, its dynamical behavior being determined by the Euler rquations of fluid motion in a gravitational field. As Milne and McCrea [5] demonstrated, the behavior of the scale factor exactly matches that in the relativistic Friedmann models.

There have been several investigations generalizing the above picture to cosmologies with shear and rotation. Broadly speaking, there have been two different types of approaches. Of these the approach discussed by Heckmann and Schucking [3, 4] and Raychaudhuri [7] will be discussed first.

22.2.1 The Potential Function Approach

Heckmann and Schucking [3, 4] wrote the gravitational force in the form

$$F_\mu = \phi_{;\mu}, \tag{4}$$

where ϕ is the gravitational potential satisfying the Poisson equation

$$\nabla^2\phi = 4\pi G\rho. \tag{5}$$

Equation (5) could also be written as $\nabla\dot{\mathbf{F}} = 4\pi G\rho$, provided we recognize that $\mathbf{F}$ is the gradient of a scalar, so that the integrability conditions

$$\text{curl}\mathbf{F} = 0 \tag{6}$$

must be satisfied.

Thus we have the following three equations to describe the general behavior of a homogeneous but anisotropic universe:

$$\text{Equation of continuity}: \quad \dot{\rho}/\rho + \nabla\dot{\mathbf{v}} = 0, (7) \tag{7}$$

$$\text{Equation of motion}: \quad \dot{\mathbf{v}} + (\mathbf{v}.\nabla)\mathbf{v} = -\phi, \tag{8}$$

$$\text{Equation of gravitation}: \quad \nabla^2\phi + \lambda = 4\pi G\rho. \tag{9}$$

It should be remembered in (7) and (8) that the density ρ function of t only, while in (9) the λ-term has also been included. The condition (6) can be rewritten in the form

$$\phi_{,\mu\nu} = \phi_{,\nu\mu}. \tag{10}$$

In this approach therefore of the 9 components of the tensor there are really only 6 algebraically independent components by virtue of (10). Equation (9) reduces the number further by 1. Hence Heckmann and Schucking had effectively five unknown quantities still to be determined. Physically we may link the arbitrariness of the problem to the arbitrariness of the choice of accelerated frames at each fundamental observer. Such frames lead to fictitious gravitational forces *without sources*.

As in relativity, the velocity vector can be used to define tensors for shear $q_{\lambda\mu}$ and spin $\omega_{\lambda\mu}$ and the spin vector ω_λ. Writing the velocity-distance relation as in

(1), we get

$$\omega_{\mu\nu} = \frac{1}{2}(H_{\mu\nu} - H_{\nu\mu}),$$
$$q_{\mu\nu} = \frac{1}{2}\left(H_{\mu\nu} + H_{\nu\mu}\right) - \frac{1}{3}\delta_{\mu\nu}H_{\lambda\lambda},$$
$$\omega_\lambda = \frac{1}{2}\epsilon_{\lambda\mu\nu}\omega_{\mu\nu}. \tag{11}$$

From these definitions and the equations (8,9), Heckmann and Schucking deduced the following relations:

$$\begin{aligned} R(t) &= \exp\left(\int \frac{1}{3}H_{\lambda\lambda}\,dt\right), \\ \frac{4}{3}\pi\rho R^3 &= \mathcal{M}, \\ \frac{d}{dt}(R^2\omega_\lambda) &= R^2\omega_\mu q_{\mu\lambda}, \\ \ddot{R} - \frac{R}{3}(\lambda - q_{\mu\nu}q_{\mu\nu}) - \frac{2}{3}R\omega_\mu\omega_\mu + \frac{G\mathcal{M}}{R^2} &= 0. \end{aligned} \tag{12}$$

Now utilizing the indeterminacy of the potential function, Heckmann and Schucking looked for a special class of solutions which are *nonsingular*, i.e, in which $R \not\to 0$. Their hope was that if such solutions exist in the Newtonian framework, we may look for their analogues in relativistic cosmology.

The situation in relativity is that the shear tensor *does not* help in this project. In Newtonian framework, however, it is possible (in the above approach) to set

$$q_{\mu\nu} = 0. \tag{13}$$

Then we have 5 additional equations to complete the solution of the problem. Following this approach, Heckmann and Schucking gave the solution

$$R^2\omega_\lambda = \text{constant}. \tag{14}$$

At small R the rotation term dominates, and we get a minimum of R for $R > 0$. It is also possible to obtain for $\lambda \neq 0$ the analogue of Gödel's model in the Newtonian framework.

We will return to this approach after discussing the second approach which is somewhat more restrictive and permits fewer models than in the present case.

22.2.2 The Gravitational Force Approach

This approach was initiated by Narlikar [6] and explored in detail by Davidson and Evans [1] and is essentially based on a generalization of the way Milne and McCrea obtained the gravitational force formula in the Newtonian framework.

Basically this approach makes use of the inverse square law of Newton. The universe at any given time t is uniformly dense and can be divided into concentric spherical shells with the origin of the coordinates as the center. From the inverse square law, the net force of gravitation due to any particular shell is zero at an interior point of the shell. Thus the force at a typical point P with coordinate $\mathbf{r}$ is due only to the matter contained in the sphere with the origin as the centre and surface passing through P. This force can be computed as was done by Milne and McCrea and is given by

$$\mathbf{F} = -\frac{4\pi G\rho}{3}\mathbf{r}. \tag{15}$$

Thus we begin by substituting (15) into the Euler equation of motion but take for $\mathbf{v}$ the general anisotropic Hubble relation (1) derived earlier. Differentiating (1) with respect to t and using $v_\mu = \dot{x}_\mu$, we get

$$\ddot{x}_\mu = A_{\mu\nu}x_\nu, \tag{16}$$

where

$$A_{\mu\nu} = \dot{H}_{\mu\nu} + H_{\mu\lambda}H_{\lambda\nu}. \tag{17}$$

The Euler equation of motion and the equation of continuity give under the present assumptions the following two relations respectively:

$$\begin{aligned} A_{\mu\nu} \equiv \dot{H}_{\mu\nu} + H_{\mu\lambda}H_{\lambda\nu} &= -\frac{4\pi G\rho}{3}\delta_{\mu\nu}, \\ \dot{\rho} + \rho H_{\lambda\lambda} &= 0. \end{aligned} \tag{18}$$

To integrate these, we follow a procedure similar to that of Milne and McCrea. Write the solution of (1) in the form

$$x_\mu = a_{\mu\nu}(t)x_\nu^0, \tag{19}$$

where at some specified time $t = t_0$, $x_\mu = x_\mu^0$, i.e.,

$$a_{\mu\nu}(t_0) = \delta_{\mu\nu}. \tag{20}$$

We then have

$$\dot{a}_{\mu\nu} = H_{\mu\lambda}a_{\lambda\nu}. \tag{21}$$

Writing

$$\Delta = \det||a_{\mu\nu}||, \tag{22}$$

we get $\dot{\Delta}/\Delta = H_{\lambda\lambda}$, so that (18) integrates to

$$\rho\Delta = \text{constant} = \rho_0 \tag{23}$$

with $\rho_0 = \rho(t_0)$. Finally, the equation of motion reduces to

$$\Delta\ddot{a}_{\mu\nu} = -\frac{4\pi G}{3}\rho_0 a_{\mu\nu}. \tag{24}$$

The general Newtonian problem is therefore contained in the solution of (23) and (24). Narlikar showed that it is possible to eliminate *all* $a_{\mu\nu}$ and obtain a fourth-order nonlinear differential equation for Δ. Writing a dimensionless time coordinate

$$\tau = \left(\frac{4\pi G}{3}\rho_0\right)^{1/2} t \tag{25}$$

and denoting $d/d\tau$ by a dash, this equation is

$$\Delta^2\Delta'''' + 7\Delta\Delta'' - 4\Delta'^2 + 9\Delta = 0. \tag{26}$$

Narlikar used this equation to investigate whether there exists a Newtonian singularity, i.e., an epoch when $\rho \to \infty$, $\Delta \to 0$. With the help of a series of transformations

$$\Delta = F^2, \quad \left(\frac{dF}{d\tau}\right)^2 = X(F), \quad F = e^U, \quad \frac{dX}{dU} = Y(X),$$

$$\text{i.e.,} \quad X = (F')^2, \quad Y = 2FF'', \tag{27}$$

the fourth-order equation (26) can be reduced to a second-order equation:

$$XY^2\frac{d^2Y}{dX^2} + XY\left(\frac{dY}{dX}\right)^2 + \left(X + \frac{Y}{2}\right)\frac{dY}{dX} \bigg/ \left(x + \frac{Y}{Z}\right) Y\frac{dY}{dX} + Y^2$$
$$- 2XY + 7Y - 2X + 9 = 0. \tag{28}$$

Since (26) has the property that if $\Delta(t)$ is a solution, so is $A^{-2}\Delta(A\tau + B)$ for arbitrary constants A and B, it turns out that all solutions of this family are characterized by one curve in the (X, Y) plane. Along a typical curve C joining points P_1 and P_2 the Δ-values values at P_1 and P_2 are related by

$$\frac{\Delta_2}{\Delta_1} = \exp\left(\int_{P_1}^{P_2} \frac{2dX}{Y}\right). \tag{29}$$

In the curves investigated by Narlikar asymptotically $Y = -2X$, so that the integral in (29) is negatively infinite as P_2 recedes to infinity: i.e., $\Delta_2 \to 0$ and a singularity results. It is also interesting to note that there is an exceptional curve, which is the straight line ξ:

$$Y = \frac{2}{3}X - 3 \tag{30}$$

and an exceptional point (E), $(9/2, 0)$, which are singular solutions. The point corresponds to the Einstein-de Sitter solution.

22.3 The Work of Davidson and Evans

Davidson and Evans have pointed out that there exist solutions of (28) which Narlikar had missed, in which Δ moves from a minimum $[X = 0, Y > 0]$ to a maximum $[X = 0, Y < 0]$ before eventually moving to a singularity. Since the curves could be continued to the past ($\tau < 0$) in a symmetrical manner, such a curve would represent a minimum with a maximum and a singularlity of Δ on either side. There are also curves, on the other hand, which have only a maximum and two singularities. Davidson and Evans considered the solutions of (24) in detail rather than analyze the fourth-order differential equation for Δ. To this end they defined the shear and spin tensor by

$$q_{\mu\nu} = H_{(\mu\nu)} - \frac{1}{3} H_{\lambda\lambda}\delta_{\mu\nu}, \tag{31}$$

so that

$$H_{\mu\nu} = q_{\mu\nu} + \omega_{\mu\nu} + \frac{1}{3}\frac{\dot{\Delta}}{\Delta}\delta_{\mu\nu}, \tag{32}$$

and

$$H_{\mu\nu}H_{\mu\nu} = q^2 - 2\omega^2 + \frac{1}{3}\frac{\dot{\Delta}^2}{\Delta^2}, \tag{33}$$

where $q^2 = q_{\mu\nu}q_{\mu\nu}$ and $2\omega^2 = \omega_{\mu\nu}\omega_{\mu\nu}$. (Thus ω is the magnitude of the spin vector $\omega_\lambda = \frac{1}{2}\epsilon_{\lambda\mu\nu}\omega_{\mu\nu}$.) The equation (17) in contracted form then leads to

$$3\frac{\ddot{R}}{R} = 2\omega^2 - q^2 - \frac{4\pi G}{R^3}\rho_0, \tag{34}$$

with $R^3 = \Delta$. This is the same equation as derived by Heckmann and Schucking [3] from their approach and is the Newtonian analogue of the Raychaudhuri equation in relativistic cosmology. A little manipulation with these equations also gives

$$\frac{d}{dt}(R^2\omega_\lambda) = R^2 q_{\lambda\mu}\omega_\mu. \tag{35}$$

It is now possible to contrast the two approaches described so far. In the Heckmann and Schucking approach the Poisson equation replaces the force equation (24). In this there is an extra freedom available, corresponding to the solutions of $\nabla^2\phi = 0$. The forces $\nabla\phi$ due to the sourceless potential equation may be interpreted as arising from arbitrarily chosen accelerated frames at different fundamental observers. To fix this part, Heckmann and Schucking assumed that the five independent components of $q_{\mu\nu}$ could be assigned arbitrarily. They therefore looked for solutions with $q_{\mu\nu} = 0$, although $\omega_{\mu\nu} \neq 0, \dot{\Delta} \neq 0$. From (34) we see that such a situation may result in preventing the singular situation $R \to 0$. And with this analogy they conjectured that nonsingular solutions might exist in relativistic cosmology.

In the Narlikar-Davidson-Evans approach the force is fixed and $q_{\mu\nu}$ is determined by the dynamical equations (24) with given initial conditions. It is not therefore possible to have $q_{\mu\nu} = 0$; indeed even if $q_{\mu\nu} = 0$ at $t = t_0$, say, subsequently $q_{\mu\nu} \neq 0$ and shear develops. Thus whereas in the Heckmann-Schucking case $q_{\mu\nu} = 0$ in (35) leads to $\omega_\nu \propto R^{-2}$ and $\omega^2 \propto R^{-4}$, so that as $R \to 0$ spin dominates and prevents a singularity (vis-a-vis the singularity-conducting gravitational term $\propto R^{-3}$), in the second approach $q_{\mu\nu} \neq 0$ does not permit the solution $\omega_\lambda \propto R^{-2}$. The second approach is therefore more analogous to the theory of relativity, where a similar conclusion was reached: *shear-free spinning universes are not possible.*

Davidson and Evans have analyzed several specific solutions of (24) and shown how they are analogous to the solutions in relativistic cosmology. For example, a model spinning about the x_3 axis and axisymmetric about it will have

$$a_{\mu\nu} = \begin{bmatrix} x & -u & 0 \\ u & x & 0 \\ 0 & 0 & z \end{bmatrix}, \quad H_{\mu\nu} = \begin{bmatrix} H & -\omega & 0 \\ \omega & H & 0 \\ 0 & 0 & K \end{bmatrix}. \tag{36}$$

The behavior of x, y, z, H, ω, K can be calculated in terms of the initial conditions. In all cases $\Delta \to 0$ eventually, i.e., a singularity is reached. Although shear is initially zero, it develops soon and eventually dominates over the spin term as the singularity is reached.

With the help of this solution Davidson and Evans have given the following ingenious argument as to why $\Delta \to 0$ in all cases of Narlikar's equation (26). In this equation if at any given instant $\tau = \tau_0$, the values $\Delta_0, \Delta_0', \Delta_0''$, and Δ_0''' are given, then the function $\Delta(\tau)$ for $\tau \geq \tau_0$ is uniquely determined. Now suppose that Δ does not reach a zero, i.e., there is no singularity. Then there must exist a nonzero lower bound of Δ. Without loss of generality set $\Delta = \Delta_0 > 0$ at this minimum and let $\tau = \tau_0$ at this instant. Then for a minimum at τ_0 we need $\Delta_0' = 0$, $\Delta_0'' > 0$. However, if $\Delta_0'' = 0$, then a minimum requires $\Delta_0''' = 0$, $\Delta_0'''' > 0$, which from (26) is impossible. Hence the minimum can only have the following range of parameters:

$$\infty > \Delta_0 > 0, \qquad \Delta' = 0, \qquad \infty > \Delta_0'' > 0, \qquad \infty > \Delta_0''' > -\infty. \tag{37}$$

Now in (36), a particular case of the axisymmetric solution, we have three general parameters H_0, ω_0, and K_0, apart from the scaling parameter Δ_0 (which may be put equal to 1 for convenience). Given any Δ_0', Δ_0'', and Δ_0''' we can always find an axisymmetric solution with the initial values equal to them. Thus for (37) we need

$$\Delta_0' = 2H_0 + K_0 = 0, \qquad \Delta_0'' = 2\omega_0^2 - \frac{3}{2}K_0^2 - 3 > 0, \tag{38}$$

and

$$\Delta_0''' = 6K_0\left(\omega_0^2 + \frac{K_0^2}{4}\right). \tag{39}$$

However, for this axisymmetric solution it can be explicitly shown that $\Delta \to 0$ at some epoch $\tau > \tau_0$. By the uniqueness argument the general case will also have this property ($\Delta \to 0$), even though the individual functions $a_{\mu\nu}$ may behave differently from the axisymmetric case. Thus the original assumption of no singularity leads to a contradiction.

Thus the second approach gives results more analogous to the results from relativity than the first approach. Davidson and Evans have calculated specific cases of models with shear and no rotation and have shown similarities with the Bianchi type I models with pancake singularities. There exist many close parallels between Newtonian models and the relativistic models of Bianchi types I, II, V, and IX.

22.4 Concluding Remarks

To the orthodox general relativist, these discussions may be of academic value only in the sense that the Newtonian framework is considered suspect for discussing cosmology. Nevertheless, because of the close parallel between the Newtonian and relativistic cosmology of isotropic universes, it is worth making comparisons between the two frameworks for the anisotropic cases. This is especially so as explicit solutions for the anisotropic cases are very difficult in the relativistic framework and the Newtonian analogies may provide a deeper insight into the problem.

References

[1] Davidson, W. and Evans, A. B., (1973), *Int.J.Theor.Phys.* **7**, 353
[2] Gödel, K. (1949), *Rev, Mod. Phys.* **21**, 447.
[3] Heckmann, O. and Schucking, E. L. (1955), *Z.Astrophys.* **38**, 95.
[4] Heckmann, O. and Schucking, E. L. (1956), *Z.Astrophys.* **40**, 81.
[5] Milne, E. A. and McCrea, W. H. (1934), *Quar. Jour. Math* **5**, 73.
[6] Narlikar, J. V. (1963), *Mon. Not. R. Astron. Soc.* **126**, 203.
[7] Raychaudhuri, A. K. (1955), *Phys. Rev.* **98**, 1123.

23

Relativistic Gravitational Fields with Close Newtonian Analogs

Pawel Nurowski
Engelbert Schucking
Andrzej Trautman

ABSTRACT Given a Newtonian velocity field $\mathbf{v}(\mathbf{x}, t)$, one considers the manifold $\mathbf{R}^4$ with the Lorentz metric $g = (\mathrm{d}\mathbf{x} - \mathbf{v}\,\mathrm{d}t)^2 - \mathrm{d}t^2$. The Riemann tensor is computed and used to characterize flat space-times with g of this form. Among nonflat solutions of Einstein's equations for such a g there are some cosmological models, the Schwarzschild and Kasner metrics and their generalizations to include matter fields and the cosmological constant. If $|\mathbf{v}| = 1$, then the vector field $\partial/\partial t$ is null and has vanishing divergence; it is geodetic and shear-free if and only if $\partial\mathbf{v}/\partial t$ is parallel to $\mathbf{v}$.

23.1 Introduction

The relation between the Einstein theory of general relativity and the Newtonian theory is usually discussed for slow motions and weak gravitational fields and described in terms of suitable approximation methods; see, for example, [2, 5]. There is, however, a class of Newtonian fields and motions with close and exact relativistic analogs [1, 7]. For those special motions, one can construct, in a simple manner, Lorentzian metrics satisfying the Einstein field equations. Among the metrics, which can be so obtained, are the Schwarzschild, Kasner, and some cosmological solutions. Recently, there has been a renewal of interest in this approach because of its relation to the dimensional reduction of a multidimensional gravitational field admitting a null Killing vector field; see [3] and the references given there.

In this paper, we take up the method outlined in [7] and prove a few new facts about Lorentz metrics constructed from a Newtonian velocity field $\mathbf{v}(\mathbf{x}, t)$. In particular, we characterize the motions that lead, in this manner, to flat space-times. We present known solutions, such as the Kasner metric, in a new form, and show how a field $\mathbf{v}$, such that[1] $|\mathbf{v}| = 1$, can be used to construct a Lorentz metric with a shear-free congruence of nondiverging null geodesics.

23.2 Notation

We use the standard notation of general relativity theory [5]. Our model of space-time is the manifold $\mathbf{R}^4$ with the Cartesian coordinates (x^μ). The Minkowski metric tensor has components $(\eta_{\mu\nu})$ such that $\eta_{ij} = \delta_{ij}$, $\eta_{i4} = 0$, and $\eta_{44} = -1$, where $i, j = 1, 2, 3$. This tensor, and its inverse $(\eta^{\mu\nu})$, are used to lower and raise the Greek indices μ, ν, etc. $= 1, \dots, 4$. We put $x^4 = t$ and often use the notation of vector calculus in $\mathbf{R}^3$. Thus, for example, if $\mathbf{v} = (v^1, v^2, v^3)$ and $\mathbf{w} = (w^1, w^2, w^3)$, then $\mathbf{v}\cdot\mathbf{w} = v_i w_i$ is their scalar product and the ith component of the vector product $\mathbf{v}\times\mathbf{w}$ is $\epsilon_{ijk} v^j w^k$, where $\epsilon_{ijk} = \epsilon_{[ijk]}$ and $\epsilon_{123} = 1$. There is no need to distinguish between the covariant and contravariant position of Latin indices, $v^i = v_i$. The radius vector is $\mathbf{x} = (x_1, x_2, x_3)$ and its length is denoted by r.

All maps are assumed to be smooth. If

$$\mathbf{v} : \mathbf{R}^4 \to \mathbf{R}^3 \tag{1}$$

is a *velocity field*, then $\partial v_i / \partial x_j = v_{i,j} = v_{(i,j)} + v_{[i,j]}$, and

$$v_{[i,j]} = \mathrm{curl}_k \mathbf{v}\, \epsilon_{kji}.$$

Note that $\mathrm{curl}_i \mathbf{curl}\, \mathbf{v} = v_{[j,i]j}$. We also write $\mathrm{div}\, \mathbf{v} = v_{i,i}$. If V is a vector space and $f : \mathbf{R}^4 \to V$, then

$$\dot{f} = \partial f/\partial t + v_i\, \partial f/\partial x^i \quad \text{and} \quad \Delta f = f_{,ii}.$$

23.3 The Metric, the Curvature, and the Ricci Tensors

Given a velocity field (1), one constructs the Lorentzian metric on $\mathbf{R}^4$,

$$g = (\mathrm{d}\mathbf{x} - \mathbf{v}\, \mathrm{d}t)^2 - \mathrm{d}t^2. \tag{2}$$

Introducing the globally defined orthonormal coframe (e^μ), $\mu = 1, \dots, 4$,

$$e^i = \mathrm{d}x^i - v^i \mathrm{d}t, \qquad i = 1, 2, 3, \qquad e^4 = \mathrm{d}t,$$

and the associated connection coefficients $\omega_{\mu\nu}$, using

$$\omega_{\mu\nu} + \omega_{\nu\mu} = 0 \quad \text{and} \quad \mathrm{d}e^\mu + \omega^\mu{}_\nu \wedge e^\nu = 0,$$

one obtains

$$\omega_{ij} = v_{[j,i]} e^4 \quad \text{and} \quad \omega_{i4} = v_{(i,j)} e^j.$$

---

[1] We assume that the system of physical units is chosen so that the velocity of light and the gravitational constant are both equal to 1.

The curvature 2-form $\Omega_{\mu\nu} = d\omega_{\mu\nu} + \omega_\mu{}^\rho \wedge \omega_{\rho\nu} = \frac{1}{2} R_{\mu\nu\rho\sigma} e^\rho \wedge e^\sigma$ has the following components:

$$R_{ijkl} = v_{(j,l)} v_{(i,k)} - v_{(i,l)} v_{(j,k)}, \tag{3}$$

$$R_{ijk4} = v_{[j,i]k}, \tag{4}$$

$$R_{4ij4} = \dot{v}_{(i,j)} + v_{(i,k)} v_{k,j} + v_{(k,j)} v_{[k,i]}. \tag{5}$$

The components of the Ricci tensor $R_{\mu\nu} = R^\rho{}_{\mu\rho\nu}$ are

$$R_{ij} = \dot{v}_{(i,j)} + \operatorname{div}\mathbf{v}\, v_{(i,j)} + \frac{1}{2}(v_{k,i} v_{k,j} - v_{i,k} v_{j,k}), \tag{6}$$

$$R_{4i} = \operatorname{curl}_i \mathbf{curl}\, \mathbf{v} \tag{7}$$

$$R_{44} = -\operatorname{div}\dot{\mathbf{v}} - v_{(i,j)} v_{i,j}. \tag{8}$$

23.4 The Comoving Coordinate System

Any metric of the form (2) can be transformed, at least locally, to comoving coordinates. Consider the system of three ordinary differential equations for the functions $\mathbf{x}(\mathbf{y}, t)$,

$$\frac{d\mathbf{x}}{dt} = \mathbf{v}(\mathbf{x}, t), \tag{9}$$

with the initial conditions $\mathbf{x}(\mathbf{y}, 0) = \mathbf{y}$. For every $\mathbf{y} \in \mathbf{R}^3$ there is a neighborhood of $(\mathbf{y}, 0)$ in $\mathbf{R}^4$ such that the map $(\mathbf{y}, t) \mapsto (\mathbf{x}(\mathbf{y}, t), t)$, is a diffeomorphism of the neighborhood on its image. In other words, solutions of (9) provide local coordinate transformations in $\mathbf{R}^4$. Since (9) gives

$$dx^i = \frac{\partial x^i}{\partial y^j} dy^j + \frac{\partial x^i}{\partial t} dt = \frac{\partial x^i}{\partial y^j} dy^j + v^i dt,$$

the metric in the *comoving coordinates* $(\mathbf{y}, t)$ is

$$\frac{\partial \mathbf{x}}{\partial y^i} \cdot \frac{\partial \mathbf{x}}{\partial y^j} dy^i dy^j - dt^2.$$

23.5 Flat Space-Times

If $R_{\mu\nu\rho\sigma} = 0$, then Eq. (4) gives $\mathbf{curl}\, \mathbf{v} = \mathbf{a}$, where $\mathbf{a}$ is a vector-valued function of t only. Therefore, there exists a "potential" $f : \mathbf{R}^4 \to \mathbf{R}$ such that $\mathbf{v} = \mathbf{grad}\, f + \frac{1}{2}\mathbf{a} \times \mathbf{x}$. From (3) it follows that the matrix $(f_{,ij})$ is of rank no larger than 1: there exists $\mathbf{b} : \mathbf{R}^4 \to \mathbf{R}^3$ such that $f_{,ij} = b_i b_j$, and (5) reduces to

$$b_i c_j + c_i b_j = 0, \quad \text{where} \quad c_i = \dot{b}_i + \mathbf{b}^2 b_i - (\mathbf{a} \times \mathbf{b})_i.$$

The vector $\mathbf{a}$ can be reduced to 0 by a rotation of the axes (x^1, x^2, x^3) with the angular velocity $\mathbf{a}$. After this has been achieved, the flatness condition $R_{\mu\nu\rho\sigma} = 0$

for (2) is equivalent to the existence of a function $f : \mathbf{R}^4 \to \mathbf{R}$ such that $\mathbf{v} = \mathbf{grad}\, f$, the rank of $(f_{,ij})$ is < 2 and

$$\dot{b}_i + \mathbf{b}^2 b_i = 0, \quad \text{where} \quad b_i b_j = f_{,ij}.$$

An example of a locally flat solution of this form is

$$(\mathrm{d}x - \sqrt{x}\,\mathrm{d}t)^2 + \mathrm{d}y^2 + \mathrm{d}z^2 - \mathrm{d}t^2.$$

23.6 Nontrivial Solutions

23.6.1 *Equations for a Perfect Fluid*

In this section we assume that the metric (2) satisfies the Einstein equations

$$R_{\mu\nu} - \frac{1}{2} g_{\mu\nu} R + \Lambda g_{\mu\nu} = T_{\mu\nu}, \tag{10}$$

with a cosmological constant Λ and with the energy-momentum tensor

$$T_{\mu\nu} = (\mu + p) u_\mu u_\nu + p g_{\mu\nu} \tag{11}$$

of a perfect fluid. The fluid is characterized by the four-velocity vector (u^μ), with

$$u^i = 0 \quad \text{for } i = 1, 2, 3, \text{ and } \quad u^4 = 1, \tag{12}$$

the energy density μ and the pressure p. Using (10)–(12), one rewrites the Einstein equations in the form

$$R_{ij} = \left(\Lambda + \frac{1}{2}(\mu - p)\right) \delta_{ij}, \tag{13}$$

$$R_{i4} = 0, \tag{14}$$

$$R_{44} = \frac{1}{2}(\mu + 3p) - \Lambda. \tag{15}$$

23.6.2 *Spherically Symmetric Spaces*

If the velocity field $\mathbf{v}$ is of the form $\mathbf{v} = \mathbf{grad}\, f$, where f is a function of t and r only, then the metric (2) has spherical symmetry. The dependence of such a $\mathbf{v}$ on t and r can be easily determined if one assumes the Einstein equations (13)–(15) with μ and p also depending on t and r only. One obtains

$$\mathbf{v}(\mathbf{x}, t) = \mathbf{x} \sqrt{\frac{1}{3}\Lambda + r^{-3}\left(2m(t) + \int^r r'^2 \mu(r', t)\,\mathrm{d}r'\right)}.$$

If, in addition, one assumes that both μ and p are independent of r, then the general solution to the Einstein equations (13)–(15) is given by the following two classes of velocity fields. Either

$$\mathbf{v} = \mathbf{x}\sqrt{\frac{1}{3}(\Lambda + \mu)}, \tag{16}$$

and

$$\dot{\mu} + \sqrt{3(\Lambda + \mu)}(\mu + p) = 0, \qquad \mu = \mu(t), \qquad p = p(t), \tag{17}$$

or

$$\mathbf{v} = \mathbf{x}\sqrt{2mr^{-3} + \frac{1}{3}(\Lambda + \mu)}, \tag{18}$$

and

$$\mu = -p = \text{const.}, \qquad m = \text{const.} \tag{19}$$

The equations (16)–(17) should be supplemented by an equation of state $F(p, \mu) = 0$ and, in particular cases, may be solved explicitly for μ and p (e.g., for the polytropes, when $p = (\gamma - 1)\mu$, $\gamma = \text{const.}$). Substituting $\mathbf{v}$ into (2), one gets Einstein metrics with a perfect fluid; in particular, the following well-known metrics are obtained as special cases:

(DS) If $\mu = p = 0$, then $\Lambda > 0$ and $\mathbf{v} = \mathbf{x}\sqrt{\Lambda/3}$ corresponds to the de Sitter metric,

(F) If $\Lambda = 0$, then $\mathbf{v} = \mathbf{x}\sqrt{\mu/3}$ with $\dot{\mu} + \sqrt{3\mu}(\mu + p) = 0$ defines the Friedmann universe with $K = 0$. In the case of a polytrope the explicit expressions for the energy density and the pressure read

$$\mu = \frac{4}{3\gamma^2 t^2}, \qquad p = \frac{4(\gamma - 1)}{3\gamma^2 t^2}, \qquad \gamma = \text{const.} \neq 0,$$

so that the velocity field can be written as

$$\mathbf{v} = \frac{2}{3|\gamma|t}\mathbf{x}.$$

The equations (19) imply that the velocity field (18) describes the Einstein metric with the energy-momentum tensor of the cosmological-constant type. Thus, without losing generality, we can restrict to the case $\mu = p = 0$. The ensuing velocity field

$$\mathbf{v} = \mathbf{x}\sqrt{2mr^{-3} + \frac{1}{3}\Lambda}, \qquad m = \text{const.},$$

corresponds to metrics which, as special cases, include the Schwarzschild solution ($\Lambda = 0$) and the de Sitter space ($m = 0$).

23.6.3 The Kasner Solution

Let us now return to the case of empty space-times. Consider a velocity field linear in the coordinates (x^i),

$$v_i(\mathbf{x}, t) = A_{ij}(t)x^j + B_i(t).$$

The vector field B_i can be eliminated by the coordinate transformation $x_i \mapsto x_i - \int B_i(t)\,dt$. Similarly, the antisymmetric part of A_{ij} can be reduced to 0 by a suitable, time-dependent rotation in $\mathbf{R}^3$. Since the matrix $A = (A_{ij})$ depends on t only, one has $\dot{A} = \mathrm{d}A/\mathrm{d}t$. Denoting by $\operatorname{tr} A$ the trace of the matrix A, assuming that $B_i = 0$ and that A is symmetric, $A^T = A$, we obtain from $R_{44} = 0$ and $R_{ij} = 0$ the equations

$$\operatorname{tr} \dot{A} + \operatorname{tr} A^2 = 0 \quad \text{and} \quad \dot{A} + (\operatorname{tr} A)A = 0.$$

By integration one obtains $A(t) = \alpha/t$, where α is a constant, symmetric matrix subject to

$$\operatorname{tr} \alpha = \operatorname{tr} \alpha^2 = 1. \tag{20}$$

By a (time-independent) rotation the matrix α can be brought to the diagonal form, $\alpha = \operatorname{diag}(p_1, p_2, p_3)$, and conditions (20) are equivalent to

$$p_1 + p_2 + p_3 = p_1^2 + p_2^2 + p_3^2 = 1.$$

Since $\mathrm{d}x_i - p_i x_i t^{-1} \mathrm{d}t = t^{p_i} \mathrm{d}(x_i t^{-p_i})$, the metric reduces to the classical Kasner form; see §11.3 in [4]. The solution is nonflat if and only if the matrix α is of maximal rank, so that $p_1 p_2 p_3 \neq 0$.

23.6.4 Perfect Fluid Generalizations of the Kasner Solution

Using the velocity field

$$v_i(\mathbf{x}, t) = A_{ij}(t) x_j, \qquad A_{ij}(t) = A_{(ij)}(t)$$

of the same form as in the Kasner case, and assuming that μ and p are functions of t only, we construct solutions to the Einstein equations for a perfect fluid (13)–(15). We consider two cases: (A) when $\mu = p = 0$ and (B) when $\Lambda = 0$.

Case (A). All solutions in this case are given by the velocity field $\mathbf{v}$ of the form

$$\mathbf{v} = \sqrt{\frac{\Lambda}{3}} \cot(t\sqrt{3\Lambda})\mathbf{x} - \frac{\sqrt{3\Lambda}}{\sin(t\sqrt{3\Lambda})} B\mathbf{x}, \tag{21}$$

where $B\mathbf{x}$ is a vector with components $B_{ij} x_j$ and the constant matrix $B = (B_{ij})$ satisfies

$$B = B^T, \qquad \operatorname{tr} B = -\frac{2}{3} + \operatorname{tr} B^2 = 0. \tag{22}$$

The solution given by (21) and (22) is meaningful also for $\Lambda \leq 0$. If $\Lambda = 0$, then $\mathbf{v} = (\mathbf{x} - 3B\mathbf{x})/3t$ and the solution reduces to the Kasner solution of Section 24.6.3. In the general case, the solution provides a known generalization of the Kasner metric to the case of vacuum Einstein equations with a cosmological constant Λ of either sign; see §11.3.2 in [4].

Case (B). In this case the general solution is given in terms of the matrix A of the form

$$A = \pi(t)I + \tau(t)B,$$

where I is the identity matrix and the matrix B is constant, symmetric, and traceless. The following two cases are worth distinguishing.

(B1) $\tau = 0$. In this case, any function $\pi(t)$ generates a solution to the equations. This is a special case ($\Lambda = 0$) of solutions of Section 24.6.2 and we do not comment on it any further.

(B2) $\tau \neq 0$. Then any nonvanishing function $\tau(t)$ generates a solution to the equations provided that

$$\pi = -\dot{\tau}/3\tau.$$

It follows that in both cases (B1) and (B2), the energy density and the pressure of the fluid can be written as

$$\mu = 3\pi^2 - \frac{1}{2}\tau^2 \operatorname{tr} B^2, \quad p = -2\dot{\pi} - 3\pi^2 - \frac{1}{2}\tau^2 \operatorname{tr} B^2.$$

These relations need to be supplemented by an equation of state. The simplest polytrope equation $p = (\gamma - 1)\mu$ applied to the nonspherically symmetric case (B2) leads to the following equation for the function τ:

$$\frac{2}{3}\frac{\mathrm{d}^2}{\mathrm{d}t^2}(\log\tau) - \frac{\gamma}{3}\left(\frac{\mathrm{d}}{\mathrm{d}t}(\log\tau)\right)^2 + \frac{\gamma-2}{2}\tau^2 \operatorname{tr} B^2 = 0. \tag{23}$$

The general solution of this equation in the case of $\gamma = 2$ generates the Einstein space-time associated with the velocity field of the form

$$\mathbf{v} = \frac{\mathbf{x} - 3B\mathbf{x}}{3t}, \qquad B = B^T, \qquad \operatorname{Tr} B = 0, \qquad \operatorname{Tr} B^2 \leq \frac{2}{3},$$

and for which

$$\mu = p = \frac{2 - 3\operatorname{tr} B^2}{6t^2}.$$

This again provides a generalization of the Kasner solution.

If $\gamma \neq 2$, then the substitution $\tau = w\sqrt{2/(2-\gamma)\operatorname{tr} B^2}$ transforms (23) into the equation

$$\ddot{w} = \frac{\gamma+2}{2}\frac{\dot{w}^2}{w} + \frac{3}{2}w^3$$

for w; it has a first integral of the form

$$\dot{w}^2 = \frac{3}{2-\gamma}w^4 + 2\sqrt{3}cw^{\gamma+2}, \qquad c = \text{const.} \tag{24}$$

For some values of the parameter γ one can solve the above equation explicitly. In particular, if $\gamma = 1$, then the general solution of (24) gives the following pure dust solution of the Einstein equations:

$$\mathbf{v} = \frac{2}{3}\frac{c^2 t}{c^2 t^2 - 1}\mathbf{x} + \sqrt{\frac{2}{3\mathrm{tr}\, B^2}}\frac{2c}{c^2t^2-1}B\mathbf{x}, \qquad \mu = \frac{4}{3}\frac{c^2}{c^2t^2-1}, \qquad p = 0.$$

where B is symmetric and traceless; see §12.4 in [4].

23.7 Congruences of Null Geodesics

If $|\mathbf{v}| = 1$, then the metric (2) reduces to

$$g = d\mathbf{x}^2 - 2\mathbf{v} \cdot d\mathbf{x}\, dt \tag{25}$$

and the 4-dimensional vector field $k = \partial/\partial t$ is null. Its 4-divergence vanishes, and the Lie derivative of g with respect to k is

$$-2(\partial \mathbf{v}/\partial t) \cdot d\mathbf{x}\, dt.$$

Therefore, the congruence of null curves, generated by k, is geodetic and shear-free if and only if [6]

$$\mathbf{v} \times \partial \mathbf{v}/\partial t = 0.$$

The 1-form associated with k by g is

$$\lambda = g(k) = -\mathbf{v} \cdot d\mathbf{x}.$$

The form λ is integrable ("k is hypersurface-orthogonal") if and only if $\mathbf{curl}\, \mathbf{v} = 0$. If this is so, then the metric (25) is flat.

Acknowledgments

This paper owes much to the stimulating discussions we have had at Washington Square in Manhattan. The research has been supported in part by the Foundation for Polish-German cooperation with funds provided by the Federal Republic of Germany, and by the Polish Committee on Scientific Research (KBN) under grant No. 2P03B 017 17.

References

[1] Heckmann, O. H. L. and Schucking, E. In *Encycl. of Physics* **LIII**, 489 (Springer, Berlin, 1959).
[2] Infeld, L. and Plebański, J. *Motion and Relativity* (PWN, Warszawa, 1960).
[3] Julia, B. and Nicolai, H. "Null-Killing vector dimensional reduction and Galilean geometrodynamics," Preprint, LPTENS 94/21 and DESY (1994). 94–156.
[4] Kramer, D., Stephani, H., MacCallum, M., and Herlt, E. *Exact Solutions of Einstein's Equations* (VEB Deutscher Verlag der Wiss., Berlin, 1980).

[5] Landau, L. D. and Lifshitz, E. M. *The Classical Theory of Fields*, Fourth Revised English Edition (Pergamon Press, Oxford, 1975).

[6] Robinson, I. and Trautman, A. In *Proc. of the Fourth Marcel Grossmann Meeting on General Relativity*, R. Ruffini, ed. (Elsevier Science Publ., 1986).

[7] Trautman, A. In *Perspectives in Geometry and Relativity*, B. Hoffmann, ed. (Indiana University Press, Bloomington, 1966).

24

Working with Engelbert

István Ozsváth

ABSTRACT My cooperation with Engelbert is long-standing and wide-ranging. It goes back to almost four decades and encompasses much of our work. In this paper, I summarize some of the old results we obtained together, mention some that I found as a consequence of our cooperation, and describe the work we have undertaken recently.

24.1 Introduction

It is a great pleasure to participate in the celebration of the seventieth birthday of Engelbert. Originally, I wanted to describe in some details Engelbert's life work. But I soon realized that one man alone cannot do justice to this task. Engelbert's interests are enormously broad, while the space alloted to me is quite narrow. Hence, I restrict my remarks to a few results we obtained together, finding consolation in the fact that Engelbert had in the course of his professional life many other collaborators who could complete my account.

I met Engelbert in 1957 in the Hamburg Observatory. As a former graduate student in Budapest, I had the good fortune to end up in Hamburg, one of the centers of the German astronomy at the time. After obtaining my Ph.D., I became the assistant to the personal assistant to the director of the observatory: Otto Heckmann. The personal assistant was *Herr Schucking.*

24.2 Exact Solutions

Our collaboration started when Engelbert showed me Kurt Gödel's paper *Rotating Universes in General Relativity Theory* [1]. I would like to cite the relevant parts of that paper.

24.2.1 Remarks by Kurt Gödel

The solution is to be homogeneous in space. ... This reduces the problem to a system of *ordinary* differential equations.

Moreover, this system of differential equations can be derived from a Hamiltonian principle ...

The symmetric case, by means of the integrals of momentum, can be reduced to a problem with three degrees of freedom (g_1, g_2, g_3), whose Lagrangian function reads as follows:

$$\left\{\sum_{i<k} \frac{\dot{g}_i \dot{g}_k}{g_1 g_k} + \frac{1}{g}\left[2\sum_i g_i^2 - \left(\sum_i g_i\right)^2\right] + \frac{V^2}{g_1(g_2-g_3)^2}\right\} g^{1/2} + 2\left(1+\frac{V^2}{g_1}\right)^{1/2},$$

where $g = g_1 g_2 g_3$ and V is a constant which determines the velocity of rotation.

...

There exist rotating stationary homogeneous solutions with finite space, no closed timelike lines, and $\Lambda > 0$ ...

24.2.2 The Schucking Equations

Inspired by those remarks, Engelbert then developed the field equations for spatially homogeneous spaces:

The metric is given by

$$ds^2 = dt^2 + \gamma_{ab}(t)\,\omega^a\omega^b.$$

The field equations read as

$$\frac{1}{2}\left\{\frac{1}{4}\dot{\gamma}_{ab}\gamma^{bc}\dot{\gamma}_{cd}\gamma^{ad} - \frac{1}{4}\left(\dot{\gamma}_{ab}\gamma^{ab}\right)^2 - R^*\right\} + \Lambda = -\kappa\rho(u_0)^2$$

$$\frac{1}{2}\dot{\gamma}_{ab}\gamma^{ad}(C^b{}_{dc} - \delta^b_c C^f{}_{fd}) = -\kappa\rho p_c u_0$$

$$\frac{1}{2}\ddot{\gamma}_{cd} - \frac{1}{2}\dot{\gamma}_{ca}\gamma^{ab}\dot{\gamma}_{bd} + \frac{1}{4}\dot{\gamma}_{cd}\dot{\gamma}_{ab}\gamma^{ab} + R^*_{cd} = (\Lambda + \frac{\kappa\rho}{2})\gamma_{cd} - \kappa\rho p_c p_d, \quad (1)$$

where

$$R^* = \gamma^{cd} R^*_{cd}$$

is the Ricci scalar of the space part of the model κ is the constant of gravitation $\rho = \rho(t)$ is the density of the incoherent matter, and

$$u_0 = u_0(t), \qquad p_c = p_c(t) \qquad c = 1, 2, 3$$

are the components of the 4-velocity. Finally, the 1-forms

$$\omega^a, \qquad a = 1, 2, 3$$

satisfy the relations

$$d\omega^a = \frac{1}{2}C^a{}_{bc}\,\omega^b \wedge \omega^c.$$

24.2.3 The Schucking Solution

The first solutions of these equations are constructed by Engelbert and are described by the following expressions:

$$\gamma_{11} = -t^{2/3[1+2\sin\alpha]}\left(\frac{9GM}{2}t - a\right)^{2/3[1-2\sin\alpha]}$$

$$\gamma_{11} = -t^{2/3[1+2\sin(\alpha+2\pi/3)]}\left(\frac{9GM}{2}t - a\right)^{2/3[1-2\sin(\alpha+2\pi/3)]}$$

$$\gamma_{11} = -t^{2/3[1+2\sin(\alpha+4\pi/3)]}\left(\frac{9GM}{2}t - a\right)^{2/3[1-2\sin\alpha+4\pi/3)]}$$

$$\gamma_{12} = \gamma_{13} = \gamma_{23} = 0, \qquad p_c = 0,$$

$$\frac{4\pi}{3}\rho\sqrt{-\gamma} = M = \text{const.} \geq 0,$$

$$a = \text{const.}, \qquad \alpha = \text{const.}$$

For $a = 0$, this solution coincides with the Friedmann solution for the case of vanishing curvature. For $M = 0$, one obtains the empty space solution found by A. Taub for Bianchi type 1. The solutions for a nonvanishing cosmological term are also easy to obtain [2].

The vast literature of the spatially homogeneous models started with the works of *Kurt Gödel, Abraham Taub, and Engelbert Schucking.*

24.2.4 The Finite Rotating Universe

This is the *rotating stationary homogeneous solution with finite space, no closed timelike lines, and* $\Lambda > 0$ as mentioned by Gödel.

This rotating universe defies Mach's principle of the relativity of rotation. It demonstrates that Einstein's wish to turn the rotation of a body into a relative motion with respect to the cosmos is *not* a consequence of his general relativity.

The anti-Mach universe is given by the line element

$$ds^2 = -dt^2 - R\sqrt{1-2k^2}\,\omega^3\,dt + \frac{R^2}{4}\{(1-k)(\omega^1)^2 + (1+k)(\omega^2)^2 + (1+2k^2)(\omega^3)^2\},$$

with parameters k and R such that

$$|k| \leq 1/2, \qquad R > 0.$$

This line element is a solution of the Einstein field equations with incoherent matter and $\Lambda > 0$. The quantities $\kappa\rho$ and Λ are given in terms of the parameters k and R by the following expressions [3]:

$$\frac{\kappa\rho}{2\Lambda} = 1 - 4k^2, \qquad \Lambda = \frac{1}{R^2(1-k^2)}.$$

24.2.5 The Anti-Mach Metric

This is a vacuum solution, free of singularities, geodesically complete, and curved, which demonstrates that one can have a genuine gravitational field without sources. The line element is given by

$$ds^2 = (\omega^1)^2 + 2\,\omega^2\,\omega^3 + (\omega^4)^2,$$

with the -forms

$$\begin{aligned}
\omega^1 &= dx^1 - x^4\,dx^3,\\
\omega^2 &= -x^4\,dx^1 + dx^2 + \frac{(x^4)^2}{2}\,dx^3,\\
\omega^3 &= dx^3,\\
\omega^4 &= dx^4.
\end{aligned}$$

The maximal group is 6-parametric [4].

By a suitable change of coordinates, we obtain

$$ds^2 = dy^2 + dz^2 + 2\,du\,dv + 2H(du)^2,$$

with

$$H = \frac{1}{2}(y^2 - z^2)\cos 2u - y\,z\sin 2u.$$

This is the well-known Robinson wave.

24.3 More on Exact Solutions

Engelbert joined the physics department at NYU; I joined the Southwest Center for Advanced Studies in Dallas. I continued working on homogeneous solutions for a while; but then I took a different approach. Previously we selected a specific Lie group and asked whether there is a solution of the field equation invariant with respect to that Lie group. Now I asked the question in a different way: are there homogeneous solutions to a specific set of field equations? If there are, I would set out to find all of them. I would like to present a few results of this work.

24.3.1 All Homogeneous Vacuum Solutions with Λ term

$$ds^2 = -dt^2 + \exp(2\sqrt{\Lambda/3}\,t)\{dx^2 + dy^2 + dz^2\}$$

W. de Sitter:

$$ds^2 = dx^2 + \exp(2\sqrt{-\Lambda/3}\,x)\{-dt^2 + dy^2 + dz^2\}$$

Anti-de Sitter:

$$\begin{aligned}
ds^2 = &-\exp(2\sqrt{-\Lambda/3}\,x)\,dt^2 + dx^2\\
&+\exp(2\sqrt{-\Lambda/3}\,z)\,dy^2 + dz^2
\end{aligned}$$

B. Bertotti (1959):

$$ds^2 = dx^2 + \exp(2\sqrt{-\Lambda/3}\,x)\{dy^2 + 2du\,dv\} + f\exp(-\sqrt{-\Lambda/3}\,x)\,dv^2 \quad (2)$$

Type N, I. Ozsváth (1962), M. Cahen (1964):

$$\begin{aligned} ds^2 = {} & dx^2 + \exp(2\sqrt{-\Lambda/3}\,x)\{dy^2 + 2du\,dv\} \\ & + f\exp(-\sqrt{-\Lambda/3}\,x)\{-2\sqrt{2}\,dy + f\exp(-3\sqrt{-\Lambda/3}\,x)\,dv\}\,dv \end{aligned}$$

Type III, I. Ozsváth:
The list is complete [5].

24.3.2 All Type N Vacuum Solutions with Λ term Vanishing Shear and Expansion

During a faculty meeting, I gave the metric

$$ds^2 = dx^2 + \exp(2\sqrt{-\Lambda/3}\,x)\{dy^2 + 2du\,dv\} + f\ \exp(-\sqrt{-\Lambda/3}\,x)\,dv^2$$

to Ivor Robinson. He exclaimed "you found the plane-fronted gravitational wave against the anti-de Sitter background!"

During a long and extremely hot summer, Ivor and I constructed all Type N solutions of the vacuum field equations with nonvanishing Λ term and vanishing shear and expansion. The structure of the metric is given by

$$ds^2 = ds_1^2 - p^{-1}qH\,d\sigma^2,$$

where

$$ds_1^2 = -2q^2p^{-2}d\sigma\left[dp + \left(-\frac{\kappa\rho}{2} + l\right)\rho\,d\sigma\right] + 2p^{-2}d\zeta\,d\bar{\zeta},$$

with

$$p = 1 + \frac{\Lambda}{6}\zeta\bar{\zeta}, \qquad q = (1 - \frac{\Lambda}{6}\zeta\bar{\zeta})\alpha + \zeta\bar{\beta} + \bar{\zeta}\beta, \qquad \kappa = \frac{\Lambda\alpha^2}{3} + 2\beta\bar{\beta},$$

where α real and η complex function of σ, and

$$l = \frac{\partial \ln|q|}{\partial\sigma}.$$

In case of a pure gravitational wave, the empty space equation is

$$H_{\zeta\bar{\zeta}} + \frac{\Lambda}{3}p^{-2}H = 0.$$

Krzysztof Rózga joined us, added electromagnetic radiation, and completed the paper with a nice discussion of the results [6].

24.3.3 Finite Rotating Universe Revisited

I figured out eventually how Gödel might have arrived at the symmetric case he mentioned in the cited quotation:

The metric

$$ds^2 = -dt^2 + A\,(\omega^1)^2 + 2F\,\omega^1\omega^2 + B\,(\omega^2)^2 + C\,(\omega^3)^2$$

with

$$d\omega^1 = -\,\omega^2 \wedge \omega^3, \quad d\omega^2 = -\,\omega^3 \wedge \omega^1, \quad d\omega^3 = -\,\omega^1 \wedge \omega^2,$$

describes the symmetric case; and the transformation

$$\begin{aligned} A &= \frac{1}{2}\{a + b + (a - b)\cos 2w\}, \\ B &= \frac{1}{2}\{a + b - (a - b)\cos 2w\}, \\ F &= \frac{1}{2}(a - b)\sin 2w, \\ C &= c, \end{aligned}$$

reduces the mechanical problem to three degrees of freedom.

The Hamiltonian function describing this mechanical problem is given by

$$\begin{aligned} H = {} & \frac{1}{2\sqrt{abc}}\{2(a^2p^2 + b^2q^2 + c^2s^2) - (ap + bq + cs)^2 \\ & \qquad + \frac{r^2}{(a-b)^2c}abc\} \\ & + \frac{1}{2\sqrt{abc}}\{2(a^2 + b^2 + c^2) - (a + b + c)^2\} \\ & + 2\Lambda\sqrt{abc} + 2\kappa\rho_0\sqrt{1 + V^2/c} \quad (= 0), \end{aligned}$$

leading to the field equations in canonical form

$$\begin{aligned} \dot{a} &= \frac{\partial H}{\partial p}, & \dot{b} &= \frac{\partial H}{\partial q}, & \dot{c} &= \frac{\partial H}{\partial s}, & \dot{w} &= \frac{\partial H}{\partial r}, \\ \dot{p} &= -\frac{\partial H}{\partial a}, & \dot{q} &= -\frac{\partial H}{\partial b}, & \dot{s} &= -\frac{\partial H}{\partial c}, & \dot{r} &= -\frac{\partial H}{\partial w}. \end{aligned}$$

Since

$$\frac{\partial H}{\partial w} = 0,$$

we have

$$\dot{w} = \frac{r}{(a-b)^2c}\sqrt{abc}, \qquad \dot{r} = 0, \qquad r = 2\kappa\rho_0 V.$$

It is very easy to work out these equations explicitly. One sees immediately that they have the following trivial solution:

$$a = \text{const.}, \quad b = \text{const.}, \quad c = \text{const.}, \quad p = 0, \quad q = 0, \quad r = 0,$$

leading to the Finite Rotating Universe. The line element has the form

$$
\begin{aligned}
ds^2 = -dt^2 &+ \frac{1}{2}\{a + b + (a-b)\cos 2w\}\,(\omega^1)^2 \\
&+ (a-b)\sin 2w\,\omega^1\omega^2 \\
&+ \frac{1}{2}\{a + b - (a-b)\cos 2w\}\,(\omega^2)^2 \\
&+ \frac{(a+b)^2 + 2(a-b)^2}{2(a+b)}\,(\omega^3)^2
\end{aligned}
$$

with

$$
w = \sqrt{\frac{(a+b)^2 - 2(a-b)^2}{(a+b)[(a+b)^2 + 2(a-b)^2]}} \times t
$$

and

$$
d\omega^1 = -\,\omega^2 \wedge \omega^3, \qquad d\omega^2 = -\,\omega^3 \wedge \omega^1, \qquad d\omega^3 = -\,\omega^1 \wedge \omega^2.
$$

exhibiting the fact that the finite rotating universe is a special case of Gödel's symmetric case, and that the *time dependence* in this form of the metric is given by the transformations mediating the reduction of the degrees of freedom of the Hamiltonian.

Solving the indicated field equations, we obtain for the quantities $\kappa\rho_0$, Λ, and V^2 explicit expressions in terms of the parameters

$$
a > 0, \qquad b > 0.
$$

These expressions impose restrictions to the range of these parameters.

The Finite Rotating Universe appears here as a stationary point of the Hamiltonian. Studying the literature of the Hamiltonian system, one could probably decide whether this fact would be useful in finding more general solutions [7].

24.4 Embedding Problems

The 3-spaces with the line element

$$
ds^2 = a\,(\omega^1)^2 + b\,(\omega^2)^2 + c\,(\omega^3)^2,
$$

where a, b, c are parameters, and the 1-forms satisfy the relations

$$
d\omega^1 = -\,\omega^2 \wedge \omega^3, \quad d\omega^2 = -\,\omega^3 \wedge \omega^1, \quad d\omega^3 = -\,\omega^1 \wedge \omega^2,
$$

play a significant role in relativistic cosmology. These spaces are used as spacelike hypersurfaces in important cosmological models. And they also are interesting in their own right as homogeneous spaces. They are S^3-s endowed with the most general left-invariant metric. Using the terms of Engelbert, I shall call them dantes. We started to study the dantes' global properties by embedding them into higher dimensional Euclidean spaces. I shall describe some of the results.

24.4.1 *Embedding of Dantes into S^4*

The most general dante which can be embedded into S^4 is characterized by the line element

$$ds^2 = \frac{1}{4}(1+\lambda)^2\omega_1^2 + \frac{1}{4}(1-\lambda)^2\omega_2^2 + \omega_3^2.$$

The embedding into the R^5 is given by

$$\begin{aligned} y_1 + iy_2 &= \frac{1}{2}[z_1^4 + z_2^4] - \lambda z_1^2 z_2^2, \\ y_3 + iy_4 &= z_1^3\bar{z}_2 - \bar{z}_1 z_2^3 + \lambda z_1 z_2(|z_1|^2 - |z_2|^2), \\ y_5 &= \frac{\sqrt{3}}{2}[z_1^2\bar{z}_2^2 + \bar{z}_1^2 z_2^2 - \frac{\lambda}{3}[(|z_1|^2 - |z_2|^2)^2 - 2|z_1|^2|z_2|^2]], \end{aligned}$$

$$|z_1|^2 + |z_2|^2 = 1 \quad \text{and} \quad \lambda \quad \text{arbitrary} \quad (|\lambda| \neq 1).$$

$$y_1^2 + y_2^2 + y_3^2 + y_4^2 + y_5^2 = \frac{1}{4}(1 + \lambda^2/3).$$

This shows that the embedded manifold fits into an S^4. It is easily checked that the substitutions

$$(z_1 \to iz_1, z_2 \to -iz_2) \quad \text{and} \quad (z_1 \to z_2, \quad z_2 \to -z_1)$$

generate the quaternion group H_8, which leaves the map invariant.

The map is isometric and injective on Poincaré's cubicle.

24.4.2 *A Special Case for $\lambda = 0$*

This dante is characterized by the line element

$$ds^2 = \frac{1}{4}(\omega_1^2 + \omega_2^2) + \omega_3^2,$$

and the embedding into S^4 is given by

$$\begin{aligned} y_1 + iy_2 &= \frac{1}{2}[z_1^4 + z_2^4], \\ y_3 + iy_4 &= z_1^3\bar{z}_2 - \bar{z}_1 z_2^3, \\ y_5 &= \frac{\sqrt{3}}{2}[z_1^2\bar{z}_2^2 + \bar{z}_1^2 z_2^2]. \end{aligned}$$

This results constitute Liland Sapiro's Ph. D. thesis, dedicated to Engelbert on the occasion of his 60th birthday [8,9].

Engelbert suggested the use of the complex numbers $z_1, z_2, |z_1|^2 + |z_2|^2 = 1$, to describe the results.

24.4.3 *Embedding of Dantes into S^5*

The most general dante which can be embedded into S^5 is characterized by the line element

$$ds^2 = (\cos\phi)^2\,\omega_1^2 + (\sin\phi)^2\,\omega_2^2 + \omega_3^2.$$

The embedding is given by the following expressions:

$$\begin{aligned}
\xi &= \cos\phi\,(z_1^2 - z_2^2),\\
\eta &= \sin\phi\,(z_1\,\bar{z}_2 + \bar{z}_1 z_2),\\
\zeta &= \cos\phi\,(z_1^2 + z_2^2),\\
\kappa &= \sin\phi\,(z_1\,\bar{z}_2 - \bar{z}_1 z_2).,
\end{aligned}$$

In the special case of

$$\phi = \pi/4, \qquad \cos\phi = \sin\phi = \frac{1}{\sqrt{2}},$$

these expressions become simpler [10]:

$$\begin{aligned}
\xi &= \frac{1}{\sqrt{2}}(z_1^2 - z_2^2),\\
\eta &= \frac{1}{\sqrt{2}}(z_1\,\bar{z}_2 + \bar{z}_1\,z_2),\\
\zeta &= \frac{1}{\sqrt{2}}(z_1^2 + z_2^2),\\
\kappa &= \frac{1}{\sqrt{2}}(z_1\bar{z}_2 - \bar{z}_1\,z_2).
\end{aligned}$$

24.4.4 *Embedding of Dantes into S^8*

I succeeded eventually in enbedding the dante characterized by the line element

$$ds^2 = a^2\,\omega_1^2 + b^2\,\omega_2^2 + c^2\,\omega_3^2$$

with the following restrictions on the parameters:

$$a^2 + b^2 - c^2 \geq 0, \qquad a^2 - b^2 + c^2 > 0, \qquad -a^2 + b^2 + c^2 > 0$$

into S^8.

The embedding is described by the expressions

$$\xi_1 = 2\sqrt{(a^2 + b^2 - c^2)/2}\, z_1 z_2,$$

$$\xi_2 = 2\sqrt{(a^2 + b^2 - c^2)/2}\, z_2 \bar{z}_2,$$

$$\xi_3 = \sqrt{(a^2 - b^2 + c^2)/2}\,(z_1^2 - z_2^2),$$

$$\xi_4 = \sqrt{(a^2 - b^2 + c^2)/2}\,(z_1 \bar{z}_2 + \bar{z}_1 z_2),$$

$$\xi_5 = \sqrt{(-a^2 + b^2 + c^2)/2}\,(z_1^2 + z_2^2),$$

$$\xi_6 = \sqrt{(-a^2 + b^2 + c^2)/2}\,(z_1 \bar{z}_2 - \bar{z}_1 z_2).$$

This is general enough to embed the space sections of the Finite Rotating Universe [11].

24.4.5 Embedding the General Dantes into S^8

Engelbert succeeded in describing this embedding in terms of quaternions and removed the last restriction.

Let

$$n_j, \qquad j = 1, 2, 3$$

denote three quaternionic units (like i, j, k) with

$$n_j \cdot n_k = -\delta_{jk} + \epsilon_{jkl}\, n_l,$$

and let

$$\xi \quad \text{with} \quad \xi \cdot \bar{\xi} = 1$$

be a unit quaternion representing a point of S^3.

The embedding map is given by the three vectorial quaternions

$$q_j = r_j\, \xi \cdot n_j \cdot \bar{\xi} = \sum_{k=1}^{3} x_{jk} n_k \qquad j = 1, 2, 3,$$

where

$$r_1 > 0, \qquad r_2 > 0, \qquad r_3 \geq 0.$$

An elegant calculation verifies that we have an embedding map

$$P^3 \longrightarrow S^8,$$

and the metric on P^3 is

$$ds^2 = a^2\, \omega_1^2 + b^2\, \omega_2^2 + c^2\, \omega_3^2$$

without any restriction on the parameters.

Each point of P^3 project on three different S^2s which are embedded into E^9 and orthogonal to each other about the origin of E^9 [12].

24.5 The SU_3 Group

The Kobayashi-Maskawa matrix C:

$$\begin{pmatrix} c_2c_3 & c_2s_3 & s_2\exp(-iw) \\ -c_1s_3 - s_1s_2c_3\exp(iw) & c_1c_3 - s_1s_2s_3\exp(iw) & s_1c_2 \\ s_1s_3 - c_1s_2c_3\exp(iw) & -s_1c_3 - c_1s_2s_3\exp(iw) & c_1c_2 \end{pmatrix}$$

gives a model of the weak interactions in particle physics [13]. Here we use the notation

$$\begin{aligned} c_1 &= \cos x & s_1 &= \sin x \\ c_2 &= \cos y & s_2 &= \sin y \\ c_3 &= \cos z & s_3 &= \sin z. \end{aligned}$$

One obtains this matrix by splitting the general element G of the SU_3 group as

$$G = G_1 \cdot C \cdot G_2,$$

where the matrices G_1 and G_2 are given by

$$G_1 = \begin{pmatrix} \exp(2ip/\sqrt{3}) & 0 & 0 \\ 0 & \exp i(-p/\sqrt{3}+q) & 0 \\ 0 & 0 & \exp -i(p/\sqrt{3}+q) \end{pmatrix}$$

and

$$G_2 = \begin{pmatrix} \exp i(r/\sqrt{3}+t) & 0 & 0 \\ 0 & \exp i(r/\sqrt{3}-t) & 0 \\ 0 & 0 & \exp(-2ir/\sqrt{3}) \end{pmatrix}.$$

In order to study that splitting, Engelbert computed the left-invariant metric on SU_3 defined by

$$ds^2 = \mathrm{tr}\{dG \cdot dG^+\},$$

or

$$ds^2 = \sum_{j=1}^{3}\sum_{k=1}^{3} du_{jk} d\bar{u}_{jk}$$

with

$$G = (u_{jk}).$$

(The second form is more adequate for a brute-force calculation with the computer.)

The amazing result is that the Kobayashi-Maskawa part is a manifold with a metric

$$ds^2 = 2\{dx^2 + dy^2 + dz^2 + 2\sin y\ \cos w\, dx\, dz + \sin^2 y\, dw^2\}.$$

The maximal group of this metric is a 3-parametric Abelian group.

By suitable change of the coordinates, the line element can be written as

$$ds^2 = 2\{du^2 + dx^2 + dz^2 + 2\cos u\, dx\, dz + \sin^2 u\, dv^2\},$$

or as

$$ds^2 = (\omega^1)^2 + (\omega^2)^2 + (\omega^3)^2 + (\omega^4)^2,$$

where the 1-forms are given by

$$\begin{aligned} \omega^1 &= \sqrt{2}\, du, \\ \omega^2 &= \sqrt{2}\ \cos u/2 \times (dx + dz), \\ \omega^3 &= \sqrt{2}\ \sin u/2 \times (-dx + dz), \\ \omega^4 &= \sqrt{2}\ \sin u\, dv. \end{aligned}$$

Continuing this research, we now would like to undertake an in-depth study of SU_3.

24.6 In Closing

I think it is not easy to find a wife with whom one can have a happy life, but it is even harder to find a colleague with whom cooperation is always a pleasure. I have been lucky on both accounts.

HAPPY BIRTHDAY ENGELBERT
bis-Hundert-und-Zwanzig

References

[1] Kurt Gödel, Rotating Universes in General Relativity, *Proc. Intern. Cong. Math.* **1** (1950), 175–181.

[2] Otto Heckmann and Engelbert L. Schucking, Relativistic Cosmology in *Gravitation: an introduction to current research*, Louis Witten ed., John Wiley and Sons, New York-London 1962.

[3] István Ozsváth and Engelbert L. Schucking, Finite Rotating Universe, *Nature* **193** (1962), 1168–1169;
Annals of Physics **55** (1969), 166–204

[4] István Ozsváth and Engelbert L. Schucking, An Anti-Mach Metric, in *Recent Developments in General Relativity*, Infeld Festschrift Pergamon Press 1962.

[5] István Ozsváth, All Homogeneous Solutions of Einstein's Vacuum Field Equations with a non-vanishing Cosmological Term, in *Gravitation and Geometry*, Robinson Festschrift, Wolfgang Rindler, and And Andrzej Trautman eds., Bibliopolis, Naples 1987.

[6] István Ozsváth, Ivor Robinson, and Krzysztof Rózga, Plane-fronted Gravitational and Electromagnetic Waves in Spaces with Cosmological Constant, *J. Math. Phys.* **26** (1985) 1755–1761.

[7] István Ozsváth, Spatially Homogeneous Rotating World Models, *J. Math. Phys.* **12** (1971), 1078-1082

[8] István Ozsváth and Leland Sapiro, The Schucking Problem, *J. Math. Phys.* **28** (1987), 2066–2073.

[9] István Ozsváth, An Embedding Problem, *J. Math. Phys.* **29** (1987), 825–834.

[10] István Ozsváth, Embedding Problem II, *J. Math. Phys.* **33** (1992), 229–247.

[11] István Ozsváth and Engelbert Schucking, *The World Viewed from Outside*, in print, J. Geometry and Physics.

[12] István Ozsváth and Engelbert Schucking, Isometric Embedding for Homogeneous Compact 3-Manifolds, *General Relativity and Gravitation* **28** (1996), 999–1011.

[13] David Griffiths, *Introduction to Elementary Particles*, John Wiley and Sons, New York-Chichester-Bribane-Toronto-Singapore, 1987.

25

Some Remarks on Twistor Theory

Roger Penrose

ABSTRACT The influence of Engelbert Schucking on the development of twistor theory is pointed out, particularly with regard to conformal invariance, the positive-frequency condition, and complexification. The current status of the problem of encoding the Einstien field equations into twistor geometry is also outlined.

25.1 Historical Comments

It is a pleasure to have this opportunity to be able to express my gratitude and appreciation to Engelbert Schucking for certain ideas that were central to the early stages in the development of twistor theory. I can clearly recall some three crucial influences that were, indeed, quite fundamental to that theory, and I wish to relate these here. They date largely back to a period in the first half of the year 1961, when Engelbert and I shared an office in Steele Hall at Syracuse University. I was visiting there as part of my stay in the USA for a period under a NATO Fellowship, held mainly at Princeton, during the academic years 1959/60 and 1960/61. It was at a time when Peter Bergmann's group at Syracuse was particularly active, and there were a great many of the then up-and-coming young workers in the field of general relativity, visiting temporarily.

One of Engelbert's profound influences upon me arose from the fact that he had stressed the importance of conformal invariance for massless free fields. He was certainly aware, at the time, of the fact that not only the wave equation (at least in flat space-time) but also the free Maxwell equations constituted conformally invariant systems. He knew of the early papers of Cunningham (1910) and Bateman (1910) establishing these respective facts, but, as I recall it, he must also have known of McLennan's (1956) work, which established conformal invariance for massless fields of general spin. I learnt of these conformal invariance properties from Engelbert, but what was much more important for me was that Engelbert evidently regarded these properties as things of particular significance which I ought to know. He encouraged me to look at these properties from the 2-spinor point of view, a point of view that I had become accustomed to for other reasons. Indeed, this 2-spinor approach to conformal invariance proved to be ultimately extremely valuable, this invariance being very immediate when the appropriate 2-spinor framework has been set up.

Indeed, I started wondering, while at Syracuse, whether conformal invariance might provide a viewpoint whereby the asymptotic structure of gravitational radiation fields could be conveniently studied in a geometrically satisfactory way. Andrzej Trautman, who was one of those at Syracuse at the time, had impressed me with his analysis of gravitational radiation, showing that the far field had the algebraic form of a null Riemann tensor. But some of his calculations and analytic conditions looked rather too complicated for me to grasp properly and I had wondered whether a more geometrical approach might be possible, using the ideas of conformal transformations and conformal invariance. Although I did not have the proper idea of conformal infinity at that time, Engelbert's remarks had sown the initial seed.

I believe that at that time I had had no particular familiarity with the way that the conformal group acts on Minkowskian, rather than Euclidean space. In the case of Euclidean geometry, the conformal compactification with which I had long been familiar simply adjoins a point at infinity, whereas with Minkowski space, an entire light cone must be adjoined. This geometrical structure of compactified Minkowski space (in various dimensions and signature) had already been known to Bôcher (1914) and Coxeter (1936), but I had not been acquainted with these references then.

I finally appreciated the way in which one could make use of this compactification in general relativity only some time later than the Syracuse period, a while after I had returned to London. At Syracuse, I had ascertained that the point at spatial infinity, what is now referred to as i^0, must be singular when the total mass is nonzero. I think I had formed the (wrong) conclusion that the Weyl curvature therefore had to be singular throughout all of infinity, so I had been left with the impression, then, that the conformal idea was probably not very useful in general relativity. Nevertheless, this idea did have an intrinsic appeal for me, so it was with great delight that I finally realized (in the summer of 1963, I think) that, owing to the presence of an "infinite boost" at null infinity, the Weyl curvature could be nonsingular there after all.

The second important idea that Engelbert impressed upon me, during that period in Syracuse, was that a key feature of quantum field theory is the splitting of field amplitudes into positive and negative frequency parts. I do not think that this was something that had been especially emphasized by other quantum field theorists, particularly at that time. Much of the heavy formalism of quantum field theory serves to obscure the fact that this splitting is indeed one of the theory's most significant ingredients—a fact that, with characteristic insight, Engelbert had singled out for me.

The normal approach to forming this splitting is to perform a Fourier analysis on the fields, and then to separate the Fourier components of positive frequency from those of negative frequency. Something that was disturbing, from my point of view, was that there is no manifest conformal invariance in this procedure. The individual Fourier components do not behave nicely under conformal transformation. The fundamental role of the conformally invariant massless fields seemed to provide an unsatisfying conflict with this. Yet, the positive/negative splitting is conformally invariant for massless fields, although this particular fact is not manifest with the normal procedures for splitting. Considerations such as this led me to approach the question of positive/negative frequency splitting in terms of a much more geometrical description than using Fourier decompositions directly—an approach in which the positive/negative frequency splitting does become manifestly conformally invariant.

To see what this description is, consider first the case of one dimension. We shall be concerned with complex functions on the real line, i.e., with maps $\mathbb{R} \to \mathbb{C}$ We now think of the real line as "compactified" into the circle $\mathbf{S}^1$. The appropriate analogue for the conformal group on this $\mathbf{S}^1$ is (in this context) the group SL(2, $\mathbb{R}$). It acts on the northern hemisphere $\mathbf{D}^+$ of the Riemann sphere $\mathbf{S}^2$, where the given $\mathbf{S}^1$ is regarded as the equator of $\mathbf{S}^2$, i.e., the boundary of $\mathbf{D}^+$. Now, an appropriate notion of "positive" frequency, for complex functions defined on $\mathbf{S}^1$, is that such a function f shall extend as a holomorphic function to the whole of $\mathbf{D}^+$. The corresponding notion of negative frequency is then that f extend holomorphically to the southern[1] hemisphere $\mathbf{D}^-$ of $\mathbf{S}^2$. This assumes that f has appropriate smoothness properties, especially at ∞. If f is *analytic* on $\mathbf{S}^1$, then there will be a "collar" containing $\mathbf{S}^1$ (on $\mathbf{S}^2$), into which f extends holomorphically. Thus, if f is of positive frequency, its holomorphic extension to $\mathbf{D}^+$ will also extend into part of $\mathbf{D}^-$, so that the entire region on which f is holomorphic will be some open set containing $\mathbf{D}^+$, and containing $\mathbf{S}^1$ in its interior. If a positive-frequency f is merely continuous, then its values on $\mathbf{S}^1$ will be the (limiting) boundary values of a holomorphic function on $\mathbf{D}^+$. More generally, this notion of positive frequency also applies when f is not necessarily continuous, and will also apply if f is any distribution. The most general notion for which the idea applies is when f is a hyperfunction (see Bailey, Ehrenpreis, and Wells 1982). We see from this procedure that the positive/negative frequency split is invariant under SL(2, $\mathbb{R}$), i.e., conformally invariant as asserted above.

I had been (very vaguely) aware of such ideas, even while at Syracuse in 1961, but I did not make all the appropriate connections then. There was another ingredient, however, which had held a particular interest for me at that time. My concern with massless fields had led me to a study of the Kirchhoff integral formula for the wave equation in Minkowski space $\mathbb{M}$, whereby the value of the field at an arbitrary point p in $\mathbb{M}$ can be obtained by means of an explicit integral performed over the sphere of intersection of the light cone of p with an initial Cauchy hypersurface in $\mathbb{M}$. I had found a generalization of this formula which applies to massless fields of arbitrary spin, where the initial hypersurface is most naturally taken to some a null hypersurface $\mathcal{N}$. I learned later that the form of the Kirchhoff integral which applies to a null initial hypersurface is referred to as the d'Adhémar integral. The generalization to arbitrary spin is, accordingly, the "generalized Kirchhoff-

[1]There is an element of convention as to which hemisphere is deemed to be the one corresponding to positive frequency and which to negative. The significance of the choice being made here is merely ease of description.

d'Adhémar formula", as so named in Penrose and Rindler (1984) pp. 394–6. I shall refer to it the GKd'A formula here.

The GKd'A formula is remarkably compact. It has a manifest conformal invariance (when reinterpreted in the right way), and it enables the field to be obtained from a single complex quantity, called the null-datum for the field, that can be chosen essentially freely (i.e., without constraints) on the initial null hypersurface N_0. (The null-datum for the field $\varphi_{AB\ldots L}$ is the quantity $\varphi = \varphi_{AB\ldots L}\xi^A\xi^B\ldots\xi^L$, where ξ^A has its flagpole pointing along the null direction in N_0.) Some of these facts had impressed me at the time and, moreover, it had seemed to me that there was perhaps some familial resemblance between the GKd'A formula and the Cauchy integral formula. I had some sort of feeling that in some sense yet to be determined, there might be a close kinship between the propagation of massless fields and the process of analytic continuation for holomorphic functions. For this to make any sense, there would have to be some kind of complex space which generalized or, in some appropriate sense, took over the role of space-time. Then, the positive-frequency condition of quantum field theory and the field propagation equations for massless fields would both emerge as different aspects of complex analyticity in this new space.

It seems that Engelbert may have had some ideas of his own which closely paralleled this kind of thinking, and he may have had other complementary motivations which he never fully revealed to me. In any case, for whatever reasons, he stressed to me on a number of different occasions the need for finding such a complex space. I recall this particularly a few years later, when I spent the academic year 1963/4 at the University of Texas in Austin. Engelbert had an office next to mine, and he brought the matter up forcefully with me early in the year. For my own motivational reasons, I had felt it to be important that this sought-for space should share with the Riemann sphere $\mathbf{S}^2$ the property that it would be naturally divided into two halves by the real part of space (analogous to the circle $\mathbf{S}^1$, in the Riemann sphere case), the positive frequency functions extending into the upper half and the negative-frequency ones into the lower half.

I was not satisfied with the straightforward complexification of Minkowski space $\mathbb{M}$ to the real 8-dimensional space $\mathbb{CM}$ (of points with complex position vectors). It might have seemed that the forward and backward tubes $\mathbb{M}^+$ and $\mathbb{M}^-$, respectively, (points with position vectors with past-timelike imaginary parts, and with future-timelike imaginary parts, respectively) serve roles closely analogous to $\mathbb{D}^+$ and $\mathbb{D}^-$. For positive frequency fields are, indeed, those which extend holomorphically into $\mathbb{M}^+$ and those of negative frequency, into $\mathbb{M}^-$. However, $\mathbb{M}^+$ and $\mathbb{M}^-$ do not form two halves of $\mathbb{M}$ (i.e., $\mathbb{M}$ is not the closure of a disjoint union of $\mathbb{M}^+$ and $\mathbb{M}^-$), unlike the case of $\mathbf{S}$, because there is a third open region (points with spacelike imaginary parts) required to make up $\mathbb{M}$. This difficulty is tied up with the fact that the massless field equations enjoy no pride of place in this complexification $\mathbb{CM}$. The positive or negative frequency requirements do not themselves exhaust the possibilities because past and future timelike vectors (and their limiting cases) do not exhaust the possible kinds of vector. My requirement was, instead, for something whereby null vectors play a special role. The past and future null vectors (and their common limiting case) do exhaust the null cone. If a complex space could be provided which directly related to null things in Minkowski space, then the hope of seeing the GKd'A formula as some reflection of analytic continuation, as motivated above, might tie in with the desired positive/negative splitting ideas.

The key realization turned out to be something that sprung from Ivor Robinson's ideas for finding singularity-free null solutions of Maxwell's equations. He had shown how to find certain shear-free congruences of light rays (henceforth rays) which generalize the family of rays meeting a given ray. These particular congruences are now called Robinson congruences. I recall the occasion in December 1963 when I realized that the Robinson congruences in $\mathbb{M}$ (together with their limiting cases) constitute a 6-dimensional space—which is naturally a 3-dimensional complex manifold, now referred to as $\mathbb{PT}$ or projective twistor space (see Penrose 1987). The space of all rays in $\mathbb{M}$ is a 5-dimensional space - now called $\mathbb{PN}$, or projective null twistor space. (I am including the "rays at infinity"—the generators of null infinity $\mathcal{I}$—in this space. They must be included if conformal invariance is to apply.) Thus, $\mathbb{PT}$ constitutes a kind of mild complexification of $\mathbb{PN}$, in that only one real dimension needs to be added to $\mathbb{PN}$ in order to get the complex manifold $\mathbb{PT}$. The basic thing about $\mathbb{PT}$ is that the manifestly physically "real" part of it, namely $\mathbb{PN}$, now does divide it into two parts, $\mathbb{PN}^-$, and $\mathbb{PN}^+$, just as $\mathbf{S}^1$ divides $\mathbf{S}^2$ into $\mathbb{D}^-$ and $\mathbb{D}^+$.

It took me more than two years to realize (in 1966) the kind of way in which massless fields in $\mathbb{M}$ are to be elegantly represented in $\mathbb{PT}$. Basically, a holomorphic function f is specified on $\mathbb{PT}$, and the space-time field is obtained from f by means of contour integration. At that time, I had not known of the much earlier work of Whittaker (1903), for solving the 3-dimensional Laplace equation, of Bateman (1904), for solving the 4-dimensional wave equation, and of Bateman (1944), for solving the free Maxwell equations. They provided definite-integral expressions which gave (locally) the general analytic solution of these equations. These expressions are easily converted into contour integrals and are then seen to provide special cases of my own formula (see Penrose 1969), for the cases of spin 0 and 1.

Even having these expressions, it took me *ten more* years before I understood the mathematical nature of the transformation properties involved in the "gauge freedom" of these expressions. As it finally emerged (see Penrose 1979, Eastwood, Penrose, and Wells 1981), the "twistor function" is not to be thought of as an ordinary function defined on (some region in) $\mathbb{PT}$ at all, but as an *element of* 1st *sheaf cohomology*. The holomorphic function f that appeared in the contour integral expressions turns out to be what is called a *Čech representative* of this cohomology element. I call an element of 1st sheaf cohomology a *1-function* (where an ordinary function would be a 0-function).

Only with this interpretation does it become true that the splitting of $\mathbb{PT}$ into two parts, $\mathbb{PT}^-$ and $\mathbb{PT}^+$, by the real hypersurface $\mathbb{PN}^0$ is correctly analogous to the splitting of $\mathbf{S}^2$ into $\mathbb{D}^-$ and $\mathbb{D}^+$. A twistor 1-function $\mathbf{f}$, defined at $\mathbb{PN}$, extends as a holomorphic 1-function into $\mathbb{PT}^+$ iff $\mathbf{f}$ is of positive frequency, and into $\mathbb{PT}^-$ iff it is of negative frequency. Moreover, the conformal invariance of this frequency splitting is now completely manifest—for the restricted (i.e., nonreflective) conformal group on $\mathbb{M}$ is realized as the group of holomorphic transformations of $\mathbb{PT}^+$ (or of $\mathbb{PT}^-$). Each of $\mathbb{PT}^+$ and $\mathbb{PT}^-$ is preserved by this group, so the splitting is preserved. Thus the three motivational strands that had originated in Engelbert's suggestions to me had finally all come together.

This provides a taste of some of Engelbert's particular skills. Not only does he have great erudition, but also an especially noteworthy ability to select, from his vast store of knowledge, just those facts that are of key importance to the problem at hand. In addition, his instincts are generated from his own personal deep understandings, and they do not at all derive from whatever might be the prevailing fashionable view as to how to proceed in physics. In other words, he has not only knowledge and judgement, but a profound originality.

25.2 Twistors and the Einstein Equations

Much has happened since these origins of twistor theory—a theory which started out, in the 1960s, as an attempt to provide a distinctly different approach to basic physics from the usual one, aimed specifically at the problem of unifying space-time structure with the mathematics of quantum theory. However, most of the current twistor-related activity (as judged by publication numbers, at least) is in pure mathematics rather than physics.

The three main pure-mathematical areas where twistor theory has proved valuable are differential geometry, representation theory, and integrable systems. In differential geometry, most current interest has its roots in a twistor construction, sometimes called *the nonlinear graviton construction*, that I found in 1975 (Penrose 1976) for describing the (anti-)self-dual solutions of the Einstein vacuum equations. These are (complex-) Riemannian 4-manifolds $\mathcal{M}$ which are Ricci-flat and for which the Weyl tensor satisfies a self-dual or anti-self-dual condition, i.e., ${}^*C_{abcd} = iC_{abcd}$ or ${}^*C_{abcd} = -iC_{abcd}$ respectively, where C_{abcd} is the Weyl conformal tensor and ${}^*C_{abcd} = \frac{1}{2}C_{ab}{}^{pq}e_{cdpq}$ is its dual, e_{abcd} being the alternating tensor. When the signature of the metric of the "space-time" $\mathcal{M}$ is taken to be positive definite rather than Lorentzian, then $\mathcal{M}$ can be real rather than complex. This is one reason that this construction has proved valuable in pure mathematics. However, I shall not be concerned with such pure-mathematical applications of this construction here.

The construction is given in terms of a curved version of twistor space $\mathbb{T}$, which is a complex 4-manifold $\mathcal{T}$ (or complex 3-manifold $\mathbb{P}\mathcal{T}$, in the projective description) having the same local holomorphic structure, in the appropriate sense, as does $\mathbb{T}$ (or $\mathbb{PT}$, respectively). This local structure of $\mathcal{T}$ contains no information as to which specific "space-time" $\mathcal{M}$ is being described by $\mathcal{T}$. There is, of course, local information in $\mathcal{M}$, namely that which defines its metric and curvature, but this is stored *nonlocally* in the twistor space.

In order to understand how this construction works, let us first return to consideration of *flat* twistor space. There is a standard way of representing a flat-space twistor Z^α in terms of 2-component spinors:

$$Z^\alpha = (\omega^A, \pi_{A'}) ,$$

where the 2-spinors ω^A and $\pi_{A'}$ are called the *spinor parts* of Z^α. The spinor ω^A is the *primary* part of Z^α and $\pi_{A'}$ is its *projection* part.

The condition that a (complex) point x^a be what is called *incident* with the twistor Z^α is that the relation

$$\omega^A = ix^{AA'}\pi_{A'} ,$$

hold. In terms of projective twistor space $\mathbb{PT}$, this incidence relation serves to define the locus within $\mathbb{PT}$ which represents the point x of $\mathbb{CM}$ whose position vector is x^a. We note that if x^a is held fixed, then the incidence relation represents two linear equations connecting ω^A with $\pi_{A'}$. This is a pair of linear relations on Z^α, so Z lies on a specific projective straight line in PT which we can label **x**. The line **x** provides a twistor representation, within the projective twistor space $\mathbb{PT}$, of the (complex) space-time point x. Thus, a geometrical interpretation of the above incidence relation is that the point Z of $\mathbb{PT}$ lies on the line **x**.

In the nonlinear graviton construction, this interpretation of a (complex) space-time point $x \in \mathcal{M}$ is closely similar. Now, there is a *holomorphic* curve $\mathbf{x} \subset \mathbb{P}\mathcal{T}$ which represents the point x. There is in general no notion of "straight line" in $\mathbb{P}\mathcal{T}$, and it is somewhat remarkable that the mere restriction that the line **x** be holomorphic, together with a requirement that it belong to the correct topological class, is sufficient for what is required. (This assumes that the complex structure of $\mathbb{P}\mathcal{T}$ is appropriate, which will indeed be the case if $\mathbb{P}\mathcal{T}$ does not differ by "too much" from a tubular neighbourhood of a line in $\mathbb{CP}^3$.) It follows from the appropriate theorems of Kodaira (1962, 1963) and of Kodaira and Spencer (1958) that the lines **x** constitute a complex-4-dimensional system, so we can construct the space-time 4-manifold $\mathcal{M}$ as the abstract space whose points represent these lines.

The location of this system of holomorphic curves is not determined by the local geometry of $\mathbb{P}\mathcal{T}$, but by aspects of its global structure. In fact, as indicated above, the local geometry of $\mathcal{T}$ is indistinguishable from that of $\mathbb{T}$, so the information defining the local geometry of $\mathcal{M}$ is stored nonlocally in $\mathcal{T}$. This is closely analogous to what happens with a twistor 1-function **f**, defined on some region in $\mathbb{T}$. A 1-function is also a nonlocal object. The physical field at a space-time point is obtained by contour integration of a holomorphic function f representing **f**, and the particular value that f might happen to have at a point of $\mathbb{T}$ has no significance. That is because any particular Čech representative f of **f** is equivalent to a whole family of other functions (which differ from it by functions which are holomorphic in appropriate regions).

In fact, the construction of the curved twistor space $\mathcal{T}$ may be legitimately regarded as providing a nonlinear analogue of a twistor 1-function **f**. Suppose, for the moment, that $\mathcal{T}$ differs only infinitesimally from flat twistor space $\mathbb{T}$. Then $\mathcal{T}$

may be constructed by the glueing together of a number of portions of flat twistor space $\mathbb{T}$, where each is displaced from the others by an infinitesimal amount. There are functions describing the relative displacements of each of the patches from each of those others which overlap it. These provide the Čech representatives of a 1-function.

In order that the complex spacetime be not only anti-self-dual (i.e., have an anti-self-dual Weyl curvature) but also *vacuum* (vanishing Ricci tensor), some further conditions on $\mathcal{T}$ are needed. Essentially what is required is that there be a projection from $\mathcal{T}$ to the space $\overline{\mathbb{S}^*} - \{0\}$ of nonzero $\pi_{A'}$-spinors. In terms of local structure, this projection is determined by 2-form $\boldsymbol{\tau}$ which is *simple* (i.e., $\boldsymbol{\tau} \wedge \boldsymbol{\tau} = 0$) and closed (i.e., $\mathrm{d}\boldsymbol{\tau} = 0$) so that it vanishes in the direction of a foliation of $\mathcal{T}$ by 2-surfaces. The projection $\mathcal{T} \to \overline{\mathbb{S}^*} - \{0\}$ is achieved by collapsing each of these 2-surfaces down to a point.

The required conditions that $\boldsymbol{\tau}$ be simple and closed follow from the existence of a 1-form $\boldsymbol{\iota}$ for which $\boldsymbol{\iota} \wedge \mathrm{d}\boldsymbol{\iota} = 0$, where $\boldsymbol{\tau} = \mathrm{d}\boldsymbol{\iota}$. The 1-form determines a foliation of the complex 4-dimensional twistor space $\mathcal{T}$ by complex 3-surfaces (restricted to which the 1-form $\boldsymbol{\iota}$ vanishes). In addition to the 1-form $\boldsymbol{\iota}$, we require $\mathcal{T}$ to possess a 3-form $\boldsymbol{\Theta}$, subject to the condition

$$\boldsymbol{\Theta} \wedge \boldsymbol{\iota} = 0\,.$$

In flat twistor space $\mathbb{T}$ these forms are given by the expressions Z^{α}

$$\begin{aligned}
\boldsymbol{\iota} &= \varepsilon^{AB'}\pi_{A'}\mathrm{d}\pi_{B'} = I_{\alpha\beta}Z^{\alpha}\mathrm{d}Z^{\beta}\,,\\
2\boldsymbol{\tau} &= \varepsilon^{A'B'}\,\mathrm{d}\pi_{A'}\,\mathrm{d}\pi_{B'} = I_{\alpha\beta}\,\mathrm{d}Z^{\alpha}\,\mathrm{d}Z^{\beta}\,,\\
\boldsymbol{\Theta} &= Z^0\mathrm{d}Z^1 \wedge \mathrm{d}Z^2 \wedge \mathrm{d}Z^3 - Z^1\mathrm{d}Z^0 \wedge \mathrm{d}Z^2 \wedge \mathrm{d}Z^3\\
&\quad + Z^2\mathrm{d}Z^0 \wedge \mathrm{d}Z^1 \wedge \mathrm{d}Z^3 - Z^3\mathrm{d}Z^0 \wedge \mathrm{d}Z^1 \wedge \mathrm{d}Z^2\,,\\
4\boldsymbol{\Phi} &= \frac{1}{6}\varepsilon_{\alpha\beta\gamma\delta}\mathrm{d}Z^{\alpha} \wedge \mathrm{d}Z^{\beta} \wedge \mathrm{d}Z^{\gamma} \wedge \mathrm{d}Z^{\delta}\,.
\end{aligned}$$

(Here we employ the *infinity twistors* $I_{\alpha\beta}$ and $I^{\alpha\beta}$, dual to one another, with spinor parts $I_{\alpha\beta} = (0,0;0,\varepsilon^{A'B'})$ and $I^{\alpha\beta} = (\varepsilon^{AB},0;0,0)$, respectively. The infinity twistors are the objects which serve to reduce the conformal group symmetry inherent in flat twistor space $\mathbb{T}$ down to the Poincaré group.)

The 3-form $\boldsymbol{\Theta}$ determines a foliation of $\mathcal{T}$ by complex curves, the tangent vectors to which annihilate $\boldsymbol{\Theta}$. I refer to the foliation by curves, determined by $\boldsymbol{\Theta}$, as the *Euler* fibration of $\mathcal{T}$. To pass from the nonprojective space $\mathcal{T}$ to the projective space $\mathbb{P}\mathcal{J}$, we factor $\mathcal{T}$ out by the Euler fibration. The condition $\boldsymbol{\Theta} \wedge \boldsymbol{\iota} = 0$, tells us that the curves of the Euler fibration all lie within the 3-surfaces determined by $\boldsymbol{\iota}$. Hence, the factoring of $\mathcal{T}$ by the Euler fibration provides us with a foliation of $\mathbb{P}\mathcal{T}$ by 2-surfaces, determined by the direction where $\boldsymbol{\iota}$ vanishes. In the case of ordinary flat projective twistor space $\mathbb{PT}$, these 2-surfaces are simply planes through the line $\mathbb{P}\mathbf{I} \subset \mathbb{PT}$, where $\mathbb{P}\mathbf{I}$ represents "infinity" (defined by $\pi_{A'} = 0$). (The line $\mathbb{P}\mathbf{I}$ is specified by the infinity twistor $I^{\alpha\beta}$.)

The lines of the Euler fibration are the integral curves of the "Euler vector field" $\boldsymbol{\Upsilon}$, where in flat space we would have

$$\boldsymbol{\Upsilon} = Z^\alpha \partial/\partial Z^\alpha .$$

This is the ordinary Euler homogeneity operator, whence the name "Euler vector field." The actual 3-form Θ determines this vector field (and not merely its direction field) in the sense that

$$\boldsymbol{\Upsilon} = \boldsymbol{\Theta} \div \boldsymbol{\Phi}, \quad \text{where} \quad \boldsymbol{\Phi} = \frac{1}{4}\mathrm{d}\boldsymbol{\Theta}.$$

The 4-form $\boldsymbol{\Phi}$ is the 4-volume form of twistor space, which in the flat case is

$$\boldsymbol{\Phi} = \mathrm{d}Z^0 \wedge \mathrm{d}Z^1 \wedge \mathrm{d}Z^2 \wedge \mathrm{d}Z^3 .$$

The Euler vector field $\boldsymbol{\Upsilon}$ serves to define the concept of *homogeneity* for a quantity ξ on $\mathcal{T}$. We say that ξ has homogeneity degree n if

$$\pounds_{\boldsymbol{\Upsilon}}\xi = n\xi.$$

It turns out that, with the definition of $\boldsymbol{\Upsilon}$ given above in terms of the 3-form $\boldsymbol{\Theta}$ and its exterior derivative Φ, the forms $\boldsymbol{\Theta}$ and $\boldsymbol{\Phi}$ are each automatically homogeneous of degree 4. However, there is one further local condition that must be satisfied by ι and $\boldsymbol{\Theta}$ in order that ι and τ should each have the (required) homogeneity degree 2. This can be stated simply as $\pounds_{\Upsilon}\iota = 2\iota$, but there is an alternative way of phrasing this condition which employs a differential operator which will also be useful for us for another reason, so it will be convenient to describe this next.

Consider expressions of the form

$$\Gamma = \boldsymbol{\alpha} \otimes \boldsymbol{\beta}, \quad \text{where} \quad \boldsymbol{\alpha} \wedge \boldsymbol{\beta} = 0,$$

$\boldsymbol{\alpha}$ being a 3-form and $\boldsymbol{\beta}$ a 1-form. There is an invariant operator $\mathbf{D}$ which acts on such quantities, defined by

$$\mathbf{D}\Gamma = \mathrm{d}\boldsymbol{\alpha} \otimes \boldsymbol{\beta} - \boldsymbol{\alpha} \oslash \mathrm{d}\boldsymbol{\beta},$$

where the operation $\oslash$ is bilinear and, for scalars p, q is given by

$$\boldsymbol{\alpha} \oslash (\mathrm{d}p \wedge \mathrm{d}q) = (\boldsymbol{\alpha} \wedge \mathrm{d}q) \otimes \mathrm{d}p - (\boldsymbol{\alpha} \wedge \mathrm{d}p) \otimes \mathrm{d}q .$$

The essential property that is possessed by the operator $\mathbf{D}$ is that it actually does act on the tensor product $\boldsymbol{\alpha} \otimes \boldsymbol{\beta}$ (so long as $\boldsymbol{\alpha} \wedge \boldsymbol{\beta} = 0$), rather than on $\boldsymbol{\alpha}$ and $\boldsymbol{\beta}$ separately; i.e.,

$$\mathbf{D}((\lambda\boldsymbol{\alpha}) \otimes \boldsymbol{\beta}) = \mathbf{D}(\boldsymbol{\alpha} \otimes (\lambda\boldsymbol{\beta})),$$

It turns out that the condition that ι (and consequently also τ, is homogeneous of degree 2 can now be written

$$\mathbf{D}(\boldsymbol{\Theta} \otimes \iota) = \frac{3}{2}\mathrm{d}\boldsymbol{\Theta} \otimes \iota .$$

The main reason for discussing these forms at this stage is that in the twistor construction of anti-self-dual vacuums, the forms provide the local structure of the twistor space $\mathcal{T}$ In fact, the 3-form Θ and the 1-form ι (subject to $\Theta \wedge \iota = 0$, $\iota \wedge d\iota = 0$, and the above relation) together specify the entire local structure that is needed. We recall that the projection to the π-space $\overline{\mathbb{S}^*} - \{0\}$ is determined by this structure, and we note also that each of the fibres (2-manifolds along which τ vanishes) has a symplectic (area) 2-form defined on it given by $\mu = \Phi \div \tau$. The metric of the space-time $\mathcal{M}$ is determined by these forms. In fact this metric turns out to be automatically Ricci-flat; moreover *any* Ricci-flat anti-self-dual 4-metric can be obtained in this way (see Penrose 1976). We find that it is essentially τ which defines the "$\varepsilon_{A'B'}$" of $\mathcal{M}$ and μ which defines its "ε_{AB}."

If we piece $\mathcal{T}$ together out of patches of twistor space, we need to make sure that the above forms are preserved. In the case where $\mathcal{T}$ is only infinitesimally deformed—corresponding to a *linearized* solution of the (anti-self-dual) Einstein equations—we can directly see how to achieve this deformation using a twistor 1-function, homogeneous of degree 2 (in accordance with the general discussion given below, the helicity being -2). Let us consider the case when only two patches are needed. This is sufficient for our purposes. The 1-function $\mathbf{f}$ is then determined by a single holomorphic function f defined on the overlap region between the two patches. The glueing may now be defined by the relation

$$X^\alpha = Z^\alpha + \epsilon I^{\alpha\beta} \partial f / \partial Z^\beta ,$$

(ϵ being an infinitesimal) where X^α and Z^α are corresponding twistors in each of the two patches and where $I^{\alpha\beta}$ is the infinity twistor, as described above. We can think of a curved twistor space which differs by a *finite* amount from the flat case as being obtained by exponentiation of this expression.

In terms of the spinor parts $(\xi^A, \eta_{A'})$ of X^α and $(\omega^A, \pi_{A'})$ respectively, this infinitesimal relation takes the form

$$\xi^A = \omega^A + \epsilon\varepsilon^{AB} \partial f / \partial \omega^B, \qquad \eta_{A'} = \pi_{A'} .$$

Note that the projection parts of the twistors are the same for each patch. Thus, there indeed is a canonical projection from the entire curved twistor space $\mathcal{T}$ to the (dual conjugate) spin-space $\overline{\mathbb{S}^*} - \{0\}$ of $\pi_{A'}$.

In order to preserve the homogeneity of the above expression, we require that f be homogeneous of degree +2. This has a particular significance because in the twistor description of a left-handed linearized graviton (given, in Minkowski space, by a massless field of helicity -2), the twistor 1-function has homogeneity degree +2. (This is the homogeneity degree of the representative holomorphic function(s) of the 1-function.) Thus, appropriately, the twistor description of a *linear* left-handed graviton corresponds exactly the *infinitesimal* version of the (seemingly quite different) twistor description of a "nonlinear graviton," which is an anti-self-dual solution of the full vacuum Einstein equations.

In fact, the homogeity degree for the twistor 1-function encountered here is part of a general pattern. We find (see Penrose and Rindler 1986) that for a massless field of helicity $n/2$ in $\mathbb{M}$ the corresponding twistor 1-function has homogeneity degree $-n-2$. Here the helicity is -2, so $n = -4$ and the homogeneity degree is indeed +2. This raises the question of the twistor description of a *right*-handed graviton, for which the helicity would be +2, so $n = -4$ and we have a twistor 1-function of homogeity degree -6. This indeed gives a massless field in $\mathbb{M}$ which describes the linearized limit of a Ricci-flat (complex) space-time with a self-dual Weyl curvature, as it should, but the real problem is to find a way of deforming twistor space so as to encode the full *non*-linear Weyl curvature of a self-dual Ricci-flat (complex) space-time. This has been termed the *googly* problem (see Penrose 1979), because, in the terminology of the game of cricket, a googly is a ball bowled so that it spins in a right-handed sense, whereas the bowling action has the appearance of that which would normally impart a left-handed spin. Twistor space (as opposed to dual twistor space) readily describes left-handed gravitons, but it is a considerable challenge to make it describe right-handed ones.

In fact, some genuine progress appears to have been recently made on this gravitational googly problem. It would not be appropriate, here, to give the details of this, but I can at least recount the essential points of my latest thinking on this problem. To begin with, I should point out that there is Poincaré-invariant exact sequence for flat twistor space

$$0 \to \mathbb{S} \to \mathbf{T} \to \overline{\mathbb{S}^*} \to 0.$$

Here, $\mathbb{S}$ stands for spin-space (the space of ω^A) and $\overline{\mathbb{S}^*}$ stands for the conjugate dual spin-space (the space of $\pi_{A'}$). The two middle maps are given, respectively, by

$$\omega^A \mapsto (\omega^A, 0) \quad \text{and} \quad (\omega^A, \pi_{A'}) \mapsto \pi_{A'}\,.$$

A space-time point x can be interpreted as a "splitting" of the sequence, according to which we have a map in the opposite direction from each of these two, given, respectively, by

$$(\omega^A, \pi_{A'}) \mapsto \omega^A - ix^{AA'}\pi_{A'} \quad \text{and} \quad \pi_{A'} \mapsto (ix^{AA'}\pi_{A'}, \pi_{A'})\,.$$

Note that the incidence relation features in each of these expressions, in two different guises.

In the original (left-handed) version of the nonlinear graviton construction, as described above, we have a deformed twistor space $\mathcal{T}$ taking the place of $\mathbf{T}$. Recall, from above, that we have a projection $\mathcal{T} \to \overline{\mathbb{S}^*} - \{\mathbf{0}\}$. This takes the place of the canonical map $\mathbf{T} \to \overline{\mathbb{S}^*}$. In this construction, the "space-time points" are interpreted as holomorphic curves in $\mathbf{P}\mathcal{T}$ which, in the nonprojective space $\mathcal{T}$, turn out to be cross sections of the fibration $\mathbf{T} \to \overline{\mathbb{S}^*}$, generalizing the map $\pi_A \mapsto (ix^{AA'}\pi_{A'}, \pi_{A'})$ above. For the googly construction, we require a deformed version $\mathbb{S} - \{\mathbf{0}\} \to \mathcal{T}$ of the injection $\mathbb{S} \to \mathbf{T}$. The image of this "deformed injection" turns out to be a singular region in $\mathbf{T}$, which in the flat case would be the 2-space $\mathbf{I} - \{\mathbf{0}\}$ (corresponding to the line $\mathbf{PI}$ referred to above). This (singular) way in which this $\mathbf{I} - \{\mathbf{0}\}$ is attached to $\mathcal{T}$ is to characterize the required reverse maps that correspondingly generalize $(\omega^A, \pi_{A'}); \mapsto \omega^A - ix^{AA'}\pi_{A'}$ (and which achieve a projection $\mathcal{T} \to \mathbb{S} - \{\mathbf{0}\}$).

To be able to describe this singular structure, it is appropriate to "blow up" the region $\mathbf{I} - \{\mathbf{0}\}$. In the flat case, the line $\mathbf{PI}$ is blown up to become a nonsingular quadric surface **PQ**. This is achieved, in the projective space, by the replacement

$$(\omega^A, \pi_{A'}) \mapsto (\omega^A \pi_{B'}, \pi_{A'}\pi_{B'}) .$$

However, for our present purposes, we must consider this nonprojectively, and then the complex 3-manifold **Q** (which is a complex line bundle over $\mathbb{CP}^1 \times \mathbb{CP}^1$) becomes adjoined to T, where **Q** consists of the quantities $(\omega^A \pi_{B'}, 0)$, obtained as the limits, when $\lambda \to 0$, of twistors $(\lambda^{-1}\omega^A, \lambda\pi_{A'})$. The manifold consisting of $\mathbb{T}$ with **Q** adjoined in this way is the *exploded* twistor space $\mathrm{T}^{\#}$. We find that the quantities Θ and Φ both diverge at **Q**, with simple poles there, whereas ι has a simple zero at **Q**. Hence the quantities $\boldsymbol{\Theta} \otimes \iota$ and $\Phi \otimes \iota$ are regular at **Q**.

In the case of curved twistor space $\mathcal{T}$ the construction is similar. We adjoin a space analogous to **Q** to $\mathcal{T}$ to obtain the exploded space $\mathcal{T}^{\#}$, just as before. But now we assign a structure to $\mathcal{T}^{\#}$ which consists of the quantity

$$\boldsymbol{\Sigma} = \boldsymbol{\Phi} \otimes \iota ,$$

together with the Euler fibration (but *not* the specific vector field Υ). We can *locally* integrate the relation $\mathbf{D}\boldsymbol{\Theta} \otimes \iota = 6\boldsymbol{\Phi} \otimes \iota$, from which $\boldsymbol{\Upsilon}$ can be locally obtained, but neither $\boldsymbol{\Theta} \otimes \iota\boldsymbol{\Upsilon}$ exists globally in general. The obstruction to the global existence of $\boldsymbol{\Theta} \otimes \iota$ (or $\boldsymbol{\Upsilon}$) can be specified by a 1-function of homogeneity -6. It is proposed that this is where the required "googly information" resides.

To find the appropriate "googly maps" (deformed versions of the flat-space maps $(\omega^A, \pi_{A'}) \mapsto \omega^A - ix^{AA'}\pi_{A'})$ which are to represent the "space-time points" defining a self-dual Ricci-flat complex 4-manifold, we use foliations of $\mathcal{T}^{\#}$ by curves to achieve these projections. In this construction, each googly map is a projection down to **Q** in the first instance (this being nonsingular), before we finally collapse **Q** down to $\mathbb{S}-\{\mathbf{0}\}$ to give us the deformed analogue of $\mathcal{T} \mapsto \mathbb{S}-\{\mathbf{0}\}$ (this latter being singular). These foliations are represented, locally, by 3-forms $\boldsymbol{\xi}$ (rather than vector fields), regular on **Q** and subject to

$$\mathrm{D}\boldsymbol{\xi} = 0, \qquad \boldsymbol{\xi} = \Theta + \eta \wedge \tau \quad \text{for some } \eta .$$

In the flat case, the point with position vector x^a would be represented by the 3-form

$$\boldsymbol{\xi} = \frac{1}{2}\mathrm{d}\omega_A \wedge \mathrm{d}\omega^A \wedge \iota - \omega_A \mathrm{d}\omega^A \wedge \tau + 2ix_A{}^{A'}\pi_{A'}\mathrm{d}\omega^A \wedge \tau .$$

The idea is that in the (googly) curved case, the condition of regularity at **Q** serves to single out the appropriate 4-parameter family of $\boldsymbol{\xi}$s. This is work still in progress.

Finally, a *third* definition of "space-time point" is required, which is to be some kind of average (or combination) of the two procedures described above, so that the resulting space-time is neither anti-self-dual nor self-dual, in general, but is to satisfy the Ricci-flatness condition, providing *generic* solutions of the Einstein vacuum equations. The most promising route towards achieving this seems to be via massless (test) fields of helicity $\frac{3}{2}$. Such fields may be said to mediate between twistors and the Einstein equations. For the Einstein equations are precisely the *consistency conditions* for such fields (written in potential form), whereas in flat space-time $\mathbb{M}$, the space of *charges* for such fields is precisely twistor space (see Penrose 1992). This also is work in progress.

References

[1] Bailey, T. N., Ehrenpreis, L. and Wells, R. O. Jr. (1982). Weak solutions of the massless field equations, *Proc. Roy. Soc. London* **A384**, 403–425.

[2] Bateman, H. (1904). The solution of partial differential equations by means of definite integrals, *Proc. Lond. Math. Soc.* 2 (**1**), 451–458.

[3] Bateman, H. (1910). The transformation of the electrodynamical equations, *Proc. Lond. Math. Soc.* 2 (**8**), 223–264.

[4] Bateman, H. (1944). *Partial Differential Equations of Mathematical Physics* (Dover, New York).

[5] Bocher, M. (1914). The infinite regions of various geometries, *Bull. Amer. Math. Soc.* **20**, 185–200.

[6] Coxeter, H. S. M. (1936). The representation of conformal space on a quadric, *Ann. Math.* **37**, 416–426.

[7] Cunningham, E. (1910). The principle of relativity in electrodynamics and an extension thereof. *Proc. Lond. Math. Soc.* 2 **8**, 77–98.

[8] Eastwood M. G., Penrose, R., and Wells, R. O., Jr. (1981). Cohomology and massless fields, *Comm. Math. Phys.* **78**, 305–351.

[9] Kodaira, K. (1962). A theorem of completeness of characteristic systems for analytic submanifolds of a complex manifold, *Ann. Math.* (**2**) 75, 146–162.

[10] Kodaira, K. (1963). On stability of compact submanifolds of complex manifolds, *Am. J. Math.* **85**, 79–94.

[11] Kodaira, K. and Spencer, D. C. (1958). On deformations of complex analytic structures I, II, *Ann. Math.* **67**, 328–401, 403–466.

[12] McLennan, J. A., Jr. (1956). Conformal invariance and conservation laws for relativistic wave equations for zero rest mass, *Nuovo. Cim.* **3**, 1360–1379.

[13] Penrose, R. (1967). Twistor algebra, *J. Math. Phys.* **8**, 345–366.

[14] Penrose, R. (1969). Solutions of the zero rest-mass equations, *J. Math. Phys.* **10**, 38–39.

[15] Penrose, R. (1976). Non-Linear gravitons and curved twistor theory, *Gen. Rel. Grav.* **7**, 31–52.

[16] Penrose, R. (1979). On the twistor description of massless fields, in Complex Manifold Techniques in *Theoretical Physics*, eds. D. E. Lerner and P. D. Sommers (Pitman, San Francisco).

[17] Penrose, R. (1987). On the origins of twistor theory, in *Gravitation and Geometry*: a volume in honour of I. Robinson. eds. W. Rindler and A. Trautman (Bibliopolis, Naples).

[18] Penrose, R. (1992). Twistors as spin 3/2 charges, in *Gravitation and Modern Cosmology* (P.G.Bergmann's 75th birthday vol.) eds. A. Zichichi, N. de Sabbata, and N.Sánchez (Plenum Press, New York).

[19] Penrose, R. (1996). Twistor theory, the Einstein equations, and quantum mechanics, in *Quantum Gravity:* International School of Cosmology and Gravitation XIV Course (80th Birthday Dedication to Peter G. Bergmann), eds. P. G. Bergmann, V. de Sabbata and H.-J. Treder (World Scientific, Singapore).

[20] Penrose, R. and Rindler, W. (1984). *Spinors and Space-Time*, Vol. 1: *Two-Spinor Calculus and Relativistic Fields* (Cambridge University Press, Cambridge).
[21] Penrose, R. and Rindler, W. (1986). *Spinors and Space-Time*, Vol. 2: *Spinor and Twistor Methods in Space-Time Geometry* (Cambridge University Press, Cambridge).
[22] Whittaker, E. T. (1904). *A Treatise on the Analytical Dynamics of Particles and Rigid Bodies* (Cambridge University Press, Cambridge).

26

Critique of the Wheeler-DeWitt Equation

Asher Peres

ABSTRACT The Wheeler-DeWitt equation is based on the use of canonical quantization rules that may be inconsistent for constrained dynamical systems, such as minisuperspaces subject to Einstein's equations. The resulting quantum dynamics has no classical limit and it suffers from the infamous "problem of time." In this article, it is shown how a dynamical time (an internal "clock") can be constructed by means of a Hamilton-Jacobi formalism, and then used for a consistent canonical quantization, with the correct classical limit.

26.1 Introduction

Classical field theories describe physical phenomena by means of field variables subject to partial differential equations. It often happens that the number of field variables exceeds that of the physical degrees of freedom; there is no unambiguous way of prescribing the values of the field variables that correspond to a given physical situation. In the mathematical structure of the theory, this property is reflected by the existence of a gauge group that allows transformations of the field variables while the physical situation remains unchanged.

The peculiar feature of Einstein's theory of gravitation which sets it quite apart from ordinary field theories is that its gauge group consists of arbitrary distortions of the space-time coordinates, and thus cannot be disentangled from the structure of space-time itself. In particular, the time evolution of the gravitational field is locally indistinguishable from a gauge transformation—namely a local distortion of the space-time coordinates. As a consequence, the Hamiltonian density $\mathcal{H}$, which generates the time evolution of the field variables, is *weakly* equal to zero [1]; namely, although $\mathcal{H}$ is a nontrivial function of the field variables (so that it can generate a nontrivial evolution), its *numerical* value is constrained to vanish.

This Hamiltonian constraint does not cause any difficulty in the classical canonical theory. The numerical value of the Hamiltonian is only an initial value constraint. It is the *functional* form of the Hamiltonian that is needed for deriving the equations of motion. In quantum theory, however, if the gravitational field equations are quantized according to the standard canonical rule, namely $\pi^{mn} = -i\hbar\delta/\delta g_{mn}$, the resulting Wheeler-DeWitt equation [2, 3] leads to a dilemma known as "the problem of time." The difficulty is that when the constraint $H\Psi = 0$ is imposed on the state vector Ψ, the latter is "frozen." There cannot be wavepackets moving along classical trajectories, in accordance with Ehrenfest's theorem [4]. That is, *the quantum equations do not lead to the expected classical limit*.

The problem was investigated long ago by Arnowitt, Deser, and Misner [5], who showed that there are an infinite number of possible coordinate conditions that may be used to put the theory in canonical form. The imposition of these coordinate conditions is equivalent to the introduction of "intrinsic" coordinates, defined by the dynamical variables of the physical system. The ADM method [5] and similar ones [6] provide, in principle, a completely general solution to the problem. Unfortunately, it is difficult to actually implement such a solution for a given specific physical situation.

In particular, many authors have been interested in the properties of highly symmetric cosmological models, for which there is a reasonable hope of obtaining an explicit solution. The trouble is that these symmetric situations have fewer (if any) dynamical degrees of freedom to which the ADM conditions can be applied, and the problem of time arises again. It would be impossible to mention here all the attempts that were made to solve that problem. Only a few randomly chosen references are listed below [7–11], with apologies to the authors of many similar works.

Why is general relativity special? The reason is the totalitarian nature of Einstein's theory: *all* physical phenomena are coupled to gravitation, for the simple reason that all phenomena occur in space-time, and the properties of space-time are determined by the gravitational field. Any stresses, or any other physical forces, are themselves sources of the gravitational field, by virtue of the Einstein equations. (Electromagnetic theory, as a counterexample, is compatible with the existence of forces of non-electromagnetic nature, that have no electromagnetic field of their own, and can cause electric charges to move in arbitrary ways.)

On the other hand, quantum theory, unlike general relativity, is *not* a "theory of everything" [12]. Its mathematical formalism can be given a consistent physical interpretation only by arbitrarily dividing the physical world into two parts: the system under study, represented by vectors and operators in a Hilbert space, and the observer (and the rest of the world), for which a classical description is used. This point was emphasized long ago by Bohr [13]:

> The necessity of discriminating in each experimental arrangement between those parts of the physical system which are to be treated as measuring instruments and those which constitute the objects under investigation may be said to form a *principal distinction between classical and quantum-mechanical description of physical phenomena* ... The place within each measuring procedure where this discrimination is made is largely a matter of convenience.

The consistency of this hybrid quantum-classical formalism can formally be proved, under suitably restrictive assumptions on the properties of the classical world [14]. That is, as foreseen by Bohr, the precise location of the boundary between the classical and quantum parts of the system is irrelevant for well-posed problems.

General relativity and quantum theory therefore appear to be fundamentally incompatible. Nevertheless, as will be shown in this article, they can be combined in a consistent way, by a careful choice of the dynamical variables. This is shown below by means of two simple examples, with a few degrees of freedom. Each example starts by specifying a Lagrangian. Indeed it is known that canonical commutation relations are compatible with specified equations of motion only if the latter are equivalent to the Euler-Lagrange equations derived from some Lagrangian [15]. In both examples, the Lagrangians are chosen in such a way that the resulting dynamics are afflicted by the infamous "problem of time," just as in canonical quantum gravity. Yet, canonical quantization is possible, provided that one degree of freedom is kept classical, so that it can be used as a clock.

The general method for solving this type of problem, for an arbitrary Lagrangian, is briefly discussed. It involves the solution of a first-order partial differential equation, similar to the Hamilton-Jacobi equation, but with a very different physical meaning.

26.2 A Simple Example of Constrained Dynamics

As our first example, consider a dynamical system with three degrees of freedom, x, y, z, and with a Lagrangian

$$L = \frac{1}{2}\left(\frac{\dot{x}^2}{z} - zx^2 - \frac{\dot{y}^2}{z} + zy^2\right). \tag{1}$$

The canonical momenta are $p_x = \dot{x}/z,\ p_y = -\dot{y}/z$, and $p_z = 0$. The equations of motion are thus

$$\dot{p}_x = d(\dot{x}/z)/dt = \partial L/\partial x = -zx, \tag{2}$$

$$\dot{p}_y = d(-\dot{y}/z)/dt = \partial L/\partial y = zy, \tag{3}$$

$$\dot{p}_z = 0 = \partial L/\partial z = \frac{1}{2}(-p_x^2 - x^2 + p_y^2 + y^2). \tag{4}$$

Equations of motion with a similar behavior occur in a cosmological model with a homogeneous scalar field, as in the Weinberg-Salam theory, but without a Higgs potential [16].

It is now convenient to introduce an auxiliary time, τ, by means of $d\tau = zdt$, so that $p_x = dx/d\tau$ and $p_y = -dy/d\tau$. Together with this new time, we also have a new Lagrangian, given by $L_\tau d\tau = Ldt$, so that the action remains the same. That is,

$$L_\tau = L/z = \frac{1}{2}\left[\left(\frac{dx}{d\tau}\right)^2 - x^2 - \left(\frac{dy}{d\tau}\right)^2 + y^2\right]. \tag{5}$$

The corresponding Hamiltonian is

$$H_\tau = \frac{1}{2}(p_x^2 + x^2 - p_y^2 - y^2), \tag{6}$$

and it is easy to derive from it the equations of motion of the two harmonic oscillators, x and y. It follows from Eq. (4) that $H_\tau = 0$. This causes no difficulty at the classical level. The Hamilton equations of motion are derived from the functional form of the Hamiltonian, irrespective of its numerical value, and they are equivalent to Eqs. (2) and (3) above.

Trouble arises, however, if we attempt to quantize such a system by introducing a wave function $\psi(x, y)$ that satisfies $H_\tau\psi$=0. Separation of variables readily leads to the general solution

$$\psi = \sum_n c_n\, u_n(x)\, u_n(y), \tag{7}$$

where the c_n are arbitrary constants, and the u_n are harmonic oscillator eigenfunctions corresponding to energy $E_n = (n + \frac{1}{2})\hbar$. Obviously, this state is time independent. Nothing moves. If we try to get a semiclassical solution by using large values of n, we find that the amplitude of the wave function is large in the vicinity of the four corners of a square, $x,\ y \simeq \pm\sqrt{2E_n}$, and it is of course fixed in time.

On the other hand, there is no such difficulty with the Heisenberg equations of motion, for example, $dx/d\tau = [x, H_\tau]/i\hbar$. These are formally identical to the classical oscillator equations of motion, and they lead to a nontrivial motion of the Heisenberg operators. We thus see that the Ehrenfest theorem [17] and, more generally, the correspondence principle, are not valid for such a dynamical system.

In order to find a quantum counterpart to the dynamical system that is represented in classical physics by the Hamiltonian (6), we must proceed more carefully. One of the harmonic oscillators, for example y, will serve us as a clock, and then the other one can be quantized in the usual way. We thus perform, still at the classical stage, a canonical transformation from y and p_y to new canonical variables,

$$Q^0 = \tan^{-1}(p_y/y), \tag{8}$$

$$P_0 = -(p_y^2 + y^2)/2. \tag{9}$$

There is no corresponding unitary transformation in quantum mechanics (since the spectrum is not invariant), but in classical mechanics, such a canonical transformation is perfectly possible.

It is easily seen that $[Q^0, H_\tau]_{\rm PB} = 1$, so that

$$dQ^0/d\tau = 1. \tag{10}$$

We can thus write

$$H_\tau = P_0 + H = 0, \tag{11}$$

where $H = \frac{1}{2}(p_x^2 + x^2)$ is the ordinary Hamiltonian of the x-oscillator. Its equations of motion are $dx/d\tau = p_x$ and $dp_x/d\tau = -x$. Thanks to Eq. (10), we can also write them as

$$dx/dQ^0 = p_x, \tag{12}$$

$$dp_x/dQ^0 = -x. \tag{13}$$

Finally, if we replace p_x by $-i\hbar\partial/\partial x$, and P_0 by $-i\hbar\partial/\partial Q^0$, as usual, Eq. (11) becomes the standard Schrödinger equation for a harmonic oscillator.

However, at this point, we must be careful: the wave function $\psi(x, Q^0)$ should be normalized according to

$$\int |\psi(x, Q^0)|^2 \, dx = 1, \tag{14}$$

without any further integration $\int \cdots dQ^0$. This follows from our decision of keeping the clock time Q^0 classical, so that it can play the ordinary role of time in Schrödinger's equation. In this way, we have obtained a simple, consistent formalism, with the correct classical limit. Obviously, there are many other possible consistent formalisms that are not equivalent to each other and yet give the same classical limit. Quantization is possible, but it is not a unique process.

26.3 Definition of a Dynamical Time

It will be now be shown how a similar quantization process can be performed for any classical dynamical system with a constrained Hamiltonian,

$$H_\tau(q, p) = 0. \tag{15}$$

Here, q and p, without indices, mean $\{q^1 \ldots q^n\}$ and $\{p_1 \ldots p_n\}$, respectively, and τ is an arbitrary, convenient time parameter, in terms of which the problem has been formulated. In the case of Einstein's gravitational field equations, there is an infinite number of dynamical variables and of constraints (there are twelve canonical variables, g_{mn} and π^{mn}, and four constraints per space point). In the present article, however, my main interest is in solving the "problem of time" for minisuperspaces. I shall therefore assume that there is only a finite number n of degrees of freedom, and a single Hamiltonian constraint, Eq. (15).

Following the method illustrated in the preceding section, let us seek a canonical transformation from $\{q^k\}$ and $\{p_k\}$ to new canonical variables, $\{Q^\mu\}$ and $\{P_\mu\}$, with $\mu = 0, \ldots, n-1$, such that

$$dQ^0/d\tau = [Q^0, H_\tau]_{\text{PB}} = 1. \tag{16}$$

It follows from Eq. (16) that

$$H_\tau = P_0 + H(Q^0 \dots Q^{n-1}, P_1 \dots P_{n-1}). \tag{17}$$

The latter equation defines an effective Hamiltonian H. Note that H does not depend on P_0 so that after we replace P_0 by $-i\hbar\partial/\partial Q^0$, the classical equation $H_\tau = 0$ becomes a Schrödinger equation for the new time Q^0 and the $(n-1)$ dynamical variables, $Q^1 \dots Q^{n-1}$.

The first step thus is to find a suitable clock time $Q^0(q, p)$. This can easily be done, as least in a restricted domain of phase space, as shown in Fig. 27.1. We start with an arbitray $(2n-1)$-dimensional hypersurface $\mathcal{K}$, oriented in such a way that the flow lines $dq/d\tau = \partial H_\tau/\partial p$ and $dp/d\tau = -\partial H_\tau/\partial q$ are nowhere tangent to $\mathcal{K}$. That is, all the flow lines lie on the same side of $\mathcal{K}$, for τ positive and short enough. Then, at least for some finite time, these flow lines will not intersect—as long as they do not reach a critical point—and they will not reenter $\mathcal{K}$ from the other side (however, if the motion is bounded, for example if it is periodic, reentry must obviously happen after enough time has elapsed). Anyway, for a finite time, each flow line that originates from $\mathcal{K}$ ascribes a unique set of q and p to each value of τ and, conversely, in a finite domain of phase space, there is a unique τ for each set of $\{q, p\}$, say $\tau = f(q, p)$.

There still is here a formidable technical difficulty, because generic dynamical problems are not integrable: the number of constants of motion is usually less than the number of degrees of freedom, and the function $f(q, p)$ defined above cannot be obtained in closed form and does not exist globally. For that reason, some authors [18] take the liberty of "fixing the gauge" by an arbitrary choice of the function $f(q, p)$, leading to a form which is convenient for further work. It is not clear to me why this is permitted. This is also not necessary, because there do exist approximation methods for performing a sequence of canonical transformations which reduce the Hamiltonian to a normal form [19, 20]. These give approximate constants of motion, which are represented in phase space by "vague tori" and are useful for describing the dynamics over extended time periods. These tori remnants [21, 22] become important in quantum theory because if their missing parts are small compared to $2\pi\hbar$, the quantum system behaves as if it were regular, with ordinary selection rules. Let us now return to the "problem of time." What we need is a canonical transformation such that $Q^0 = f(q, p)$ is a prescribed function. As explained above, $f(q, p)$ is chosen in such a way that

$$[f(q, p), H_\tau]_{\mathrm{PB}} = 1, \tag{18}$$

and therefore $dQ^0/d\tau = 1$. At this point, it is natural to ask what would happen if we had chosen another initial hypersurface, say $\mathcal{K}'$, leading us to a different time function, $f'(q, p)$, say. We would then have

$$[f(q, p) - f'(q, p), H_\tau]_{\mathrm{PB}} = 0, \tag{19}$$

so that $(f - f')$ has to be a constant of the motion. Either it is a function of H_τ or, if there are other, nontrivial constants of the motion, $(f - f')$ can be a function of them. In particular, if $H_\tau = 0$ is the only constant of motion, $(f - f')$ can

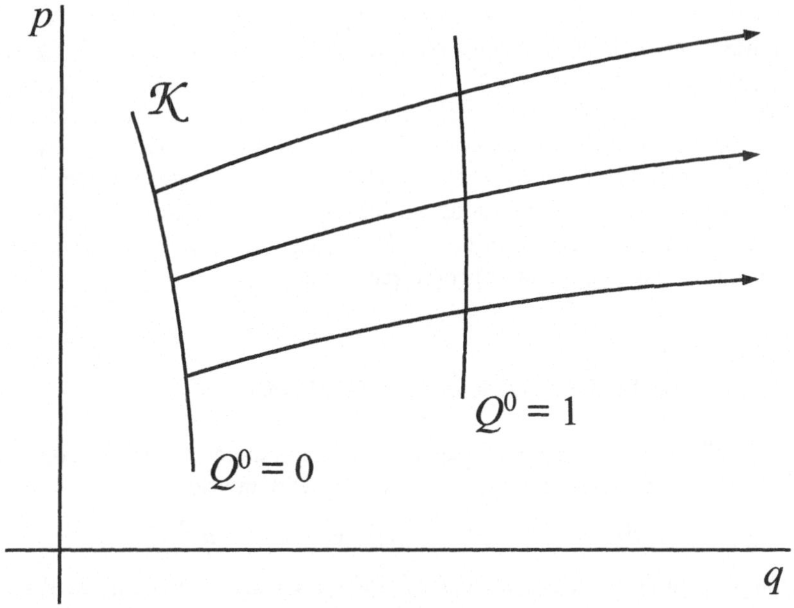

Figure 26.1 Orbits in phase space, starting on the hypersurface $\mathcal{K}$. Each one of these orbits defines an internal clocktime, $Q^0(q, p)$.

only be a mere number. Anyway, it does not matter for the sequel whether $f(q, p)$ is uniquely defined, up to a numerical constant, or can be modified by adding a nontrivial constant of the motion (thus effectively giving a different version of the theory).

Our task thus is is to find explicitly a canonical transformation that leads to the decomposition (17). This can be done, in principle, by the solution of a first-order partial differential equation of the same type as the Hamilton-Jacobi equation. It is easiest to use a generating function [23] of type F_1, which we shall write as $S(q, Q)$. We have

$$p_k = \partial S/\partial q^k, \tag{20}$$

$$P_\mu = -\partial S/\partial Q^\mu. \tag{21}$$

Since S is time independent (there is no explicit appearance of τ in S), the new Hamiltonian is numerically equal to the old one, as in Eq. (17).

To obtain S explicitly, we have to solve

$$Q^0 = f\left(q, \frac{\partial S(q, Q)}{\partial q}\right). \tag{22}$$

The various Q^μ, with $\mu > 0$, are unspecified integration constants in the solution of Eq. (22). As in the Hamilton-Jacobi case, there is no guarantee that (22) has well-behaved global solutions. However, it is always possible to achieve arbitrarily close approximations in a finite domain. An example is given in the next section.

Once we have obtained $S(q, Q)$, we get $P_\mu(q, Q)$ from Eq. (21). We can then invert these equations, in principle, and find $q(Q, P)$. Likewise, Eq. (20) gives us $p = p(q, Q)$, and since $q(Q, P)$ is already known, this gives $p = p(Q, P)$. All these results are then substituted in H_τ so as to obtain the explicit form of Eq. (17). Finally, that equation can be quantized in the usual way, replacing Q^0 by a new variable, t (recall that $dQ^0/d\tau = 1$), and P_0 by $-i\hbar\partial/\partial t$. Note, however, that the new parameter t is not a function of the space-time coordinates: it is a function, $f(q, p)$, of the phase-space coordinates. This is a meaningful dynamical time, not a meaningless (gauge-dependent) coordinate-time.

26.4 Quantization of a Minisuperspace

Let us finally return to general relativity. As the simplest example, consider a spatially flat Friedmann-Lemaitre universe [24], with metric

$$ds^2 = N^2(t)\,dt^2 - a^2(t)\,(dx^2 + dy^2 + dz^2). \tag{23}$$

The matter source is a massless scalar field ϕ, for which the energy density and pressure are

$$\rho = p = \frac{1}{2}\dot{\phi}^2, \tag{24}$$

where natural units have been used: $c = 8\pi G = 1$.

The Einstein field equations for the above metric and sources become ordinary differential equations for the three variables $N(t)$, $a(t)$, and $\phi(t)$. In order to obtain a quantum version of this theory, the above differential equations must be obtainable as the Euler-Lagrange equations resulting from a Lagrangian [15]. It is easily found that a suitable Lagrangian, giving the correct equations, is

$$L = \frac{1}{N}\left(\frac{1}{2}a^3\,\dot{\phi}^2 - 3a\,\dot{a}^2\right). \tag{25}$$

Note that $\dot{N}$ does not appear in L, so that

$$p_N \equiv \partial L/\partial \dot{N} = 0, \tag{26}$$

and therefore

$$\dot{p}_N = \partial L/\partial N = -L/N^2 = 0. \tag{27}$$

The fact that $L = 0$ is an initial-value constraint imposed on the variables a, $\dot{a}$, and $\dot{\phi}$.

Likewise, we have

$$p_\phi \equiv \partial L/\partial \dot{\phi} = a^3\,\dot{\phi}/N. \tag{28}$$

This is a constant of the motion, because $\partial L/\partial \phi = 0$. Finally,

$$p_a \equiv \partial L/\partial \dot{a} = -6\,a\,\dot{a}/N, \tag{29}$$

and

$$\dot{p}_a = \frac{\partial L}{\partial a} = \frac{3}{N}\left(\frac{1}{2}a^2\dot{\phi}^2 - \dot{a}^2\right). \tag{30}$$

As in Section 27.2, it is convenient to introduce an auxiliary time τ by means of $d\tau = Ndt$. We then have a new Lagrangian, given by $L_\tau d\tau = Ldt$, so that the action remains invariant. Furthermore, it is convenient to introduce, instead of the radial scale variable a, a new variable $v(t) = a^3(t)$ which scales the volume element. We then have

$$L_\tau = \frac{v}{2}\left(\frac{d\phi}{d\tau}\right)^2 - \frac{1}{3v}\left(\frac{dv}{d\tau}\right)^2, \tag{31}$$

from which we obtain

$$p_\phi = v\,\frac{d\phi}{d\tau}, \tag{32}$$

$$p_v = -\frac{2}{3v}\,\frac{dv}{d\tau}. \tag{33}$$

The corresponding Hamiltonian is

$$H_\tau \equiv p_\phi\frac{d\phi}{d\tau} + p_v\frac{dv}{d\tau} - L_\tau = \frac{1}{2v}\,{p_\phi}^2 - \frac{3v}{4}\,{p_v}^2. \tag{34}$$

Note that both L_τ and H_τ vanish weakly, as a consequence of (27). The non-essential dynamical variable $N(t)$ has thus been eliminated, but it has left a remnant, which is the initial value constraint $H_\tau = 0$. The equations of motion resulting from the new Hamiltonian are $p_\phi =$ const., Eq. (33), and

$$\frac{dp_v}{d\tau} = -\frac{\partial H_\tau}{\partial v} = \frac{{p_\phi}^2}{2v^2} + \frac{3\,{p_v}^2}{4}. \tag{35}$$

Our task now is to find a dynamical time function $Q^0 = f(v, p_v, p_\phi)$ such that $dQ^0/d\tau = 1$. (Obviously, f is not a function of the cyclic variable ϕ, since the latter does not appear explicitly in the equations of motion.) In other words, we want a function $f(v, p_v, p_\phi)$ that satisfies

$$[f(v, p_v, p_\phi), H_\tau]_{\text{PB}} = 1. \tag{36}$$

For this, we have to solve the equations of motion explicitly.

Substitution of (33) into the right-hand side of (34) gives, after some rearrangement,

$$\frac{1}{2}\left(\frac{dv}{d\tau}\right)^2 + \frac{3}{2}\,H_\tau v = \frac{3}{4}\,{p_\phi}^2. \tag{37}$$

This looks like the elementary energy equation for free fall of a particle of unit mass, height v, and total energy $\frac{3}{4}\,{p_\phi}^2$, in a gravity field $g = \frac{3}{2}H_\tau$. The solution is

$$v = -\frac{3}{4}\,H_\tau\,\tau^2 \pm \sqrt{\frac{3}{2}}\,p_\phi\,\tau, \tag{38}$$

where the integration constant was set so that $v = 0$ when $\tau = 0$ (in other words, the $\mathcal{K}$ hypersurface is given by $v = 0$). Since by definition $v \geq 0$, the $\pm$ sign in (38) has to be the same as the sign of of $p_\phi \tau$. Note that we are not allowed to set $H_\tau = 0$ at this stage; consistency of the method that was proposed in the preceding section requires that the equations of motion be valid for the entire phase space, not only for the orbits with initial conditions that satisfy $H_\tau = 0$.

It is possible to solve directly (38) for τ, and then to substitute (34) in the result. However, it is simpler to proceed as follows. From (38), we have

$$\frac{dv}{d\tau} = -\frac{3}{2}H_\tau \tau \pm \sqrt{\frac{3}{2}}p_\phi, \tag{39}$$

whence, thanks to Eq. (33),

$$vp_v = -\frac{2}{3}\frac{dv}{d\tau} = H_\tau \tau \mp \sqrt{\frac{2}{3}}p_\phi. \tag{40}$$

Thus, (38) becomes

$$v = -\frac{3}{4}\tau(H_\tau \tau \mp \sqrt{\frac{8}{3}}p_\phi) = -\frac{3}{4}\tau(vp_v \mp \sqrt{\frac{2}{3}}p_\phi), \tag{41}$$

and therefore

$$\tau \equiv f(v, p_v, p_\phi) = \frac{v}{-\frac{3}{4}vp_v \pm \sqrt{\frac{3}{8}}\,p_\phi}. \tag{42}$$

It is easy to verify directly that Eq. (36) indeed holds.

The next step is to find explicitly the transformation from the original canonical coordinates to the new ones, which include Q^0 and P_0. We have, from Eqs. (22) and (42),

$$Q^0 = \frac{v}{-\frac{3}{4}v\frac{\partial S}{\partial v} \pm \sqrt{\frac{3}{8}}\frac{\partial S}{\partial \phi}}, \tag{43}$$

where $S = S(v, \phi, Q^0, Q^1)$. An obvious way for obtaining a solution is to separate variables, namely,

$$S = \phi Q^1 + S'(v, Q^0, Q^1), \tag{44}$$

so that

$$Q^1 = p_\phi. \tag{45}$$

Rearranging Eq. (43), we obtain

$$\frac{1}{Q^0} = -\frac{3}{4}\frac{\partial S'}{\partial v} \pm \sqrt{\frac{3}{8}}\frac{Q^1}{v}, \tag{46}$$

whose solution is

$$S' = \frac{4}{3}\left(-\frac{v}{Q^0} \pm \sqrt{\frac{3}{8}}Q^1 \ln v\right). \tag{47}$$

We thus have

$$p_v = \frac{\partial S'}{\partial v} = \frac{3}{4}\left(-\frac{1}{Q^0} \pm \sqrt{\frac{3}{8}\frac{p_\phi}{v}}\right), \tag{48}$$

in agreement with (41).

Note that

$$P_0 = -\frac{\partial S'}{\partial Q^0} = -\frac{4}{3}\frac{v}{(Q^0)^2}, \tag{49}$$

whence

$$v = -\frac{3}{4}\,P_0\,(Q^0)^2. \tag{50}$$

When these equations for v and p_v are substituted into (34), we obtain

$$H_\tau = P_0 \pm \sqrt{\frac{8}{3}\frac{p_\phi}{Q^0}}. \tag{51}$$

The reduced Hamiltonian H, defined by Eq. (17), is thus

$$H = \pm\sqrt{\frac{8}{3}\frac{p_\phi}{Q^0}}. \tag{52}$$

Recall that the $\pm$ sign in H is the same as the sign of p_ϕ/Q^0. Note that if Q^0 is considered as equivalent to the time τ, the number of degrees of freedom has been reduced by 2: the variable N disappeared in the transformation from t to τ, and the v and p_v variables have been absorbed in the dynamical definition of a "clock time" Q^0.

Here, we must be careful and avoid expressing Q^0, in Eq. (52), by means of the right-hand side of (42). This would give

$$H = \frac{{p_\phi}^2}{v} \mp \sqrt{\frac{3}{2}}\,p_\phi\,p_v \qquad \text{(wrong)}. \tag{53}$$

Such a way of writing the Hamiltonian is not correct: it would give the true equations of motion for v and p_v only if the initial conditions are set in such a way that $H_\tau = 0$ in Eq. (34), namely

$${p_\phi}^2 = \frac{3}{2}\,(vp_v)^2. \tag{54}$$

Indeed, we have from (53)

$$\frac{dv}{d\tau} = [v, H]_{\text{PB}} = \mp\sqrt{\frac{3}{2}}\,p_\phi \qquad \text{(wrong)}, \tag{55}$$

and

$$\frac{dp_v}{d\tau} = [p_v, H]_{\text{PB}} = \frac{{p_\phi}^2}{v^2} \qquad \text{(wrong)}, \tag{56}$$

and these agree with Eqs. (35) and (37) only if (54) is satisfied. Therefore H in Eq. (53) is not a valid, unconstrained Hamiltonian for this problem. Only H

given by (52) is acceptable. (More generally, the reader may easily verify that $P_0 = H_\tau - H$ has vanishing Poisson brackets with all the canonical variables only on the hypersurface $H_\tau = 0$.)

We thus remain with the reduced Hamiltonian (52), and we may now replace in it Q^0 by τ. The only nontrivial equation of motion is

$$\frac{d\phi}{d\tau} = [\phi, H]_{\rm PB} = \pm\sqrt{\frac{8}{3}}\,\frac{1}{\tau}, \tag{57}$$

whence $\phi = \phi_0 \pm \sqrt{\frac{8}{3}}\ln\tau$. Quantization is trivial: the wave function $\psi(p_\phi, \tau)$ satisfies a Schrödinger equation,

$$i\hbar\frac{\partial\psi}{\partial\tau} = \pm\sqrt{\frac{8}{3}}\,\frac{p_\phi}{\tau}\,\psi, \tag{58}$$

so that

$$\psi = F(p_\phi)\exp\left(\mp\frac{i}{\hbar}\sqrt{\frac{8}{3}}\,p_\phi\ln\tau\right), \tag{59}$$

where $F(p_\phi)$ is an arbitrary function that takes care of normalization.

Obviously, only a superspace with a larger number of degrees of freedom can give an interesting theory. Unfortunately, "interesting" also means "nonintegrable": the function $f(q, p)$ is not in general well behaved (it is not "isolating"), and approximation methods must be used [19–22].

Finally, the question must be raised whether the notion of a minisuperspace is a valid approximation for studying quantum gravity [25]. The arbitrary imposition of symmetry constraints on the gravitational field freezes almost all its dynamical degrees of freedom in a way that appears to be incompatible with the existence of quantum fluctuations. A similar dilemma arises in elementary classical mechanics, when we impose mundane mechanical constraints, such as restricting the motion of a mass to a 2-dimensional surface. Classically, such a system is well defined. However, its quantization is not unique and it essentially depends on the nature of the constraining forces [26]. I hope to return to this problem in a future publication.

Acknowledgments

It is a pleasure to dedicate this article to Englebert Schucking, on the occasion of his 70th birthday. I am grateful to Stanley Deser for clarifying comments. This work was supported by the Gerard Swope Fund, and the Fund for Encouragement of Research.

References

[1] P. A. M. Dirac, *Proc. Roy. Soc.* London **A246** (1958) 333.

[2] J. A. Wheeler, in *Battelle Rencontres: 1967 Lectures on Mathematical Physics* (Benjamin, New York, 1968).
[3] B. S. DeWitt, *Phys. Rev.* **160** (1967) 1113.
[4] T. Brotz and C. Kiefer, *Nucl. Phys.* **B475** (1996) 339.
[5] R. Arnowitt, S. Deser, and C. W. Misner, in *Gravitation: an Introduction to Current Research*, ed. by L. Witten (Wiley, New York, 1962).
[6] A. Peres, *Phys. Rev.* **171** (1968) 1335.
[7] W. G. Unruh, *Phys. Rev. D* **40** (1989) 1048.
[8] W. G. Unruh and R. M. Wald, *Phys. Rev. D* **40** (1989) 2598.
[9] C. G. Torre, *Phys. Rev. D* **46** (1992) 3231.
[10] R. M. Wald, *Phys. Rev. D* **48** (1993).
[11] J. D. Brown and K. Kuchař, *Phys. Rev. D* **51** (1995) 5600.
[12] A. Peres and W. H. Zurek, *Am. J. Phys.* **50** (1982) 807.
[13] N. Bohr, *Phys. Rev.* **48** (1935) 696.
[14] A. Peres, *Quantum Theory: Concepts and Methods* (Kluwer, Dordrecht, 1993) p. 376.
[15] S. A. Hojman and L. C. Shepley, *J. Math. Phys.* **32** (1991) 142.
[16] V. N. Pervushin and V. I. Smirichinski, report JINR E2-97-155 (e-print gr-qc/9704078).
[17] P. Ehrenfest, *Z. Phys.* **45** (1927) 455.
[18] M. Cavaglià, V. de Alfaro, and A. T. Filippov, *Int. J. Mod. Phys.* **A10** (1995) 611.
[19] G. Contopoulos, *Astrophys. J.* **138** (1963) 1297.
[20] F. G. Gustavson, *Astronom. J.* **71** (1966) 670.
[21] C. Jaffé and W. P. Reinhardt, *J. Chem. Phys.* **77** (1982) 5191.
[22] R. B. Shirts and W. P. Reinhardt, *J. Chem. Phys.* **77** (1982) 5204.
[23] H. Goldstein, *Classical Mechanics* (Addison-Wesley, Reading, MA, 1980) p. 382.
[24] C. W. Misner, K. S. Thorne, and J. A. Wheeler, *Gravitation* (Freeman, San Francisco, 1973) Chapt. 27.
[25] K. V. Kuchař and M. P. Ryan, Jr., *Phys. Rev. D* **40** (1989) 3982.
[26] N. G. van Kampen and J. J. Lodder, *Am. J. Phys.* **52** (1984) 419.

27

A New Version of the Heavenly Equation

Jerzy F. Plebański
Maciej Przanowski

ABSTRACT A new version of the heavenly equation constructed with the use of two expanding congruences of anti-self-dual null strings is presented. Some simple examples of the heavenly metrics defined by this new equation are also given.

Dedicated to Professor Engelbert Schucking

27.1 Introduction

Self-dual gravity seems to be one of the most interesting parts of the mathematical relativity. Nowadays there exist many approaches to self-dual gravity. However, in fact, they all are some variations of two main approaches. The essence of the first of them consists in the analysis of the Einstein equations describing the self-dual spacetime and in a reasonable simplification of these equations [1]–[4]. The second one is the twistor technique [5,6].

In this paper we deal with the first approach. It has been shown in [1] that every self-dual, Ricci-flat complex spacetime, i.e., the $\mathcal{H}$ space, appears to be a complexified Kähler manifold and Einstein equations can be reduced to one second-order nonlinear partial differential equation for the complex Kähler function. This equation is called the *first heavenly equation.* The aim of the present work is to reduce the self-dual vacuum Einstein equations employing only the complexified Hermitian but non-Kählerian structure. In Section 27.2 we examine such structures, and it is proved that they exist for every $\mathcal{H}$ space in complex relativity (CR). Then in Section 27.3 the analogous result concerning Hermitian but non-Kählerian structures for real $\mathcal{H}$-space in Euclidean relativity (ER) is obtained.

Finally, Section 27.4 is devoted to the reduction of the self-dual vacuum Einstein equations. As is shown, the reduction leads to one second-order nonlinear partial differential equation for one function (the potential). This equation can be considered to be a *new version of the heavenly equation.*

We hope that the equation obtained gives a deeper insight into the structure of self-dual spacetime.

27.2 An Expanding Congruence of Null Strings

First we recall the main points of the null tetrad and spinor formalisms [1,2,7–11]. Let (M, ds^2) be a complex spacetime i.e., a 4-dimensional complex analytic manifold M endowed with a holomorphic symmetric nondegenerate covariant tensor field of second order ds^2

$$ds^2 = g_{ij}dx^i \otimes dx^j\,, \qquad g_{ij} = g_{ij}(x^k),$$
$$g_{ij} = g_{ji}\,, \qquad \det(g_{ij}) \neq 0\,; \qquad i, j, k = 1, \ldots, 4, \tag{1}$$

where $x^i$are local coordinates in M. As usual, ds^2 will be called the *metric on M.* Let U be an open set of M. *A null tetrad of 1-forms on U* is defined to be a set of four holomorphic 1-forms $e^a \epsilon \Lambda^1(U)$, $a = 1, \ldots, 4$, on U such that the metric ds^2 on U reads

$$ds^2 = g_{ab}e^a \otimes e^b,$$

$$(g_{ab}) = \begin{pmatrix} 0 & 1 & 0 & 0 \\ 1 & 0 & 0 & 0 \\ 0 & 0 & 0 & 1 \\ 0 & 0 & 1 & 0 \end{pmatrix} = (g^{ab})\,, \qquad a, b = 1, \ldots, 4. \tag{2}$$

Then the *null tetrad of tangent vector fields on U* is defined to be the four holomorphic tangent vector fields $e_a \epsilon T(U)$, $a = 1, \ldots, 4$, on U dual to the basis $e^a \epsilon \Lambda^1(U)$, $a = 1, \ldots, 4$, i.e.,

$$e^a(e_b) = \delta^a{}_b, \qquad a, b = 1, \ldots, 4. \tag{3}$$

An orientation on $U \subset M$ is defined by choosing the null tetrad $e^a \epsilon \Lambda^1(U)$ on U. If U is oriented, then one can define the *Hodge* $*$ *- operation* as follows:

$$* : \quad \Lambda^r(U) \to \Lambda^{4-r}(U)$$

$$* : \quad \omega = \frac{1}{r!}\omega_{a_1 \ldots a_r} e^{a_1} \wedge \cdots \wedge e^{a_r} \mapsto$$
$$*\omega = -\frac{1}{r!(4-r)!} \exp[\frac{i\pi}{2} r(4-r)] \epsilon_{a_1 \ldots a_r b_1 \ldots b_{4-r}} \omega^{a_1 \ldots a_r} e^{b_1} \wedge \cdots \wedge e^{b_{4-r}}$$
$$** = id \tag{4}$$

The Cartan structure equations read

$$De^a = de^a + \Gamma^a{}_b \wedge e^b = 0 \tag{5}$$

$$\mathcal{R}^a{}_b = d\Gamma^a{}_b + \Gamma^a{}_c \wedge \Gamma^c{}_b, \tag{6}$$

where $\Gamma^a{}_b \epsilon \Lambda^1(U)$ are the Riemannian connection 1-forms and $\mathcal{R}^a{}_b \epsilon \Lambda^2(U)$ stand for the Riemannian curvature 2-forms. It is well known that the following relations hold:

$$\Gamma_{ab} = -\Gamma_{ba}. \tag{7}$$

It is also well known that by writing Γ_{ab} in the form

$$\Gamma_{ab} = \Gamma_{abc} e^c \tag{8}$$

and e^a and e_a in the form

$$e^a = e^a{}_i dx^i, \qquad e_a = e_a{}^i \frac{\partial}{\partial x^i}, \tag{9}$$

one finds Γ_{abc} (the *Ricci coefficients*) to be

$$\Gamma_{abc} = -e_{ai;j} e_b{}^i e_c{}^j, \tag{10}$$

where the semicolon ";" means the covariant derivative.

Now we define the spinor 1-form on $U \subset M$

$$(g^{A\dot{B}}) = \begin{pmatrix} e^4 & e^2 \\ e^1 & -e^3 \end{pmatrix}, \qquad A = 1, 2; \quad \dot{B}=\dot{1}, \dot{2} \tag{11}$$

In terms of $g^{A\dot{B}}$ the metric ds^2 reads

$$ds^2 = -g_{A\dot{B}} \otimes g^{A\dot{B}}, \tag{12}$$

where the spinorial indices are to be manipulated according to the rule

$$\epsilon_{AC}\Psi^C = \Psi_A, \qquad \epsilon_{\dot{A}\dot{C}}\Psi^{\dot{C}} = \Psi_{\dot{A}},$$
$$\epsilon^{CA}\Psi_C = \Psi^A, \qquad \epsilon^{\dot{C}\dot{A}}\Psi_{\dot{C}} = \Psi^{\dot{A}},$$
$$(\epsilon_{AB}) = (\epsilon^{AB}) = \begin{pmatrix} 0 & 1 \\ -1 & 0 \end{pmatrix} = (\epsilon^{\dot{A}\dot{B}}) = (\epsilon_{\dot{A}\dot{B}}). \tag{13}$$

In fact the object

$$g^{A\dot{B}} = g^{A\dot{B}}{}_a e^a \tag{14}$$

can be considered to be a cross section of the vector fiber bundle

$$S(U) \otimes \dot{S}(U) \otimes \Lambda^1(U),$$

where $S(U)$ and $\dot{S}(U)$ are the *vector fiber bundles of undotted* and *dotted spinors over U* , respectively, and $\Lambda^1(U)$ denotes the *holomorphic cotangent bundle over U* . [Note that in the present section we deal only with holomorphic bundles. Moreover, the bundles and the set of bundle cross sections are denoted by the same symbol.]

The components of the undotted and dotted spinors transform according to the rule

$$\Psi'^{A} = l'^{A}{}_{B}\Psi^{B}, \qquad (l'^{A}{}_{B})\epsilon SL(2;C),$$
$$\Psi'^{\dot{A}} = l'^{\dot{A}}{}_{\dot{B}}\Psi^{\dot{B}}, \qquad (l'^{\dot{A}}{}_{\dot{B}})\epsilon SL(2;C), \tag{15}$$

where there is no relation between $l'^{A}{}_{B}$ and $l'^{\dot{A}}{}_{\dot{B}}$.

One can define the cross section of the bundle $S^{*}(U)\otimes \dot{S}^{*}(U)\otimes T(U)$ by

$$\partial_{A\dot{B}} := -g_{A\dot{B}}{}^{a}e_{a}. \tag{16}$$

It is easy to check that

$$g^{A\dot{B}}(\partial_{E\dot{F}}) = \delta^{A}{}_{E}\delta^{\dot{B}}{}_{\dot{F}}, \tag{17}$$

i.e.,

$$(\partial_{A\dot{B}}) = \begin{pmatrix} e_4 & e_2 \\ e_1 & -e_3 \end{pmatrix}. \tag{18}$$

Then we define the objects which play the crucial role in our further considerations. Namely

$$S(U)\otimes S(U)\otimes \Lambda^2(U) \ni S^{AB} := \epsilon_{\dot{E}\dot{F}}g^{A\dot{E}}\wedge g^{B\dot{F}} = S^{BA}, \tag{19}$$

$$\dot{S}(U)\otimes \dot{S}(U)\otimes \Lambda^2(U) \ni S^{\dot{A}\dot{B}} := \epsilon_{EF}g^{E\dot{A}}\wedge g^{F\dot{B}} = S^{\dot{B}\dot{A}}. \tag{20}$$

One quickly finds that

$$* S^{AB} = S^{AB} \quad \text{and} \quad * S^{\dot{A}\dot{B}} = -S^{\dot{A}\dot{B}}. \tag{21}$$

Therefore (S^{11}, S^{12}, S^{22}) is the basis of the space of self-dual 2-forms $\Lambda^2_+(U)$ and $(S^{\dot{1}\dot{1}}, S^{\dot{1}\dot{2}}, S^{\dot{2}\dot{2}})$ is the basis of the space of anti-self-dual 2-forms $\Lambda^2_-(U)$. Of course

$$\Lambda^2(U) = \Lambda^2_+(U)\oplus\Lambda^2_-(U). \tag{22}$$

The first Cartan structure equations in the spinor formalism read

$$Dg^{A\dot{B}} = dg^{A\dot{B}} + \Gamma^{A}{}_{C}\wedge g^{C\dot{B}} + \Gamma^{\dot{B}}{}_{\dot{C}}\wedge g^{A\dot{C}} = 0, \tag{23}$$

where $\Gamma^{A}{}_{B}$ and $\Gamma^{\dot{B}}{}_{\dot{C}}$ are the connection 1-forms in the vector fiber bundles $S(U)$ and $\dot{S}(U)$, respectively. As $\Gamma^{A}{}_{B}, \Gamma^{\dot{A}}{}_{\dot{B}}\epsilon sL(2;C)$, one has

$$\Gamma^{A}{}_{A} = 0 \;\Rightarrow\; \Gamma_{AB} = \Gamma_{BA},$$
$$\Gamma^{\dot{A}}{}_{\dot{A}} = 0 \;\Rightarrow\; \Gamma_{\dot{A}\dot{B}} = \Gamma_{\dot{B}\dot{A}}. \tag{24}$$

Then from (5), (11), and (23), it follows that

$$\begin{aligned}\Gamma_{AB} &= -\frac{1}{2} g_A{}^{\dot{C}a} g_{B\dot{C}}{}^{b}\Gamma_{ab} = -\frac{1}{4} S_{AB}{}^{ab}\Gamma_{ab},\\ \Gamma_{\dot{A}\dot{B}} &= -\frac{1}{2} g^{C}{}_{\dot{A}}{}^{a} g_{C\dot{B}}{}^{b}\Gamma_{ab} = -\frac{1}{4} S_{\dot{A}\dot{B}}{}^{ab}\Gamma_{ab}.\end{aligned} \tag{25}$$

Consequently

$$\Gamma_{AB} = \begin{pmatrix} -\Gamma_{42} & -\frac{1}{2}(\Gamma_{12}+\Gamma_{34}) \\ -\frac{1}{2}(\Gamma_{12}+\Gamma_{34}) & -\Gamma_{31} \end{pmatrix},$$

$$\Gamma_{\dot{A}\dot{B}} = \begin{pmatrix} -\Gamma_{41} & \frac{1}{2}(\Gamma_{12}-\Gamma_{34}) \\ \frac{1}{2}(\Gamma_{12}-\Gamma_{34}) & -\Gamma_{32} \end{pmatrix}. \tag{26}$$

The second Cartan structure equations take the form

$$\begin{aligned}\mathcal{R}^{A}{}_{B} &= d\Gamma^{A}{}_{B} + \Gamma^{A}{}_{C} \wedge \Gamma^{C}{}_{B},\\ \mathcal{R}^{\dot{A}}{}_{\dot{B}} &= d\Gamma^{\dot{A}}{}_{\dot{B}} + \Gamma^{\dot{A}}{}_{\dot{C}} \wedge \Gamma^{\dot{C}}{}_{\dot{B}},\end{aligned} \tag{27}$$

where $\mathcal{R}^{A}{}_{B}$ and $\mathcal{R}^{\dot{A}}{}_{\dot{B}}$ are the curvature 2-forms of the connections $\Gamma^{A}{}_{B}$ and $\Gamma^{\dot{A}}{}_{\dot{B}}$, respectively. $\mathcal{R}^{A}{}_{B}$ and $\mathcal{R}^{\dot{A}}{}_{\dot{B}}$ can be decomposed into irreducible parts as follows:

$$\begin{aligned}\mathcal{R}_{AB} &= -\frac{1}{2} C_{ABCD} S^{CD} + \frac{R}{24} S_{AB} + \frac{1}{2} C_{AB\dot{C}\dot{D}} S^{\dot{C}\dot{D}},\\ \mathcal{R}_{\dot{A}\dot{B}} &= -\frac{1}{2} C_{\dot{A}\dot{B}\dot{C}\dot{D}} S^{\dot{C}\dot{D}} + \frac{R}{24} S_{\dot{A}\dot{B}} + \frac{1}{2} C_{CD\dot{A}\dot{B}} S^{CD},\end{aligned} \tag{28}$$

where

$$\begin{aligned}C_{ABCD} &= C_{(ABCD)} := \frac{1}{16} S_{AB}{}^{ab} C_{abcd} S_{CD}{}^{cd},\\ C_{\dot{A}\dot{B}\dot{C}\dot{D}} &= C_{(\dot{A}\dot{B}\dot{C}\dot{D})} := \frac{1}{16} S_{\dot{A}\dot{B}}{}^{ab} C_{abcd} S_{\dot{C}\dot{D}}{}^{cd},\\ C_{AB\dot{C}\dot{D}} &= C_{(AB)\dot{C}\dot{D}} = C_{AB(\dot{C}\dot{D})} := \frac{1}{2} g_{A\dot{C}a} g_{B\dot{D}b} C^{ab},\end{aligned} \tag{29}$$

and C_{abcd} is the Weyl tensor, C_{ab} is the traceless Ricci tensor and R stands for the scalar curvature.

In the convention adopted in the present paper, the spinors C_{ABCD} and $C_{\dot{A}\dot{B}\dot{C}\dot{D}}$ correspond to the self-dual and the anti-self-dual part of the Weyl tensor, respectively.

From the first Cartan structure equations (23) one easily infers that the exterior covariant derivatives of S^{AB} and $S^{\dot{A}\dot{B}}$ vanish, i.e.,

$$DS^{AB} = dS^{AB} + \Gamma^{A}{}_{C} \wedge S^{CB} + \Gamma^{B}{}_{C} \wedge S^{AC} = 0,$$
$$DS^{\dot{A}\dot{B}} = dS^{\dot{A}\dot{B}} + \Gamma^{\dot{A}}{}_{\dot{C}} \wedge S^{\dot{C}\dot{B}} + \Gamma^{\dot{B}}{}_{\dot{C}} \wedge S^{\dot{A}\dot{C}} = 0. \tag{30}$$

Now we are prepared to consider the concept of the null string (the twistor surface) [2, 6, 10, 12]. An *anti-self-dual null string* (the *twistor surface*) in $U \subset M$ is defined to be a complex connected 2-surface $N \subset U$ such that
(i) for each point $p \in N$ and each vector v tangent to N at p, i.e., $v \in T_p(N)$,

$$ds^2(v, v) = 0.$$

(ii) for each point $p \in N$ any 2-form Σ at p orthogonal to $T_p(N)$ is anti-self-dual, i.e., $*\Sigma = -\Sigma$. Note that by (i) N is also totally geodesic.
Analogously one defines the self-dual null string, but now $*\Sigma = \Sigma$.

Hereafter we assume that $U \subset M$ is oriented, connected, and sufficiently small. This is also assumed in all parts of the paper.

The *congruence of anti-self-dual null strings in U* is defined to be a set $\mathcal{N}$ of anti-self-dual null strings in U such that for every point $p \in U$ there exists one and only one $N \in \mathcal{N}$ such that $p \in N$ and moreover, there exists a holomorphic 2-form Σ over U such that $*\Sigma = -\Sigma$ and for every $p \in U$, Σ is orthogonal to $T_p(N)$, where $p \in N \in \mathcal{N}$. The 2-form Σ is said to *define the congruence* $\mathcal{N}$. As is well known, Σ defines a congruence of anti-self-dual null strings if and only if

$$(a) \quad \Sigma \wedge \Sigma = 0\,, \qquad (b) \quad *\,\Sigma = -\Sigma,$$
$$(c') \quad d\Sigma = \omega \wedge \Sigma, \tag{31}$$

where $\omega \in \Lambda^1(U)$. If U is sufficiently small, then without any loss of generality one can choose Σ so that

$$\omega = 0. \tag{32}$$

In what follows, we assume that (32) holds; i.e., we have

$$(c) \quad d\Sigma = 0. \tag{33}$$

Similarly, assuming $*\Sigma = \Sigma$, one defines the congruence of self-dual null strings in U. In the present paper we deal with congruences of anti-self-dual null strings. Let $\mathcal{N}$ be such a congruence in $U \subset M$ and let Σ be the holomorphic 2-form on U defining $\mathcal{N}$ and satisfying the conditions (a), (b), and (c). Then one can easily show that there exist coordinates $\{q^A,\ q^{\dot{A}}\}$ such that

$$g^{A\dot{2}} = dq^A \tag{34}$$

and (see (20))

$$\Sigma = S^{\dot{2}\dot{2}} = \epsilon_{EF} g^{E\dot{2}} \wedge g^{F\dot{2}} = 2g^{1\dot{2}} \wedge g^{2\dot{2}} = 2dq^1 \wedge dq^2. \tag{35}$$

Consequently, from (30) and (35) we get

$$\Gamma_{\dot{1}\dot{1}} \wedge S^{\dot{1}\dot{2}} + \Gamma_{\dot{1}\dot{2}} \wedge S^{\dot{2}\dot{2}} = 0. \tag{36}$$

Finally, (20) and (36) yield

$$\Gamma_{\dot{1}\dot{1}A\dot{1}} g^{A\dot{1}} = 0 \tag{37}$$

and

$$\Gamma_{\dot{1}\dot{1}1\dot{2}} = 2\Gamma_{\dot{1}\dot{2}1\dot{1}}, \qquad \Gamma_{\dot{1}\dot{1}2\dot{2}} = 2\Gamma_{\dot{1}\dot{2}2\dot{1}}. \tag{38}$$

The relation (37) is consistent with the Frobenius theorem. In fact (37) is the necessary and sufficient condition for the Pfaff system

$$g^{1\dot{2}} = g^{2\dot{2}} = 0 \tag{39}$$

to be completely integrable. The vector fields $\partial_{1\dot{1}}$ and $\partial_{2\dot{1}}$ span at every point $p \in U$ the vector space tangent to the anti-self-dual null string passing through p.

In terms of the null tetrad formalism (see (18)),

$$\partial_{1\dot{1}} = e_4 \quad \text{and} \quad \partial_{2\dot{1}} = e_1 \tag{40}$$

and (37) mean that (see(27.2) with (11))

$$\Gamma_{414} = 0 = \Gamma_{411}. \tag{41}$$

Then assuming (34) and (35), we define the *expansion 1-form θ of the congruence $\mathcal{N}$ to be* [2][10]

$$\theta := \Gamma_{\dot{1}\dot{1}A\dot{2}} g^{A\dot{2}}. \tag{42}$$

The congruence $\mathcal{N}$ is called *nonexpanding* if $\theta = 0$ and $\mathcal{N}$ is called *expanding* if $\theta \neq 0$. In the null tetrad language we have

$$\theta = 0 \quad \text{iff} \quad \Gamma_{412} = 0 \ \text{and} \ \Gamma_{413} = 0. \tag{43}$$

One can quickly show that the condition $\Gamma_{412} = 0 = \Gamma_{413}$, under (41), is *independent of the choice of the basis (e_4, e_1) spanning the vector spaces tangent to $\mathcal{N}$.*

Now let us define an $\mathcal{H}$ space (the heavenly space) in CR.

An $\mathcal{H}$ *space* is a complex space-time (M, ds^2) for which

$$\mathcal{R}^{\dot{A}}{}_{\dot{B}} = 0. \tag{44}$$

From (28) one infers that

$$\mathcal{R}^{\dot{A}}{}_{\dot{B}} = 0 \ \Leftrightarrow \ C_{\dot{A}\dot{B}\dot{C}\dot{D}} = C_{AB\dot{C}\dot{D}} = R = 0. \tag{45}$$

If the curvature 2-form $\mathcal{R}^{\dot{A}}{}_{\dot{B}} = 0$, then for every point $p \in M$ there exists an open neighborhood U of p and a dotted spinor basis on U such that

$$\Gamma^{\dot{A}}{}_{\dot{B}} = 0. \tag{46}$$

Consequently, from the Frobenius theorem it follows that for the dotted spinor basis leading to (46), the Pfaff systems (39) and

$$g^{1\dot{1}} = g^{2\dot{1}} = 0 \tag{47}$$

are completely integrable. Therefore there exist local coordinates $\{q^A, q^{\dot{A}}\}$ such that (compare with (34))

$$\begin{gathered} g^{A\dot{2}} = dq^A\,, \qquad g^{A\dot{1}} = \Phi^A{}_{\dot{B}}dq^{\dot{B}}, \\ \Phi^A{}_{\dot{B}} = \Phi^A{}_{\dot{B}}(q^C, q^{\dot{C}}). \end{gathered} \tag{48}$$

The equations

$$q^A = \text{const.} \tag{49}$$

or

$$q^{\dot{A}} = \text{const.} \tag{50}$$

define two congruences of anti-self-dual null strings consisting of integrable manifolds of the Pfaff systems (39) or (47), respectively. By (46) it is evident that these two congruences are nonexpanding.

From (12) and (48) one gets the metric ds^2 to read

$$ds^2 = \Phi_{A\dot{B}}(dq^A \otimes dq^{\dot{B}} + dq^{\dot{B}} \otimes dq^A). \tag{51}$$

Then from (30) with (46) we find that

$$dS^{\dot{1}\dot{1}} = 0 \tag{52}$$

and

$$dS^{\dot{1}\dot{2}} = 0. \tag{53}$$

Equation (52) under (48) leads to the conclusion that without any loss of generality we can choose the coordinates $q^{\dot{A}}$ so that

$$\det(\Phi_{A\dot{B}}) = \frac{1}{2}\Phi_{A\dot{B}}\Phi^{A\dot{B}} = 1. \tag{54}$$

Finally, (53) and (54) yield

$$\begin{gathered} \Phi_{A\dot{B}} = \frac{\partial^2\Omega}{\partial q^A \partial q^{\dot{B}}} =: \Omega,_{A\dot{B}}\,, \\ \det(\Omega,_{A\dot{B}}) = 1\,, \qquad \Omega = \Omega(q^A, q^{\dot{A}}). \end{gathered} \tag{55}$$

This is the *first heavenly equation* [1,2]. From the considerations leading to the first heavenly equation it is seen that this equation is in a sense founded on two *nonexpanding* congruences of the anti-self-dual null strings. The main intention of the present paper is to find the heavenly equation as founded on two expanding congruences of anti-self-dual null strings. Therefore the question is if such two congruences exist in every $\mathcal{H}$ space. The rest of this section is devoted to this question.

Denote by $(e_{\dot{1}}, e_{\dot{2}})$ the dotted spinor basis for which (46) holds. Then for any spinor basis $(e'_{\dot{1}}, e'_{\dot{2}})$ we have

$$e'_{\dot{A}} = e_{\dot{B}} l^{\dot{B}}{}_{\dot{A}}, \qquad \det(l^{\dot{B}}{}_{\dot{A}}) = 1, \tag{56}$$

and $l^{\dot{B}}{}_{\dot{A}}$ are holomorphic functions on U.

The null tetrad of 1-forms $g'^{A\dot{B}}$ as defined by this new basis $e'_{\dot{A}}$ are related to $g^{A\dot{B}}$ defined by (48) for the basis $e_{\dot{A}}$, as follows:

$$g^{A\dot{B}} = l^{\dot{B}}{}_{\dot{C}} g'^{A\dot{C}}. \tag{57}$$

The connection 1-forms $\Gamma'^{\dot{A}}{}_{\dot{B}}$ read

$$\Gamma'^{\dot{A}}{}_{\dot{B}} = l^{-1\dot{A}}{}_{\dot{C}} dl^{\dot{C}}{}_{\dot{B}} = -l_{\dot{C}}{}^{\dot{A}} dl^{\dot{C}}{}_{\dot{B}}, \tag{58}$$

where

$$l_{\dot{C}}{}^{\dot{A}} = \epsilon_{\dot{C}\dot{B}} \epsilon^{\dot{D}\dot{A}} l^{\dot{B}}{}_{\dot{D}}. \tag{59}$$

According to the Frobenius theorem, one finds that the Pfaff systems

$$g'^{1\dot{2}} = g'^{2\dot{2}} = 0 \tag{60}$$

and

$$g'^{1\dot{1}} = g'^{2\dot{1}} = 0 \tag{61}$$

are completely integrable if and only if

$$\Gamma'_{\dot{1}\dot{1}A\dot{1}} g'^{A\dot{1}} = 0 \tag{62}$$

and

$$\Gamma'_{\dot{2}\dot{2}A\dot{2}} g'^{A\dot{2}} = 0 \tag{63}$$

respectively.

Employing (57) and (58), we can rewrite the condition (62) in the form

$$l_{\dot{B}}{}^{\dot{2}} l^{\dot{C}}{}_{\dot{1}} (\partial_{A\dot{C}} l^{\dot{B}}{}_{\dot{1}}) g'^{A\dot{1}} = 0 \tag{64}$$

and the condition (63) in the form

$$l_{\dot{B}}{}^{\dot{1}} l^{\dot{C}}{}_{\dot{2}} (\partial_{A\dot{C}} l^{\dot{B}}{}_{\dot{2}}) g'^{A\dot{2}} = 0, \tag{65}$$

where $\partial_{A\dot{C}}$ is defined by (16).

By some simple manipulations one finds that assuming $l^{\dot{1}}{}_{\dot{1}} \neq 0$, condition (64) is equivalent to the system of differential equations

$$\begin{aligned} \Gamma'_{\dot{1}\dot{1}1\dot{1}} = 0 \;&\Leftrightarrow\; \partial_{1\dot{1}}\rho + \rho\partial_{1\dot{2}}\rho = 0, \\ \Gamma'_{\dot{1}\dot{1}2\dot{1}} = 0 \;&\Leftrightarrow\; \partial_{2\dot{1}}\rho + \rho\partial_{2\dot{2}}\rho = 0, \end{aligned} \tag{66}$$

where

$$\rho := \frac{l^{\dot{2}}{}_{\dot{1}}}{l^{\dot{1}}{}_{\dot{1}}}. \tag{67}$$

Analogously for $l^{\dot{2}}{}_{\dot{2}} \neq 0$, condition (65) can be equivalently written as the system of the following differential equations:

$$\begin{aligned} \Gamma'_{\dot{2}\dot{2}2\dot{2}} = 0 \;&\Leftrightarrow\; \partial_{2\dot{2}}\sigma + \sigma\partial_{2\dot{1}}\sigma = 0, \\ \Gamma'_{\dot{2}\dot{2}1\dot{2}} = 0 \;&\Leftrightarrow\; \partial_{1\dot{2}}\sigma + \sigma\partial_{1\dot{1}}\sigma = 0, \end{aligned} \tag{68}$$

where

$$\sigma := \frac{l^{\dot{1}}{}_{\dot{2}}}{l^{\dot{2}}{}_{\dot{2}}}. \tag{69}$$

From (48) and (54) one finds $\partial_{A\dot{B}}$ to be

$$\partial_{A\dot{1}} = -\Phi_A{}^{\dot{B}}\frac{\partial}{\partial q^{\dot{B}}}, \tag{70}$$

$$\partial_{A\dot{2}} = \frac{\partial}{\partial q^{A}}, \tag{71}$$

where $\Phi_A{}^{\dot{B}} = \epsilon_{AC}\epsilon^{\dot{D}\dot{B}}\,\Phi^C{}_{\dot{D}}$.

Inserting (70) into (66) and (68), we obtain the systems of differential equations

$$\left(\Phi_A{}^{\dot{B}}\frac{\partial}{\partial q^{\dot{B}}} - \rho\frac{\partial}{\partial q^A}\right)\rho = 0 \tag{72}$$

and

$$\left(\Phi^A{}_{\dot{B}}\frac{\partial}{\partial q^A} + \sigma\frac{\partial}{\partial q^{\dot{B}}}\right)\sigma = 0 \tag{73}$$

respectively. Observe that for $\sigma \neq 0$, Eq. (73) can be rewritten in the form of (72), i.e.,

$$\left(\Phi_A{}^{\dot{B}}\frac{\partial}{\partial q^{\dot{B}}} - \frac{1}{\sigma}\frac{\partial}{\partial q^A}\right)\frac{1}{\sigma} = 0. \tag{74}$$

Observe also that without any loss of generality one can put

$$l^{\dot{1}}{}_{\dot{1}}\, l^{\dot{2}}{}_{\dot{2}} \neq 0. \tag{75}$$

Hereafter we assume that (75) holds.

One can easily check that the first heavenly equation (55) guarantees the complete integrability of the systems (72) and (73). This is the crucial point of our considerations. (As is well known, this is a crucial point in the twistor constructions [6,13,14]. Compare also with the paper by Robinson and Rózga [15].)

Choose any Cauchy data for Eqs. (72)

$$\rho_0 = \rho_0(q^A) \tag{76}$$

on the 2-surface in U defined by

$$q^{\dot{1}} = q^{\dot{2}} = 0. \tag{77}$$

If U is sufficiently small, we get (by the Cauchy-Kovalevskaya theorem) the unique analytic solution $\rho = \rho(q^A, q^{\dot{A}})$ to the system (72) on U satisfying the Cauchy data (76)

$$\rho(q^A, 0) = \rho_0(q^A) \tag{78}$$

Then we choose the Cauchy data for the system (73)

$$\sigma_0 = \sigma_0(q^{\dot{A}}). \tag{79}$$

on the 2-surface in U

$$q^1 = q^2 = 0 \tag{80}$$

to be such that

$$\sigma_0(q^{\dot{A}})\rho(0, q^{\dot{A}}) \neq 1 \tag{81}$$

for every point of the 2-surface (80). The condition (81) follows from the definitions (67) and (69) of ρ and σ, and from the fact that $\det(l^{\dot{A}}{}_{\dot{B}}) = 1$.

By the Cauchy-Kovalevskaya theorem one finds the unique analytic solution $\sigma = \sigma(q^A, q^{\dot{A}})$ of the system (73) on U (where U is sufficiently small) and by (81)

$$\sigma\rho \neq 1 \tag{82}$$

everywhere on U. The solutions ρ and σ define two congruances of anti-self-dual null strings $\mathcal{N}_{\dot{1}}$ and $\mathcal{N}_{\dot{2}}$, respectively, in U and $\mathcal{N}_{\dot{1}} \cap \mathcal{N}_{\dot{2}} = \emptyset$. The vector fields

$$\partial'_{A\dot{1}} = l^{\dot{B}}{}_{\dot{1}}\partial_{A\dot{B}}, \quad \text{i. e.,} \quad (e'_4, e'_1) \tag{83}$$

and

$$\partial'_{A\dot{2}} = l^{\dot{B}}{}_{\dot{2}}\partial_{A\dot{B}}, \quad \text{i. e.,} \quad (e'_2, -e'_3) \tag{84}$$

are tangent to $\mathcal{N}_{\dot{1}}$ and $\mathcal{N}_{\dot{2}}$, respectively. Consider now whether $\mathcal{N}_{\dot{1}}$ and $\mathcal{N}_{\dot{2}}$ are expanding or not. According to (42) we should examine the object $\Gamma'_{\dot{1}\dot{1}A\dot{2}}$ for the case of $\mathcal{N}_{\dot{1}}$ and the object $\Gamma'_{\dot{1}\dot{1}A\dot{1}}$ for $\mathcal{N}_{\dot{2}}$.

Straightforward calculations lead to the following results:

$$\Gamma'_{\dot{1}\dot{1}A\dot{2}} = l^{2}{}_{\dot{2}}(l^{1}{}_{\dot{1}})^2(\partial_{A\dot{2}} + \sigma\partial_{A\dot{1}})\rho \tag{85}$$

and

$$\Gamma'_{\dot{2}\dot{2}A\dot{1}} = -l^{1}{}_{\dot{1}}(l^{2}{}_{\dot{2}})^2(\partial_{A\dot{1}} + \rho\partial_{A\dot{2}})\sigma. \tag{86}$$

From (66), (82), and (85) it follows that

$$\Gamma'_{\dot{1}\dot{1}A\dot{2}} = 0 \quad \text{iff} \quad \partial_{A\dot{B}}\rho = 0, \tag{87}$$

Analogously, from (68), (82), and (86) one infers that

$$\Gamma'_{\dot{2}\dot{2}A\dot{1}} = 0 \quad \text{iff} \quad \partial_{A\dot{B}}\sigma = 0, \tag{88}$$

Therefore we arrive at the conclusion that the *congruence* $\mathcal{N}_{\dot{1}}$ *is nonexpanding iff* $\rho = const$ *and* $\mathcal{N}_{\dot{2}}$ *is nonexpanding iff* $\sigma = const$.

Moreover, one concludes also that taking the Cauchy data ρ_0 and σ_0 to be such that $\mid \frac{\partial\rho_0}{\partial q^{\dot{1}}} \mid + \mid \frac{\partial\rho_0}{\partial q^{\dot{2}}} \mid \neq 0$ for every point of the 2-surface (77) and $\mid \frac{\partial\sigma_0}{\partial q^{1}} \mid + \mid \frac{\partial\sigma_0}{\partial q^{2}} \mid \neq 0$ for every point of the 2-surface (80), assuming also that $\rho_0 \cdot \sigma_0 \neq 1$ (see (81)), one gets *both congruences* $\mathcal{N}_{\dot{1}}$ *and* $\mathcal{N}_{\dot{2}}$ *to be expanding for sufficiently small* U (where the point $(q^A = 0, q^{\dot{A}} = 0) \epsilon U$) *and* $\mathcal{N}_{\dot{1}} \cap \mathcal{N}_{\dot{2}} = \emptyset$. Then there exist coordinates $\{z^A, z^{\dot{A}}\}$ in U such that the equations

$$z^A = \text{const.} \tag{89}$$

define the congruence $\mathcal{N}_{\dot{1}}$ and the equations

$$z^{\dot{A}} = \text{const.} \tag{90}$$

define $\mathcal{N}_{\dot{2}}$.

Finally, the corresponding null tetrad of 1-forms reads

$$g'^{A\dot{2}} = f^A{}_B dz^B, \qquad \det(f^A{}_B) \neq 0 \tag{91}$$

and

$$g'^{A\dot{1}} = f^A{}_{\dot{B}} dz^{\dot{B}}, \qquad \det(f^A{}_{\dot{B}}) \neq 0, \tag{92}$$

where $f^A{}_B$ and $f^A{}_{\dot{B}}$ are holomorphic functions on U. It is evident that without any loss of generality one can choose the gauge such that

$$f^A{}_B = \delta^A{}_B. \tag{93}$$

Hence the metric takes the form of (51)

$$ds^2 = f_{A\dot{B}}(dz^A \otimes dz^{\dot{B}} + dz^{\dot{B}} \otimes dz^A). \tag{94}$$

The components $f_{A\dot{B}}$ will be found in Section 27.4.

27.3 Hermitian and Kählerian Structures on $\mathcal{H}$ Space in Euclidean Relativity.

It is an easy matter to carry over the results of the previous section to the case of "Ultrahyperbolic Relativity" (UR), i.e., when M is a 4-dimensional real manifold and ds^2 is the metric on M of the signature (++ - -). To achieve this, one should consider all the objects of the previous section to be real and $(l^A{}_B)\epsilon SL(2; R)$, $(l^A{}_{\dot{B}})\epsilon SL(2; R)$.
In the present section we deal with "Euclidean Relativity" (ER). Here M is a 4-dimensional real manifold endowed with the matric ds^2 of the signature (++++). *A null tetrad of 1-forms on* $U \subset M$ is now defined to be a set of four 1-forms $e^a \epsilon \Lambda^1(U)$, $a = 1, \ldots, 4$, on U such that the metric ds^2 on U has the form (2) and, moreover,

$$\overline{e^1} = e^2 \quad \text{and} \quad \overline{e^3} = e^4, \tag{95}$$

where the overbar stands for complex conjugation. Consequently, for the *null tetrad of tangent vector fields on* U defined by (3), one finds the following relation to hold:

$$\overline{e_1} = e_2 \quad \text{and} \quad \overline{e_3} = e_4. \tag{96}$$

The connection 1-forms $\Gamma^a{}_b \epsilon \Lambda^1(U)$ and the Riemannian curvature 2-forms $\mathcal{R}^a{}_b \epsilon \Lambda^2(U)$ are defined as before. It is evident by (95) and (96) that

$$\overline{\Gamma^a{}_b} = \Gamma_a{}^b \quad \text{and} \quad \overline{\mathcal{R}^a{}_b} = \mathcal{R}_a{}^b. \tag{97}$$

Analogously as before (see (11)), we define the spinor 1-form $g^{A\dot{B}}$ on U. But now, by (95), one obtains the following relations

$$\overline{g^{A\dot{B}}} = -g_{A\dot{B}} \quad \text{for} \quad A = 1, 2; \quad \dot{B} = \dot{1}, \dot{2}, \tag{98}$$

Consequently the components of undotted and dotted spinors transform as follows:

$$\begin{aligned} \Psi'^A &= l'^A{}_B \Psi^B, \qquad (l'^A{}_B)\,\epsilon\, SU(2), \\ \Psi'^{\dot{A}} &= l'^{\dot{A}}{}_{\dot{B}} \Psi^{\dot{B}}, \qquad (l'^{\dot{A}}{}_{\dot{B}})\,\epsilon\, SU(2), \end{aligned} \tag{99}$$

and, as before, $l'^A{}_B$ and $l'^{\dot{A}}{}_{\dot{B}}$ are unrelated one to another. Finally, one quickly finds the formulas

$$\overline{\partial_{A\dot{B}}} = -\partial^{A\dot{B}}, \tag{100}$$

$$\overline{S^{AB}} = S_{AB}, \quad \overline{S^{\dot{A}\dot{B}}} = S_{\dot{A}\dot{B}}, \tag{101}$$

$$\overline{\Gamma^{AB}} = \Gamma_{AB}, \quad \overline{\Gamma^{\dot{A}\dot{B}}} = \Gamma_{\dot{A}\dot{B}}, \tag{102}$$

$$\overline{\mathcal{R}^{AB}} = \mathcal{R}_{AB}, \quad \overline{\mathcal{R}^{\dot{A}\dot{B}}} = \mathcal{R}_{\dot{A}\dot{B}}, \tag{103}$$

$$\overline{C^{ABCD}} = C_{ABCD}, \quad \overline{C^{\dot{A}\dot{B}\dot{C}\dot{D}}} = C_{\dot{A}\dot{B}\dot{C}\dot{D}}$$

$$\overline{C^{AB\dot{C}\dot{D}}} = C_{AB\dot{C}\dot{D}}. \tag{104}$$

where the above objects are defined analogously as in Section 27.2.

It is obvious that as the metric ds^2 is positive definite we have no real null vector. Therefore there exist no totally null 2-surfaces, i.e., null strings, in M. The question arises if one can give any geometrical interpretation of the 2-form Σ on $U \subset M$ satisfying the conditions (a), (b), and (c) (see (31) and (33)), i.e., to recall those conditions

$$(a) \quad \Sigma \wedge \Sigma = 0\,, \qquad (b) \quad * \Sigma = -\Sigma, \quad \text{and} \quad (c) \quad d\Sigma = 0. \tag{105}$$

From (105), by the complex conjugation we get

$$(\overline{a}) \quad \overline{\Sigma} \wedge \overline{\Sigma} = 0\,, \qquad (\overline{b}) \quad * \overline{\Sigma} = -\overline{\Sigma}, \quad \text{and} \quad (\overline{c}) \quad d\overline{\Sigma} = 0. \tag{106}$$

Consequently, from (a), (b), $(\overline{a})$, and $(\overline{b})$ one infers that there exists a null tetrad of 1-forms $g^{A\dot{B}} \in \Lambda^1(U)$ such that

$$\Sigma = h^{-1} S^{\dot{2}\dot{2}} = 2h^{-1} g^{1\dot{2}} \wedge g^{2\dot{2}} = 2h^{-1} e^3 \wedge e^2,$$

$$\overline{\Sigma} = \overline{h^{-1}} S^{\dot{1}\dot{1}} = 2\overline{h^{-1}} g^{1\dot{1}} \wedge g^{2\dot{1}} = 2\overline{h^{-1}} e^4 \wedge e^1, \tag{107}$$

where h is some function on U. Finally, the conditions (c) and $(\overline{c})$ are equivalent to the following ones:

$$\Gamma_{\dot{1}\dot{1}A\dot{1}} g^{A\dot{1}} = 0 \quad \text{and} \quad \Gamma_{\dot{2}\dot{2}A\dot{2}} g^{A\dot{2}} = 0. \tag{108}$$

To find the geometrical meaning of (108), consider the following tensor field on U [16,17]

$$\mathrm{End}(T(U)) \ni J := i(e_3 \otimes e^3 + e_2 \otimes e^2 - e_4 \otimes e^4 - e_1 \otimes e^1)$$

$$= 2i\,\partial_{A(\dot{1}} \otimes g^{A}{}_{\dot{2})}. \tag{109}$$

One quickly shows that

$$J^2 = -id_{T(U)} \tag{110}$$

and

$$ds^2(JX, JY) = ds^2(X, Y). \tag{111}$$

for every $X, Y \in T(U)$.

Hence J is an *almost complex* structure on U and the metric ds^2 is an Hermitian metric with respect to J. In another word (U, ds^2, J) is an *almost Hermitian manifold.* Now the main result which has been proved in [16,17] says that

J is integrable iff the conditions (108) *hold.* Thus we are led to the following theorem:

Theorem 27.3.1 *Let Σ be a 2-form defined on an open set $U \subset M$ and let Σ satisfy the conditions (a), (b), and (c) of (105). Then (U, ds^2, J) is a Hermitian manifold, where J is the complex structure on U defined by (107) and (109).*

(Evidently, the inverse theorem also holds).

Note that from the considerations leading from (105) to (108) one concludes that the *existence of the 2-form Σ on U fulfilling the conditions* (105) *is equivalent to the existence of the null tetrad of 1-forms $g^{A\dot{B}} \in \Lambda^1(U)$ which fulfill* (108).

So let (108) be satisfied. Hence, by Theorem 27.3.1 J defined by (109) is seen to be a complex structure on U and ds^2 is the Hermitian metric with respect to J. Locally, on some $V \subset U$, there exist complex coordinates $\{q^A, q^{\dot{A}}\}$, $q^{\dot{A}} = \overline{q^A}$, and the functions $h^A{}_B$ and $h^A{}_{\dot{B}}$ such that

$$g^{A\dot{2}} = h^A{}_B dq^B \quad \text{and} \quad g^{A\dot{1}} = h^A{}_{\dot{B}} dq^{\dot{B}} \quad \text{on } V. \tag{112}$$

The relations (98) yield

$$\overline{h^A{}_B} = h_{A\dot{B}}. \tag{113}$$

Then from (12), (112), and (113) we find the Hermitian metric ds^2 to be

$$ds^2 = \Phi_{A\dot{B}}(dq^A \otimes dq^{\dot{B}} + dq^{\dot{B}} \otimes dq^A), \tag{114}$$

$$\Phi_{A\dot{B}} = h^C{}_A h_{C\dot{B}} = \sum_{C=1}^{2} h^C{}_A \overline{h^C{}_B}.$$

For $S^{\dot{1}\dot{1}}$ and $S^{\dot{2}\dot{2}}$ one gets

$$S^{\dot{1}\dot{1}} = 2\bar{h} dq^{\dot{1}} \wedge dq^{\dot{2}}, \qquad S^{\dot{2}\dot{2}} = 2h dq^1 \wedge dq^2, \tag{115}$$

where $h = \det(h^A{}_B) \neq 0$ (compare this with (107)). Finally, $S^{\dot{1}\dot{2}}$ reads

$$S^{\dot{1}\dot{2}} = \Phi_{A\dot{B}} dq^A \wedge dq^{\dot{B}}. \tag{116}$$

The Kähler form ϕ on U is defined as follows [16,17,18]:

$$\phi(X, Y) := ds^2(X, JY)\,; \qquad X, Y \in T(U). \tag{117}$$

Thus from (109), (112), (114), and (116) one infers that

$$\phi = -i\Phi_{A\dot{B}} dq^A \wedge dq^{\dot{B}} = -iS^{\dot{1}\dot{2}}. \tag{118}$$

Recall that the Hermitian manifold (U, ds^2, J) is called a *Kählerian manifold* if [16,17,18]

$$d\phi = 0. \tag{119}$$

Consequently, by (30) and (118) we conclude that (U, ds^2, J) is the Kählerian manifold iff [16,17]

$$\Gamma_{\dot{1}\dot{1}A\dot{2}}\, g^{A\dot{2}} = 0 \quad \text{and} \quad \Gamma_{\dot{2}\dot{2}A\dot{1}}\, g^{A\dot{1}} = 0. \tag{120}$$

Comparing (120) with (42), one quickly finds that the existence of the Kählerian structure on the Hermitian manifold (U, ds^2, J) in ER is the analogue of the existence in CR of two nonexpanding congruences of anti-self-dual null strings $\mathcal{N}_{\dot{1}}$ and $\mathcal{N}_{\dot{2}}$ such that $\mathcal{N}_{\dot{1}} \cap \mathcal{N}_{\dot{2}} = \emptyset$. Gathering all that, we see that the following theorem holds.

Theorem 27.3.2 *Let (M_C, ds^2) be a complex space-time and let $U_C \subset M_C$ be an open set of M_C admitting the congruence of anti-self-dual null strings $\mathcal{N}_{\dot{1}}$. Then the Euclidean slice (U, ds^2), if it exists, admits the complex structure J defined by (109) such that (U, ds^2, J) is an Hermitian manifold. Moreover, if $\mathcal{N}_{\dot{1}}$ is nonexpanding, then (U, ds^2, J) is a Kählerian manifold.*

Consider now an $\mathcal{H}$ space in ER, i.e., a real space-time (M, ds^2) such that the signature of ds^2 is $(++++)$ and $\mathcal{R}^{\dot{A}}{}_{\dot{B}} = 0$ $(\Leftrightarrow C_{\dot{A}\dot{B}\dot{C}\dot{D}} = C_{AB\dot{C}\dot{D}} = R = 0)$.

As before, for every point $p \in M$ there exists an open neighborhood U of p and a dotted spinor basis on U such that

$$\Gamma^{\dot{A}}{}_{\dot{B}} = 0 \quad \text{on } U. \tag{121}$$

Therefore (U, ds^2, J) with J defined by (109) is a Kählerian manifold. Locally, on some $V \subset U$, the null tetrad of 1-forms $g^{A\dot{B}}$ is defined by (112), but now without any loss of generality one can put

$$h := \det(h^A{}_B) = 1 = \det(h^{\dot{A}}{}_{\dot{B}}) = \bar{h}. \tag{122}$$

Finally, from (30), (114), (116), (121), and (122) we are led to the first heavenly equation (55) with $\Omega = \Omega(q^A, q^{\dot{A}})$ being now a real function. Let $(e_{\dot{1}}, e_{\dot{2}})$ be the dotted spinor basis on U for which (121) holds. Then any dotted spinor basis $(e'_{\dot{1}}, e'_{\dot{2}})$ on U can be obtained by

$$e'_{\dot{A}} = e_{\dot{B}}\, l^{\dot{B}}{}_{\dot{A}}, \qquad (l^{\dot{B}}{}_{\dot{A}}) \in SU(2). \tag{123}$$

The corresponding almost-complex structure J' on U such that (U, ds^2, J') is an almost-Hermitian manifold reads

$$J' = 2i\,\partial'_{A(\dot{1}} \otimes g'^{A}{}_{\dot{2})} = a_1 J + a_2 K + a_3 L,$$

$$a_1 = | l^{\dot{1}}{}_{\dot{1}} |^2 - | l^{\dot{1}}{}_{\dot{2}} |^2 , \qquad a_2 = -(l^{\dot{1}}{}_{\dot{1}} l^{\dot{1}}{}_{\dot{2}} + \overline{l^{\dot{1}}{}_{\dot{1}} l^{\dot{1}}{}_{\dot{2}}}),$$

$$a_3 = -i(l^{\dot{1}}{}_{\dot{1}} l^{\dot{1}}{}_{\dot{2}} - \overline{l^{\dot{1}}{}_{\dot{1}} l^{\dot{1}}{}_{\dot{2}}}),$$

$$J = 2i\partial_{A(\dot{1}} \otimes g^{A}{}_{\dot{2})} , \qquad K = i(\partial_{A\dot{2}} \otimes g^{A}{}_{\dot{2}} - \partial_{A\dot{1}} \otimes g^{A}{}_{\dot{1}}),$$

$$L = -(\partial_{A\dot{2}} \otimes g^{A}{}_{\dot{2}} + \partial_{A\dot{1}} \otimes g^{A}{}_{\dot{1}}). \tag{124}$$

It is an easy matter to prove the following relations:

$$J^2 = K^2 = L^2 = -id_{T(U)},$$

$$JK = L, \qquad LJ = K, \qquad KL = J, \tag{125}$$

and

$$a_1^2 + a_2^2 + a_3^2 = 1. \tag{126}$$

With the use of ρ and σ defined by (67) and (69), J' can be rewritten as follows:

$$J' = \frac{1 - |\rho|^2}{1 + |\rho|^2} J + 2\frac{\Re\rho}{1 + |\rho|^2} K + 2\frac{\Im\rho}{1 + |\rho|^2} L. \tag{127}$$

[Note that in (124) and (127) we use the fact that $(l^{\dot{B}}{}_{\dot{A}}) \in SU(2)$, i.e.,

$$l^{\dot{2}}{}_{\dot{2}} = \overline{l^{\dot{1}}{}_{\dot{1}}}, \qquad l^{\dot{2}}{}_{\dot{1}} = -\overline{l^{\dot{1}}{}_{\dot{2}}}, \qquad | l^{\dot{1}}{}_{\dot{1}} |^2 + | l^{\dot{1}}{}_{\dot{2}} |^2 = 1;$$

If one assumes only that $(l^{\dot{B}}{}_{\dot{A}} \in SL(2, K)$ then by making the substitutions in the formulas (124)–(127)

$$\overline{l^{\dot{1}}{}_{\dot{1}}} \to l^{\dot{2}}{}_{\dot{2}}, \qquad \overline{l^{\dot{1}}{}_{\dot{2}}} \to -l^{\dot{2}}{}_{\dot{1}} \quad \text{and} \quad \overline{\rho} \to -\sigma$$

one gets the corresponding relations in CR.]

Employing the results of the previous section, one infers that (U, ds^2, J') is an Hermitian manifold iff ρ satisfies the following system of differential equations (see (66):

$$(\partial_{A\dot{1}} + \rho\partial_{A\dot{2}})\rho = 0. \tag{128}$$

Moreover, by (87), (88), and(120), remembering also that in ER $\sigma = -\overline{\rho}$, we conclude that (U, ds^2, J') is a *Kählerian manifold iff ρ =const. on U.*

In particular,

$$\rho = 0 \Rightarrow J' = J; \qquad \rho = 1 \Rightarrow J' = K; \qquad \rho = i \Rightarrow J' = L \tag{129}$$

Hence, *J, K, and L are three complex structures on U satisfying the quaternionic relations* (125) *and defining the Kählerian structures on* (U, ds^2). It means that the $\mathcal{H}$ space (U, ds^2) in ER is a hyper-Kähler manifold [19].

Finally, any ρ fulfilling the system of differential equations (128) and such that (see(85))

$$(\partial_{1\dot{2}} - \overline{\rho}\partial_{1\dot{1}})\rho \neq 0 \quad \text{or} \quad (\partial_{2\dot{2}} - \overline{\rho}\partial_{2\dot{1}})\rho \neq 0 \tag{130}$$

for every point $p \in U$, defines according to (127) the complex structure J' on U such that (U, ds^2, J') is Hermitian but non-Kählerian manifold. One quickly finds that (128) and (130) are satisfied iff ρ satisfies the system (128) and also

$$(\partial_{1\dot{1}}\rho \neq 0 \quad \text{and} \quad \partial_{1\dot{2}}\rho \neq 0) \quad \text{or} \quad (\partial_{2\dot{2}}\rho \neq 0 \quad \text{and} \quad \partial_{2\dot{1}}\rho \neq 0) \tag{131}$$

everywhere on U.

Observe that in terms of the local coordinates $\{q^A, q^{\dot{A}}\}$ the system (128) takes the form of (72). Therefore, using the arguments similar to those given in Section 27.2, we arrive at the conclusion that if U is sufficiently small then in any (analytic) $\mathcal{H}$ space (U, ds^2) in ER there exist complex structures J' such that (U, ds^2, J') is an Hermitian but non-Kählerian manifold (with respect to J').

In the light of the results obtained in this section, one can easily interpret the considerations of Section 27.2 in terms of a *complexified Hermitian structure* and a *complexified Kählerian structure* [16,20].

In the next section we use this result and the analogous result in CR obtained in Section 27.2, and we find a new version of the heavenly equation.

27.4 The Heavenly Equation

Let $\mathcal{N}_{\dot{1}}$ and $\mathcal{N}_{\dot{2}}$, $\mathcal{N}_{\dot{1}} \cap \mathcal{N}_{\dot{2}} = \emptyset$, be two *expanding* congruences of anti-self-dual null strings in $U \subset M$, where (M, ds^2) is a complex $\mathcal{H}$ space. Employing the results of Section 27.2 (see (89)–(94)) we choose the coordinates $\{z^A, z^{\dot{A}}\}$ on U such that, omitting the primes "'", one has

$$g^{A\dot{2}} = dz^A, \qquad g^{A\dot{1}} = f^A{}_{\dot{B}}\, dz^{\dot{B}} \tag{132}$$

and the equations z^A =const. or $z^{\dot{A}}$ =const. define the congruences $\mathcal{N}_{\dot{1}}$ or $\mathcal{N}_{\dot{2}}$, respectively. Then the metric is defined by (94). Straightforward calculations lead to the following dotted spinorial connection forms (see (23) with (24)):

$$\begin{aligned}
\Gamma_{\dot{1}\dot{1}} &= \frac{1}{f} f_{A[\dot{2},\dot{1}]}\, dz^A, \\
\Gamma_{\dot{1}\dot{2}} &= \frac{1}{2} f^{\dot{B}A} \{ f_{A[\dot{B},\dot{C}]}\, dz^{\dot{C}} + f_{(A|\dot{B}|,C)}\, dz^C \}, \\
\Gamma_{\dot{2}\dot{2}} &= f_{[2|\dot{A}|,1]}\, dz^{\dot{A}},
\end{aligned} \tag{133}$$

where $f^{\dot{B}A}$ is the inverse matrix to $f_{A\dot{B}}$, i.e.,

$$f^{\dot{B}A} f_{A\dot{C}} = \delta^{\dot{B}}{}_{\dot{C}}\,, \qquad f_{A\dot{C}} f^{\dot{C}B} = \delta^B{}_A \tag{134}$$

and $f = \det(f_{A\dot{B}})$; moreover

$$
\begin{aligned}
f_{A[\dot{B},\dot{C}]} &= \frac{1}{2}(f_{A\dot{B},\dot{C}} - f_{A\dot{C},\dot{B}}),\\
f_{[A|\dot{B}|,C]} &= \frac{1}{2}(f_{A\dot{B},C} - f_{C\dot{B},A}),\\
f_{(A|\dot{B}|,C)} &= \frac{1}{2}(f_{A\dot{B},C} + f_{C\dot{B},A}).
\end{aligned}
$$

The $\mathcal{H}$ space equations (44), by (27) and (133), read

$$(f_{A[\dot{2},\dot{1}]})_{,\dot{B}} - f^{\dot{C}D} f_{D(\dot{C},\dot{B})} f_{A[\dot{2},\dot{1}]} = 0, \quad (135)$$

$$(f_{1[\dot{2},\dot{1}]})_{,2} - (f_{2[\dot{2},\dot{1}]})_{,1} + f^{\dot{B}1} f_{[2|\dot{B}|,1]} f_{1[\dot{2},\dot{1}]} - f^{\dot{B}2} f_{[1|\dot{B}|,2]} f_{2[\dot{2},\dot{1}]} = 0, \quad (136)$$

$$(f_{[2|\dot{A}|,1]})_{,B} - f^{\dot{D}C} f_{(C|\dot{D}|,B)} f_{[2|\dot{A}|,1]} = 0, \quad (137)$$

$$(f_{[2|\dot{1}|,1]})_{,\dot{2}} - (f_{[2|\dot{2}|,1]})_{,\dot{1}} + f^{\dot{1}B} f_{B[\dot{2},\dot{1}]} f_{[2|\dot{1}|,1]} - f^{\dot{2}B} f_{B[\dot{1},\dot{2}]} f_{[2|\dot{2}|,1]} = 0, \quad (138)$$

$$(f^{\dot{B}A} f_{A[\dot{B},\dot{C}]})_{,\dot{D}} - (f^{\dot{B}A} f_{A[\dot{B},\dot{D}]})_{,\dot{C}} = 0, \quad (139)$$

$$(f^{\dot{A}B} f_{[B|\dot{A}|,C]})_{,D} - (f^{\dot{A}B} f_{[B|\dot{A}|,D]})_{,C} = 0, \quad (140)$$

$$(f^{\dot{B}A} f_{(A|\dot{B}|,D)})_{,\dot{C}} - (f^{\dot{B}A} f_{A[\dot{B},\dot{C}]})_{,D} + \frac{2}{f} f_{D[\dot{2},\dot{1}]} f_{[2|\dot{C}|,1]} = 0. \quad (141)$$

[Note that Eqs. (133) as well as Eqs. (135)–(141) correspond to the equations found in [21]. Also the further considerations are very similar to those of [21]].

Equations (139) imply the existence of a function $G = G(z^A, z^{\dot{A}})$ such that

$$f^{\dot{B}A} f_{A[\dot{B},\dot{C}]} = -\frac{1}{2} G_{,\dot{C}}. \tag{142}$$

As

$$f^{\dot{B}A} f_{A\dot{B},\dot{C}} = (\ln f)_{,\dot{C}}, \tag{143}$$

from (142) and (143) one gets

$$f^{\dot{B}A} f_{A\dot{C},\dot{B}} = (G + \ln f)_{,\dot{C}}. \tag{144}$$

Analogously, from (140) we obtain

$$
\begin{aligned}
f^{\dot{A}B} f_{[B|\dot{A}|,C]} &= -\frac{1}{2}\tilde{G}_{,C},\\
\tilde{G} &= \tilde{G}(z^A, z^{\dot{A}}),
\end{aligned}
\tag{145}
$$

and

$$f^{\dot{A}B} f_{C\dot{A},B} = (\tilde{G} + \ln f)_{,C}. \tag{146}$$

It is assumed that both congruences $\mathcal{N}_{\dot{1}}$ and $\mathcal{N}_{\dot{2}}$ are expanding. Consequently

$$\Gamma_{\dot{1}\dot{1}} \neq 0 \quad \Leftrightarrow \quad f_{A[\dot{2},\dot{1}]} dz^A \neq 0 \tag{147}$$

and

$$\Gamma_{\dot{2}\dot{2}} \neq 0 \quad \Leftrightarrow \quad f_{[2|\dot{A}|,1]} dz^{\dot{A}} \neq 0 \tag{148}$$

Assume that $f_{1[\dot{2},\dot{1}]} \neq 0$ and $f_{2[\dot{2},\dot{1}]} \neq 0$. Then from (135), by (143) and (144) one has

$$\left[\ln f_{A[\dot{2},\dot{1}]} - \left(\ln f + \frac{1}{2} G \right) \right]_{,\dot{B}} = 0, \tag{149}$$

i.e., $f_{A[\dot{2},\dot{1}]}$ is of the form

$$f_{A[\dot{2},\dot{1}]} = g_A(z^B) \cdot L(z^C, z^{\dot{C}}). \tag{150}$$

Consequently, we can choose coordinates z^A so that

$$f_{1[\dot{2},\dot{1}]} = 0 \quad \text{and} \quad f_{2[\dot{2},\dot{1}]} \equiv P \neq 0. \tag{151}$$

Analogously, employing (137), (146), and the formula $f^{\dot{A}B} f_{B\dot{A},C} = (\ln f)_{,C}$, one quickly shows that without any loss of generality we can put

$$f_{[2|\dot{1}|,1]} = 0 \quad \text{and} \quad f_{[2|\dot{2}|,1]} \equiv \tilde{P} \neq 0. \tag{152}$$

In what follows, we assume that (151) and (152) hold. Hence, (142) with (149) and (151) yields

$$f_{1\dot{B}} = \left(\frac{f}{P} \right)_{,\dot{B}}. \tag{153}$$

Similarly, one easily finds that

$$f_{B\dot{1}} = \left(\frac{f}{\tilde{P}} \right)_{,B}. \tag{154}$$

Comparing (153) with (154) for $\dot{B} = \dot{1}$ and $B = 1$, we conclude that there exists a function $\Psi = \Psi(z^A, z^{\dot{A}})$ (the *potential*) such that

$$\frac{f}{P} = \Psi_{,1} \quad \text{and} \quad \frac{f}{\tilde{P}} = \Psi_{,\dot{1}}. \tag{155}$$

Thus we have

$$f_{1\dot{B}} = \Psi_{,1\dot{B}} \quad \text{and} \quad f_{B\dot{1}} = \Psi_{,B\dot{1}}. \tag{156}$$

Employing (151), (152), (155), and (156), one quickly shows that Eqs. (136) read

$$K_{,1} = 0 \ , \qquad K := \ln \left(\frac{\Psi_{,1} \Psi_{,\dot{1}}}{f} \right) \tag{157}$$

and Eqs. (138) read

$$K_{,\dot{1}} = 0. \tag{158}$$

Hence

$$K = K(z^2, z^{\dot{2}}). \tag{159}$$

By simple manipulations using (151), (152), (155), (156), and (159), we find that Eqs. (141) are equivalent to the Liouville equation

$$K_{,2\dot{2}} - 2\,e^{-K} = 0. \tag{160}$$

The general solution of Eq. (160) has the form

$$K = K(z^2, z^{\dot{2}}) = \ln\left[\frac{(1+\alpha\tilde{\alpha})^2}{\alpha_{,2}\,\tilde{\alpha}_{,\dot{2}}}\right], \tag{161}$$

where $\alpha = \alpha(z^2)$ and $\tilde{\alpha} = \tilde{\alpha}(z^{\dot{2}})$ are arbitrary functions of their arguments.

Finally, one easily checks that by (151), (152), (155), and (156) Eqs. (135) and (137) are fulfilled. Thus we are left with the following equations:

$$f_{2[\dot{2},\dot{1}]} = \frac{f}{\Psi_{,1}}, \tag{162}$$

$$f_{[2|\dot{2}|,1]} = \frac{f}{\Psi_{,\dot{1}}}, \tag{163}$$

$$\frac{\Psi_{,1}\,\Psi_{,\dot{1}}}{f} = \frac{(1+\alpha\tilde{\alpha})^2}{\alpha_{,2}\,\tilde{\alpha}_{,\dot{2}}}, \tag{164}$$

and (156).

From (162) and (164), using also (156), we get

$$f_{2\dot{2},\dot{1}} = \frac{2\alpha_{,2}\,\tilde{\alpha}_{,\dot{2}}}{(1+\alpha\tilde{\alpha})^2}\Psi_{,\dot{1}} + \Psi_{,2\dot{1}\dot{2}}\,. \tag{165}$$

Integration of (165) yields

$$f_{2\dot{2}} = \frac{2\alpha_{,2}\,\tilde{\alpha}_{,\dot{2}}}{(1+\alpha\tilde{\alpha})^2}\Psi + \Psi_{,2\dot{2}} + \beta(z^2, z^{\dot{2}}), \tag{166}$$

where $\beta = \beta(z^2, z^{\dot{2}})$ is some function.

From (155) and (156) it is evident that the potential Ψ is defined with precision up to the following transformation:

$$\Psi \;\rightarrow\; \Psi + \gamma(z^2, z^{\dot{2}}), \tag{167}$$

where $\gamma = \gamma(z^2, z^{\dot{2}})$ is an arbitrary function of z^2 and $z^{\dot{2}}$. Hence one can choose the function γ so that the function β in (166) vanishes. Consequently, without any generality lost we have

$$f_{2\dot{2}} = \frac{2\alpha_{,2}\,\tilde{\alpha}_{,\dot{2}}}{(1+\alpha\tilde{\alpha})^2}\Psi + \Psi_{,2\dot{2}}\,. \tag{168}$$

Finally, it is also evident that without any loss of generality one can put

$$\alpha = z^2 \quad \text{and} \quad \tilde{\alpha} = z^{\dot{2}}. \tag{169}$$

Thus

$$f_{2\dot{2}} = \frac{2}{(1+z^2 z^{\dot{2}})^2}\Psi + \Psi_{,2\dot{2}}\,. \tag{170}$$

Employing (156), (164), with (169) and (170), one quickly finds that Eq. (163) is satisfied. Consequently, we are left with (164), (169), (156), and (170). As $f = f_{1\dot{1}} f_{2\dot{2}} - f_{1\dot{2}} f_{2\dot{1}}$, one obtains the following nonlinear partial differential equation on the potential Ψ:

$$\Psi,_{1\dot{1}} \Psi,_{2\dot{2}} - \Psi,_{1\dot{2}} \Psi,_{2\dot{1}} + \frac{1}{(1+z^2 z^{\dot{2}})^2}(2\Psi\Psi,_{1\dot{1}} - \Psi,_1 \Psi,_{\dot{1}}) = 0. \tag{171}$$

This is our *new version of the heavenly equation.* The heavenly metric is given by (94) with

$$f_{1\dot{1}} = \Psi,_{1\dot{1}}\ , \qquad f_{1\dot{2}} = \Psi,_{1\dot{2}}\ , \qquad f_{2\dot{1}} = \Psi,_{2\dot{1}}\ ,$$
$$f_{2\dot{2}} = \frac{2}{(1+z^2 z^{\dot{2}})^2}\Psi + \Psi,_{2\dot{2}}\ . \tag{172}$$

[To carry over these results to UR, one should assume that Ψ is a real function and $\{z^A, z^{\dot{A}}\}$ are real coordinates. Then assuming Ψ to be real function and $z^{\dot{1}} = \overline{z^1}$, $z^{\dot{2}} = \overline{z^2}$, we obtain the heavenly equation (171) and the heavenly metric (172) in ER.]

We conclude this paper with some simple examples of searching for solutions of Eq. (171).

(i) First assume that

$$\Psi_{,\dot{2}} = 0. \tag{173}$$

Consequently from (171) one gets the following equation to be satisfied:

$$2\Psi\Psi_{,1\dot{1}} - \Psi_{,1}\Psi_{,\dot{1}} = 0,$$
$$\Psi = \Psi(z^1, z^{\dot{1}}, z^2). \tag{174}$$

The general solution of (174) reads

$$\Psi = (\tau(z^1, z^2) + \dot{\tau}\,(z^{\dot{1}}, z^2))^2, \tag{175}$$

where $\tau = \tau(z^1, z^2)$ and $\dot{\tau} = \dot{\tau}\,(z^{\dot{1}}, z^2)$ are arbitrary functions of their arguments. Finally, according to (172), the heavenly metric corresponding to the solution (175) takes the form

$$f_{1\dot{1}} = 2\tau_{,1}\,\dot{\tau},_{\dot{1}}\,, \qquad f_{2\dot{1}} = 2(\tau + \dot{\tau})_{,2}\,\dot{\tau}_{,\dot{1}} + 2(\tau + \dot{\tau})\,\dot{\tau}_{,2\dot{1}},$$
$$f_{1\dot{2}} = 0, \qquad f_{2\dot{2}} = \frac{2(\tau + \dot{\tau})^2}{(1+z^2 z^{\dot{2}})^2}. \tag{176}$$

In particular, assuming also

$$\Psi_{,2} = 0, \tag{177}$$

one finds that $\tau = \tau(z^1)$ and $\dot\tau = \dot\tau\,(z^{\dot 1})$. Therefore, performing the coordinate transformation

$$\sqrt{2}\tau(z^1) \mapsto z^1, \qquad \sqrt{2}\,\dot\tau\,(z^{\dot 1}) \mapsto z^{\dot 1}, \tag{178}$$

we obtain the following simple heavenly metric:

$$ds^2 = dz^1 \otimes dz^{\dot 1} + dz^{\dot 1} \otimes dz^1 + \frac{(z^1 + z^{\dot 1})^2}{(1 + z^2 z^{\dot 2})^2}(dz^2 \otimes dz^{\dot 2} + dz^{\dot 2} \otimes dz^2). \tag{179}$$

It is evident that the metric (179) has a Euclidean slice with

$$z^{\dot 1} = \overline{z^1}, \qquad z^{\dot 2} = \overline{z^2}.$$

(ii) In this example we deal with the heavenly metric admitting the Killing vector field ξ

$$\xi = i\left(\frac{\partial}{\partial z^1} - \frac{\partial}{\partial z^{\dot 1}}\right). \tag{180}$$

Consequently, the solution of Eq. (171) is of the following form:

$$\Psi = \Psi(w, z^2, z^{\dot 2}), \qquad w := z^1 + z^{\dot 1}. \tag{181}$$

The heavenly equation (171) reads now

$$\Psi,_{ww}\Psi,_{2\dot 2} - \Psi,_{w\dot 2}\Psi,_{2w} + \frac{1}{(1 + z^2 z^{\dot 2})^2}(2\Psi\Psi,_{ww} - \Psi,_w^2) = 0, \tag{182}$$

where, of course, $\Psi,_w := \frac{\partial\Psi}{\partial w}$, etc. In terms of differential forms the equation (182) can be rewritten as follows:

$$d\Psi - \Psi,_w dw - \Psi,_2 dz^2 - \Psi,_{\dot 2} dz^{\dot 2} = 0, \tag{183}$$

$$d\Psi,_w \wedge d\Psi,_{\dot 2} \wedge dz^{\dot 2} + \frac{1}{(1 + z^2 z^{\dot 2})^2}(2\Psi d\Psi,_w \wedge dz^2 \wedge dz^{\dot 2} - \Psi,_w^2 dw \wedge dz^2 \wedge dz^{\dot 2})$$
$$= 0 \tag{184}$$

We write (183) in the form

$$d(\Psi - \Psi,_w w) - (-w)d\Psi,_w - \Psi,_2 dz^2 - \Psi,_{\dot 2} dz^{\dot 2} = 0. \tag{185}$$

Define

$$\begin{aligned} x &:= \Psi,_w \;\Rightarrow w = w(x, z^2, z^{\dot 2}), \\ \chi' &= \chi'(x, z^2, z^{\dot 2}) := \Psi(w(x, z^2, z^{\dot 2}), z^2, z^{\dot 2}) - x w(x, z^2, z^{\dot 2}). \end{aligned} \tag{186}$$

Consequently, (183) and (184) with (185) and (186) lead to the relations

$$\begin{aligned} &d\chi' - \chi',_x dx - \chi',_2 dz^2 - \chi',_{\dot 2} dz^{\dot 2} = 0 \\ &\chi',_x = -w\,, \qquad \chi',_2 = \Psi,_2 \quad \text{and} \quad \chi',_{\dot 2} = \Psi,_{\dot 2}\,, \end{aligned} \tag{187}$$

and

$$\begin{aligned} &dx \wedge d\chi'_{,\dot{2}} \wedge dz^{\dot{2}} \\ &\quad + \frac{1}{(1+z^2z^{\dot{2}})^2}[2(\chi' - x\chi'_{,x})dx \wedge dz^2 \wedge dz^{\dot{2}} + x^2 d\chi'_{,x} \wedge dz^2 \wedge dz^{\dot{2}}] \\ &= 0 \end{aligned} \tag{188}$$

Thus we get the following equation for χ':

$$\chi'_{,2\dot{2}} + \frac{1}{(1+z^2z^{\dot{2}})^2}(2\chi' - 2x\chi'_{,x} + x^2\chi'_{,xx}) = 0. \tag{189}$$

Finally, one quickly finds that by the substitution

$$\chi := \frac{\chi'}{x} \tag{190}$$

the equation (189) gives

$$\chi_{,2\dot{2}} + \frac{x^2}{(1+z^2z^{\dot{2}})^2}\chi_{,xx} = 0. \tag{191}$$

Concluding: *If the heavenly metric admits the Killing vector field* (180), *then Eq.* (171) *can be brought to the linear partial differential equation* (191).

Simple manipulations show that in terms of the function χ the metric reads

$$\begin{aligned} f_{1\dot{1}} &= -\frac{1}{(x\chi)_{,xx}}, \quad f_{1\dot{2}} = -\frac{(x\chi)_{,x\dot{2}}}{(x\chi)_{,xx}}, \\ f_{2\dot{1}} &= -\frac{(x\chi)_{,x2}}{(x\chi)_{,xx}}, \quad f_{2\dot{2}} = -\frac{2x^2}{(1+z^2z^{\dot{2}})^2}\chi_{,x} + x\chi_{,2\dot{2}} - \frac{(x\chi)_{,x2}\,(x\chi)_{,x\dot{2}}}{(x\chi)_{,xx}}. \end{aligned} \tag{192}$$

An interesting question is what the connection is between our result and the considerations concerning the symmetries of the $\mathcal{H}$ spaces given in references [22,23].

Acknowledgments

We are grateful to Sebastian Formański for valuable discussions and for preparing the present text for publication. We are also indebted to Daniel Finley for reading the manuscript and many useful critical comments and improvements.

References

[1] J. F. Plebański, *J. Math. Phys.* **16**, 2395 (1975).

[2] C. P. Boyer, J. D. Finley III, and J. F. Plebański, *Complex General Relativity, $\mathcal{H}$ and $\mathcal{HH}$ Spaces - A Survey of One Approach,* in *General Relativity and Gravitation,* Einstein's memorial volume, edited by A. Held (Plenum, New York, 1980), Vol. 2, pp. 241–281.

[3] M. Ko, M. Ludvigsen, E. T. Newman, and K. P. Tod, *Phys. Rep.* **71**, 51 (1981).
[4] A. Ashtekar, T. Jacobson and L. Smolin, *Commun. Math. Phys.* **115**, 631 (1988).
[5] R. Penrose, *Gen. Relativ. Grav.* **7**, 31 (1976).
[6] R. Penrose and R. S. Ward, *Twistors for Flat and Curved Space-Time* in *General Relativity and Gravitation*, Einstein's memorial volume, edited by A. Held (Plenum, New York, 1980), Vol. 2, pp. 283–328.
[7] J. F. Plebański, *Spinors, Tetrads and Forms*, unpublished monograph of CINVESTAV (México 14, D. F.,1974).
[8] J. F. Plebański, *The Spinorial and Helicity Formalisms of Riemannian Structures in Complex or Real Four Dimensions*, preprint of CINVESTAV (México 14, D. F., 1980).
[9] R. Penrose and W. Rindler, *Spinors and Space-Time*, Vol.1 (Cambridge University Press, Cambridge, 1984).
[10] J. F. Plebański and K. Rózga, *J. Math. Phys.* **25**, 1930 (1984).
[11] J. F. Plebański and M. Przanowski, *Acta. Phys. Pol.* **B19**, 805 (1988).
[12] J. F. Plebański and S. Hacyan, *J. Math. Phys.* **16**, 2403 (1975).
[13] E. T. Newman, J. R. Porter, and K. P. Tod, *Gen. Relativ. Grav.* **9**, 1129 (1978).
[14] I.A. B. Strachan, *Class. Quantum Grav.* **10**, 1417 (1993).
[15] I. Robinson and K. Rózga, *J. Math. Phys.* **25**, 1941 (1984).
[16] E. J. Flaherty, *Hermitian and Kählerian Geometry in Relativity*(Springer-Verlag Berlin-Heidelberg, 1976).
[17] M. Przanowski and B. Broda, *Acta Phys. Pol.* **B14**, 637 (1983).
[18] S. Kobayashi and K. Nomizu, *Foundations of Differential Geometry*, Vol. II. (Interscience Publishers, New York 1969).
[19] E. Calabi, *Ann. Sci. École. Norm. Sup.* **12**, 269 (1979).
[20] C. P. Boyer and J. F. Plebański, *J. Math. Phys.* **26**, 229 (1985).
[21] M. Przanowski and B. Baka, *Gen. Relativ. Grav.* **16**, 797 (1984).
[22] J. D. Finley III and J. F. Plebański, *J. Math. Phys.* **20**, 1938 (1979).
[23] C. P. Boyer and J. D. Finley III, *J. Math. Phys.* **23**, 1126 (1981).

28

A Plain Man's Guide to Bivectors, Biquaternions, and the Algebra and Geometry of Lorentz Transformations

Wolfgang Rindler
Ivor Robinson

ABSTRACT We here present an elementary account of the algebra and geometry of bivectors and of the biquaternions they give rise to. We then proceed to show how the latter allow a detailed study of the algebra and geometry of Lorentz transformations.

28.1 Introduction

For practical problems in special relativity it is customary to write out Lorentz transformations explicitly in terms of coordinates. But this direct representation is extremely inconvenient for the study of Lorentz transformations as such. For example, even the computation of the Thomas precession then becomes a major undertaking. There is, of course, an elegant and well-known representation of Lorentz transformations by unimodular 2×2 matrices, but this requires the introduction of an auxiliary space.

Contrast this situation with the study of rotations in Euclidean 3-space by means of quaternions, where a rotation through an angle θ about a unit vector $\mathbf{d}$ is represented by a quaternion $(\alpha, \mathbf{a})$, α being the scalar $\cos \frac{\theta}{2}$ and $\mathbf{a}$ the vector $\mathbf{d} \sin \frac{\theta}{2}$. Then the result of compounding two rotations $(\alpha, \mathbf{a})$ and $(\beta, \mathbf{b})$ is neatly given by the quaternion product $(\gamma, \mathbf{c}) = (\alpha, \mathbf{a})(\beta, \mathbf{b})$.

In an analogous manner one can represent restricted Lorentz transformations by biquaternions (α, A) consisting of a *complex* scalar α and a real bivector A. The algebra of biquaternions parallels that of quaternions. We devoted an earlier paper (with J. Ehlers) [1] to that topic. If we now return to it after thirty years, it is for several reasons. We here develop bivector algebra and geometry more fully and in a more elementary way that is intended as a tutorial on the subject. We similarly develop biquaternions and stress their use as a computational tool for dealing with the algebra and geometry of Lorentz transformations. Also, we give more examples and applications, including a relatively painless derivation of the Thomas Precession.

28.2 Basic Algebra of Real Bivectors and Complex Scalars

An electromagnetic field corresponds to a pair $(\mathbf{e}, \mathbf{b})$ of real 3-vectors whose components make up the antisymmetric field tensor (bivector) F_{ij} and its dual

$$\overset{*}{F}_{ij} = \frac{1}{2}\eta_{ijkl}F^{lk} \tag{1}$$

as follows:

$$F_{ij} = \begin{pmatrix} 0 & -e_1 & -e_2 & -e_3 \\ e_1 & 0 & b_3 & -b_2 \\ e_2 & -b_3 & 0 & b_1 \\ e_3 & b_2 & -b_1 & 0 \end{pmatrix}, \quad \overset{*}{F}_{ij} = \begin{pmatrix} 0 & -b_1 & -b_2 & -b_3 \\ b_1 & 0 & -e_3 & e_2 \\ b_2 & e_3 & 0 & -e_1 \\ b_3 & -e_2 & e_1 & 0 \end{pmatrix} \tag{2}$$

This correspondence is form invariant under spatial rotations.

We shall here be concerned with bivectors in general. *Any* bivector and its dual can be associated as above with a pair of real 3-vectors.

For inner products of arbitrary tensors we shall use the following index-free notation:

$$A^{\cdots}_{\ldots i}B^{i\cdots}_{\ldots} =: A \cdot B, \tag{3}$$

$$\frac{1}{2!}A^{\cdots}_{\ldots ij}B^{ji\cdots}_{\ldots} =: A : B, \tag{4}$$

and so on. Equation (1), for example, can then be written $\overset{*}{F} = \eta : F$. For two bivectors F_{ij} and G_{ij}, corresponding, respectively, to $(\mathbf{e}, \mathbf{b})$ and $(\mathbf{e}', \mathbf{b}')$, we easily verify

$$F : G = G : F = \mathbf{b} \cdot \mathbf{b}' - \mathbf{e} \cdot \mathbf{e}', \tag{5}$$

where, however, yet another convention comes into play: Since we shall use the metric

$$ds^2 = dt^2 - dx^2 - dy^2 - dz^2 \tag{6}$$

(with $t, x, y, z \longleftrightarrow 0, 1, 2, 3$), the raising or lowering of a *spatial* index must be accompanied by a change of sign, e.g., $b_1 = -b^1$; we therefore write, *contrarily to common usage,*

$$\mathbf{a} \cdot \mathbf{b} = a_i b^i - (a_1 b_1 + a_2 b_2 + a_3 b_3). \tag{7}$$

Consider next the following association of F_{ij} with a *complex* 3-vector $\mathbf{f}$:

$$F_{ij} \sim \mathbf{f} := \mathbf{b} + i\mathbf{e}. \tag{8}$$

It can be shown that under Lorentz transformations of F_{ij}, $\mathbf{f}$ undergoes a complex rotation which preserves scalar and vector products of complex vectors. Since $\overset{*}{F}_{ij}$ results formally from the operation $\mathbf{e} \longmapsto \mathbf{b}$, $\mathbf{b} \longmapsto -\mathbf{e}$, we have

$$\overset{*}{F}_{ij} \sim i\mathbf{f} = -\mathbf{e} + i\mathbf{b}. \tag{9}$$

The above association can be used to define three important bivector operations:

$$\alpha \circ F \sim \alpha\mathbf{f}, \tag{10}$$

$$F \circ G = \mathbf{f} \cdot \mathbf{g} = G \circ F, \tag{11}$$

$$F \times G \sim \mathbf{f} \times \mathbf{g} \sim -G \times F, \tag{12}$$

where α is a *complex* scalar and $\mathbf{f} \times \mathbf{g}$ is the usual vector product whose first component is $f_2 g_3 - f_3 g_2$. Note that operations (10) and (12) yield a *real* bivector, while (11) yields a *complex* scalar. Note also that $\alpha \circ F = \alpha F$ when α is real.

Directly from these definitions follow such "obvious" properties as

$$(F + G) \circ H = F \circ H + G \circ H, \tag{13}$$

$$(F + G) \times H = F \times H + G \times H, \tag{14}$$

$$\alpha \circ (\beta \circ F) = (\alpha\beta) \circ F, \tag{15}$$

$$F \circ (\alpha \circ G) = \alpha(F \circ G), \tag{16}$$

$$\alpha \circ F = 0 \Longleftrightarrow \alpha = 0 \quad \text{or} \quad F = 0, \tag{17}$$

$$F \times G = 0 \Longleftrightarrow F = \alpha \circ G \quad (\text{for some } \alpha), \tag{18}$$

$$\alpha \circ (F \times G) = (\alpha \circ F) \times G = F \times (\alpha \circ G), \tag{19}$$

$$F \circ (G \times H) = (F \times G) \circ H = G \circ (H \times F), \tag{20}$$

$$F \circ (G \times H) = 0, G \neq 0 \neq H \Longrightarrow F = \alpha \circ G + \beta \circ H, \tag{21}$$

$$F \times (G \times H) = (F \circ G) \circ H - (F \circ H) \circ G, \tag{22}$$

$$(F \times G) \circ (H \circ J) = (F \circ J)(G \circ H) - (F \circ H)(G \circ J), \tag{23}$$

and also the lemma needed later

$$F \circ F = 0,\ G \circ G = 0,\ F \circ G = 0 \Longrightarrow F = \alpha \circ G \quad (\text{for some } \alpha). \tag{24}$$

Proof: Let $F \sim \mathbf{f} = \mathbf{b} + \mathbf{ie}$ and $G \sim \mathbf{g}$ and choose the z-axis normal to $\mathbf{e}$ and $\mathbf{b}$ so that, in accordance with $F \circ F = 0$, $\mathbf{f} = (f_1, if_1, 0)$; then $F \circ G = 0$ implies $g_1 + ig_2 = 0$, whereupon $G \circ G = 0$ gives $g_3 = 0$, and thus $\mathbf{g} = (g_1, ig_1, 0)$, which establishes the result.

We also have, because of (9), the following identities involving duals:

$$\overset{*}{F} = i \circ F, \qquad \overset{**}{F} = -F, \tag{25}$$

$$F \circ G = -i\overset{*}{F} \circ G = -\overset{*}{F} \circ \overset{*}{G}, \tag{26}$$

$$F \times G = -\overset{*}{F} \times \overset{*}{G} = -i \circ (\overset{*}{F} \times G), \tag{27}$$

$$(F \times G)^* = \overset{*}{F} \times G = F \times \overset{*}{G}, \tag{28}$$

Next we establish the explicit forms for the three operations defined in (10)–(12). They are, with $\alpha = \alpha_1 + i\alpha_2$, α_1 and α_2 real,

$$\alpha \circ F = \alpha_1 F + \alpha_2 \overset{*}{F} \tag{29}$$
$$= \frac{1}{2}\alpha(F - i\overset{*}{F}) + \frac{1}{2}\overline{\alpha}(F + i\overset{*}{F}),$$

$$F \circ G = F : G - i\overset{*}{F} : G, \tag{30}$$

$$F \times G = F \cdot G - G \cdot F. \tag{31}$$

For proof of (29), we have

$$\alpha \circ F \sim \alpha \mathbf{f} = (\alpha_1 + i\alpha_2)\mathbf{f} = a_1\mathbf{f} + \alpha_2(i\mathbf{f}) \tag{32}$$
$$\sim \alpha_1 F + \alpha_2 \overset{*}{F}.$$

To establish (30) we use (5) and (9):

$$F \circ G = \mathbf{f} \cdot \mathbf{g} = (b + ie) \cdot (b' + ie'), \tag{33}$$
$$= b \cdot b' - e \cdot e' + i(e \cdot b' + b \cdot e')$$
$$= F : G - i\overset{*}{F} : G.$$

From (30) and the symmetry of $F \circ G$ and $F : G$, we incidentally deduce

$$\overset{*}{F} : G = F : \overset{*}{G}, \tag{34}$$

which is also obvious directly.

The simplest way to prove (31) is to compute the matrix products on the right-hand side. In the same way we can justify the important identity

$$F \cdot G - \overset{*}{G} \cdot \overset{*}{F} = (\overset{*}{F} : G)\delta. \tag{35}$$

Many calculations are aided by the introduction of the *standard bivector triad* $I,\ J,\ K$, defined in terms of the right-handed triad of real unit 3-vectors $\mathbf{x},\ \mathbf{y},\ \mathbf{z}$ along the similarly named spatial axes:

$$I,\ J,\ K \sim \mathbf{x},\ \mathbf{y},\ \mathbf{z}. \tag{36}$$

Then

$$\overset{*}{I},\ \overset{*}{J},\ \overset{*}{K} \sim i\mathbf{x},\ i\mathbf{y},\ i\mathbf{z}, \tag{37}$$

and evidently $I,\ J,\ K,\ \overset{*}{I},\ \overset{*}{J},\ \overset{*}{K}$ serve as a basis for bivectors, since every complex vector like $\mathbf{f}$ in (8) can be uniquely written in the form $a\mathbf{x}+b\mathbf{y}+c\mathbf{z}+i(d\mathbf{x}+e\mathbf{y}+f\mathbf{z})$ for some real numbers $a, \ldots, f$. Directly from the definitions we have

$$I \times J = K, \qquad J \times K = I, \qquad K \times I = J \tag{38}$$

and

$$I^2 := I \circ I = -1, \text{etc.}, \quad \text{and} \quad I \circ J = 0, \text{etc.} \tag{39}$$

Since **x** is the "**b**" of I, we have, from (1), $I_{23} = -I_{32} = 1$, and all other components vanish. Consequently, I can be expressed as in the first of the following formulae; the other formulae are obtained similarly (with t denoting the unit vector along the time axis):

$$\begin{aligned} I &= yz - zy, & \overset{*}{I} &= xt - tx, \\ J &= zx - xz, & \overset{*}{J} &= yt - ty, \\ K &= xy - yx, & \overset{*}{K} &= zt - tz. \end{aligned} \tag{40}$$

Once again, we use an index-free notation for what would otherwise be written $I_{ij} = y_i z_j - z_i y_j$, etc. The vectors in (40) are to be understood as *four*-vectors: $t = (1, 0, 0, 0)$, $x = (0, 1, 0, 0)$, etc., whereas those in (36) are *three*-vectors $\mathbf{x} = (1, 0, 0)$, etc.

28.3 Basic Geometry of Bivectors

We say a bivector F is *simple* if it can be expressed as the commutator of two 4-vectors, say a and b,

$$F = ab - ba, \tag{41}$$

in which case it is said to *span* the plane of a and b. As one easily verifies, F can then be equally well expressed as the commutator of any other vector pair in the same plane, suitably scaled. One necessary and sufficient condition for the simplicity of F is

$$F : \overset{*}{F} = 0. \tag{42}$$

For proof, we first write this out in full:

$$F_{ij}F_{kl} + F_{ik}F_{lj} + F_{il}F_{jk} = 0. \tag{43}$$

Now, unless $F_{ij} = 0$ (a case of no interest), we can obviously find two vectors c and d such that $F_{ij}c^i d^j = 1$. Transvecting (43) with $c^i d^j$ then yields

$$F_{kl} = (F_{ki}c^i)(F_{lj}d^j) - (F_{kj}d^j)(F_{li}c^i), \tag{44}$$

which has the form $a_k b_l - b_k a_l$, as required. Conversely, direct substitution of (41) into the totally antisymmetric left-hand side of (43) yields 12 terms of type $\pm\, a_i a_j b_k b_l$, and thus a multiple of $a_{[i}a_j b_k b_{l]}$ which vanishes.

The symmetry $F : \overset{*}{F} = \overset{*}{F} : F$ (cf. (34)) shows that the simplicity of F implies that of $\overset{*}{F}$ and vice versa. Examples are provided by the triad bivectors I, J, K, and their duals.

We next establish the following alternative conditions for simplicity:

$$F : \overset{*}{F} = 0 \Leftrightarrow F \cdot \overset{*}{F} = 0 \Leftrightarrow F \circ F = \text{real} \Leftrightarrow \overset{*}{F} \circ \overset{*}{F} = \text{real} \Leftrightarrow \mathbf{e} \cdot \mathbf{b} = 0. \tag{45}$$

For proof of the second condition, we put $G = \overset{*}{F}$ in (34) and use (25)(ii) to find

$$F : \overset{*}{F} = \frac{1}{2}(F : \overset{*}{F})\delta = \overset{*}{F} \cdot F. \tag{46}$$

The third and fourth conditions follow from (30) and (26), while the last follows from $F \circ F = \mathbf{f} \cdot \mathbf{f}$.

A bivector F satisfying $F \circ F = 0$ is called *null*. The following are equivalent conditions for nullity:

$$F \circ F = 0 \Leftrightarrow F : F = 0 \quad \text{and} \quad F : \overset{*}{F} = 0 \Leftrightarrow \mathbf{e} \cdot \mathbf{b} = 0 \quad \text{and} \quad e^2 = b^2, \tag{47}$$

as is clear from (33); (45) shows that every *null bivector is simple*.

Two simple bivectors F and G are *orthogonal* (in the sense of spanning orthogonal planes) if and only if

$$F \cdot G = 0. \tag{48}$$

For suppose $F = ab - ba$ and $G = cd - dc$. Then (48) reads (with indices where helpful):

$$\begin{aligned} 0 &= (ab - ba) \cdot (cd - dc) \\ &= (b \cdot c)ad - (b \cdot d)ac - (a \cdot c)bd + (a \cdot d)bc \\ &= [(b \cdot c)a_i - (a \cdot c)b_i]d_j + [(a \cdot d)b_i - (b \cdot d)a_i]c_j. \end{aligned}$$

Now, for *fixed* i this is true for each j; but c and d are linearly independent, so the expressions in the square brackets vanish, e.g.,

$$(b \cdot c)a_i - (a \cdot c)b_i = 0 \tag{49}$$

But, also, a and b are linearly independent, so c is orthogonal to both a and b; and the same follows for d from the vanishing of the second square bracket. Since this proof also works in reverse, the necessity and sufficiency of the orthogonality condition (48) is established. Note in particular, from (45), that *a simple bivector is always orthogonal to its dual.* Conversely, since the plane orthogonal to a given plane is unique, it follows that *any pair of simple orthogonal bivectors must be the duals of each other, up to a numerical factor.*

We next classify simple bivectors F into three types: *timelike, null, or spacelike*, according to whether the plane they span contains two null vectors, one null vector, or no null vector. We can easily establish the following criteria (note that the reality of $F \circ F$ is guaranteed by (45)):

$$\begin{aligned} F \circ F > 0 &\Leftrightarrow \text{timelike}, \\ F \circ F = 0 &\Leftrightarrow \text{null}, \\ F \circ F < 0 &\Leftrightarrow \text{spacelike}. \end{aligned} \tag{50}$$

For suppose $F = ab - ba$. Then from (30) and the reality of $F \circ F$ we have

$$\begin{aligned} F \circ F = F : F &= (ab - ba) : (ab - ba) \\ &= (a \cdot b)^2 - (a \cdot a)(b \cdot b). \end{aligned} \tag{51}$$

In order that a null vector $n = ua + vb$ exist in the plane of F, we need two real numbers u and v to satisfy the equation

$$n \cdot n = u^2(a \cdot a) + 2uv(a \cdot b) + v^2(b \cdot b) = 0. \tag{52}$$

But the discriminant of this quadratic is just the right-hand side of (51), so that (50) is established.

From (26) it follows at once that the nullity of either F or $\overset{*}{F}$ implies that of the other, and that if one of these is spacelike, the other is timelike and vice versa.

Consider next, an arbitrary but *non-null* bivector F. Then it is easy to see that there exists a unique (up to sign) *unit timelike* bivector A ($A^2 = 1$) and a unique (up to sign) complex number α such that

$$F = \alpha \circ A. \tag{53}$$

Evidently

$$A = \alpha^{-1} \circ F, \tag{54}$$

with

$$\alpha^2 = F \circ F. \tag{55}$$

Suppose that for a second non-null bivector G we have, analogously, $G = \beta \circ B$. Let us then define (again, only up to sign) a quantity μ that looks like a cosine between F and G by the equation

$$\mu \equiv \cos(F, G) := \frac{F \circ G}{(F \circ F)^{1/2}(G \circ G)^{1/2}} \tag{56}$$

This is invariant under $F \longmapsto \alpha^{-1} \circ F,\ G \longmapsto \beta^{-1} \circ G$, so $\mu = \pm\, A \circ B$, and we can tie the various sign ambiguities together by defining

$$\mu = A \circ B. \tag{57}$$

Now suppose

$$A = ln - nl, \qquad B = pq - qp, \tag{58}$$

where l, n, p, q are the null vectors in the planes of A and B respectively. If $A : \overset{*}{B} \neq 0$, we have

$$\eta_{ijkl} l^i n^j p^k q^l \neq 0, \tag{59}$$

which shows that the determinant of the components of l, n, p, q is nonzero; this in turn means that these four null vectors span the entire 4-space and A and B do not intersect. Here we shall rather concentrate on the case (cf. (30))

$$A : \overset{*}{B} = 0 \Leftrightarrow A \circ B = \mu = \text{real}. \tag{60}$$

This implies a linear relation between the null vectors l, n, p, q:

$$al + bn = cp + dq, \tag{61}$$

which means that the planes of A and B have exactly one line in common, unless they coincide, a case we shall exclude from our discussion.

An important corollary is that (60) *is the necessary and sufficient condition for the intersection of the planes of any two simple bivectors A and B*, regardless of their type, since we have nowhere used the nullity of l, n, p, q so far. Thus, in particular, by (26), the planes of a *null* bivector A and its dual $\overset{*}{A}$ *always* intersect. Moreover, that intersection is a null vector: for, lying in $\overset{*}{A}$, it is orthogonal to every vector in A and consequently to itself. Thus A and $\overset{*}{A}$ share their unique null vector.

The intersection of two *timelike* bivectors can be timelike, null, or spacelike. We shall now prove that these three cases correspond, respectively, to

$$\mu^2 < 1, \qquad \mu^2 = 1, \qquad \mu^2 > 1. \tag{62}$$

From (30) and (58) we find the following equivalents of the unicity requirements on A and B:

$$\begin{aligned} A \circ A = 1 \Leftrightarrow A : A = 1 \Leftrightarrow l \cdot n = \pm 1, \\ B \circ B = 1 \Leftrightarrow B : B = 1 \Leftrightarrow p \cdot q = \pm 1. \end{aligned} \tag{63}$$

We can take the products $l \cdot n$ and $p \cdot q$ to be $+1$, if necessary by replacing l by $-l$ and p by $-p$ in (58). The case when p and q coincide with l and n has been excluded, so that without loss of generality, we may assume l, n, p to be distinct. Since distinct null vectors cannot be orthogonal, we may then rescale l, n, p, q as follows:

$$\begin{aligned} l' &= \lambda(n \cdot p)l, \\ n' &= \lambda(l \cdot p)n, \\ p' &= \lambda p, \\ q' &= \lambda^{-1} q, \end{aligned} \tag{64}$$

with

$$\lambda^2 (n \cdot p)(l \cdot p) = 1. \tag{65}$$

Then A and B still have the form $l'n' - n'l'$ and $p'q' - p'q'$, respectively, but, dropping primes, we now also have

$$l \cdot n = l \cdot p = n \cdot p = p \cdot q = 1. \tag{66}$$

Since l, n, p are linearly independent by hypothesis, (61) shows that q can be expressed linearly in terms of these vectors. We therefore write, for some real number ν,

$$q = \frac{1}{2}(1 - \nu)l + \frac{1}{2}(1 + \nu)n - \frac{1}{4}(1 - \nu^2)p, \tag{67}$$

a form dictated by the requirements $q \cdot p = 1$ and $q^2 = 0$. In conjunction with (66) this gives

$$q \cdot l = \frac{1}{4}(1 + \nu)^2, \qquad q \cdot n = \frac{1}{4}(1 - \nu)^2, \tag{68}$$

which leads us to discover

$$\mu = A \circ B = A : B = (n \cdot p)(l \cdot q) - (n \cdot q)(l \cdot p) = \nu. \tag{69}$$

Thus, by rewriting (67) in the form

$$\frac{1}{2}(1-\mu)l + \frac{1}{2}(1+\mu)n = \frac{1}{4}(1-\mu^2)p + q, \tag{70}$$

we actually exhibit the vector common to the planes of A and B. Since its squared length is $\frac{1}{2}(1-\mu^2)$, our assertion (62) is established.

We shall later need the following lemma.

Lemma: *Given any non-null bivector F, there exist two null bivectors U, V such that $F = U \times V$ and U, V are unique up to $U \longmapsto \lambda \circ U,\ V \longmapsto \lambda^{-1} \circ V$.*

Proof: First, we establish the existence part of the lemma for a simple unit timelike bivector $A(A^2 = 1)$. Without loss of generality, we can take A to be the $\overset{*}{K}$ of the bivector basis $I \ldots \overset{*}{K}$ (cf. (36)–(40)), and then we have

$$A = \overset{*}{K} = \overset{*}{I} \times J = \frac{1}{\sqrt{2}}(\overset{*}{I} - J) \times \frac{1}{\sqrt{2}}(\overset{*}{I} + J) =: U \times V. \tag{71}$$

The general existence result now follows from the fact (see (53)) that every non-null bivector F is expressible in the form $F = \alpha \circ A$ for some such A as above. Accordingly $F = \alpha \circ (U \times V) = (\alpha \circ U) \times V =: \tilde{U} \times V,\ \tilde{U}^2 = V^2 = 0$.

To prove the unicity of the splitting up to scaling, suppose

$$F = U \times V = P \times Q, \qquad U^2 = V^2 = P^2 = Q^2 = 0. \tag{72}$$

Then (see (23))

$$F^2 = (U \circ V)^2 = (P \circ Q)^2, \tag{73}$$

and we also have

$$U \times (U \times V) = U \times (P \times Q), \tag{74}$$

i.e., (see (22)),

$$-(U \circ V) \circ U = (U \circ P) \circ Q - (U \circ Q) \circ P, \tag{75}$$

which when squared yields

$$0 = -2(U \circ P) \circ (U \circ Q) \circ (P \circ Q). \tag{76}$$

Since $P \circ Q = \sqrt{F^2} \neq 0$, we must have $U \circ P = 0$ or $U \circ Q = 0$, whence (see (24)) $U = \lambda \circ P$ or $U = \lambda \circ Q$ for some λ. Similarly, we find $V = \mu \circ P$ or $V = \mu \circ Q$ for some μ. But $U \circ V = P \circ Q$, so $\lambda\mu = 1$, and unicity up to this scaling is established.

Collecting our results and adding some easily established extra ones, we have

$$F = U \times V,\ U^2 = V^2 = 0$$

$$U \circ V = \sqrt{F^2} =: \alpha, \qquad F \times U = \alpha \circ U, \qquad F \times V = -\alpha \circ V,$$
$$F \circ U = 0, \quad F \circ V = 0. \tag{77}$$

In light of our earlier result (between (61) and (62)) on the intersection of a null A and its dual, our present lemma implies that uniquely associated with every non-null bivector F are two null directions, these being, respectively, the intersections of the pencils of null planes $\lambda \circ U = \lambda_1 U + \lambda_2 \overset{*}{U}$ $(\lambda = \lambda_1 + i\lambda_2)$ and $\mu \circ V$. Associated with a *null* bivector there is just *one* such null direction.

28.4 Biquaternions and the Bivector Transformations They Generate

We define a *biquaternion* $\mathcal{A}$ as a pair

$$\mathcal{A} = (\alpha, A) \tag{78}$$

consisting of a complex number α and a real bivector A. In analogy with the usual quaternion product, we define

$$\mathcal{AB} = (\alpha, \mathcal{A})(\beta, \mathcal{B}) = (\alpha\beta + \mathcal{A} \circ \mathcal{B},\ \beta \circ \mathcal{A} + \alpha \circ \mathcal{B} + \mathcal{A} \times \mathcal{B}) \tag{79}$$

and note that this product commutes if and only if $A \times B = 0$. The sum of two biquaternions is defined in the obvious way:

$$\mathcal{A} + \mathcal{B} = (\alpha + \beta,\ \mathcal{A} + \mathcal{B}), \tag{80}$$

and biquaternion multiplication is then easily checked to be distributive from the left,

$$\mathcal{A}(\mathcal{B} + \mathcal{C}) = \mathcal{AB} + \mathcal{AC}, \tag{81}$$

and similarly from the right. Scalars and bivectors can be regarded as special cases of biquaternions: $\alpha = (\alpha, 0)$, $A = (0, A)$, $\mathcal{A} = \alpha + A$. In particular, multiplication by a *real* number yields

$$a\mathcal{A} = (a\alpha, a\mathcal{A}). \tag{82}$$

It is a little harder to establish that multiplication is associative:

$$(\mathcal{AB})\mathcal{C} = \mathcal{A}(\mathcal{BC}). \tag{83}$$

We shall first establish this for a special case. Consider

$$\begin{aligned} [AB]C &= (A \circ B,\ A \times B)\, C \\ &= ((A \times B) \circ C,\ (A \circ B) \circ C + (A \times B)\ \times C) \end{aligned} \tag{84}$$

and

$$\begin{aligned} A[BC] &= A(B \circ C,\ B \times C) \\ &= (A \circ (B \times C),\ (B \circ C) \circ A + A \times\ (B \times C)). \end{aligned} \tag{85}$$

We see from (20) that the *scalars* on the right-hand sides of (84) and (85) are equal. The equality of the bivectors is established by repeated use of (22):

$$\begin{aligned}(A \circ B) \circ C + (A \times B) \times C &= (A \circ B) \circ C - C \times (A \times B) \\ &= (A \circ B) \circ C - (C \circ A) \circ B + (C \circ B) \circ A \\ &= A \times (B \times C) + (B \circ C) \circ A.\end{aligned} \tag{86}$$

General associativity would follow if

$$[(\alpha + A)(\beta + B)](\gamma + C) = (\alpha + A)[(\beta + B)(\gamma + C)]$$

were true. But, from the distributive law and the already established associativity of the product ABC, this indeed *is* true.

We define the *transpose* $\mathcal{A}'$ of a biquaternion $\mathcal{A}$ as follows:

$$(\alpha, A)' = (\alpha, -A). \tag{87}$$

Then we have

$$\begin{aligned}[(\alpha, A)(\beta, B)]' &= (\alpha\beta + A \circ B, -\beta \circ A - \alpha \circ B - A \times B) \\ &= (\beta, -B)(\alpha, -A),\end{aligned}$$

i.e.,

$$(\mathcal{A}\mathcal{B})' = \mathcal{B}'\mathcal{A}', \tag{88}$$

just as for matrices. We note

$$(\alpha, A)(\alpha, A)' = (\alpha, A)'(\alpha, A) = (\alpha^2 - A \circ A, 0), \tag{89}$$

and we define the *norm* of (α, A) as

$$N(\alpha, A) = \alpha^2 - A \circ A. \tag{90}$$

Regarding this as a purely scalar biquaternion, we can also write

$$N(\mathcal{A}) = \mathcal{A}\mathcal{A}' = \mathcal{A}'\mathcal{A}. \tag{91}$$

It then follows (by appeal to associativity) that

$$\begin{aligned}N(\mathcal{A}\mathcal{B}) &= \mathcal{A}\mathcal{B}\mathcal{B}'\mathcal{A}' = \mathcal{A}N(\mathcal{B})\mathcal{A}' \\ &= N(\mathcal{A})N(\mathcal{B}).\end{aligned} \tag{92}$$

We call $\mathcal{A}$ *normed* if

$$N(\mathcal{A}) = \mathcal{A}\mathcal{A}' = \alpha^2 - A \circ A = 1. \tag{93}$$

It then follows from (92) that *normed biquaternions form a subgroup under multiplication.*

Now consider, for some normed $\mathcal{A} = (\alpha, A)$, the following transformation of a bivector X:

$$X \mapsto \hat{X} = \mathcal{A}'X\mathcal{A} = (\alpha, -A)(0, X)(\alpha, A). \tag{94}$$

This transformation, which will play a key role in our work, will ultimately be recognized as the most general proper orthochronous Lorentz transformation on bivectors. Multiplying out the right-hand side of (94) and using the normalization condition of (93), we find the explicit formula

$$\hat{X} = \mathcal{A}'X\mathcal{A} = X - 2\alpha \circ (A \times X) + 2A \times (A \times X), \tag{95}$$

which by use of (22) can also be written in the form

$$\hat{X} = (2\alpha^2 - 1) \circ X - 2\,(A \circ X) \circ A - 2\alpha \circ (A \times X). \tag{96}$$

An important property of (94) is that it preserves bivector products:

$$\hat{X} \circ \hat{Y} = X \circ Y. \tag{97}$$

This follows from the biquaternion identity $\hat{X}\hat{Y} = \mathcal{A}' \times \mathcal{A}\mathcal{A}'Y\mathcal{A} = \mathcal{A}'XY\mathcal{A}$, whose scalar part is (97).

If we apply two transformations of type (94) in succession, say

$$\hat{X} = \mathcal{A}'X\mathcal{A}, \qquad \hat{\hat{X}} = \mathcal{B}'\hat{X}\mathcal{B}, \tag{98}$$

we get

$$\hat{\hat{X}} = \mathcal{B}'\mathcal{A}'X\mathcal{A}\mathcal{B} = (\mathcal{A}\mathcal{B})'X(\mathcal{A}\mathcal{B}), \tag{99}$$

which shows that the result of applying *first* $\mathcal{A}$, *then* $\mathcal{B}$ is equivalent to applying $\mathcal{A}\mathcal{B}$. In particular, since $\mathcal{A}\mathcal{A}' = \mathcal{A}'\mathcal{A} = 1$, $\mathcal{A}$ and $\mathcal{A}'$ generate inverse transformations. These *transformations* thus form a group. Evidently, $\pm\mathcal{A}$ define the same transformation. Thus, since $\mathcal{A}' = (\alpha, -A)$ represents the inverse of $\mathcal{A}$, so does $(-\alpha, A)$.

It is clear from (95) that a transformation by $\mathcal{A} = (\alpha, A)$ leaves invariant any bivector $X = \lambda \circ A$, i.e., any real linear combination of A and $\overset{*}{A}$. Let us inquire into the *most general* bivector left invariant up to scale: $X \mapsto aX$, a real. The invariance of $X \circ X$ implies that $a = \pm 1$ or $X \circ X = 0$. The case $a = -1$ can occur only for very special transformations $\mathcal{A}$, since $\mathcal{A}'X\mathcal{A} = -X$ is equivalent to $X\mathcal{A} = -\mathcal{A}X$, i.e.,

$$(X \circ A, \alpha \circ X + X \times A) = -(X \circ A, \alpha \circ X + A \times X), \tag{100}$$

which implies (remember, A is normed)

$$\alpha = 0, \qquad A \circ A = -1, \qquad A \circ X = 0. \tag{101}$$

As we shall see later, this represents a 180° rotation in the spacelike plane of A (cf. (149) below).

When $a = 1$ ($X \mapsto X$), we have a positive sign in front of the right-hand parenthesis in (100), which then implies $A \times X = 0$ and thus (cf. (18)) $X = \lambda \circ A$. This is the "obvious" set of invariant bivectors already spotted.

Lastly, suppose $X^2 = 0$, $a \neq \pm 1$. With now an a in front of the right-hand parenthesis in (100) we find, to start with,

$$X \circ A = 0. \tag{102}$$

If A is null, this leads again (cf. (24)) to $X = \lambda \circ A$, i.e., to the "obvious" set of invariant bivectors; so in this case, the only null bivectors left invariant up to real scaling are in fact left invariant. If A is non-null, let us split it according to (77):

$$A = U \times V, \qquad U^2 = V^2 = 0, \qquad A^2 = (U \circ V)^2 = \alpha^2 - 1 \neq 0. \tag{103}$$

Then (101) implies (cf. (21))

$$X = \lambda \circ U + \mu \circ V. \tag{104}$$

Squaring this gives

$$0 = (\lambda\mu) \circ (U \circ V). \tag{105}$$

But since $U \circ V \neq 0$, we must have $\lambda = 0$ or $\mu = 0$, i.e.,

$$X = \lambda \circ U \ \text{ or } \ X = \mu \circ V. \tag{106}$$

Suppose $X = \lambda \circ U$ and substitute this into the right-hand side of equation (96). For that right-hand side to be a real multiple of $\lambda \circ U$, we then need

$$2\alpha^2 - 1 \mp 2\alpha\sqrt{\alpha^2 - 1} = \text{real} \tag{107}$$

with the upper sign, while (107) with the lower sign applies to $\mu \circ V \mapsto a\,(\mu \circ V)$. Putting $\alpha = \cosh\phi$, we convert (107) into

$$\cosh\frac{\phi}{2} \mp \sinh\frac{\phi}{2} = e^{\mp\,\phi/2} = \text{real}, \tag{108}$$

which requires ϕ and with it, α, to be real and thus from (103) A to be simple and timelike. In this case, therefore, and only in this case, these are two families of null bivectors, $\lambda \circ U$ and $\mu \circ V$, each individual member of which maps into a real multiple of itself. We shall see the geometric interpretation of this state of affairs in Section 28.6 (after (157)).

We have already noted that $\pm\mathcal{A}$ generate the same transformation. We can now prove that no other biquaternion does. Suppose, contrariwise, that $\mathcal{A}'X\mathcal{A} = \mathcal{B}'X\mathcal{B}$, i.e., $X = (\mathcal{B}\mathcal{A}')'X(\mathcal{B}\mathcal{A}')$ for all X. Then X is of the form $\lambda \circ C$, C being the bivector part of $\mathcal{B}\mathcal{A}'$. But such a relation cannot hold for all X, so $C = 0$ and $\mathcal{B}\mathcal{A}'$ is a scalar γ. The normalization condition finally requires $\gamma = \pm 1$, i.e., $\mathcal{B} = \pm A$.

28.5 Lorentz Matrices

We begin this section by establishing two preliminary equations. If we subtract from both sides of equation (35) the product $G \cdot F$, we obtain

$$F \times G - \overset{*}{G} \cdot \overset{*}{F} = (F : G)\delta - G \cdot F, \tag{109}$$

which, on interchanging F and G, first yields

$$F \cdot G - \overset{*}{F} \cdot \overset{*}{G} = (F : G)\,\delta + (F \times G), \tag{110}$$

and then, on writing $\overset{*}{G}$ for G and using (25) and (28),

$$F \cdot \overset{*}{G} + \overset{*}{F} \cdot G = (F : \overset{*}{G})\delta + (F \times G)^*. \tag{111}$$

Next, corresponding to each biquaternion $\mathcal{A} = (\alpha, A)$, we define two complex tensors:

$$^{\oplus}\mathcal{A} = {}^{\oplus}(\alpha, A) = \alpha\delta + A - i\overset{*}{A} = \alpha\delta + 2\,{}^{+}A, \tag{112}$$

$$^{\ominus}\mathcal{A} = \overline{^{\oplus}\mathcal{A}} = \overline{\alpha}\delta + A + i\overset{*}{A} = \overline{\alpha}\delta + 2\,{}^{-}A, \tag{113}$$

where we have introduced, for later use, the standard notation

$$^{+}A = \frac{1}{2}(A - i\overset{*}{A}), \qquad {}^{-}A = \frac{1}{2}(A + i\overset{*}{A}). \tag{114}$$

Note that for the first time in this paper, we contemplate *complex* multiples of real bivectors, i.e., we are working in *complexified bivector space.*

The operations $\oplus$ and $\ominus$ act on biquaternions (or on scalars or bivectors regarded as special biquaternions) and make complex tensors out of them. They clearly satisfy

$$^{\oplus}(\mathcal{A}') = ({}^{\oplus}\mathcal{A})', \qquad {}^{\oplus}X = 2\,{}^{+}X, \tag{115}$$

and similarly for $\ominus$. The property that makes them important is the following:

$$^{\oplus}(\mathcal{A}\mathcal{B}) = {}^{\oplus}\mathcal{A}.{}^{\oplus}\mathcal{B}, \qquad {}^{\ominus}(\mathcal{A}\mathcal{B}) = {}^{\ominus}\mathcal{A}.{}^{\ominus}\mathcal{B}, \tag{116}$$

where $\mathcal{A}\mathcal{B}$ is the *biquaternion* product and ${}^{\oplus}\mathcal{A}.{}^{\oplus}\mathcal{B}$, ${}^{\ominus}\mathcal{A}.{}^{\ominus}\mathcal{B}$ are *tensor* products. Equation (116)(i) is a little tricky to establish, but equation (116)(ii) then follows immediately by taking conjugates. One key lemma in the proof of (116)(i) is the relation

$$\beta\,{}^{+}\mathcal{A} = {}^{+}(\beta \circ A), \tag{117}$$

derivable at once from (29) by use of the identities

$$({}^{+}A)^* = i\,{}^{+}A, \qquad ({}^{-}A)^* = -i\,{}^{-}A, \tag{118}$$

which follow from the definitions (114). A second needed lemma is the identity

$$4\,{}^{+}A.{}^{+}B = (A \circ B)\delta + 2\,{}^{+}(A \times B). \tag{119}$$

Its proof rests on (110) and (111):

$$\begin{aligned} \text{l.h.s.} &= (A - i\overset{*}{A}).(B - i\overset{*}{B}) \\ &= (A.B - \overset{*}{A}.\overset{*}{B}) - i\,(\overset{*}{A}.B + A.\overset{*}{B}) \\ &= (A : B)\delta + A \times B - i[(A : \overset{*}{B})\delta + (A \times B)^*] \\ &= \text{r.h.s.} \end{aligned} \tag{120}$$

We are now ready to establish (116)(i). We have

$$\begin{aligned} {}^{\oplus}\mathcal{A}.{}^{\oplus}\mathcal{B} &= (\alpha\delta + 2^{+}A)\,.\,(\beta\delta + 2^{+}B) \\ &= \alpha\beta\delta + 2\beta^{+}A + 2\alpha^{+}B + 4^{+}A.^{+}B, \end{aligned} \tag{121}$$

which, by virtue of (117) and (119), becomes

$$\begin{aligned} &= \alpha\beta\delta + 2^{+}(\beta \circ A) + 2^{+}(\alpha \circ B) + (A \circ B)\delta + 2^{+}(A \times B) \\ &= {}^{\oplus}(\mathcal{AB}), \end{aligned} \tag{122}$$

thus completing our proof.

The next important identity states that every ${}^{\oplus}\mathcal{A}$ commutes with every ${}^{\ominus}\mathcal{B}$:

$$ {}^{\oplus}\mathcal{A}.{}^{\ominus}\mathcal{B} ={}^{\ominus} \mathcal{B}.{}^{\oplus}\mathcal{A}. \tag{123}$$

For proof, we expand both sides in terms of ${}^{+}A$ and ${}^{-}B$, using the definitions (112) and (113), and subtract the one from the other, which results in $4({}^{+}A.{}^{-}B \;-\; {}^{-}B.{}^{+}A)$. Now this equals $4({}^{+}A \times^{-} B)$, since we can obviously extend the definition (31) of the vector product $F \times G$ to *complex* bivectors. Such identities as (28) must then continue to hold, since they must be algebraically derivable from (31) without regard to the reality or otherwise of the components. To finish our proof, we need to show ${}^{+}A \times^{-} B = 0$. Using (28) and (118), we have, on the one hand,

$$ ({}^{+}A \times^{-} B)^{*} = ({}^{+}A)^{*} \times^{-} B = i({}^{+}A \times^{-} B) \tag{124}$$

and, on the other hand,

$$ ({}^{+}A \times^{-} B)^{*} = {}^{+}A \times ({}^{-}B)^{*} = -i({}^{+}A \times^{-} B), \tag{125}$$

so that, for every two bivectors A and B,

$$ {}^{+}A \times^{-} B = 0. \tag{126}$$

This establishes (123). Of course, since ${}^{+}A$ and ${}^{-}B$ are not real, our earlier lemma (18) is not applicable here.

We next associate with any biquaternion $\mathcal{A} = (\alpha, A)$ a real tensor L defined as

$$ L\,(\mathcal{A}) ={}^{\oplus} \mathcal{A}.{}^{\ominus}\mathcal{A}. \tag{127}$$

Its reality follows from the definition (113) and the commutativity (123):

$$ \overline{L(\mathcal{A})} = \overline{{}^{\oplus}\mathcal{A}.{}^{\ominus}\mathcal{A}} = {}^{\ominus}\mathcal{A}.{}^{\oplus}\mathcal{A} = {}^{\oplus}\mathcal{A}.{}^{\ominus}\mathcal{A} = L(\mathcal{A}). \tag{128}$$

We also have, again using the commutativity (123) and the property (116),

$$\begin{aligned} L(\mathcal{A}).L(\mathcal{B}) &= {}^{\oplus}\mathcal{A}.{}^{\ominus}\mathcal{A}.{}^{\oplus}\mathcal{B}.{}^{\ominus}\mathcal{B} \\ &= {}^{\oplus}\mathcal{A}.{}^{\oplus}\mathcal{B}.{}^{\ominus}\mathcal{A}.{}^{\ominus}\mathcal{B} = {}^{\oplus}(\mathcal{AB}).{}^{\ominus}(\mathcal{AB}) \\ &= L\,(\mathcal{AB}). \end{aligned} \tag{129}$$

To get the explicit form of L, we calculate

$$\begin{aligned} L(\alpha, \mathcal{A}) &= (\alpha\delta + 2^{+}A).(\overline{\alpha}\delta + 2^{-}A) \\ &= \alpha\overline{\alpha}\delta + (2\overline{\alpha}^{+}A + 2\alpha^{-}A) + 4^{+}A.^{-}A. \end{aligned} \tag{130}$$

The term in parenthesis is recognized, via (114) and (29) as $2\overline{\alpha} \circ A$, while the last term, evaluated as after (88), is $A.A + \overset{*}{A}.\overset{*}{B}$. Hence

$$L(\alpha, A) = \alpha\overline{\alpha}\delta + A.A + \overset{*}{A}.\overset{*}{A} + 2\overline{\alpha} \circ A. \tag{131}$$

An important special case of (131) arises when A is simple. According to (45), $A \circ A$ is then real and consequently equal to $A : A$; cf. (30). If we add to or subtract from (131), the equation

$$0 = A.A - \overset{*}{A}.\overset{*}{A} - (A : A)\delta, \tag{132}$$

which follows from (110), we get, respectively,

$$L = (\alpha\overline{\alpha} - A \circ A)\delta + 2\overline{\alpha} \circ A + 2A.A \tag{133}$$

and

$$L = (\alpha\overline{\alpha} + A \circ A)\delta + 2\overline{\alpha} \circ A + 2\overset{*}{A}.\overset{*}{A}. \tag{134}$$

These equations simplify if we further require $\mathcal{A} = (\alpha, A)$ to be normed (cf. (93)), i.e., to satisfy

$$\alpha^2 - A \circ A = 1. \tag{135}$$

With A simple, this implies that α^2 is real and consequently that α is either real or purely imaginary,

$$\alpha = a, \quad \text{or} \quad \alpha = ib, \tag{136}$$

a and b being real. Applying (135) in these two cases to (133) and (134), respectively, yields the two equations

$$L = \delta + 2aA + 2A.A \quad (A \circ A \geq -1) \tag{137}$$

and

$$L = -\delta - 2bA^* + 2A^*.A^* \quad (A \circ A \leq -1), \tag{138}$$

which will be needed presently.

But first, we note from (131) that the *transpose* L' of L ($L_{ij} \mapsto L_{ij}$) corresponds to $A \mapsto -A$, whence

$$L'(\mathcal{A}) = L(\mathcal{A}'). \tag{139}$$

It therefore follows from (129) that

$$L'(\mathcal{A}).L(\mathcal{A}) = L(\mathcal{A}).L'(\mathcal{A}) = L(\mathcal{A}\mathcal{A}'), \tag{140}$$

and, in particular, when $\mathcal{A}$ is *normed*,

$$L'(\mathcal{A}).L(\mathcal{A}) = L(1) = \delta, \tag{141}$$

by (131). But this implies that $L(\mathcal{A})$ *is a Lorentz matrix!* For a Lorentz matrix $L = L^i{}_j$ is characterized by the property

$$\eta_{ij} = \eta_{pq} L^p_{\ i} L^q_{\ j}, \tag{142}$$

where η_{ij} are the coefficients of the Minkowski metric (6); and raising the index i in (142) yields

$$\delta^i_j = L^{\ i}_q L^q_{\ j}, \tag{143}$$

i.e., equation (141).

What we have shown here is that matrices generated as in (117) by normed biquaternions $\mathcal{A}$ are Lorentz matrices. For the converse, namely that every (restricted) Lorentz matrix can be so generated, we refer the reader to our earlier paper, [1] (after equation (52)).

The Lorentz transformations of a 4-vector x and *a* bivector X, respectively, can be written in an index-free way as

$$x \mapsto x.L, \qquad X \mapsto L'.X.L. \tag{144}$$

An interesting and important result is that the second of the transformations (144) is *identical* to our earlier transformation (94), $X \mapsto \mathcal{A}'X\mathcal{A}$. For proof, we have

$$\begin{aligned} L'.{}^{+}X.L &= {}^{\oplus}\mathcal{A}'.{}^{\ominus}\mathcal{A}'.\frac{1}{2}\,{}^{\oplus}X.{}^{\ominus}\mathcal{A}.{}^{\oplus}\mathcal{A} \\ &= \frac{1}{2}\,{}^{\oplus}\mathcal{A}'.{}^{\oplus}X.{}^{\ominus}\mathcal{A}'\,{}^{\ominus}\mathcal{A}.{}^{\oplus}\mathcal{A}. \\ &= \frac{1}{2}\,{}^{\oplus}\mathcal{A}'.{}^{\oplus}X.{}^{\oplus}\mathcal{A} \\ &= \frac{1}{2}\,{}^{\oplus}(\mathcal{A}'X\mathcal{A}) ={}^{+} (\mathcal{A}'X\mathcal{A}), \end{aligned} \tag{145}$$

and similarly

$$L'.{}^{-}X.L = {}^{-}(\mathcal{A}'X\mathcal{A}). \tag{146}$$

Addition of these two results then yields the advertized identity

$$L'.X.L = \mathcal{A}'X\mathcal{A}. \tag{147}$$

As one might expect, $\pm\mathcal{A}$ generate the *same* transformation matrix L, which can be seen, for example, from (131).

28.6 The Geometry of Lorentz Transformations

As we have just seen in Section 28.5, every normed biquaternion $\mathcal{A} = (\alpha, A)$ determines a unique Lorentz transformation (LT). By the normalization requirement on $\mathcal{A}$, therefore, a bivector A alone determines *two* LT's having opposite scalars,

$\pm\,\alpha$, which, as we have seen (in the first paragraph after (99)), correspond to mutually inverse transformations. Apart from this ambiguity, we may thus consider *bivectors* to generate LT's.

We saw at the send of Section 28.4 that *every* LT, generated by a bivector A, leaves invariant precisely the bivector family $\lambda \circ A$, i.e., A, $\overset{*}{A}$, and all real linear combinations of these. Transformations generated by a simple *timelike* bivector A additionally leave invariant *up to scale* the two families of *null* bivectors $\lambda \circ U$ and $\mu \circ V$ determined by the splitting $A = U \times V$.

Now we know from (77) that every *non-null* A can be uniquely expressed as $A = \beta \circ B$, $B^2 = 1$, so that the linear combinations of A and $\overset{*}{A}$ are identical to those of B and $\overset{*}{B}$. Moreover, by the unicity of B (up to sign), B and $\overset{*}{B}$ are the *only* simple bivectors in $\lambda \circ A$. In the case of a *null* bivector A, every bivector $\lambda \circ A$ is null and therefore simple.

Geometrically, therefore, since simple bivectors correspond to 2-planes, we see that transformations generated by a *generic* bivector A leave invariant precisely two 2-planes B and $\overset{*}{B}$, one timelike and one spacelike. Those generated by a timelike bivector $A = U \times V$ additionally leave invariant the two entire pencils of null 2-planes generated by U and V, and transformations generated by a null bivector A leave invariant the entire pencil of null 2-planes $\lambda \circ A$.

Moreover, in the case when A is *simple* and $A \circ A \geq -1$ (which includes the null case), the plane $\overset{*}{A}$ is left *vectorwise* invariant. For if a vector x lies in $\overset{*}{A}$, it is orthogonal to A and therefore satisfies $x \cdot A = 0$. With that, the transformation equations (144), (137) yield $\hat{x} = x \cdot L = x$, thereby establishing our assertion. Similarly, if $A \circ A \leq -1$, the plane A itself is left vectorwise invariant.

We shall show at the end of this section that the most general transformation, i.e., one generated by a nonsimple bivector, can be reduced to the product of two "simple" transformations—our term for LT's generated by simple bivectors. It will therefore suffice to study the latter, i.e., LT's with generators satisfying the condition $A^2 =$ real. However, even among these, the class with $A^2 < -1$ can be reduced to products of transformations with

$$A^2 \geq -1. \tag{148}$$

For suppose

$$A = \sqrt{1+\gamma^2}C, \qquad C^2 = -1, \qquad \gamma = \text{real}.$$

It is then easily verified that the corresponding normed biquaternion $\mathcal{A} = (\alpha, A)$ splits as follows:

$$\mathcal{A} = (\alpha, A) = (\sqrt{1+\gamma^2}, \gamma\overset{*}{C})(0, C) \tag{149}$$

into two components satisfying the restriction (148); they will presently be recognized as a boost in the plane of $\overset{*}{C}$ (same as the plane of $\overset{*}{A}$) and a 180°-rotation in the plane of C (i.e., of A).

Let us, therefore, concentrate on the LT's generated by A's satisfying not only the simplicity condition A^2 = real but also condition (148), $A^2 = -1$. And let us at first exclude null A. We can then write

$$A = \sinh \frac{\phi}{2} \circ B, \qquad B \circ B = 1, \tag{150}$$

thereby defining ϕ. For timelike A, ϕ is real, while for spacelike A subject to (148) we can set $\phi = i\theta$, $\sinh \frac{\phi}{2} = i \sin \frac{\theta}{2}$, θ = real.

The normalization requirement then yields

$$\alpha = \pm \cosh \frac{\phi}{2}, \tag{151}$$

and we arbitrarily choose of the corresponding mutually inverse LT's that with the positive sign. Then the transformation equation (95) reads

$$\hat{X} = X - \sinh \phi_\circ (B \times X) + 2 \sinh^2 \frac{\phi}{2} \circ B \times (B \times X). \tag{152}$$

Thus, in particular, for the null bivector U associated with B via $B = U \times V$, we have (cf. (77))

$$\hat{U} = U - \sinh \phi \circ U + 2 \sinh^2 \frac{\phi}{2} \circ U = e^{-\phi} \circ U,$$

while for V we have $\hat{V} = e^{\phi} \circ V$. The LT generated by a simple non-null bivector subject to (148) can thus be reduced to the form

$$\hat{U} = -e^{-\phi} \circ U, \qquad \hat{V} = e^{\phi} \circ V. \tag{153}$$

If we choose coordinates so that $B = \overset{*}{K}$ (cf. (71)), we have

$$\begin{aligned} \sqrt{2}\, U &= \overset{*}{I} - J = x(t+z) - (t+z)x, \\ \sqrt{2}\overset{*}{U} &= -I - \overset{*}{J} = -y(t+z) + (t+z)y, \\ \sqrt{2}\, V &= \overset{*}{I} + J = x(t-z) - (t-z)x, \\ \sqrt{2}\overset{*}{V} &= -I + \overset{*}{J} = y(t-z) - (t-z)y. \end{aligned} \tag{154}$$

For *timelike* A, ϕ is real, the vector x lies in $K = -\overset{*}{B} = -(\sinh \phi)^{-1} \overset{*}{A}$, and *all* vectors in $\overset{*}{A}$ remain fixed. Consequently (153) with (154) implies

$$\hat{t} + \hat{z} = e^{-\phi}(t+z), \qquad \hat{t} - \hat{z} = e^{\phi}(t+z), \tag{155}$$

while the constancy of K corresponds to

$$\hat{x} = x, \qquad \hat{y} = y. \tag{156}$$

In (155), (156) we recognize an active boost in the direction of z through a rapidity ϕ. Since this transformation was generated by the biquaternion $(\cosh \frac{\phi}{2}, \sinh \frac{\phi}{2} B)$, $B = \overset{*}{K} \sim i\mathbf{z}$, we see that, in general, the biquaternion

$$\mathcal{A}_{\phi,\mathbf{d}} = \left(\cosh \frac{\phi}{2}, \sinh \frac{\phi}{2} B\right), \qquad B \sim i\mathbf{d}, \tag{157}$$

generates a ϕ-boost in the direction **d**.

The two pencils of invariant null planes $\lambda \circ U$ and $\mu \circ V$, whose existence in the case of timelike generators we have already discussed in Section 29.4 (after (108)), are now recognized geometrically: they are all those planes formed by either of the two invariant null directions in A ($t \pm z$ in the coordinatization of (154)) together with any one of the invariant vectors in $\overset{*}{A}$.

For a *spacelike* A, (153) can conveniently be written

$$\hat{U} = U\cos\theta - \sin\theta \overset{*}{U}, \qquad \hat{\overset{*}{V}} = \overset{*}{V}\cos\theta - V\sin\theta. \tag{158}$$

Since now all vectors in the plane of B $(= \overset{*}{K})$ remain invariant, z and t remain invariant. With that, and with (154), the equations (158) are equivalent to

$$\hat{x} = x\cos\theta + y\sin\theta, \; \hat{y} = -x\sin\theta + y\cos\theta, \tag{159}$$

and we have already observed that

$$\hat{z} = z, \qquad \hat{t} = t. \tag{160}$$

We recognize in this transformation, regarded as active, a pure rotation of the 3-space of **x**, **y**, **z** about the **z**-axis through an angle θ. The transformation was generated by the biquaternion $(\cos\frac{\theta}{2}, \sin\frac{\theta}{2}\overset{*}{B})$, $\overset{*}{B} = -K \sim -\mathbf{z}$. In general, therefore, a θ-rotation about a unit vector **d** in **x**, **y**, **z** space is generated by

$$\mathcal{A}_{\theta,\mathbf{d}} = \left(\cos\frac{\theta}{2}, \; \sin\frac{\theta}{2}C\right), \qquad C \sim -\mathbf{d}. \tag{161}$$

As the last of the simple generators, let us now suppose A is null; the normalization requirement (93) then implies $\alpha = \pm 1$ and we shall choose $\alpha = 1$. We shall jointly discuss the entire ray of generators cA, $c =$ real. Recall that A and $\overset{*}{A}$ intersect in their unique null vector, say l. Since A and $\overset{*}{A}$ are invariant, so is l. Pick two more vectors x and y (necessarily spacelike and orthogonal) in $\overset{*}{A}$ and A, respectively. Since $\overset{*}{A}$ is vectorwise invariant, x is invariant. Let us define a null vector n uniquely by the requirements

$$n \cdot n = n \cdot x = n \cdot y = 0, \qquad n \cdot l = 1, \tag{162}$$

and we already have

$$x.y = x.l = y.l = l.l = 0, \qquad x.x = y.y = -1. \tag{163}$$

By suitably adjusting the scale of l, we can write

$$A = ly - yl. \tag{164}$$

Then $A \cdot A = ll$ and, with $\mathcal{A} = (1, cA)$, the form (137) of L yields, for y,

$$\hat{y} = y \cdot L = y + 2cy \cdot (ly - yl) + 2c^2 y \cdot ll = y + 2cl, \tag{165}$$

and similarly we obtain the $\hat{n}$ in the following collected results:

$$\begin{aligned} \hat{x} &= x, \\ \hat{y} &= y + 2cl, \\ \hat{l} &= l, \\ \hat{n} &= \hat{n} + 2cy + 2c^2 l. \end{aligned} \tag{166}$$

This transformation represents what is called a *null rotation* "about" the (null) plane of x and l, i.e., of $\overset{*}{A}$, which is left vectorwise invariant. All other planes containing l and a vector from the x, y plane transform into themselves. The tetrad of vectors x, y, l, n is related to the usual orthonormal tetrad t, x, y, z as follows:

$$z = (l - n)/\sqrt{2}, \qquad t = (l + n)/\sqrt{2}. \tag{167}$$

The remaining task now is to establish our earlier claim (above (148)) that every LT generated by a nonsimple, and therefore non-null bivector A is the product of two *simple* LT's. In fact, we shall demonstrate that every such LT is *uniquely* the product of a boost and a rotation in orthogonal planes, these planes corresponding to the unique simple bivectors B and $\overset{*}{B}$ in the family $\lambda \circ A$.

Since we assume $A^2 \neq 0$, we can once more set $A = \beta \circ B,\ B^2 = 1$, and then we have, because of the normalization of $\mathcal{A} = (\alpha, A)$,

$$A^2 = \beta^2, \qquad \alpha^2 = 1 + \beta^2. \tag{168}$$

Let us define real parameters θ and ϕ by the first of the following equations, the second then following from (168):

$$\alpha = \cosh\frac{1}{2}(\phi + i\theta) = \cosh\frac{\phi}{2}\cos\frac{\theta}{2} + i\sinh\frac{\phi}{2}\sin\frac{\theta}{2}, \tag{169}$$

$$\beta = \sinh\frac{1}{2}(\phi + i\theta) = \sinh\frac{\phi}{2}\cos\frac{\theta}{2} + i\cosh\frac{\phi}{2}\sin\frac{\theta}{2}. \tag{170}$$

If we further define two normed biquaternions,

$$\mathcal{A}_\theta = \left(\cos\frac{\theta}{2}, \sin\frac{\theta}{2}\overset{*}{B}\right), \ \mathcal{A}_\phi = \left(\cosh\frac{\phi}{2}, \sinh\frac{\phi}{2}B\right), \tag{171}$$

which induce, respectively, a θ-rotation in the plane of $\overset{*}{B}$ and a ϕ-boost in the plane of B, we can check at once, by use of (169) and (170), that

$$\mathcal{A}_\phi\mathcal{A}_\theta = \mathcal{A}_\theta\mathcal{A}_\phi = \mathcal{A}, \tag{172}$$

thereby establishing the existence part of our claim. The unicity follows from the already established fact that *exactly* two non-null 2-planes are left invariant by any non-null LT.

Incidentally, we have already seen a special case of the splitting (172), when in (149) we treated the case of a simple bivector A with $A^2 < -1$. Then α is purely imaginary, so from (169) $\theta = \pi$, and the rotation component $\mathcal{A}_\theta$ is a π-rotation.

28.7 t-Real Biquaternions, t-Rotations, and t-Boosts

Although the reality or pure-imaginariness of the 3-vector $\mathbf{f} = \mathbf{b} + i\mathbf{e}$ in the association $F_{ij} \sim \mathbf{f}$ (cf. (8)) has no frame-invariant significance (an LT generally scrambles $\mathbf{e}$ and $\mathbf{b}$), yet it *is* a useful concept *relative* to a given time axis (or timelike 4-vector) t. We shall say that F_{ij} is t-real or t-imaginary if

$$e_i = F_{i0} = F.t = -t \cdot F = 0, \tag{173}$$

or

$$b_i = \overset{*}{F}_{i0} = \overset{*}{F}.t = -t \cdot \overset{*}{F} = 0, \tag{174}$$

respectively. For example, from (36) and (37) we see that $I,\ J,\ K$ are t-real, while $\overset{*}{I},\ \overset{*}{J},\ \overset{*}{L}$ are t-imaginary.

Products $F \circ G$ or $F \times G$ of t-real or t-imaginary bivectors are t-real or t-imaginary in the obvious way via (11) and (12). Dualizing, $F \mapsto \overset{*}{F}$, changes t-reality into t-imaginariness and vice versa.

Any t-real bivector is simple and spacelike, while a t-imaginary bivector is simple and timelike (recall (1.7)!):

$$A \sim \mathbf{b} \Longrightarrow A \circ A = -{b_1}^2 - {b_2}^2 - {b_3}^2 < 0, \tag{175}$$

$$B \sim i\mathbf{e} \Longrightarrow B \circ B = {e_1}^2 + {e_2}^2 + {e_3}^2 > 0. \tag{176}$$

From (2) we see directly that for a t-imaginary bivector $B \sim i\mathbf{e}$, we have $B = et - te$, and so t lies in the plane of B. The converse is also true: if t lies in the plane of a simple non-null bivector B, it is orthonormal to all vectors in the plane of $\overset{*}{B}$, whence $t \cdot \overset{*}{B} = 0$, and so $\overset{*}{B}$ is t-real and B is t-imaginary.

If the specific t is not of immediate interest, we may simply speak of "real" and "imaginary" bivectors, without risk of confusion. Evidently every bivector F can be uniquely expressed (given t) as the sum of its real and imaginary parts:

$$F = F_b + F_e, \tag{177}$$

where F_b in matrix form (2) has the e-terms zero, while F_e has the b-terms zero.

We call a biquaternion real if its scalar and bivector parts are real, and imaginary if the parts are imaginary. As we see from inspection of the product rule (79), products of real biquaternions are real. Every biquaternion (α, A) can be split into its real and imaginary parts,

$$(\alpha_1 + i\alpha_2, A_b + A_e) = (\alpha_1, A_b) + i(\alpha_2, -\overset{*}{A}_e), \tag{178}$$

where i on the right-hand side is regarded as a biquaternion and we have replaced A_e by $-\, i \circ \overset{*}{A}_e$; both parentheses on the right-hand side are real.

Real biquaternions $\mathcal{U} = (\alpha, U)$ have positive norm (cf. (175)):

$$\mathcal{U}\mathcal{U}' = \alpha^2 - U^2 > 0. \tag{179}$$

Now suppose $\mathcal{A} = (\alpha, A)$ is general but normed, and split it according to (178),

$$\mathcal{A} = \mathcal{U} + i\mathcal{V}, \tag{180}$$

where $\mathcal{U}$, $\mathcal{V}$ are real. Then

$$\mathcal{A}\mathcal{A}' = (\mathcal{U} + i\mathcal{V})(\mathcal{U}' + i\mathcal{V}') = \mathcal{U}\mathcal{U}' - \mathcal{V}\mathcal{V}' + i(\mathcal{V}\mathcal{U}' + \mathcal{U}\mathcal{V}') = 1 \tag{181}$$

from which we deduce

$$0 = \mathcal{V}\mathcal{U}' + \mathcal{U}\mathcal{V}' = \mathcal{V}\mathcal{U}' + (\mathcal{V}\mathcal{U}')', \tag{182}$$

whence $\mathcal{V}\mathcal{U}'$ is a bivector,

$$\mathcal{V}\mathcal{U}' =: B. \tag{183}$$

Moreover, it is a *real* bivector, being the product of real biquaternions. (Similarly, starting with $\mathcal{A}'\mathcal{A}$, we find $\mathcal{U}'\mathcal{V}$ to be a real bivector.)

Next, if $u = \sqrt{\mathcal{U}\mathcal{U}'}$, define

$$\mathcal{R} = \frac{1}{u}\mathcal{U} =: (\beta, R), \tag{184}$$

which is real and normed, $\mathcal{R}\mathcal{R}' = 1$. Since $\mathcal{R}^2$ is real and negative (cf. (175)), $\mathcal{R}$ represents a rotation (cf. (158)), which in fact leaves t invariant, since t lies in the plane $\overset{*}{\mathcal{R}}$ left pointwise invariant by $\mathcal{R}$. Accordingly, we call $\mathcal{R}$ a *t-rotation*.

Returning to the $\mathcal{A}$ of (180), consider now

$$\mathcal{A}\mathcal{R}' = (\mathcal{U} + i\mathcal{V})\,\mathcal{U}'u^{-1} = (u + u^{-1}i\mathcal{V}\mathcal{U}') = (u, u^{-1}i \circ B), \tag{185}$$

where we have made use of (183). Since B is real, $u^{-1} \circ B$ is imaginary and therefore simple and timelike (cf. (176)). Also, since $\mathcal{A}$ and $\mathcal{R}$ are normed, so is $\mathcal{A}\mathcal{R}'$. This biquaternion, therefore, represents a pure boost (cf. after (154)); call it $\mathcal{B}_1 : \mathcal{A}\mathcal{R}' = \mathcal{B}_1$. Postmultiplying by $\mathcal{R}$ gives

$$\mathcal{A} = \mathcal{B}_1\mathcal{R}, \tag{186}$$

an equation which shows that a general LT can be split into a t-boost followed by a t-rotation about any preassigned timelike vector t. The term "t-boost" shall denote any pure boost that sends a 2-plane containing t into itself; evidently $\mathcal{B}_1$ is of this type, since its generating bivector $u^{-1}\overset{*}{B}$ sends the plane of $\overset{*}{B}$ into itself, and that plane contains t because B is real. Note that the present splitting is quite different from that of (149), since the invariant 2-planes of $\mathcal{A}$ may not contain t.

We can alternatively split $\mathcal{A}$ into *first* a t-rotation and *then* a t-boost as follows. Consider

$$\mathcal{R}'\mathcal{A} = u^{-1}\mathcal{U}'\,(\mathcal{U} + i\mathcal{V}) = (u + u^{-1}i\,\mathcal{U}'\mathcal{V}) =: \mathcal{B}_2, \tag{187}$$

where $\mathcal{B}_2$ evidently is a t-boost, whence, as claimed,

$$\mathcal{A} = \mathcal{R}\mathcal{B}_2. \tag{188}$$

Given t, the splittings (186) and (188) are, in fact, unique: Suppose, for example, that contrariwise

$$\mathcal{BR} = \tilde{\mathcal{B}}\tilde{\mathcal{R}}, \ \therefore \ \mathcal{R}\tilde{\mathcal{R}}' = \tilde{\mathcal{B}}\mathcal{B}'. \tag{189}$$

But the product of two t-rotations, i.e., of two real normed biquaternions, is clearly a t-rotation [this corresponds to the well-known fact that the product of 2 rotations in Euclidean 3-space is a rotation] and the product of two t-boosts is (less obviously) a boost, though not necessarily a t-boost, as we shall presently establish. Given that, both sides of (189)(ii) must equal the identity, whence $\tilde{\mathcal{R}} = \mathcal{R}$ and $\tilde{\mathcal{B}} = \mathcal{B}$ and so (186) is unique. Quite analogously we establish the unicity of (188).

To show that the product of two t-boosts $\mathcal{B}$ and $\tilde{\mathcal{B}}$ is a boost, let us write

$$\mathcal{B} = (C, SB), \qquad \tilde{\mathcal{B}} = (\tilde{C}, \tilde{S}\tilde{B}), \tag{190}$$

$$B^2 = 1, \qquad \tilde{B}^2 = 1, \qquad B \circ \tilde{B} = \cos\theta = c^2 - s^2 \tag{191}$$

where $c = \cos(\theta/2)$, $s = \sin(\theta/2)$, $C = \cosh(\phi/2)$, $S = \sinh(\phi/2)$ and similarly for $\tilde{C}, \tilde{S}$; $\mathcal{B}, \tilde{\mathcal{B}}$ are the unit bivectors determining the boost planes, and their intersection along a timelike vector t is guaranteed by $(B \circ \tilde{B})^2 < 1$ (cf. (62)). So $\mathcal{B}, \tilde{\mathcal{B}}$ are t-boosts. For the scalar β of $\mathcal{B}\tilde{\mathcal{B}}$ we then find

$$\begin{aligned} \beta &= C\tilde{C} + S\tilde{S}(c^2 - s^2) = (c^2 + s^2)C\tilde{C} + (c^2 - s^2)S\tilde{S} \\ &= c^2(C\tilde{C} + S\tilde{S}) + s^2(C\tilde{C} - S\tilde{S}) \\ &= c^2 \cosh(\varphi + \tilde{\varphi})/2 + s^2 \cosh(\varphi - \tilde{\varphi})/2 > c^2 + s^2 = 1. \end{aligned} \tag{192}$$

But, $\mathcal{B}\tilde{\mathcal{B}}$ is normed, so its bivector part P satisfies

$$P^2 = \beta^2 - 1 \geq 0, \tag{193}$$

and is therefore timelike, whence $\mathcal{B}\tilde{\mathcal{B}}$ is a boost as claimed.

The converse of the above result, namely that the product of two boosts $\mathcal{B}$ and $\tilde{\mathcal{B}}$ is a boost *only* if their invariant timelike planes intersect in a timelike vector, is false. Suppose the planes of B and $\tilde{B}$ in (190) intersect in a spacelike or null vector; then we write $i\theta$ for θ in (191) so that (192) reads

$$\beta = \cosh^2\frac{\theta}{2}\cosh\frac{\varphi + \tilde{\varphi}}{2} - \sinh^2\frac{\theta}{2}\cosh\frac{\varphi - \tilde{\varphi}}{2}, \tag{194}$$

and this number exceeds unity if and only if $\mid \varphi + \tilde{\varphi} \mid > \mid \varphi - \tilde{\varphi} \mid$. On the other hand, if the invariant timelike planes of $\mathcal{B}, \tilde{\mathcal{B}}$ do *not* intersect, $\mathcal{B}\tilde{\mathcal{B}}$ *cannot* be a boost. For then $B \circ \tilde{B}$, and with it β is complex and β^2 cannot be real and positive.

Also, whereas products of rotations in Euclidean 3-space are rotations, this is *not* necessarily true in Minkowski space. If the two invariant spacelike 2-planes do *not* intersect (for example, $B \sim \sqrt{2}x + iy$, $\tilde{B} = \sqrt{2}y + ix$) the scalar β of the product is complex, so that β^2 cannot be real and positive and the product cannot be a rotation. (See end of Section 28.6.)

As an example of the splitting (186), let us so split a typical null rotation $\mathcal{A} = (1, A)$, $A^2 = 0$ (cf. above (162)), relative to a given t. Without loss of generality we can pick the reference tetrad so that t coincides with the time axis and

$$A \sim u(x+iy), \qquad A = u(I + \overset{*}{J}), \tag{195}$$

for some real u. Then (180) reads

$$\mathcal{A} = (1,\ uI) + i(0,\ uJ) := \mathcal{U} + i\mathcal{V}, \tag{196}$$

and now $\mathcal{U}\mathcal{U}' = 1 + u^2$. According to (184), therefore,

$$\mathcal{R} = (1+u^2)^{-1/2}(1, uI). \tag{197}$$

Furthermore, $\mathcal{V}\mathcal{U}' = uJ + u^2K$, which when substituted in (185) gives

$$\mathcal{B}_1 = \mathcal{A}\mathcal{R}' = ((1+u^2)^{1/2},\ (1+u^2)^{-1/2}(uJ + u^2\overset{*}{K})). \tag{198}$$

This represents a boost in the timelike plane of the simple bivector $B = u\overset{*}{J} + u^2\overset{*}{K}$ which intersects the rotation plane I (we easily check that $B \circ B > 0$, $I \circ B = 0$). By reference to (195)(ii) and (40) we see how the choice of t-axis determines the parameter u in (197) and (198):

$$A \cdot t = uy, \tag{199}$$

and reference to (171) shows how this u determines the θ and ϕ of $\mathcal{R}$ and $\mathcal{B}_1$ respectively, i.e.,

$$\cos\frac{\theta}{2} = (1+u^2)^{-1/2},\ \cosh\frac{\phi}{2} = (1+u^2)^{1/2}, \tag{200}$$
$$u = \tan\frac{\theta}{2}, \qquad u = \sinh\frac{\phi}{2}.$$

28.8 The Thomas Precession

Within our present formalism it turns out that one can relatively simply, and entirely without tedious computation, derive the notorious *Thomas Precession* formula, which shows how the resultant of two boosts in general involves a space rotation. Of special interest is the case where the second boost is infinitesimal. Let us specifically consider a particle moving along a worldline l having equation $x^i = x^i(\tau)$ relative to a given inertial frame S, τ being S-time. The particle carries a reference tetrad $T(\tau)$ (corresponding to its instantaneous inertial rest frame) with time axis tangent to l. At some instant $\tau = 0$ let $T(0)$ be so oriented that it arises out of the permanent reference tetrad $T = (t, x, y, z)$ of S by a t-boost $\mathcal{B}_o$. Thereafter, it is to be transported along l by a succession of infinitesimal boosts ("Fermi-Walker" transport). Now *any* Lorentz transformation away from T can be uniquely split into a t-rotation followed by a t-boost. Let $\mathcal{R}(\tau)$ and $\mathcal{B}(\tau)$ be the t-rotation and t-boost, respectively, whose product $\mathcal{R}\mathcal{B}$ sends T to $T(\tau)$. Then

$$\mathcal{R}(0) = 1, \qquad \mathcal{B}(0) = \mathcal{B}_o, \tag{201}$$

and we shall write

$$\mathcal{R}(d\tau) =: \mathcal{R}_1, \qquad \mathcal{B}(d\tau) =: \mathcal{B}_1. \tag{202}$$

If, furthermore, $\mathcal{C}_o$ is the boost sending $T(0)$ to $T(d\tau)$, we have

$$\mathcal{R}_1\mathcal{B}_1 = \mathcal{B}_o\mathcal{C}_o. \tag{203}$$

We now construct a new biquaternion $\mathcal{C}$ as follows:

$$\mathcal{C} := \mathcal{B}_o\mathcal{C}_o\mathcal{B}_o' = \mathcal{R}_1\mathcal{B}_1\mathcal{B}_o', \tag{204}$$

where we have made use of (203). The key to our main argument will be that $\mathcal{C}$ is a t-boost: Since $\mathcal{C}_o$ is an infinitesimal boost, say through a rapidity $d\phi$, we have, from (171), that for some infinitesimal bivector C_o,

$$\mathcal{C} = \mathcal{B}_o(1, C_o)\mathcal{B}_o' = (1, C), \qquad C = \mathcal{B}_o C_o \mathcal{B}_o', \tag{205}$$

to first order in $d\phi$. Thus C is the transform of C_o under the reverse boost $\mathcal{B}_o'$, which also carries t_o, the tangent vector to l at $\tau = 0$, into t, the time axis of S. Then, since the dual of the transform is the transform of the dual, and since products are preserved, we have

$$\overset{*}{C} \cdot t = \overset{*}{C}_o \cdot t_o \tag{206}$$

But C_o is a t_o-boost, whence $\overset{*}{C}_o \cdot t_o = 0$. Thus $\overset{*}{C} \cdot t = 0$ and C is a t-boost, as claimed.

From (174) we get $\mathcal{C}' = \mathcal{B}_o\mathcal{B}_1'\mathcal{R}_1'$, and then, since $\mathcal{R}_1$ is presumed normed,

$$\mathcal{C}'\mathcal{R}_1 = \mathcal{B}_o\mathcal{B}_1'. \tag{207}$$

Let us write (by reference to (171))

$$\mathcal{R}_1 = \left(1, \frac{1}{2}d\theta R_1\right), \tag{208}$$

where $d\theta$ is the angle of the rotation and R_1 represents its plane. Also, let us write

$$\mathcal{B}_o = (\beta_o, i \circ B_o), \qquad \mathcal{B}_1 = \mathcal{B}(0) + \dot{\mathcal{B}}(0)\, d\tau. \tag{209}$$

Then by use of (205), (208), (209), equation (207) becomes

$$(1, i \circ \overset{*}{C})\left(1, \frac{1}{2}d\theta R_1\right) = (\beta_o, i \circ B_o)(\beta_o, -i \circ (B_o + \dot{B}_o d\tau), \tag{210}$$

where *all* the letters (except i) represent real numbers or t-real bivectors. Equating the t-real bivector parts of both sides of (210) then yields

$$\frac{1}{2}\, d\theta R_1 = B_o \times \dot{B}_o d\tau. \tag{211}$$

By means of the correspondence (8) between bivectors and complex 3-vectors, and by reference to (157) and (161), we can rewrite (211) in vector form. But first we need to convert from rapidity ϕ to velocity v by means of the basic relation $\cosh\phi = \gamma := (1 - v^2/c^2)^{-1/2}$. Then,

$$B_0 \sim \sqrt{\frac{1}{2}(\gamma - 1)}\,\mathbf{d} = \left\{\frac{\sqrt{\frac{1}{2}(\gamma - 1)}}{v}\right\}\mathbf{v}, \qquad \mathbf{v} := v\mathbf{d}, \tag{212}$$

$$\dot{B}_o \sim \{\}\dot{\mathbf{v}} + \{\}\dot{\mathbf{v}}, \tag{213}$$

$$\frac{1}{2}\,d\theta R_1 \sim -\frac{1}{2}\,d\theta\mathbf{d}_1 =: -\frac{1}{2}\mathbf{d}\theta. \tag{214}$$

With these equivalences, (211) finally yields the well-known Thomas Precession formula

$$\frac{d\theta}{d\tau} = \frac{-\frac{1}{2}\left[(1 - v^2/c^2)^{-1/2} - 1\right]}{v^2}\,\mathbf{v} \times \dot{\mathbf{v}}. \tag{215}$$

Envoi

It is with great pleasure that we dedicate this work to our good friend and distinguished colleague Engelbert. We are grateful to him for thirty-seven years of wonderful friendship.

References

[1] J. Ehlers, W. Rindler, and I. Robinson, *Quaternions, Bivectors and the Lorentz Group*, in *Perspectives in Geometry and Relativity* (Hlavaty Festschrift), B. Hoffmann, ed., Indiana University Press, 1966.

29

Leon Lichtenstein's Work on Rotating Fluids

Bernd Schmidt[1]

29.1 Introduction

Leon Lichtenstein was professor of mathematics in Leipzig from 1922 until he died in 1933. J. Schauder and E. Kähler were among his students [1]. In 1933, he published a monograph [2] with the title "Gleichgewichtsfiguren rotierender Flüssigkeiten" — "Equilibrium figures of rotating fluids" — in which were collected all the work he had done on this topic between 1918 and 1933. His contributions to the theory of rotating fluids have been forgotten almost completely. The purpose of this paper is to bring Lichtenstein back to the stage and show that his work still influences today's research.

29.2 Rigidly Rotating Fluids in Newtonian Theory

In comoving coordinates the equations of motion for self-gravitating fluid bodies rotating rigidly around the z-axis are given by

$$\begin{aligned} -\omega^2 x &= \frac{\partial V}{\partial x} - \frac{1}{f}\frac{\partial p}{\partial x}, \\ -\omega^2 y &= \frac{\partial V}{\partial y} - \frac{1}{f}\frac{\partial p}{\partial y}, \\ 0 &= \frac{\partial V}{\partial z} - \frac{1}{f}\frac{\partial p}{\partial z}. \end{aligned} \tag{1}$$

[1] Dedicated to Engelbert Schucking.

Here f is the density of the fluid, assumed to be constant in this paper, and p is the pressure. The constant ω is the angular velocity. The gravitational potential is

$$V(\vec{x}) = \kappa \int_T \frac{f(\vec{x}')\,dx'\,dy'\,dz'}{|\vec{x} - \vec{x}'|}\,. \tag{2}$$

In this paper I shall deal with the following two cases:

1. One axisymmetric fluid ball rotating around the z-axis.
2. Two fluid balls in circular motion around their center of mass.

The support T of the density is connected in the first case and has two components in the second.

Equations (1, 2) hold at all points of T. Then (1) can easily be integrated, with the result equivalent to (1)

$$-\frac{1}{2}\omega^2(x^2 + y^2) = V(\vec{x}) - \frac{1}{f}p(\vec{x}) + s \tag{3}$$

at each point of the body. The constant s may have different values on the components of T. In particular, at the boundaries S of the bodies where the pressure vanishes we have

$$-\frac{1}{2}\omega^2(x^2 + y^2) = V(\vec{x}) + s, \tag{4}$$

which determines the value of s. The sum of the centrifugal and gravitational potential is constant at the surface.

Suppose, conversely, that we find domains T such that at the boundaries S of T Eq. (4) holds, with V given by (2). We can then use (3) to define the pressure inside T, and we obtain a solution of (1). The pressure inside the body is positive, provided $2\omega^2 < \kappa 4\pi f$. This follows from the maximum principle if we act on (3) with the Laplace operator. Hence in the constant density case the search for a rigidly rotating fluid configuration is reduced to the problem of finding shapes T in Euclidian space such that (4) holds for the Poisson integral (2).

Following work by Liapunoff and Poincaré, Lichtenstein [2] demonstrated the existence of such solutions "near known solutions" or " near approximately known solutions." I shall describe his method by explaining two particular cases:

1. Starting from a nonrotating spherical constant density solution T_0, one obtains "slowly rotating" configurations. This is the simplest case.
2. The case of two fluid balls rotating around their center of mass is much more sophisticated. Existence can be shown when the two bodies are "sufficiently far apart."

Let us discuss case 1: To find slowly rotating solutions, one describes this solution by a displacement function ζ which is defined on the spherical surface S_0 of the nonrotating body T_0. A displacement of the amount ζ in the direction of the outwards-directed normal $\vec{n}$ defines a new configuration T. If we write Eq. (4) for this configuration using the variable ζ, we obtain

$$\frac{1}{2}\omega^2\left((x+\zeta(\vec{x})n_x)^2+(y+\zeta(\vec{x})n_y)^2\right)$$
$$+\kappa f\int_T \frac{dx'\,dy'\,dz'}{|\vec{x}+\zeta(\vec{x})\vec{n}(\vec{x})-\vec{x}'|}+s=0 \tag{5}$$

at the boundary of T, i.e., for $\vec{x}\in S_0$. A solution of this equation defines a new equilibrium configuration.

If we expand (5) in ζ, we find for $\vec{x}\in S$

$$\frac{1}{2}\omega^2\left(x^2+y^2\right)+\kappa f\int_{T_0}\frac{dx'\,dy'\,dz'}{|\vec{x}-\vec{x}'|}+s+\omega^2\zeta(xn_x+yn_y)$$
$$-\frac{4\pi}{3}\kappa fR\zeta+\kappa f\int_S\frac{\zeta(\vec{x}')\,d\sigma'}{|\vec{x}-\vec{x}'|}+F(\zeta)=0, \tag{6}$$

where R is the radius of the static reference configuration. Lichtenstein showed that the term $F(\zeta)$, which is $O(\zeta^2)$, can be written as an infinite sum of terms

$$V^{(n)}=\int_S\frac{K^{(n)}[\zeta(\vec{x}')]}{|\vec{x}-\vec{x}'|^n}\,d\sigma', \qquad n>1, \tag{7}$$

where $K^{(n)}$ is a nth-order polynomial in ζ and its first derivatives.

Rewriting (6) slightly and putting the value $\frac{2\pi}{3}f\,R^2$ of the volume integral into the constant s we obtain

$$-\frac{4\pi}{3}\kappa fR\zeta+\kappa f\int_S\frac{\zeta'\,d\sigma'}{|\vec{x}-\vec{x}'|}+\frac{1}{2}\omega^2\left(x^2+y^2\right)$$
$$+s'+\omega^2\zeta(xn_x+yn_y)+F(\zeta)=0. \tag{8}$$

Suppose that the linear integral equation

$$G(\zeta):=-\frac{4\pi}{3}\kappa fR\zeta+\kappa f\int_S\frac{\zeta'\,d\sigma'}{|\vec{x}-\vec{x}'|}=h \tag{9}$$

has a solution ζ for any right-hand side h. Then there is a natural iteration for finding solutions: choose some small ω and s', and solve

$$G(\zeta^{(1)})=-\frac{1}{2}\omega^2\left(x^2+y^2\right)-s', \tag{10}$$

$$G(\zeta^{(n)})=-\frac{1}{2}\omega^2\left(x^2+y^2\right)-s'-\omega^2\zeta^{(n-1)}(xn_x+yn_y)-F(\zeta^{(n-1)}). \tag{11}$$

In our simple case of a spherical nonrotating reference configuration, the operator G is well understood: For a spherical harmonic Y_{lm},

$$\frac{4\pi}{3}RY_{lm}(\vec{x})=\frac{2l+1}{3}\int_S\frac{Y'_{lm}}{|\vec{x}-\vec{x}'|}\,d\sigma', \qquad |\vec{x}|=R. \tag{12}$$

Hence the $l = 1$ spherical harmonics solve the equation $G(\zeta) = 0$. For any h which contains only spherical harmonics with $l \neq 1$, we can solve Eq. (9). The right-hand side of (10) is invariant under rotations around the z-axis and against the reflection $z \to -z$. All functions with these symmetries do not contain $l = 1$ spherical harmonics. The same is true for the iterate ζ^2, because $xn_x + yn_y$ also has this symmetry. Products of functions with these symmetries again show these symmetries, hence all $\zeta^{(n)}$ are well defined. Furthermore, Lichtenstein has shown that for sufficiently small ω and s' the sequence converges. In fact, it converges to a Maclaurin ellipsoid.

Let us now turn to the more complicated case of two bodies in a circular orbit around the z-axis. (The case of n bodies moving on a circular orbit was the topic of E. Kähler's Ph.D. thesis in Leipzig 1928.[2])

As an initial configuration, we take two equal spherical bodies of constant density of radius a which move along the point particle orbit with radius ρ. Then we have $\omega^2 = \frac{M}{2\rho^3}$. Imagine that the centers of the bodies are in the plane $z = 0$ on the x-axis of the comoving coordinate system. This configuration does not satisfy the equations (1). Now let ${}^{(i)}\zeta$ for $i = 1, 2$ be displacement fields on the boundaries of the two spheres such that the shapes defined by the displacements satisfy the equilibrium conditions (4). The displacement fields satisfy nonlinear integral equations of the form

$$-\frac{4\pi}{3} fa\, {}^{(i)}\zeta + f \int\limits_{S_i} \frac{{}^{(i)}\zeta' d\sigma'}{|\vec{x} - \vec{x}'|} = {}^{(i)}F\,(\rho, {}^{(1)}\zeta, {}^{(2)}\zeta). \tag{13}$$

Again the idea of the existence proof is the obvious iteration. But in this case it is more difficult to "deal with the kernel" of the linear integral operator.

Denote by (u_i, v_i, w_i) for $i = 1, 2$ two Cartesian coordinate systems centered at the reference bodies such that the (u_i, v_i, w_i) axes are parallel to the x, y, z-axes. Then the functions (u_i, v_i, w_i) at the surfaces of the bodies are the $l = 1$ spherical harmonics.

The integral kernel

$$N = \frac{1}{|\vec{x} - \vec{x}'|} - \frac{uu' + vv' + ww'}{a^3} \tag{14}$$

defines an integral operator

$$H(\zeta) = \frac{4\pi}{3} f\zeta - \int\limits_{S_i} fN\zeta' d\sigma' \tag{15}$$

with trivial kernel. The integral equation $G(\zeta) = F$ becomes

$$H(\zeta) = F + \frac{1}{a^3}(Au + Bv + Cw) \tag{16}$$

---

[2] I follow essentialy a paper by Kähler [3] which simplifies Lichtenstein's treatment.

provided

$$A = \int_S \zeta' u' \, d\sigma', \qquad B = \int_S \zeta' v' \, d\sigma', \qquad C = \int_S \zeta' w' \, d\sigma'. \tag{17}$$

If F depended on some parameters which could be determined such that $A = B = C = 0$, one would just have to solve $H(\zeta) = F$.

To create such parameters, Lichtenstein and Kähler displace the centers of the reference configuration in the x-y-plane relative to the point particle orbits by an amount α_i in the x-direction. The integral equations then become

$$H({}^{(i)}\zeta) = F_i(\rho, \alpha_i, {}^{(1)}\zeta, {}^{(2)}\zeta), \tag{18}$$

and by the obvious iteration it can be shown that they have a solution for all sufficiently small $\frac{1}{\rho}$, α_i.

The essential point is finally to check that for certain values of these parameters $\frac{1}{\rho}$, α, the conditions (17) are satisfied. This leads to the so-called "bifurcation equation," which in the case of two bodies always has a solution provided the two trial bodies are sufficiently far apart.

Besides the two cases discussed above, Lichtenstein also treats in his monograph "Gleichgewichtsfiguren rotierender Flüssigkeiten" [2] the following cases: from known homogeneous or inhomogeneous, rigidly or differentially rotating, solutions one can always obtain nearby solutions with different rotation or equations of state.

Known approximate solutions can be used to prove existence for ring-type fluids with or without central bodies, provided the diameter of the ring is sufficiently large for fixed radius; several rings may also occur. Solutions of the Newtonian point particle n-body problem, in which the particles rotate rigidly around their center of mass, give rise to a similar configuration with sufficiently small bodies. Examples are n bodies placed symmetrically on a circle or a line.

29.3 Rigidly Rotating Fluids in Einstein's Theory of Gravity

There are various, rather different, numerical methods to calculate rigidly rotating fluid solutions in general relativity. However, up to now there is only one theorem in general relativity which demonstrates the existence of a certain class of rotating fluid bodies. It was proven by U. Heilig [4] in 1995, and the structure of the proof is quite interesting.

The proof uses the field equations of the "frame theory" developed by Ehlers [5]. The field equations of this theory contain a parameter $\lambda = c^{-2}$ such that for $\lambda \neq 0$ one has Einstein's field equation; for $\lambda = 0$ one obtains a system equivalent to the equations of Newtonian theory. In general, the limit $\lambda \to 0$ is singular in the sense that if Einstein's equations are hyperbolic in some coordinate system, one will obtain in general in the Newtonian limit a mixed hyperbolic-elliptic system.

Hence it is extremely hard to show the existence of λ-families of solutions with a Newtonian limit.

This is different in a time-independent situation. Because of the absence of gravitational radiation, there are formulations of the relativistic field equations which are elliptic and have a well-behaved limit for $\lambda \to 0$ [5]. This fact is exploited by Heilig, when he constructs via the implicit function theorem (for a restricted class of equations of state) families of rigidly rotating, axisymmetric solutions $g(\lambda, \omega)$, starting from $g(0, 0)$, a static, nonrotating Newtonian fluid solution.

In comparison to the Newtonian proof, one meets many extra complications. For example, it is not possible to formulate the problem just as a problem of the body. One has to take along the outside field, and to control the conditions of asymptotic flatness and the coordinate conditions (harmonicity) which make the equations elliptic.

This result shows that thinking about the Newtonian limit of Einstein's theory is not only desirable from a conceptual point of view, but may even give insight that leads to new results within general relativity.

29.4 Speculations

Does there exist a solution in Einstein's theory which describes two bodies in circular motion in analogy to the Newtonian system treated in section 29.2? As such a system has to emit gravitational radiation, it can only stay on the "same orbit" when the outgoing radiation is compensated by the right amount of incoming radiation.

Consider the similar case of an "electromagnetic two-body problem" treated by Schild [6]. Two point particles of opposite charge can move with constant magnitude of their velocity on a circular orbit, if each particle is moving according to the Lorentz force of the 1/2(retarded + advanced) field of the other particle. In this sum the tangential component of the force on the particle cancels. If we take as the field the sum of the retarded and advanced fields of both particles, we have a solution which is invariant under the Killing field $\partial_t + \omega(y\partial_x - x\partial_y)$, which descibes a rigid rotation.

From this example we can learn two more things: if one tries to calculate the radiation field at null infinity of Minkowski space, it turns out that it does not exist! Furthermore, the energy of the field contained in the outside of a big ball is infinite. Since the charges are in an accelerated motion for all times, there is too much radiation energy in the system.

What can we expect in Einstein's theory? It is not even clear how to characterize such a solution. Clearly the solution should be invariant under a Killing vector which describes rigid rotation. In analogy with $\partial_t + \omega(y\partial_x - x\partial_y)$, we expect it to be timelike near the bodies and spacelike near infinity. The asymptotic properties of such a system are also unclear. If the system admitted null infinity with the usual properties, we could demand that the Killing field be a rigid rotation near infinity.

However, we expect this solution to be only asymptotically flat in a very weak sense. Most likely the ADM mass will be infinite.

If we write the field equations in coordinates adapted to the Killing field, they should be elliptic near the bodies because the Killing field is timelike. Near infinity, however, they should be hyperbolic. Perhaps one can demonstrate the existence of a solution in the stationary part along the lines of Heilig's construction. Such a solution will not be unique, because we have to choose boundary values for the elliptic equations. All these solutions will be analytic, and we can try to extend them into the time-dependent regime. Hopefully only one solution will admit a singularity free extension which is asymptotically flat in some weak sense.

Without having really good reasons, I conjecture that each Newtonian 2-body solution as described by Lichtenstein defines a unique λ-family of solutions in Einstein's theory. It is a very challenging partial differential equations problem to prove or reject this conjecture.

References

[1] S. Gottwald, H. Ilgauds, K. Schlote, *Lexikon bedeutender Mathematiker*, Bibliographisches Institut, Leipzig, 1990.

[2] L. Lichtenstein, *Gleichgewichtsfiguren rotierender Flüssigkeiten*, Springer, Berlin, 1933.

[3] E. Kähler, *Math. Zeitschrift,* **28** (1928), 220–237.

[4] U. Heilig, *Commun. Math. Phys.*, **166** (1995), 457–493.

[5] J. Ehlers, The Newtonian limit of general relativity, in *Classical mechanics and relativity: relationships and consistency*, ed. G. Ferrarese, Naples, 1991.

[6] A. Schild, *Phys. Rev.*, **131** (1963), 2762–2766.

30

Decaying Neutrinos and the Flattening of the Galactic Halo

Dennis W. Sciama

ABSTRACT The recently constructed Dehnen-Binney set of mass models for the galaxy is used to show that the decaying neutrino theory for the ionization of the interstellar medium (Sciama 1990a, 1993) requires the neutrino halo of the galaxy to be as flattened as is observationally permitted (axial ratio $q = 0.2$ or shape E8). The argument involves an evaluation of the contribution of red-shifted decay photons from the cosmological distribution of neutrinos to the extragalactic diffuse background at 1500Å. This contribution must be as large as is observationally permitted. These two requirements depend on the decay lifetime τ in potentially conflicting ways. For consistency to be achieved, τ must lie within 30% of 10^{23} seconds.

30.1 Introduction

It gives me great pleasure to dedicate this paper to Engelbert Schucking, whose subtle mind has illuminated many problems in cosmology and general relativity. I hope that he enjoys the way in which, because of the pervasiveness of neutrinos, cosmology and galactic astronomy become interdependent in the decaying neutrino theory for the ionization of the interstellar medium (Sciama 1990a, 1993). I shall give an example of this interdependence here. My discussion also exemplifies two other useful features of the decaying neutrino theory. First, the theory leads to specific and testable predictions concerning the configuration of various matter distributions. For example, it predicts that if the ionization of hydrogen in an opaque region of the galaxy is mainly due to decay photons, then the resulting electron density will be independent of the neutral hydrogen density in the region. There is observational evidence in support of this prediction (Sciama 1990b, 1997). The second feature is that various observations have reduced the domain of validity of the theory to a small region of its parameter space. For example, the energy E_γ of a decay photon in the rest frame of the decaying neutrino is constrained with a precision of 1% ($E_\gamma = 13.7 \pm 0.1$ eV). While this feature may eventually lead to the demise of the theory, it has so far managed to survive.

In this paper I provide another example of these two features, which is based on a new set of comprehensive mass models of the galaxy (Dehnen and Binney 1997). I will demonstrate that if the decaying neutrino theory is correct, the neutrino halo of our galaxy must be flattened to the maximum extent allowed by these models, that is, with an axial ratio $q = 0.2$, corresponding to a shape factor E8 (where q is related to En by $q = 1 - n/10$). Associated with this result is a constraint imposed on the neutrino decay lifetime τ, whose value is required to be close to 10^{23} seconds. The argument leading to these conclusions is based on the following observational results in addition to those underlying the Dehnen-Binney models:

(i) Pulsar dispersion data imply that the free election density in the intercloud medium within one kiloparsec of the sun lies in the range 0.04-0.06 cm^{-3} (Reynolds 1990, Sciama 1990b) and these free electrons have a scale height ~1 kpc (Reynolds 1991, Nordgren et al. 1992, Taylor and Cordes 1993).

(ii) Hα data imply that there are 4×10^6 hydrogen ionizations per cm^2 per sec along a column at the sun perpendicular to the galactic plane (Reynolds 1984).

(iii) The isotropic extragalatic photon flux at 1500Å ~ 300 ± 80 photons cm^{-2} sec^{-1} ster^{-1} Å^{-1} (continuum units or CU) (Henry and Murthy 1993, Witt and Petersohn 1994). This value is still somewhat controversial (compare Bowyer 1991 with Henry 1991). In fact a significantly smaller value would rule out the decaying neutrino theory, as we shall see.

30.2 The Neutrino Density Near the Sun

The neutrino density $n_\nu(0)$ near the sun, which we are assuming to be mainly responsible for the free electron density n_e in opaque regions of the intercloud medium, will be given in ionization equilibrium by

$$\frac{n_\nu(0)}{\tau} = \alpha n_e^2 ,$$

where α is the hydrogen recombination coefficient excluding transitions directly to the ground state. There is, however, a danger that the decaying neutrino theory may lead to a value for $n_\nu(0)$ which is larger than is permitted by the Dehnen-Binney mass models. We therefore immediately adopt the smallest observationally allowed values for α and n_e, namely $2.6 \times 10^{-13} \text{cm}^3 \text{sec}^{-1}$ (corresponding to the reasonable electron temperature of 10^4 K (Osterbrock 1989)) and 0.04 cm^{-3} respectively. Then

$$n_\nu(0) = 4.16 \times 10^7 \tau_{23} \text{ cm}^{-3} ,$$

where $\tau = 10^{23}\ \tau_{23}$ secs. We may convert this number density of neutrinos into a mass density $\rho_\nu(0)$ by using the mass derived for a decaying neutrino in the theory, namely 27.4 ± 0.2 eV (Sciama 1993). We then find that

$$\rho_\nu(0) = 2.04 \times 10^{-24}\ \tau_{23} \text{ g} \cdot \text{cm}^{-3}$$

$$= 0.03\ \tau_{23}\ \text{M}_\odot \text{ pc}^{-3} .$$

We now ask what is the largest observationally permitted value for $\rho_\nu(0)$, since we shall soon see that the extragalactic background at 1500 Å leads to a strong lower bound on τ_{23}. The mass models of Dehnen and Binney (1997) provide a detailed answer to this question (cf. also Gates et al. 1995), but a rough estimate can be derived in the following simple manner. We may obtain observational constraints on $\rho_\nu(0)$ in two ways, by considering (i) estimates for the total density $\rho(0)$ near the sun (the Oort limit) and (ii) the column densities at the sun for various matter distributions. The value of the Oort limit is controversial (e.g., Kuijken and Gilmore 1991, Bahcall et al. 1992). Recent data from the Hubble Space Telescope have placed strict limits on the contribution to $\rho(0)$ from very faint stars (e.g., Gould et al. 1996). A reasonable upper limit on $\rho_\nu(0)$, derived from observational estimates of the gravitational force due to $\rho(0)$ within $\sim$300 pcs of the plane and of the density $\rho_{obs}(0)$ of observed material, would be 0.1 $M_\odot$ pc^{-3}, which would lead to an upper limit on τ_{23} of $\sim$3. In this connexion we note that Binney et al. suggested already in 1987 that all the dark matter near the sun might be due to a flattened halo.

A more stringent upper limit on $\rho_\nu(0)$, and therefore on τ_{23}, follows from considering various column densities at the sun. According to Kuijken and Gilmore (1991, the column density $\sum_{1.1}$ of the observed and dark matter combined out to 1.1 kpc is 71 ± 6 $M_\odot$ pc^{-2}. Some authors have argued that their error estimates should be increased somewhat (e.g., Bahcall et al. 1994). We therefore follow Gates et al. (1995) and assume that $\sum_{1.1} \leq 100$ $M_\odot$ pc^{-2}. For the total column density $\sum_{\rm obs}$ of the observed material at the sun we adopt $\sum_{\rm obs} = 40\, M_\odot$ pc^{-2} (Gould et al. 1996). Hence $\sum_{\nu,1.1} \leq 60\, M_\odot$ pc^{-2}. So long as the scale height of the neutrino distribution is much greater than 1.1 kpc, we have that $0.03\ \tau_{23} \times 2.2 \times 10^3 \leq 60$ and so $\tau_{23} \leq 0.9$. In view of the uncertainties in the values we have adopted, we shall suppose that $\tau_{23} \leq 1$. Thus the upper limit on $\rho_\nu(0)$ is 0.03 $M_\odot$ pc^{-3}. When we come to consider the background at 1500 Å, we shall find that $\tau_{23} \geq 1$, so that the only consistent possibility is $\tau_{23} \sim 1$ and $\rho_\nu(0) \sim 0.03$ $M_\odot$ pc^{-3}. We therefore examine the implications of this value of $\rho_\nu(0)$ for the flattening of the neutrino halo.

Since a large $\rho_\nu(0)$ implies a flattened neutrino halo, we begin by considering the total column density $\sum_{\rm rot}$ of a flattened system required to account for the rotation velocity v_c of the galaxy at the sun's position ($R = R_\odot$). Binney and Tremaine (1987) give for this quantity

$$\sum_{\rm rot} = \frac{v_c^2}{2\pi G R_\odot}$$
$$= 210\ M_\odot\ {\rm pc}^{-2}$$

for $v_c = 220$ km sec^{-1} and $R_\odot = 8.5$ kpc. Hence

$$\sum_\nu = 170\ M_\odot\ {\rm pc}^{-2}\ ,$$

and so the scale height l_ν of the neutrino distribution is given by $l_\nu = 2.8$ kpc, which is indeed substantially greater than 1.1 kpc, so that our previous discussion is valid. To derive the implied flattening of the neutrino halo, we assume that the neutrino distribution in the plane has the form $\rho_\nu(r) = \rho_\nu(0)a^2/(a^2 + r^2)$, with $a = 8$ kpc (Sciama 1993). Then the "scale height" in the plane is $\pi a/2$ or 4π kpc, and so the flattening q is given by $q = 0.2$, corresponding to the shape E8. This simple reasoning is confirmed by the Dehnen-Binney models which Walter Dehnen kindly extended for me to include $q = 0.3$ and $q = 0.2$, for which $\rho_n u(0) = 0.03\,\mathrm{M}_\odot\,\mathrm{pc}^{-3}$ lies just on the edge of the allowed range of models (cf. also Gates et al. 1995).

30.3 τ_{23} and the Hα Data

We now show that Reynolds' (1984) Hα data, as interpreted by the decaying neutrino theory, also lead to the result $\tau_{23} \sim 1$. We consider separately the ionizations produced by decay photons in the opaque layer of free electrons lying above and below the sun out to a distance l and those produced by decaying neutrinos lying in the transparent regions outside this layer. A detailed model of the opaque layer would be complicated, because we should consider the contribution of both clouds and the intercloud medium to the opacity. For simplicity we shall assume that the opaque region corresponds to the electron layer of scale height l as derived from the pulsar dispersion data. Reynolds (1991), Nordgren et al. (1992), and Taylor and Cordes (1993) found that $l \sim 1$ kpc. Inside this opaque layer every decay photon produces an ionization in its vicinity, so along a line of sight normal to the galactic plane there will be n_l ionizations in this layer, where $n_l = 2.6 \times 10^{-13} \times (0.04)^2 \times 6 \times 10^{21}$, or 2.5×10^6. The column density of neutrinos outside the opaque layer corresponds to $115.5\,\mathrm{M}_\odot\,\mathrm{pc}^{-2}$ and so is 5×10^{29} neutrinos cm^{-2}. The number of ionizations which they produce inside the layer is reduced by the usual factor 4 which arises from an integration over solid angle related to the slab geometry. Hence they produce $5 \times 10^6/(4\tau_{23})$ ionizations cm^{-2} sec^{-1}. Thus

$$\frac{1.25 \times 10^6}{\tau_{23}} + 2.5 \times 10^6 = 4 \times 10^6 ,$$

or

$$\tau_{23} = 0.8 .$$

Given the simplicity of our model for the opaque layer, we round this result off to $\tau_{23} \sim 1$, which is just compatible with our previous result $\tau_{23} \lesssim 1$.

30.4 τ_{23} and the Extragalactic Background at 1500 Å

Some of the earliest lower limits on τ_{23} were based on observational estimates of the cosmic background in the far UV, due to red-shifted decay photons produced by the cosmological distribution of neutrinos (Stecker 1980, Kimble et al. 1981).

As mentioned in the introduction, we here adopt an observed flux of 300 ± 80CU at 1500 Å. The most recent estimate (Armand et al. 1994) for the contribution due to galaxies at 2000 Å is 40–130 CU. The red-shifted contribution from decay photons has recently been recalculated by Sciama (1991), Overduin et al. (1993), and Dodelson and Jubas (1994). The main uncertainty arises from absorption by dust in the galaxy. Allowing a factor 2 for this absorption, one obtains 400 τ_{23}^{-1} CU. Given the uncertainties, a reasonable conclusion is that

$$\tau_{23} \geq 1.$$

In conjunction with our previous discussion we arrive at a solution which is just consistent with all the observational constraints, with

$$\tau_{23} \sim 1 ,$$

$$\rho_\nu(0) \sim 0.03 \ M_\odot \ pc^{-3},$$

$$q \sim 0.2 .$$

This solution corresponds to the most flattened possible halo for our galaxy (E8).

30.5 Conclusions

We ask in conclusion whether such a large flattening is otherwise reasonable. I believe that it is. It is noteworthy that another galaxy, NGC4650A, is observed to have a highly flattened halo. This was deduced from observations of an outer ring of gas, dust, and stars which are on orbits that are nearly perpendicular to the plane of the flattened central galaxy, which rotates about its own apparent minor axis (a polar ring galaxy). These orbits delineate the gravitational potential of the galaxy outside its central plane. Sackett et al. (1994) deduced that the halo of this galaxy is flattened towards the plane of its central body. They state that whenever the data were ambiguous they attempted to err on the side of favoring rounder halos. Still they found for this galaxy that q lies between 0.3 and 0.4 (E6–E7). I therefore regard the requirement from the decaying neutrino theory, that the dark halo of our galaxy is as flat as it could possibly be, is a reasonable one.

Acknowledgments

I am very grateful to Walter Dehnen for extending the Dehnen-Binney mass models of our galaxy into the extreme regime required by the decaying neutrino theory, and for helpful discussions of the limits imposed by these mass models. I am also grateful to Geza Gyuk for his advice. This work was financially supported by the Ministero dell'Universita' e della Ricerca Scientifica.

References

[1] C. Armand, B. Milliard, and J. M. Deharveng, *Astron. and Astrophys.* **284** (1994) 12.
[2] J. N. Bahcall, C. Flynn, and A. Gould, *Astrophys. J.* **389** (1992) 234.
[3] J. N. Bahcall, C. Flynn, A. Gould, and S. Kirhakos, *Astrophys. J.* **435** (1994) L51.
[4] J. J. Binney, A. May, and J. P. Ostriker, *MNRAS* **226** (1987) 149.
[5] J. J. Binney and S. Tremaine, *Galactic Dynamics*, Princeton University Press, Princeton (1987).
[6] S. Bowyer, *Ann. Rev. Astr. Ap.* **29** (1991) 59.
[7] W. Dehnen and J. J. Binney, *MNRAS* in press (1997).
[8] S. Dodelson and J. M. Jubas, *MNRAS* **266** (1994) 886.
[9] E .I. Gates, G. Gyuk, and M.S. Turner, *Astrophys. J.* **449** (1995) L123.
[10] A. Gould, J.N. Bahcall, and C. Flynn, *Astrophys. J.* **465** (1996) 759.
[11] R. C. Henry, *Ann. Rev. Astr. Ap.* **29** (1991) 89.
[12] R. C. Henry and J. Murthy, *Astrophys. J.* **418** (1993) L17.
[13] R. Kimble, S. Bowyer, and P. Jakobsen, *Phys. Rev. Lett.* **46** (1981) 80.
[14] K. Kuijken and G. Gilmore, *Astrophys. J.* **367** (1991) L9.
[15] T. Nordgren, J. Cordes, and Y. Terzian, *Astronom. J.* **104** (1992) 465.
[16] D. E. Osterbrock *Astrophysics of Gaseous Nebulae and Active Galactic Nuclei*, University Science Books, Mill Valley, CA (1989).
[17] J. M.Overduin, P. S. Wesson, and S. Bowyer, *Astrophys. J.* **404** (1993) 1.
[18] R. J. Reynolds, *Astrophys. J.* **282** (1984) 191.
[19] R. J. Reynolds, *Astrophys. J.* **348** (1990) 153.
[20] R. J. Reynolds, in *IAU Symposium No. 144 The Interstellar Disk-Halo Connection in Galaxies*, ed. H. Bloemen, Kluwer, Dordrecht (1991) 67.
[21] P. D. Sackett, H-W. Rix, B. J. Jarvis, and K. C. Freeman, *Astrophys. J.* **436** (1994) 629.
[22] D. W. Sciama, *Astrophys. J.* **364** (1990a) 549.
[23] D. W. Sciama, *Nature* **346** (1990b) 40.
[24] D. W. Sciama, in *The Early Observable Universe from Diffuse Backgrounds*, eds. G. Rocca-Volmerange, J.M. Deharveng and J. Tran Thanh Van, Edition Frontières (1991) 127.
[25] D. W. Sciama, *Modern Cosmology and the Dark Matter Problem*, Cambridge University Press, Cambridge (1993).
[26] D. W. Sciama, *Astro-ph* **9702188** (1997).
[27] F. W. Stecker, *Phys. Rev. Lett.* **45** (1980) 1460.
[28] J. H. Taylor and J. M. Cordes, *Astrophys. J.* **411** (1993) 674.
[29] A. N. Witt and J. K. Petersohn, in *First Symposium on the Infrared Cirrus and Diffuse Interstellar Clouds* ASP Conf. Series Vol. **58**, eds. R. M. Cutri and W. B. Latter (1994).

31

The Kasner Condition and Inhomogeneous Perfect Fluid Cosmologies

Jim E.F. Skea

ABSTRACT We analyze the consequences of a power-law *ansatz* for the metric coefficients of diagonalizable perfect fluid cosmological solutions which admit a 2-dimensional isometry group. It is noted that a condition on the power-law exponents is identical to one of the constraints on the exponents of the Kasner vacuum solution. All possible solutions are classified, and those obeying the Kasner condition in two variables are studied in detail. It is shown that in general they represent tilted stiff perfect fluid solutions. The Kasner condition is applied to a metric with one Killing vector, and an apparently new perfect fluid solution obtained. Successes and difficulties in applying invariant classification to the G_2 and G_1 solutions are discussed.

31.1 Introduction

The key contributions of Professor Schucking to cosmological solutions of Einstein's field equations are well known, both in the study of vacuum metrics [1] and perfect fluid metrics [2] which admit high symmetry. I am therefore delighted at the opportunity to present this paper, which deals with space-times with markedly less symmetry, but which also links vacuum and perfect fluid cosmological solutions.

One of the best-known and most widely referred to solutions in general relativity is the vacuum solution discovered by Kasner [3]. The solution has appeared in diverse forms (an interesting account of the history of the Kasner metric is given by Harvey [4]), but all share the characteristic that the metric coefficients are simple powers of the space-time variables. The best-known form of the Kasner metric is probably

$$ds^2 = dt^2 - t^{2p_1}dx^2 - t^{2p_2}dy^2 - t^{2p_3}dz^2, \tag{1}$$

where, in order that the solution represents a vacuum space-time, the constant parameters p_1, p_2, and p_3 obey the simultaneous conditions

$$\sum_{i=1}^{3} p_i - 1 = 0, \tag{2}$$

$$\sum_{i=1}^{3} p_i^2 - 1 = 0. \tag{3}$$

Though both these conditions are associated with the Kasner solution, we shall principally be interested in condition (2), which, in an abuse of language, we shall refer to as "*the* Kasner condition." A notable particular solution is the *axisymmetric case*, which occurs when two of the exponents coincide such that, for example, $p_1 = p_2 = \frac{2}{3}$ and $p_3 = -\frac{1}{3}$. (The other case of coincident exponents, when one exponent is 1 and the others 0, is simply Minkowski space in nonstandard coordinates and is of no interest to us in this work.) In what follows, we define

$$K_n \equiv n_1 + n_2 + n_3 - 1,$$

as a shorthand for (2), where the lettered index n will be one of $\{a, b, c\}$.

As we shall see, it turns out that condition (2) plays an important rôle not only for the Kasner vacuum solution, but also for perfect fluid space-times; it is also important for space-times with isometry groups more general than the G_3 which the Kasner solution admits, with applications to space-times admitting 2-dimensional and even 1-dimensional isotropy groups. It is likely that the condition is of use in developing cosmological models which have no Killing vectors, and this subject is currently under investigation.

To see how the Kasner condition arises, we take as our starting point "power-law metrics," where all metric coefficients are products of simple powers of the space-time variables. Suspending the summation convention for this paragraph, we have

$$g_{ij} = k_{ij} t^{A_{ij}} x^{B_{ij}} y^{C_{ij}} z^{D_{ij}}, \qquad (A_{ij}, B_{ij}, C_{ij}, D_{ij}, k_{ij} \text{ constants}). \tag{4}$$

In particular we shall be interested in diagonal power-law metrics, where we can set all k_{ii} to 1 and, rescaling the coordinates suitably, write the metric as

$$\begin{aligned} ds^2 &= t^{2\alpha_0} x^{2\beta_0} y^{2\gamma_0} z^{2\delta_0} dt^2 - t^{2\alpha_1} x^{2\beta_1} y^{2\gamma_1} z^{2\delta_1} dx^2 \\ &\quad - t^{2\alpha_2} x^{2\beta_2} y^{2\gamma_2} z^{2\delta_2} dy^2 - t^{2\alpha_3} x^{2\beta_3} y^{2\gamma_3} z^{2\delta_3} dz^2, \end{aligned} \tag{5}$$

with α_i, β_i, γ_i and δ_i, $i = 0, \ldots, 3$ constants. With further redefinitions of $\{t, x, y, z\}$ these space-times can be rewritten with metric

$$\begin{aligned} ds^2 &= x^{2b_0} y^{2c_0} z^{2d_0} dt^2 - t^{2a_1} y^{2c_1} z^{2d_1} dx^2 \\ &\quad - t^{2a_2} x^{2b_2} z^{2d_2} dy^2 - t^{2a_3} x^{2b_3} y^{2c_3} dz^2, \end{aligned} \tag{6}$$

where a_i, b_i, c_i and $d_i, i = 0, \ldots, 3$ are constants. We shall be interested in metrics of the form (6) which represent perfect fluid space-times. Despite the substantial simplifications made, Einstein's field equations for (6) under this condition are still complicated. In order to gain some insight about the structure of the field equations, we therefore analyze (6) in the simpler case when there exist Killing vectors. Focusing on inhomogeneous cosmologies, we begin with the two-Killing vector case and choose ∂_y and ∂_z as the Killing vectors to obtain a cosmological solution. This leads us to the metric

$$ds^2 = x^{2b_0}\,dt^2 - t^{2a_1}\,dx^2 - t^{2a_2}x^{2b_2}\,dy^2 - t^{2a_3}x^{2b_3}\,dz^2. \tag{7}$$

In addition to the two Killing vectors, the space-time also, though somewhat less evidently, admits the homothetic vector

$$\frac{b_0-1}{a_1b_0-1}\partial_t + \frac{a_1-1}{a_1b_0-1}\partial_x + \frac{a_1b_0-a_1b_2-a_2b_0+a_2+b_2-1}{a_1b_0-1}\partial_y$$

$$+\frac{a_1b_0-a_1b_3-a_3b_0+a_3+b_3-1}{a_1b_0-1}\partial_z \quad (a_1b_0 \neq 1).$$

The space-times therefore admit a 3-dimensional homothety group, H_3. Perfect fluid cosmological solutions of this class have recently been intensively studied (see [5] for a very good summary of work), though not all solutions have, to our knowledge, been explicitly given.

The intention here, however, is not principally to obtain new cosmological solutions of Einstein's equations with a G_2 or H_3, but rather, by analyzing the structure of the field equations in this case, to derive results which we hope are applicable to more general cosmological solutions, in particular power-law metrics with lower symmetry.

The *ansatz* (5) is admittedly in part opportunistic: if we start from the premise that the metric coefficients have a power-law behavior in terms of the coordinates, after suitable transformations, Einstein's equations are polynomials in the coordinates, making the problem amenable to attack by the barrage of tools available in computer algebra systems (such as Gröbner bases, resultants, and factorization algorithms) for dealing with systems of polynomial equations. The *ansatz* is not, however, without physical motivation: the analysis of cosmologies admitting a G_3 by Jentzen, Rosquist, and Uggla [6],[7], has shown that homothetic metrics act as sinks or sources (in the sense of dynamical systems) for these classes of cosmological models. A similar behavior has been noted by Hewitt et al. [8] for G_2 cosmologies. The metric (5) also incorporates the (spatially flat) FRW cosmologies, and so can be regarded as an inhomogeneous perturbation of the same (though the imposition of the Kasner condition on any parameter set other than the a_i unfortunately removes this possibility).

With Greek indices running from 0 to 3, we assume that the energy-momentum tensor has the form of a perfect fluid (PF), namely

$$T_{\mu\nu} = (\rho + p)u_\mu u_\nu + p g_{\mu\nu},$$

where u_μ is the 4-velocity of the fluid ($u_\mu u^\mu = 1$) and ρ and p are the energy density and pressure of the fluid respectively. Over and above this, no restriction will be placed on the equation of state *a priori*, and the field equations will therefore define the equation of state.

The calculations required were carried out using both the computer algebra system MAPLE (for the manipulation of the polynomial equations) and the program CLASSI [9, 10] written by Jan Åman (to generate the field equations). One of the facilities of CLASSI is its ability to produce an invariant classification for a space-time, which it delivers in terms of a spinor decomposition of the Riemann tensor and its covariant derivatives in a canonically defined basis (see [11] for details). A second facility of CLASSI which we shall use is related to its ability to handle spinors. In spinor language, a basis-independent condition that a space-time represents a PF at the macroscopic level is the vanishing of the Plebanski-Petrov spinor

$$\chi_{ABCD} \equiv \Phi_{(AB|E'F'}\Phi_{|CD)}{}^{E'F'} = 0, \tag{8}$$

where $\Phi_{ABC'D'}$ is the Ricci spinor, parentheses denote symmetrization over the enclosed indices, and a vertical bar excludes indices from the symmetrization. As with the Weyl spinor, each spinor index of χ_{ABCD} can be only 0 or 1, and advantage is taken of this, together with the symmetry of χ_{ABCD}, to write χ_E as a shorthand, with the single index E taking values from 0 to 4 and indicating the number of times the digit 1 occurs in the indices of χ_{ABCD}. We can use the same ideas to compactify separately the primed and unprimed indices of $\Phi_{ABC'D'}$ and write $\Phi_{AB'}$, with A and B running from 0 to 2.

The great advantage of (8) is that it makes no assumption about the characteristics of the fluid flow: in particular it includes both aligned and nonaligned perfect fluids. On the down-side there are, however, two difficulties in using (8) to define the condition for a PF: firstly (8) is quadratic in terms of the components of the Ricci spinor, which considerably increases the difficulty of the equations involved; secondly, the condition is only a necessary condition for a PF, not sufficient. In fact, besides the trivial cases of vacuum space-times and empty space-times with a cosmological constant (where $\chi_A = 0$ trivially because $\Phi_{AB'} = 0$), condition (8) holds not only for perfect fluids but also for null radiation fluids (fortunately of some physical interest and worth studying in any case), and tachyon fluids (TFs), which are of dubious physical interest. (Senovilla and Mars [12] give a nice discussion of the same problem in terms of the components of the Ricci tensor.) We must therefore bear in mind that at some point we have to distinguish which types of matter field we are dealing with, and how to go about this in our particular case is one of the points we discuss in the next section.

31.2 Mathematical Background

Since we shall be working with spinors, we first identify a null tetrad in which we shall be performing most of the calculations. We define the "natural Lorentz tetrad" for (7) as

$$\omega^0 = x^{b_0}\,\mathrm{d}t, \quad \omega^1 = t^{a_1}\,\mathrm{d}x, \quad \omega^2 = t^{a_2}x^{b_2}\,\mathrm{d}y, \quad \omega^3 = t^{a_3}x^{b_3}\,\mathrm{d}z, \tag{9}$$

and the "natural null tetrad" $\{\mathbf{k}, \mathbf{l}, \mathbf{m}, \overline{\mathbf{m}}\}$ derived from this basis by

$$\mathbf{k} = \frac{1}{\sqrt{2}}(\omega^0 + \omega^1), \quad \mathbf{l} = \frac{1}{\sqrt{2}}(\omega^0 - \omega^1), \quad \mathbf{m} = \frac{1}{\sqrt{2}}(\omega^2 + i\omega^3). \tag{10}$$

In this tetrad we find that two of the components of the Weyl spinor, namely Ψ_1 and Ψ_3, vanish identically. In general, the space-time is of Petrov type I. As for the Ricci spinor, $\Phi_{AB'}$, we shall see that $\Phi_{01'} = \Phi_{12'} = 0$ identically in the natural null tetrad, and that $\Phi_{02'}$ is real. From the definition of the Plebanski-Petrov spinor, we have that

$$\chi_0 = \frac{1}{2}\,\Phi_{02'}\Phi_{00'},$$

$$\chi_4 = \frac{1}{2}\,\Phi_{02'}\Phi_{22'},$$

$$\chi_2 = \frac{1}{12}(\Phi_{00'}\Phi_{22'} + \Phi_{02'}^2 - 4\Phi_{11'}^2),$$

with χ_1 and χ_3 identically zero. The nontrivial solutions to these equations result in two possibilities: if $\Phi_{02'} = 0$, we are left with the single equation

$$\Phi_{00'}\Phi_{22'} - 4\Phi_{11'}^2 = 0. \tag{11}$$

This group includes both PF and TF solutions in their so-called canonical forms, as defined below. On the other hand, if $\Phi_{02'} \neq 0$, then (at least) one of $\Phi_{00'}$ or $\Phi_{22'}$ must vanish, and the restriction that $\chi_2 = 0$ provides us with the equation

$$\Phi_{02} = \pm 2\Phi_{11'}. \tag{12}$$

It is relatively easy to show (see, for example, [13]) that (12) is incompatible with a PF solution (though it is compatible with a TF). We are led to the conclusion that the metric (7) in the tetrad (10) can only represent a PF if $\Phi_{02'} = 0$.

From (11) we see that once we have set $\Phi_{02'}$ to zero for the models under consideration, we can write

$$\Phi_{22'} = 2\alpha\Phi_{11'}, \qquad \Phi_{00'} = 2\alpha^{-1}\Phi_{11'}, \tag{13}$$

where α, in general, will be some function of the coordinates. Now a Ricci spinor which satisfies (11) can represent either a PF or TF. These types can be distinguished if we express the Ricci spinor in its canonical basis, in which, for a PF,

$$\Phi_{00'} = 2\,\Phi_{11'} = \Phi_{22'},$$

while for a TF, in the same basis,

$$\Phi_{00'} = -2\,\Phi_{11'} = \Phi_{22'}.$$

From the nonzero components of $\Phi_{AB'}$, we see that the natural null tetrad for our metric must be related to the canonical tetrad through a dyad transformation of the form

$$\begin{pmatrix} w & 0 \\ 0 & w^{-1} \end{pmatrix}, \qquad w \in \mathbb{C}.$$

Applying this dyad transformation to (13), we find that the transformed Ricci spinor, $\hat{\Phi}_{AB'}$, has the form

$$\hat{\Phi}_{00'} = 2\alpha^{-1}(w\bar{w})^2\Phi_{11'}, \qquad \hat{\Phi}_{11'} = \Phi_{11'}, \qquad \hat{\Phi}_{22'} = 2\alpha(w\bar{w})^{-2}\Phi_{11'}.$$

To achieve the canonical form for $\hat{\Phi}_{AB'}$, we therefore require

$$(w\bar{w})^4 = \alpha^2 \Rightarrow (w\bar{w})^2 = \pm\alpha. \tag{14}$$

If we choose the positive sign, it follows that

$$\hat{\Phi}_{00'} = \hat{\Phi}_{22'} = 2\hat{\Phi}_{11'},$$

and the solution represents a perfect fluid, while with the negative sign

$$\hat{\Phi}_{00'} = \hat{\Phi}_{22'} = -2\hat{\Phi}_{11'},$$

and the solution represents a tachyon fluid. It appears that we have a Ricci spinor which can be interpreted as two different types of fluids! However, since we must have $w\bar{w} \geq 0$, it follows that the sign of $\pm\alpha$ must be positive to satisfy this condition. Thus at any given point in space-time, only one choice of sign is valid, and one type of fluid is distinguished. Globally, however, $\alpha = \alpha(x, t)$ and the sign of α will (in general) vary over the space-time. In this case, the solution represents a PF over some region of the manifold, and a tachyonic fluid over the rest. We shall refer to these solutions as mixed (or PF/TF) solutions.

Having described the general technique for obtaining and standardising perfect fluid solutions from (7), we develop in the next section, the specific expressions for that metric.

31.3 The Basic Quantities

It turns out that the expressions for the spinor components related to (7) are more compact and more easily treated if we define new variables

$$T \equiv t^{-1}x^{-b_0}, \qquad X \equiv x^{-1}t^{-a_1}.$$

In order that T and X be independent, we exclude for now the case $a_1 b_0 = 1$. With respect to (10), the nonzero components of the Weyl spinor are given by

$$\begin{aligned} 4\Psi_0 &= (b_2 - b_3)(1 + b_0 - b_2 - b_3)X^2 + (a_2 - a_3)(1 + a_1 - a_2 - a_3)T^2 \\ &\quad + 2[a_1(b_2 - b_3) + a_2(b_0 - b_2) + a_3(b_3 - b_0)]TX, \\ 12\Psi_2 &= (2b_0^2 - b_0b_2 - b_0b_3 - b_2^2 + 2b_2b_3 - b_3^2 - 2b_0 + b_2 + b_3)X^2 \\ &\quad - (2a_1^2 - a_1a_2 - a_1a_3 - a_2^2 + 2a_2a_3 - a_3^2 - 2a_1 + a_2 + a_3)T^2, \end{aligned}$$

$$4\Psi_4 = (b_2 - b_3)(1 + b_0 - b_2 - b_3)X^2 + (a_2 - a_3)(1 + a_1 - a_2 - a_3)T^2$$
$$- 2[a_1(b_2 - b_3) + a_2(b_0 - b_2) + a_3(b_3 - b_0)]TX, \qquad (15)$$

and we see explicitly that $\Psi_1 = \Psi_3 = 0$. The only nonzero components of the Ricci spinor are

$$4\Phi_{00'} = [(1 + b_0)(b_2 + b_3) - b_2^2 - b_3^2]X^2 + [(1 + a_1)(a_2 + a_3) - a_2^2 - a_3^2]T^2$$
$$+ 2[a_1(b_2 + b_3) + a_2(b_0 - b_2) + a_3(b_0 - b_3)]TX,$$
$$4\Phi_{11'} = (b_0^2 - b_0 - b_2b_3)X^2 - (a_1^2 - a_1 - a_2a_3)T^2,$$
$$4\Phi_{22'} = [(1 + b_0)(b_2 + b_3) - b_2^2 - b_3^2]X^2 + [(1 + a_1)(a_2 + a_3) - a_2^2 - a_3^2]T^2$$
$$- 2[a_1(b_2 + b_3) + a_2(b_0 - b_2) + a_3(b_0 - b_3)]TX,$$
$$4\Phi_{02'} = (b_2 - b_3)(1 - b_0 - b_2 - b_3)X^2 - (a_2 - a_3)(1 - a_1 - a_2 - a_3)T^2.$$

We turn first to the necessary PF condition $\Phi_{02'} = 0$. In the (T, X) variables it is clear that there is no possibility of satisfying this by matching like exponents of the variables, thus canceling terms. Hence the coefficients of T and X must vanish separately, implying

$$(a_2 - a_3)(1 - a_1 - a_2 - a_3) = 0 \quad \text{and} \quad (b_2 - b_3)(1 - b_0 - b_2 - b_3) = 0. \quad (16)$$

Once (16) is solved, we need only impose the condition $\chi_2 = 0$. This provides the equation

$$\chi_2 = \frac{1}{96}\Sigma_{TTTT}T^4 + \frac{1}{96}\Sigma_{XXXX}X^4 + \frac{1}{48}\Sigma_{TTXX}T^2X^2 = 0, \qquad (17)$$

where we introduce the notation that Σ_{ABCD} represents (up to a constant factor) the coefficient of the term in variables $ABCD$ in χ_2. These normalized coefficients are

$$\Sigma_{TTTT} \equiv (a_1 + a_2 + a_3 - 1)$$
$$\cdot (a_2^3 - a_2^2 - a_1a_2^2 - a_2^2a_3 - a_2a_3^2 + 2a_2a_1^2 + 2a_3a_1^2$$
$$- a_1a_3^2 - a_3^2 + a_3^3 - 2a_1^3 + 2a_1^2),$$
$$\Sigma_{XXXX} \equiv (b_0 + b_2 + b_3 - 1)$$
$$\cdot (b_3^3 - 2b_0^3 + 2b_0^2 + 2b_0^2b_2 + 2b_0^2b_3 - b_0b_2^2$$
$$- b_0b_3^2 - b_2^2b_3 - b_2b_3^2 + b_2^3 - b_2^2 - b_3^2),$$
$$\Sigma_{TTXX} \equiv Q(a_1, a_2, a_3, b_0, b_2, b_3), \qquad (18)$$

with Q a polynomial of 38 terms when fully expanded, quadratic in all its arguments.

From (16) we see the importance of the Kasner condition applied separately to the parameter sets $\{a_i\}$ and $\{b_i\}$: the condition is sufficient (though not necessary) for the vanishing of $\Phi_{02'}$ and hence is a key item in the solution of the PF field equations for these power-law models. In order to fully solve the field equations, we still of course have to satisfy (17). From (18) we see that, in addition to setting $\Phi_{02'} = 0$, the application of the Kasner condition produces the bonus that we automatically eliminate two independent terms from χ_2!

The 6 original parameters, along with the 3 algebraic restrictions imposed by the 2 Kasner conditions and the only additional condition that $\Sigma_{TTXX} = 0$, thus produce a 3-parameter class of solutions. In contrast, if we use the pair of axisymmetric conditions to set $\Phi_{02'}$ to zero, then, together with the 3 independent algebraic conditions (18), we have 5 conditions on 6 parameters and the resulting solutions have at most 1 parameter. By similar arguments we expect the use of one Kasner condition and one axisymmetric condition to result in solutions with two free parameters.

The simultaneous solution of various algebraic conditions via the Kasner conditions therefore not only simplifies the solution of the field equations by reducing the effective number of independent algebraic restrictions on the exponents, but, as a by-product, also produces a larger parameter space of cosmological models.

Returning to (16), we have four possible combinations of parameters which satisfy the equation, and these will form the basis of the broad classes of solutions:
Class A: ("doubly axisymmetric") $a_2 = a_3$ and $b_2 = b_3$.
Class B: $a_2 = a_3$ and $K_b = 0$.
Class C: $K_a = 0$ and $a_2 = a_3$.
Class D: ("doubly Kasner") $K_a = 0$ and $K_b = 0$.

In the next section we look at the remaining algebraic equations for $\chi_2 = 0$ for all classes, but focus principally on the "double Kasner" metrics.

31.4 The Cosmological Models

Class A (doubly axisymmetric). Setting $a_3 = a_2$ and $b_3 = b_2$, we discover from (18) that

$$\Sigma_{TTTT} = 2(a_2 - a_1)(a_1 + 2a_2 - 1)(a_1^2 - a_1a_2 - a_1 - a_2),$$

$$\Sigma_{XXXX} = 2(b_2 - b_0)(b_0 + 2b_2 - 1)(b_0^2 - b_0b_2 - b_0 - b_2),$$

$$\begin{aligned}\Sigma_{TTXX} = 2[&(3a_1 - 1)(a_2 - a_1)b_2^2 \\ &+ a_2b_2(3a_2b_0 - 3a_1b_0 + a_1 - a_2 + b_0 + 1) \\ &+ b_0(a_1^2b_0 - 3a_2^2b_0 - a_1^2 - a_1b_0 + a_2^2 + a_1)],\end{aligned}$$

all of which must vanish to satisfy the PF field equations. There are thus 9 different combinations of factors for which Σ_{TTTT} and Σ_{XXXX} vanish. These combinations are listed in Table 31.1, along with their consequences for the remaining condition

Table 31.1 Subdivisions of Class A. The final column is an algebraic quantity whose vanishing is equivalent to that of the coefficient Σ_{TTXX}.

Class	$a_2 = a_3$	$b_2 = b_3$	$\Sigma_{TTXX} = 0$
A1	a_1	b_0	$a_1 b_0(1 - a_1 b_0)$
A2	a_1	$\frac{1-b_0}{2}$	$a_1(-2b_0 + 3b_0^2 - 1 + 4a_1 b_0^2)$
A3	a_1	$\frac{b_0(b_0-1)}{1+b_0}$	$a_1 b_0$
A4	$\frac{1-a_1}{2}$	b_0	$b_0(-2a_1 - 1 + 3{a_1}^2 + 4a_1^2 b_0)$
A5	$\frac{a_1(a_1-1)}{1+a_1}$	b_0	$a_1 b_0$
A6	$\frac{1-a_1}{2}$	$\frac{1-b_0}{2}$	$(6a_1 - 4a_1^2 - 3)b_0^2 + (6a_1^2 - 6a_1 + 2)b_0 - a_1(3a_1 - 2)$
A7	$\frac{1-a_1}{2}$	$\frac{b_0(b_0-1)}{1+b_0}$	$b_0(a_1 b_0 - 2a_1 + 1)$
A8	$\frac{a_1(a_1-1)}{1+a_1}$	$\frac{1-b_0}{2}$	$a_1(a_1 b_0 - 2b_0 + 1)$
A9	$\frac{a_1(a_1-1)}{1+a_1}$	$\frac{b_0(b_0-1)}{1+b_0}$	$a_1 b_0$

$\Sigma_{TTXX} = 0$ (the quantity in the last column vanishes if and only if Σ_{TTXX} vanishes, and may be considered as the square-free decomposition of the primitive part of the numerator of Σ_{TTXX}).

Though we will not go into any solutions in detail, we do note that not all solutions lead to models with a maximal G_2. For example, when we take into account that $a_1 b_0 \neq 1$, the models of class A must have either $a_1 = a_2 = a_3 = 0$ or $b_0 = b_2 = b_3 = 0$, which respectively eliminate the dependence on t or x, thereby introducing a third Killing vector. It is also worth pointing out that if we identify $\{t, x, y, z\}$ with the same variables as used in the standard metric of spatially flat FRW, the candidates for perturbations of FRW will belong to class A, since the general spatially flat FRW model cannot be considered to lie in a neighborhood of the solutions with either $K_a = 0$ or $K_b = 0$.

Classes B and C. For models of class B, we set $a_3 = a_2$ and $K_b = 0$, and see that $\Sigma_{XXXX} = 0$ identically. From (18) we have that

$$\Sigma_{TTTT} = 2(a_2 - a_1)(a_1 + 2a_2 - 1)(a_1^2 - a_1 a_2 - a_1 - a_2),$$

$$\begin{aligned}\Sigma_{TTXX} = {} & 2[{a_1}^2 - (a_2 + 1)a_1 - a_2]b_2^2 - 2(b_0 - 1)(a_2 a_1 + a_2 - {a_1}^2 + a_1)b_2 \\ & - (3b_0 - 1)^2 a_2^2 + 2(b_0 - 1)(2a_1 b_0 - b_0 - a_1)a_2 \\ & + a_1(b_0 - 1)(a_1 b_0 - 2b_0 + a_1).\end{aligned}$$

Table 31.2 Subdivisions of Class B. The final column is an algebraic quantity whose vanishing is equivalent to that of the coefficient Σ_{TTXX}.

Class	$a_2 = a_3$	$\Sigma_{TTXX} = 0$
B1	a_1	$a_1(b_0^2 - b_2 - b_0 + b_0 b_2 + b_2^2 + a_1 b_0^2)$
B2	$\frac{1-a_1}{2}$	$4(3a_1+1)(-1+a_1)b_2^2 + 4(3a_1+1)(-1+a_1)(b_0-1)b_2$ $-(13a_1^2 - 22a_1 + 13)b_0^2 + (18a_1^2 + 10 - 20a_1)b_0 - (3a_1 - 1)^2$
B3	$\frac{a_1(a_1-1)}{1+a_1}$	$a_1(a_1 b_0 - 2b_0 + 1)$

We can divide these solutions into three finer classes, corresponding to the vanishing of each of the factors of Σ_{TTTT}, as outlined in Table 32.2. The consequences for the remaining parameters from the condition $\Sigma_{TTXX} = 0$ are also given. We note that, in the most general cases, there are two free parameters.

Playing the same game with $b_2 = b_3$ and $K_a = 0$, we find that we can analogously define three classes of solutions in class C, equivalent to those of class B, but with $a_1 \leftrightarrow b_0$, $a_2 \leftrightarrow b_2$, and $X \leftrightarrow T$. We shall not investigate these solutions further here, but instead consider in more detail the doubly Kasner solutions.

Class D. Imposing $\sum a_i = 1$ and $\sum b_i = 1$, the condition that the χ_A vanish reduces to the single equation for the coefficient $\Sigma_{TTXX} = 0$. Using the Kasner conditions to eliminate a_3 and b_3, we find that the remaining parameters must satisfy the (in general) quadratic equation

$$(3a_1+1)(a_1-1){b_2}^2 + 2\left(3a_1 + 3a_2 - a_1^2 - 5a_1 a_2 - 2\right) b_0 b_2$$
$$+ \left(3a_2^2 - 5a_1^2 - 2a_1 a_2 + 8a_1 - 4\right) b_0^2 + 2(3a_1 a_2 - a_1 - a_2 + 1) b_2$$
$$+ 2\left(4a_1^2 - a_2^2 + 3a_1 a_2 - 5a_1 - a_2 + 2\right) b_0 - (2a_1 + a_2 - 1)^2 = 0 \quad (19)$$

We can, of course, write the explicit solution(s) of (19) for, say, $b_2 = b_2(a_1, a_2, b_1)$, and thereby explicitly obtain the doubly Kasner perfect fluid metrics of (7). The expressions that result are not, however, more illuminating than the original ones, and we prefer to retain (19) as a condition to be imposed on the 4 remaining parameters whenever we require it. Despite the complexity of the expressions, we can, in general, obtain some useful physical information about these solutions. Probably the most important is that all the models have equation of state $p = \rho$ and therefore represent stiff perfect fluids. Explicitly,

$$p = \rho = (b_0^2 + b_2^2 + b_0 b_2 - b_0 - b_2)X^2 - (a_1^2 + a_2^2 + a_1 a_2 - a_1 - a_2)T^2$$

(with the parameters linked by (19)). We therefore see that, though these models retain many parameters, the equation of state means they are largely uninteresting from the point of view of generalizing FRW models. Also of physical interest is the fact that the solutions are, in general, tilted. The nonzero components of the fluid 4-velocity (with $u_a u^a = 1$) satisfy

$$u_0^2 = \frac{(a_1^2 + a_2^2 + a_1 a_2 - a_1 - a_2)t^{-2}}{(a_1^2 + a_2^2 + a_1 a_2 - a_1 - a_2)T^2 - (b_0^2 + b_2^2 + b_0 b_2 - b_0 - b_2)X^2},$$

$$u_1^2 = \frac{(b_0^2 + b_2^2 + b_0 b_2 - b_0 - b_2)x^{-2}}{(a_1^2 + a_2^2 + a_1 a_2 - a_1 - a_2)T^2 - (b_0^2 + b_2^2 + b_0 b_2 - b_0 - b_2)X^2},$$

showing the tilted nature, unless either

$$a_1^2 + a_2^2 + a_1 a_2 - a_1 - a_2 = 0 \quad \text{or} \quad b_0^2 + b_2^2 + b_0 b_2 - b_0 - b_2 = 0.$$

It is always nice to see some explicit examples, so we provide below some combinations of the power-law parameters which satisfy the double Kasner condition. The first of these also belongs to class B, and can be obtained by many routes: for instance it appears if we demand that (19) is linear in b_2 and, further, that the discriminant of the remaining equation, considered as a quadratic equation for b_0, is zero. Alternatively, it can be obtained as a nontilted model, axisymmetric with respect to the a_i.

31.5 Application to Space-Times Admitting a G_1

Since the Kasner conditions proved so useful in solving the field equations for diagonal power-law G_2 metrics, one might ask whether this usefulness extends to more general space-times. The obvious generalization of the work on cosmological solutions admitting a G_2 isometry group is to consider the metric

$$ds^2 = x^{2b_0} y^{2c_0} dt^2 - t^{2a_1} y^{2c_1} dx^2 - t^{2a_2} x^{2b_2} dy^2 - t^{2a_3} x^{2b_3} y^{2c_3} dz^2, \qquad (20)$$

which manifestly admits only 1 Killing vector, ∂_z. The metric also admits a homothetic Killing vector, and so the space-times will at least admit an H_2 of homotheties. Extending the calculations of Section 31.3, we define variables, τ, ξ, γ, such that

$$T \equiv x^{-b_0} y^{-c_0} t^{-1}, \qquad X \equiv t^{-a_1} y^{-c_1} x^{-1}, \qquad Y \equiv t^{-a_2} x^{-b_2} y^{-1}.$$

It is found that in terms of these variables, the field equations once again reduce to polynomial equations, and so once again the problem of solving the field equations reduces to equating a set of coefficients to zero. However, the solution of these equations is substantially more complicated: in the natural null tetrad, we find

Table 31.3 Some particular values of exponents for cosmological models of class D.

a_1	a_2	a_3	b_0	b_2	b_3
$-\frac{1}{3}$	$\frac{2}{3}$	$\frac{2}{3}$	$\frac{3}{7}$	b_2	$1 - b_2 - \frac{3}{7}$
$-\frac{1}{3}$	$\frac{5}{3}$	$-\frac{1}{3}$	$\frac{15}{19}$	$\frac{10}{19}$	$-\frac{6}{19}$
$-\frac{1}{3}$	$\frac{6+4\sqrt{3}}{9}$	$\frac{6-4\sqrt{3}}{9}$	$\frac{1}{13}$	$\frac{6-4\sqrt{3}}{13}$	$\frac{6+4\sqrt{3}}{13}$

that, in general, no component of Ψ_A or $\Phi_{AB'}$ is zero, and we cannot introduce the simplifying arguments of (11). All components of χ_A are nonzero and are quartic polynomials in the new variables. We find, however, that many of the coefficients of these quartics have a Kasner condition as one of the factors. Using the same notation for coefficients introduced in Section 31.3, we may, for instance, write

$$\begin{aligned}\chi_0 = {} & \Sigma_{TTTT}T^4 + \Sigma_{XXXX}X^4 + \Sigma_{YYYY}Y^4 \\ & + \Sigma_{TTTX}T^3X + \Sigma_{TXXX}TX^3 + \Sigma_{XXYY}X^2Y^2 \\ & + \Sigma_{TTXX}T^2X^2 + \Sigma_{TTYY}T^2Y^2 + \Sigma_{TXYY}TXY^2\,,\end{aligned}$$

(with all other coefficients identically zero). If we assume the three Kasner conditions between the a_i, b_i, and c_i, then six of these coefficients vanish, leaving only Σ_{TTYY}, Σ_{TXYY}, and Σ_{XXYY} to solve for.

In fact, after imposing the Kasner conditions, the number of nonzero coefficients in each coefficient of $\{\chi_A\}$, $A = 0, \ldots, 4$, reduces from $\{9, 6, 6, 6, 9\}$ to $\{3, 2, 3, 2, 3\}$. Though still apparently highly overdetermined, with 13 conditions on the 6 remaining parameters, many of the remaining coefficients are equal up to a sign, and it turns out that only 6 of the coefficients are linearly independent.

As a concrete example that we can determine cosmological solutions in this manner, the metric (21) represents a perfect fluid cosmological solution, which we believe to be new, that has been constructed to have only one Killing vector. using a triple Kasner condition

$$ds^2 = x^{4/7}y^{4/3}dt^2 - t^{-1/2}y^{-2/3}dx^2 - t^{3/2}x^{6/7}dy^2 - tx^{4/7}y^{4/3}dz^2. \tag{21}$$

Unfortunately it is difficult to verify whether the solution really does admit only one Killing vector, a point which will be discussed in Section 31.6. Key to that discussion is the form of the Weyl spinor for (21) in the natural null tetrad, with components

$$\begin{aligned}\Psi_0 &= \frac{1}{49}t^{1/2}x^{-2}y^{2/3} - \frac{1}{14}t^{-3/4}x^{-9/7}y^{-1/3} - \frac{1}{3}t^{-3/2}x^{-6/7}y^{-2} \\ &\quad - \frac{1}{32}t^{-2}x^{-4/7}y^{-4/3}, \\ \Psi_1 &= \Psi_3 = -\frac{1}{4}t^{-7/4}x^{-5/7}y^{-5/3}, \\ \Psi_2 &= \frac{1}{147}t^{1/2}x^{-2}y^{2/3} - \frac{1}{9}t^{-3/2}x^{-6/7}y^{-2} - \frac{17}{96}t^{-2}x^{-4/7}y^{-4/3}, \\ \Psi_4 &= \frac{1}{49}t^{1/2}x^{-2}y^{2/3} + \frac{1}{14}t^{-3/4}x^{-9/7}y^{-1/3} - \frac{1}{3}t^{-3/2}x^{-6/7}y^{-2} \\ &\quad - \frac{1}{32}t^{-2}x^{-4/7}y^{-4/3},\end{aligned}$$

while the components of the Ricci spinor factorize as

$$\Phi_{00'} = \frac{(16t^{5/4}y - 7x^{5/7})^2}{1568t^2x^2y^{4/3}},$$

$$\Phi_{11'} = \frac{(16t^{5/4}y - 7x^{5/7})(16t^{5/4}y + 7x^{5/7})}{3136t^2x^2y^{4/3}},$$

$$\Phi_{22'} = \frac{(16t^{5/4}y + 7x^{5/7})^2}{1568t^2x^2y^{4/3}}.$$

We see that the solution is of mixed PF/TF form. In fact, once again, the solution represents a stiff perfect fluid.

31.6 Aspects of Invariant Classification

For a Petrov type I space-time it is well-known (see for example [14]) that a tetrad can be found in which the Weyl spinor assumes the form

$$\Psi_0 = \Psi_4, \qquad \Psi_1 = \Psi_3 = 0, \tag{22}$$

with Ψ_2 unrestricted. All nonzero components of Ψ_A in this tetrad will, in general, be complex. For a type I space-time, condition (22) completely fixes the tetrad (up to discrete swaps of null directions), and this completely determined tetrad is known as the canonical tetrad for the space-time. Once the Riemann tensor, and its decomposition in terms of symmetric spinors, is known, the real and imaginary parts of each spinor component can be regarded as potentially new coordinates for the space-time. Examining the real and imaginary parts of all nonconstant spinor components for functional independence will therefore produce a set of n essential coordinates for the space-time and, since type I metrics have no isotropy group, the isometry group will have dimension $4 - n$.

Of course the condition (22) is not the only possible condition which completely fixes the tetrad. Among a host of other possibilities, we could well have chosen

$$\Psi_0 = \Psi_4 = 0, \qquad \Psi_1 = \Psi_3, \tag{23}$$

with Ψ_2 unrestricted. In fact, the standard form need not be initially based on the Weyl spinor—a possibility which we shall discuss in Section 31.6. However, in the given circumstances, it would be rather perverse to use (23) as a standard form, particularly since CLASSI uses (22).

For the G_2 metrics, the dyad transformation which brings Ψ_A into the standard form (22) is easy to determine (conceptually, if not algebraically). Since Ψ_1 and Ψ_3 are already zero, and Ψ_0 and Ψ_4 are real, we need only apply a boost of the form

$$\begin{pmatrix} q & 0 \\ 0 & q^{-1} \end{pmatrix}, \qquad q = \left(\frac{\Psi_4}{\Psi_0}\right)^{1/8} \in \mathbb{R},$$

to the natural null tetrad to obtain the canonical tetrad, in which the transformed Weyl spinor, $\hat{\Psi}_A$, has nonzero components

$$\hat{\Psi}_0 = \hat{\Psi}_4 = \sqrt{\Psi_0\Psi_4}, \qquad \hat{\Psi}_2 = \Psi_2.$$

For metrics of class D, one finds in general that $\hat{\Psi}_0$ and $\hat{\Psi}_2$ are functionally independent and therefore that the space-times admit at most a G_2 (Abelian).

When we turn to the G_1 metrics, things become rather more complicated. As we can see from (21), Ψ_1 and Ψ_3 are no longer nonzero, which means that the dyad transformation required to bring Ψ_A into standard form is no longer a simple spin-boost. In fact the elements of the dyad transformation depends on the roots of a quartic equation involving the components of Ψ_A, which are themselves rather lengthy algebraic expressions. Though this quartic is, in principle, solvable, the resulting expressions are, to say the least, algebraically complicated and the calculation of the spinor components in the canonical basis, and subsequent analysis of functional independence of these components, has defied attempts at computation.

The G_1 thus solutions demonstrate a characteristic that was thought would be rare in exact solutions: previously it was suspected that because exact solutions tend to be developed in a frame that simplifies the structure of the Riemann tensor, the resulting dyad transformations to bring the working tetrad into standard form would be relatively easy to calculate. However, despite the apparent simplicity of the working tetrad (10), the solutions developed here have largely been produced from algebraic considerations, with little regard to the consequences for the form of the Riemann tensor components. The fact that we can discover simple solutions such as (21), whose invariant classification defies a standard approach, poses new challenges for the invariant classification algorithms.

One subject currently under investigation is the development of canonical tetrads based primarily on the components of the Ricci spinor. We can see from (22) that, at least for the metric (21), the Ricci spinor is within a boost of its standard form. We can thus use the Ricci spinor to restrict the frame up to the SO(3) isometry group of a perfect fluid. To fix the frame completely, we need then consider only the effect of the SO(3) group on Ψ_A and not the full Lorentz group. The problem here is that the effect of SO(3) transformations on Ψ_A is rather complicated, and it is not obvious how best to tie down the components of Ψ_A. For this reason we cannot say definitively that (21) represents a perfect fluid with only one Killing vector.

31.7 Discussion

In this paper we have shown how computer algebra methods can be used to determine perfect fluid cosmological solutions with low symmetry by starting from power-law ansätze for the metric coefficients. A study of the diagonal G_2 case reveals that two types of algebraic conditions between the power-law exponents are important for the solution of the perfect fluid field equations: either the model is "axisymmetric" or the exponents obey the "Kasner condition," as defined in Section 31.1. It is hoped that some of the power-law models can be viewed as inhomogeneous generalizations of the spatially flat FRW models, though the fact that the "double Kasner" class D models represent only stiff perfect fluids makes them of little interest from this point of view.

Despite being developed from the point of view of G_2 cosmologies, it is found that both the axisymmetric and Kasner conditions bring simplifications to the diagonal G_1 power-law models, and an explicit metric, whose exponents in t, x, and y separately obey the Kasner condition has been presented here as an example. Whether the same conditions are useful in solving the field equations with no isometry remains to be seen.

Various questions have been brought to light by studying these power-law models.

(i) Can the same techniques be used to study metrics which are not reducible to diagonal metrics? This seems likely, since the introduction of off-diagonal components should not alter the polynomial form of the Einstein equations, though it may well be that the coordinates T and X used here are no longer suitable.

(ii) Can similar techniques be used to generalize spatially closed and open FRW models, by introducing trigonometric and/or hyperbolic functions of the variables in the metric?

(iii) Does the imposition of n Kasner conditions on a perfect fluid power-law metric with isometry group G_{4-n} necessarily produce a stiff fluid? Is there any way, other than by explicitly solving the field equations, that this can be seen?

(iv) Do the perfect fluid equations remain underdetermined when there is no isotropy, or are the number of constraints such that no digonal power-law solutions exist when there is no Killing vector?

(v) Are there any nonstiff perturbations of FRW models in these solutions?

(vi) How do we develop an alternative set of standard forms and canonical tetrads based on the Ricci spinor for handling "pathological" space-times where it is algebraically difficult to calculate the canonical tetrad based on the Weyl spinor?

We hope to have the answers to some of these questions in the future.

Acknowledgments

The author is grateful for comments by Filipe Paiva on an earlier version of this paper.

References

[1] I. Ozsváth and E. Schücking, (1962). An anti-Mach metric, in *Recent Developments in General Relativity*, Pergamon Press, Oxford.

[2] I. Ozsváth and E. Schücking, (1969). The finite rotating universe *Ann. Phys.*, **55**, 166.

[3] E. Kasner (1921). Geometrical theorems on Einstein's cosmological equations, *Amer. J. Math.* **43**, 217.

[4] A. Harvey, (1990). Will the real Kasner metric please stand up, *Gen. Rel. Grav.* **22**, 1433–1445.

[5] J. Carot and A. Sintes Olives, (1997). Homothetic perfect fluid space-times, *Class. Quant. Grav.* **14**, 1183–1205.

[6] C. Uggla and K. Rosquist, (1988). Asymptotic cosmological solutions; orthogonal Bianchi type-I, III, IV, VII models, *Class. Quant. Grav.* **5**, 767.

[7] C. Uggla, R. T. Jantzen, and K. Rosquist, (1995). Exact hypersurface-homogeneous solutions in cosmology and astrophysics, *Phys. Rev. D* **51**, 5522.

[8] C. G. Hewitt, J. W.ainwright, and S. W. Goode, (1988). Qualitative analysis of a class of inhomogeneous self-similar cosmological models, *Class. Quant. Grav.* **5**, 1313.

[9] J. E. Åman and A. Karlhede, (1980). *Phys. Lett.* **A80** 229.

[10] J. E. Åman, (1987). *Manual for CLASSI: classification programs for geometries in general relativity* (Third provisional edition) University of Stockholm Institute of Theoretical Physics technical report.

[11] M. A. H. MacCallum and J. E. Åman, (1986). Algebraically independent n^{th} derivatives of the Riemann curvature in general relativity, *Class. Quant. Grav.* **3**, 1133–1141.

[12] J. M. M. Senovilla and M. Mars, (1997). Non-diagonal separable G_2 perfect-fluid spacetimes, *Class. Quant. Grav.* **14**, 205–226.

[13] W. Seixas, (1991). Extensions to the computer-aided classification of the Ricci tensor *Class. Quant. Grav.* **8**, 1577–1585.

[14] R. Penrose and W. Rindler, (1986). *Spinors and space-time*, vol.2, Spinor and twistor methods in space-time geometry (Cambridge University Press, Cambridge).

32

Gravitational Screening

E.A. Spiegel

ABSTRACT Calculations of the stopping power of a medium lead to divergent integrals in gravitational theory as they do in the analogous electromagnetic problem. In the latter case, one usually introduces the Debye length as a cutoff at large distances to remove the divergence. The question of what to do in the Newtonian gravitational analogue can be answered by including the self-gravity of the medium. Then the Jeans length appears naturally as a cutoff. Despite the different sign of the coupling constant from the case of the electric plasma, the same sort of theory works in both cases and removes the need of introducing *ad hoc* cutoffs from both of them.

32.1 Gravitational Stopping Power

In calculating the total Coulomb scattering cross section, one is normally faced at the end with an integral over all scattering angles. This integral diverges, and some fix is needed. In plasma physics one usually says that the smallest scattering occurs at the largest impact parameter and that particles at sufficiently great distances do not feel the scatterer. The reason given is that the intervening medium effectively screens the electric potential for any object passing at a distance greater than the Debye length. Another example of this problem is seen in the calculation of the drag of a medium on a charged object passing through it. Again the Debye-Hückel [1] theory is called to the rescue.

In this discussion I am concerned with the second example, but mainly in the gravitational setting. The problem is how to deal with the divergence of the integral for the gravitational drag on a massive object passing through a uniform medium. A cutoff is needed, but the question is whether we can simply use the seemingly natural Jeans length as a cutoff in the purely gravitational case. As we shall see, if one takes into account explicitly the self-gravity of the ambient medium, there is no need to invoke an *ad hoc* assumption about cutoffs. The problem solves itself, and the same is true for a charged projectile. There does remain the need for a short-range cutoff, but this has to do with processes that are not closely connected with the present considerations.

The discussion here will be couched in purely Newtonian terms, but I feel that this needs no apology. After all, this volume has been assembled to honor Engelbert Schucking, one of the great heroes of the neo-Newtonian revolution in cosmology that began in the mid-thirties. What better precedent could one have? In fact, Engelbert is indirectly responsible for the work reported here.

After I had attended the first Texas Conference in Relativistic Astrophysics that Engelbert and his fellow Texans organized in 1964, I returned with ideas of how to speed up the gravitational collapse of the core of a galaxy (a problem which I had been discussing previously with M. A. Ruderman). I wrote out my notions and sent them to D. E. Osterbrock, who, in a pleasant way, raised an objection that I never overcame. However, the following calculations arose from my attempts to respond to him. My understanding of what it all meant was advanced by comments of C. W. Misner. So with thanks to Engelbert, whose actions got me onto this path of agreeable interactions, both personal and gravitational, let me next state the problem that arises when the self-gravity of the ambient medium is not included.

32.2 A Drag Crisis

When an object of mass m moves with velocity $\mathbf{U}$ through an infinite, uniform medium of density ρ_0, the fluid, initially at rest, begins to move with velocity $\mathbf{u}$ and its density is modified to ρ. The fluid dynamics is then described by these equations:

$$\partial_t(\rho\mathbf{u}) + \nabla \cdot (\rho\mathbf{u}\mathbf{u}) = -\nabla p - \rho\nabla V, \tag{1}$$

$$\partial_t\rho + \nabla \cdot (\rho\mathbf{u}) = 0, \tag{2}$$

where V is the potential of the gravitational field. The field is, in this section *only*, assumed to come from the passing object alone, so that

$$\Delta V = 4\pi Gm\delta(\mathbf{x} - \mathbf{U}t)\,, \tag{3}$$

where Δ is the Laplacian and δ is the Dirac function. We assume that the drag on the object is so weak that we may treat $\mathbf{U}$ as constant.

Let us write $\rho = \rho_0(1+\psi)$ and assume that $|\psi| << 1$ everywhere, with similar assumptions about the fluid velocity and pressure. Then we obtain linear equations for the perturbation caused by the intruding object. Though there are also some results on the nonlinear problem, I will restrict myself to the linear case here, since this illustrates nicely the point I want to raise.

With the assumption that $p = p(\rho)$, we can (in the linear case) write the pressure perturbation as $a^2\rho_0\psi$, where a is the adiabatic speed of sound, given by $a^2 = [\partial p/\partial\rho]_0$. It is then a straightforward matter to linearize the equations and boil them down to an equation for ψ. The linearized forms of (1) and (2) are

$$\partial_t\mathbf{u} = a^2\nabla\psi + \nabla V, \tag{4}$$

$$\partial_t\psi = -\nabla \cdot \mathbf{u}. \tag{5}$$

With the help of (3), we may condense this to an inhomogeneous wave equation for ψ:

$$\partial_t^2 \psi - a^2 \Delta \psi = 4\pi G m \delta(\mathbf{x} - \mathbf{U}t) \,. \tag{6}$$

The flow induced by the object is called an accretion flow, though, to study the actual accretion, one might also include a sink term on the right of (2). What I am interested in here is the so-called stopping power of the medium. In the next section we shall calculate this by evaluating the gravitational force of the disturbed medium on the object. This can be done as described by Landau and Lifshitz [2] for the standard Cherenkov problem. I postpone this to the next section, since I want to explain here why something extra is needed. That something is the self-force of the medium, which I am leaving out in this section. The problem without self-interaction has a significant literature.

The solution to the drag problem in the gravitational case (with magnetic field included) was first published by Dokuchaev [3], as I learned from G. Golytsin. F. D. Kahn (private communication) and K. H. Prendergast (private communication) also were early among those who thought about this calculation. A very simple way of solving (6) is to be found in the book of Ward [4], as S. Childress has told me. There are also papers dealing with the wakes of plasma probes [5], charged satellites [6], and massive objects in galactic disks [7] [8]. And of course, behind and beyond all this is the heritage of the nuclear and atomic literature on stopping power with contributions from Bethe, Bohr, Fermi, and other Olympians (see [9]).

The line along which the object moves is called the accretion axis [10]. In the reference frame of the object, we may find solutions which are steady and symmetric around the accretion axis [11] by the methods of linear acoustic theory [4]. To express these, we let $U = |\mathbf{U}|$ and introduce the Mach number, $M = U/a$, and the accretion radius,

$$R_A = \frac{2Gm}{a^2} \,. \tag{7}$$

We also let θ be the angle measured from the downstream accretion axis and r be the distance from the object, both in the object frame. Then, for subsonic motion ($M < 1$), we obtain the solution

$$\psi = \frac{R_A M^2}{r\sqrt{(1 - M^2 \sin^2 \theta)}} \,. \tag{8}$$

For supersonic motion ($M > 1$), the disturbance is confined to the region inside the downstream Mach cone; that is, $\psi = 0$ for $\theta > \arcsin M^{-1}$. Inside the downstream Mach cone, the solution is again (8). These solutions may be verified by direct substitution into (6).

Apart from remarking that the shock wave at $\theta = \arcsin M^{-1}$ is singular on account of the linearization and the point nature of the interloper, I forgo discussion of this solution (but see [11] for some illustrations). Of concern here are two conclusions about the drag force that may be drawn from the formulation of the problem as stated. This force is, for reasons of symmetry, along the accretion axis, and is written in the form $\mathbf{F} = -F_A\mathbf{U}/U$ where, on dimensional grounds, the accretion drag is

$$F_A = \pi R_A^2 \rho_0 U^2 C_A \,. \tag{9}$$

Here, we have introduced the accretion drag coefficient C_A that contains the crux of the problem.

The first conclusion is that the drag in the subsonic case, $M < 1$, is zero. This is a familiar result in terrestrial subsonic acoustics since, there too, subsonic flows have fore-aft symmetry. The second conclusion is that the drag formula makes no sense in the supersonic case unless we introduce some additional physics. To these rather bald statements, let me add some heuristic justification for those who may not wish to await the more extensive calculations of the next section. Those who do not care much for heuristic arguments might skip to the next section at once.

As in the Coulombic case, one may make the sudden approximation to calculate the stopping power [12]. A fluid particle going by in the object's rest frame is given a transverse kick and starts moving with velocity $v_\perp$ toward the accretion axis. This velocity is really a transverse momentum per unit mass, and we may therefore estimate it as the force (per unit mass) on the fluid element multiplied by the time it acts. That force per unit mass is Gm/b^2, where b is the impact parameter. The time of action is approximately $2b/U$. The product is then

$$v_\perp = \frac{2Gm}{bU} \,. \tag{10}$$

Energy is delivered to the medium at the rate $\mathbf{F} \cdot \mathbf{U}$. To estimate this, we note that the kinetic energy of the medium is created at the rate $\frac{1}{2}\rho_0 v_\perp^2 U$ per unit time per unit of area normal to the direction of symmetry axis. If we integrate over that area, we obtain the rate of energy delivered to the medium (or lost by the object). On dividing that by U, we find that the force has magnitude

$$F_A = 2\pi \int_{b_{min}}^{b_{max}} \frac{1}{2}\rho_0 v_\perp^2 b \, db \,, \tag{11}$$

where the integration has been limited between a minimum and a maximum impact parameter. If we put in the derived estimate for $v_\perp$, and compare the result with (9), we find that the accretion drag coefficient is

$$C_A = \log \frac{b_{\text{max}}}{b_{\text{min}}} \,. \tag{12}$$

The cutoff impact parameters have here been put in by hand so as to bring out the nature of the problem. To get a finite answer, we need to see where the present formulation breaks down and either cut off there or introduce some additional physics. For the inner cutoff, one possibility is that we may acknowledge that the object actually has a finite radius, R. Then, as Eddington showed [13], a fluid element coming in with impact parameter $b < b_0$, where

$$b_0 = \sqrt{R(R_A + R)} \,, \tag{13}$$

will hit the object. Assuming that the collision is sticky, we can make b_0 the inner cutoff. Of course, we should add a correction to the drag to account for the particles that hit the object, but I will not do that here. Our concern is really with the outer cutoff.

If the medium is expanding, that will produce a cutoff at a distance where the Hubble velocity, $H_0 r$, is comparable to $v_\perp$. With the expression just derived, $v_\perp = 2Gm/(Ub)$, this leads us a value

$$b_{\max}^2 = \frac{R_A U}{H_0} = R_A R_U , \tag{14}$$

where $R_U = U/H_0$ is the distance at the which expansion speed is the same as that of the object with respect to the medium.

However, the real issue here is not so much the value of the drag but its interest in bringing out the nature of the gravitational behavior of a fluid. In that case we want to know how faithful the analogy to the electromagnetic case is. What we have seen here is that a cutoff is needed. The question is whether the Jeans length $\pi a/\sqrt{4\pi G\rho_0}$ may be used for this cutoff if extraneous phenomena do not intervene. A positive answer is obtained in the next section.

32.3 Saved by Self-Gravity

In this section I sketch the calculation of the drag with the inclusion of the self-gravity of the medium. Apart from that addition, the approach is like that used to study Cherenkov radiation by Landau and Lifshitz [2]. The effect of self-gravity is to invalidate both of the conclusions about the drag mentioned in the previous section. When I first told these things to knowledgeable people, I found that some of them did not want accept one or the other of these results. Martin Lampe, who was sympathetic to my problem, independently repeated the calculations and verified the conclusions in what I regard as an act of remarkable kindness, even for someone who might be fond of contour integrations.

To include the effect of the self-gravity of the medium, we add the term $4\pi G\rho$ to the right-hand side of equation (4). This however introduces a problem: there is no longer a static homogeneous solution to the equations, even in the absence of the intruding object. This difficulty is one of long standing and there have been two general ways of coping with it, if we leave aside the Jeans [14] approach of pretending it isn't there. One thing to do is to give up the simplicity of the homogeneous background and look for inhomogeneous equilibria. In many such structures, the gravitational force balances the pressure gradient, and so their sizes are on the order of the Jeans length, which in this way can also get into the act. This is a less interesting direction for the present discussion than the second one, which is to modify the theory to allow the existence of a static, homogenous and infinite medium. Let us take the latter route.

In the analogous case of plasma dynamics, the physics itself provides the solution. There, in first approximation, ρ is the density of the electron gas and it is proportional to the charge density. If the medium is to be electrically neutral, one must include the positive charge density of the ions on the right-hand side of the Poisson equation. In the simplest models, this ion charge density is a constant but, in more sophisticated versions, it has a dynamics that is excited through coupling to the electrons through their joint potential. One then has a two-fluid problem.

In the gravitational case, we may similarly modify the Poisson equation, (4), to

$$\Delta V = 4\pi G m \delta(\mathbf{x} - \mathbf{U}t) + 4\pi G(\rho - \rho_\Lambda) . \tag{15}$$

where ρ_Λ is the density of negative gravitational mass. This is the approach I will follow here but, in passing, I feel it is worth asking whether one should go all the way in the analogy to the plasma dynamics and couple in some dynamical model for this dark antigravitational fluid to see where that leads.

When ρ_Λ is constant, we can absorb it into the potential with the replacement $V \to V + (4\pi/6)G\rho_\Lambda r^2$. Then ρ_Λ disappears from the Poisson equation and reappears explicitly on the right-hand side of the equation of motion, (1), in the form $\frac{1}{3}\Lambda\rho\mathbf{r}$, where $\Lambda = 4\pi G\rho_\Lambda$. This is the Newtonian analogue of Einstein's cosmological term. According to Bondi [15], the cosmological term was introduced in the Newtonian setting by Seeliger in 1895. Engelbert tells me that Seeliger in fact put in a shielding effect, not a repulsive force, and that Einstein believed that he too was doing such a thing in the relativistic case. However, Engelbert remarks, it was Eddington [16] who first noted that the cosmological term represented a repulsive force. I believe that its use is far preferable to Jeans' disregard of the problem of finding a static homogenous solution, especially as (15) is so suggestive.

Now we may proceed as before with the linear theory. We again let $\rho = \rho_0(1 + \psi)$, where this time $\rho_0 = \rho_\Lambda$. Thus the introduction of the cosmological density changes nothing that matters here, but it does permit us to expand about a uniform state in a better frame of mind than Jeans must have been in when he studied gravitational instability. On assuming that $|\psi| << 1$ as before, we obtain an inhomogeneous Klein-Gordon equation (6), namely,

$$\partial_t^2 \psi - a^2 \Delta\psi - a^2 k_J^2 \psi = 4\pi G m \delta(\mathbf{x} - \mathbf{U}t) , \tag{16}$$

where $k_J^2 = 4\pi G\rho_0/a^2$ and where ψ is real.

The solution to (16) is a linear superposition of the forced, or driven solution, and of the solution of the homogeneous equation consisting of the free waves. The latter, which is needed to solve the general initial value problem, contains the seeds of gravitational instability. In this problem we may expect this instability to produce gravitationally condensing wakes that resemble what is seen in a cloud chamber in the wake of a charged object. Such a process was invoked in the steady state theory to produce new galaxies. I have not seen any nonlinear treatment of this phenomenon and there is not much point in pursuing that line without one. I will therefore adopt initial conditions in which the free modes are not excited.

To find the forced solution of this problem, we may follow Landau, Lifshitz and many others into Fourier space, setting

$$\psi(\mathbf{x}, t) = \int e^{i\mathbf{k}\cdot\mathbf{x}} \hat{\psi}(\mathbf{k}, t) d^3\mathbf{k}. \tag{17}$$

Then we take the Fourier transform of (17), use

$$\delta(\mathbf{x} - \mathbf{U}t) = \frac{1}{(2\pi)^3} \int e^{i\mathbf{k}\cdot(\mathbf{x}-\mathbf{U}t)} d^3\mathbf{k}, \tag{18}$$

and obtain

$$\partial_t^2 \hat{\psi} + a^2(\mathbf{k}^2 - k_J^2)\hat{\psi} = \frac{Gm}{2\pi^2} e^{-i\mathbf{k}\cdot\mathbf{U}t} . \tag{19}$$

At this point, it is possible to anticipate that when we do the Fourier inversions, we shall have to pay attention to poles in $\hat{\psi}$. When these are on the real axis, we shall have to decide which way to go around them. To make this choice unambiguously, we replace m by $m \exp(\epsilon t)$, where ϵ is an arbitrarily small, positive quantity. With the introduction of this standard device, we see that back at $t = -\infty$ there was no intruding object so that the disturbance is turned on very gently without making waves.

With $\epsilon = 0^+$, we introduce

$$\omega = \mathbf{k} \cdot \mathbf{U} + i\epsilon, \tag{20}$$

and rewrite (19) as

$$\partial_t^2 \hat{\psi} + a^2(\mathbf{k}^2 - k_J^2)\hat{\psi} = \frac{Gm}{2\pi^2} e^{-i\omega t}. \tag{21}$$

The slow turn-on lets us concentrate on the forced solution to (20), which is

$$\hat{\psi}(\mathbf{k}, t) = \frac{Gm}{2\pi^2} \frac{e^{-i\omega t}}{a^2(\mathbf{k}^2 - k_J^2) - \omega^2} . \tag{22}$$

For $k_J = 0$, Fourier inversion of this result gives (8).

The drag on the object, $\mathbf{F}$, is the gravitational force of the medium on the object:

$$\mathbf{F} = -F_A \mathbf{U}/U \ = -m\nabla V_d , \tag{23}$$

where V_d is the gravitational potential of the disturbance in the medium. The third member of (23) is to be evaluated *at the object.*

We obtain V_d from the Poisson equation,

$$\Delta V_d = 4\pi G \rho_0 \psi , \tag{24}$$

which we solve by taking its Fourier transform:

$$\hat{V}_d = -\frac{4\pi G\rho_0}{\mathbf{k}^2} \hat{\psi} . \tag{25}$$

Then, we make a Fourier inversion of (24) and take its gradient, bringing an $i\mathbf{k}$ down into the integrand. With the z-axis chosen as the accretion axis, we must evaluate the resulting expression on the object, which is at $x = 0$, $y = 0$, $z = Ut$. Then, on multiplying by m and introducing (22), we obtain for the accretion drag force,

$$\mathbf{F} = \frac{2iG^2m^2\rho_0}{\pi} \int \frac{\mathbf{k}e^{\epsilon t}}{\mathbf{k}^2[a^2(\mathbf{k}^2 - k_J^2) - \omega^2]} d^3\mathbf{k} . \tag{26}$$

Scalar multiplication of (26) by $\mathbf{U}$ and comparison with (9) leads us to the following expression for the accretion drag for ϵ going to zero and t finite:

$$F_A = \frac{i R_A^2 \rho_0 U^4}{2\pi} \int \frac{k_z\, d^3\mathbf{k}}{\mathbf{k}^2[a^2(\mathbf{k}^2 - k_J^2) - (k_z U)^2 - 2i\epsilon U k_z]}, \tag{27}$$

where R_A is defined in (7) and we recall that $M = U/a$.

Let $\mathbf{q}$ be the component of $\mathbf{k}$ normal to the accretion axis so that $\mathbf{k} = (\mathbf{q}, k_z)$. Then this last result can be rewritten as

$$F_A = i R_A^2 \rho_0 M^2 U^2 \int_0^\infty q\, dq\, \mathcal{I}(q^2), \tag{28}$$

where

$$\mathcal{I}(q^2) = \int_{-\infty}^{+\infty} \frac{k_z\, dk_z}{(k_z^2 + q^2)[(1 - M^2)k_z^2 + q^2 - k_J^2 - 2i\epsilon M k_z/a]}. \tag{29}$$

When $(q^2 - k_J^2)/(1 - M^2) > 0$, the integrand in (29) has four poles on the imaginary k_z-axis and we can safely set $\epsilon = 0$. The integrand is then seen to be odd and nonsingular and we find $\mathcal{I} = 0$ in that case. Therefore, for $M < 1$, we have that $\mathcal{I} = 0$ for $q^2 < k_J^2$, while for $M > 1$, we find $\mathcal{I} = 0$ for $q^2 > k_J^2$.

In the case with the other sign, $(q^2 - k_J^2)/(1 - M^2) > 0$, there are two poles on the imaginary axis and two poles close to the real axis, displaced from it by an amount of order ϵ and in a direction determined by the sign of $1 - M^2$. We close the contour in the half-plane away from these latter poles and, for $(q^2 - k_J^2)/(1 - M^2) < 0$, use the method of residues to find

$$\mathcal{I} = \frac{\pi i}{(k_J^2 - q^2 M^2)} \operatorname{sgn}(1 - M^2). \tag{30}$$

For the subsonic case, we see that the integral over q has a nonzero integrand only for $0 < q < k_J$ and that the drag is

$$F_A = -\pi R_A^2 \rho_0\, U^2 \log \frac{1}{\sqrt{1 - M^2}}, \qquad M < 1, \tag{31}$$

where, by our sign convention, a negative F_A means acceleration.

In the supersonic case, the integral over q is now from k_J to ∞. The limit of integration at infinity corresponds to small scales, and we shall assume a cutoff at a large wavenumber q_0, say. We could, for example, take $q_0 = 1/b_0$ (see equation (13) or use $1/R$; I shall not try to deal with this issue here. The key point is that there is no need for a cutoff at small wavenumbers. The drag in this case is given by

$$F_A = \pi R_A^2 \rho_0 U^2 \log \sqrt{\frac{(M q_0/k_J)^2 - 1}{M^2 - 1}}, \qquad M < 1, \tag{32}$$

where the specification of the large q_0 will depend on the problem.

The way in which the introduction of a cutoff is averted by the inclusion of self-gravity is different in the subsonic and supersonic cases but, in each case, the usual problem is removed by the presence of the k_A^2 term. There may still be a worry in the divergence of the drag at Mach 1, but this cannot be helped by a cutoff. This problem may connected with linearization. Thus, though there is much to be resolved in this subject, I would claim that the artificial cutting off of the integrals is not needed.

32.4 The Message Is The Medium

The notion of shielding in an electric plasma is so familiar that one takes for granted a cutoff at the Debye distance. However, the introduction of a cutoff in a gravitational problem may produce raised eyebrows. In retrospect, I find the ready acceptance of the *ad hoc* cutoff in the former case as questionable as it would be in the latter. Once we have introduced the background cosmological density, the two situations are *almost* identical. In fact, the only role of the background density is to allow us to study a homogeneous medium. Why then should the electric case be more exempt from careful scrutiny than the gravitational case? I believe that, in both cases, although the introduction of the natural looking cutoffs does seem to gibe with physical intuition, the argument that often is offered for it is not complete.

In the example studied here, there is a steady solution in the reference frame of the intruding object (albeit an unstable one in the gravitational case). If we ask what force may be contributed by a Fourier component of the density perturbation, we may expect that, over a distance of many wavelengths, the positive and negative density contributions cancel. On the other hand, at any distance, there are wavelenths to match and to exceed that distance. These contributions are not self-canceling, but they may perhaps be weak. It appears, however, that they are strong enough to integrate up to a logarithm, and this is why the cutoffs are sometimes introduced. In fact, what is suggested by the calculations just outlined is that that argument is not enough. It appears that the reason that the long wavelengths are excluded is that they are evanescent and cannot reach out over great distances. Without that factor, the drag does indeed diverge. This is not the whole story, as we see from the trouble at Mach 1, but I suggest that it is not inappropriate to introduce cutoffs without realizing that this is an issue that requires some consideration.

The same calculations as those described in the previous section can be done for the one-component electric plasma. If you calculate the Coulombic drag, as has been done here for the gravitational case, and include the self-interaction of the plasma, you find a finite answer with no ad hockery. There are some slight differences in the outcome—in particular the drag is zero in the subsonic case for a charged object—but the removal of the singularity takes place nicely. It is somewhat pleasing then to see the gravitational theory offer something to the study of plasmas, since one gets the impression that the flow has been in the other direction in recent years [17], at least in the nonrelativistic case. In fact, the results given here were first stated in a *samizdat* that recorded the proceedings of a summer school on the dynamics of interstellar matter by D.E. Osterbrock at the University of Wisconsin in 1966. (In those days it was not automatic to publish such things except informally.) In any case, I did not give the derivation in those notes so am glad to be able outline it here. More recently, similar results have been obtained using the Vlasov equation [18], as X. Chen has informed me. As to the treatment of such issues in relativity, we may hope that this will all have been worked out nicely for Engelbert's 120th birthday.

REFERENCES

[1] Debye, P. and Hückel, E. *Physik. Z.* **24**, 185 (1923).
[2] Landau, L. D. and Lifshitz, E. M. *Electrodynamics of Continuous Media*, translated by J. B. Sykes and J. S. Bell (Addison-Wesley, Reading, MA, 1960), p. 344.
[3] Dokuchaev, V. P. *Sov. Atron.* **8**, 23 (1964).
[4] Ward, G. N. *Linearized Theory of High Speed Flow* (Cambridge Univ. Press, Cambridge, 1955).
[5] Lam, S. H. *Phys. Fluids* **8**, 73 (1965).
[6] Kraus, L. and Watson, K. M. *Phys. Fluids* **1**, 480 (1958).
[7] Julian, W. H. and Toomre, A. *Astrophys. J.* **146**, 810 (1966).
[8] Simkin, S. M. *Astrophys. J.* **159**, 463 (1969).
[9] Ichimaru, S. *Statistical Plasma Physics I* (Addison-Wesley, Reading, MA, 1992), p. 63.
[10] Lyttleton, R. A. *The Comets and Their Origin* (Cambridge Univ. Press, Cambridge, 1953), p. 66.
[11] Spiegel, E. A. p. 201 in *Interstellar Gas Dynamics* Int. Astr. Union, Symp. No. 39, H. Habing, ed. (D. Reidel, Dordrecht, 1970).
[12] Ruderman, M. A. and Spiegel, E. A. *Astrophys. J.* **165**, 1 (1971).
[13] Eddington, A. S. *The Internal Constitution of the Stars* (Dover Pub., New York, 1959), p. 391.
[14] Jeans, J. H. *Astronomy and Cosmogony* (Cambridge Univ. Press, Cambridge, 1928), p. 336.
[15] Bondi, H. *Cosmology* (Cambridge Univ. Press, Cambridge, 1960).
[16] Eddington, A. S. *The Mathematical Theory of Relativity* (Cambridge Univ. Press, Cambridge, 1924), p. 161.
[17] Fridman, A. M. and Polyachenko, V.L. *Physics of Gravitating Systems*, two volumes, (Springer-Verlag, New York, 1984); translated from the Russian by A. B. Aries and I. N. Poliakoff.
[18] Kukharenko, Yu., Vityazev, A., and Bashkirov, A. *Phys. Lett. A* **195**, 27 (1994).

33

On the Interpretation of the Einstein–Cartan Formalism

John Stachel[1]

33.1 Introduction

Hehl and collaborators [1] have suggested the need to generalize Riemannian geometry to a metric-affine geometry, by admitting torsion and nonmetricity of the connection field. They assume that this geometry represents the microstructure of space-time, with Riemannian geometry emerging as some sort of macroscopic average over the metric-affine microstructure. They thereby generalize the earlier approach to the Einstein-Cartan formalism of Hehl et al. [2] based on a metric connection with torsion. A particularly clear statement of this point of view is found in Hehl, von der Heyde, and Kerlick [3]: "We claim that the [Einstein-Cartan] field equations . . . are, at a classical level, the correct microscopic gravitational field equations. Einstein's field equation ought to be considered a macroscopic phenomenological equation of limited validity, obtained by averaging [the Einstein-Cartan field equations]" (p. 1067).

Adamowicz takes an alternate approach [4], asserting that "the relation between the Einstein-Cartan theory and general relativity is similar to that between the Maxwell theory of continuous media and the classical microscopic electrodynamics" (p. 1203). However, he only develops the idea of treating the spin density that enters the Einstein-Cartan theory as the macroscopic average of microscopic angular momenta in the linear approximation, and does not make explicit the relation he suggests by developing a formal analogy between quantities in macroscopic electrodynamics and in the Einstein-Cartan theory.

In this paper, I shall develop such an analogy with macroscopic electrodynamics in detail for the exact, nonlinear version of the Einstein-Cartan theory. I discuss a correspondence between linearized solutions to the Einstein-Cartan equations

---

[1]Dedicated to Engelbert Schucking on his seventieth birthday with friendship and appreciation of his many contributions to science as a scholar and as a human being.

and linearized solutions to the Einstein equations that strongly supports the interpretation of the Einstein-Cartan theory as a macroscopic theory. Finally, I discuss the prospects for extending these results to exact solutions to the Einstein-Cartan equations, and of generalizing the macroscopic approach of this paper to the full metric-affine theory.

33.2 The Analogy: Electromagnetism

I shall start with what, in order to use a terminology parallel to that used in the discussions of gravitation theory cited above [5], I shall call microscopic electrodynamics. But it should be born in mind that the term "microscopic" here is really a misnomer. It does not carry any implication that the 4-current density introduced is to be interpreted atomistically. It is assumed to be a continuous, differentiable function of its arguments, and insofar as the atomistic structure of charged matter is taken into account, some sort of averaging process is assumed to have already been done. What is implied by the term "microscopic" is merely that *all* sources of charge and current, both "free" and "bound," have been included in its evaluation.

In the 4-dimensional version of microscopic electrodynamics, two antisymmetric tensor fields are introduced: the covariant tensor or 2-form $F_{\alpha\beta}$, which incorporates the E and B fields in a particular inertial frame of reference into one 4-dimensional Lorentz-covariant field; and the contravariant tensor $G^{\alpha\beta}$, which similarly incorporates the D and H fields [6]. The first set of microscopic Maxwell field equations assert that curl F vanishes:

$$F_{[\alpha\beta,\tau]} = 0 \quad \text{or} \quad \mathbf{d}F = 0 \tag{1}$$

in the differential forms notation, i.e., F is a closed form. Locally, at least, this implies that the $F_{\alpha\beta}$ field can be derived from a 4-vector potential A_β, which incorporates the usual scalar and vector potentials ϕ and $\mathbf{A}$. Indeed, aside from topological complications (which do not occur in Minkowski spacetime), this first set of Maxwell equations (1) is equivalent to:

$$F_{\alpha\beta} = A_{\beta,\alpha} - A_{\alpha,\beta} \quad \text{or} \quad \mathbf{d}A, \tag{2}$$

i.e., the form F is exact [7]. These equations may be interpreted as postulating the absence of magnetic monopoles. Given the field F, the potentials are not unique, but only determined up to a gauge transformation:

$$A'_\alpha = A_\alpha + \phi_{,\alpha} \quad \text{or} \quad \mathbf{A}' = \mathbf{A} + \mathbf{d}\phi. \tag{3}$$

As we shall see, Eq. (1) and its consequences are actually common to both the microscopic and macroscopic versions of Maxwell's theory. So I shall refer to them just as the first set of Maxwell equations. The second set of microscopic Maxwell field equations assert that div G equals the total charge-current density:

$$G^{\alpha\beta}{}_{,\beta} = j^\alpha \tag{4}$$

Here j^α is the charge-current 4-vector density, which incorporates the usual charge and current densities p and j. As emphasized above, this includes both "bound" and "free" quantities.

In order to proceed with any applications of the two sets of microscopic Maxwell equations, it is necessary to introduce some relation between the F and G fields. In vacuum, the two fields are related by means of the metric tensor $g_{\alpha\beta}$:

$$G^{\alpha\beta} = g^{\alpha\delta} g^{\beta\tau} F_{\delta\tau} = F^{\alpha\beta} \tag{5}$$

(depending on the system of units used, a constant ϵ_0 representing the "polarization of the vacuum" may be introduced). So in vacuum, the two tensors are effectively equal [8].

Now we come to what I shall call macroscopic Maxwell theory, in which it is assumed that the charges and currents can be divided into "free" and "bound." This is done formally by introducing a third antisymmetric contravariant tensor field, the polarization tensor $p^{\alpha\beta}$, inside of matter. This tensor incorporates the usual **P** and −**M** fields, i.e., the electric and magnetic polarization vectors, respectively. Its value will depend on the properties of the matter under consideration, as well as on the electromagnetic fields to which the matter is subject. Some *Ansatz* for the form of these vectors in the rest frame at a point of the matter then must be introduced. The adequacy of the postulated relations is judged by the success of the resulting theory in explaining the observed electrodynamical properties of the matter.

It is also possible to go a step further by introducing a microscopic model of the matter, and deriving the form of the polarization tensor by some sort of averaging process over the multipole moments of the atomistic constituents of the matter. In the usual treatment of dielectrics, for example, the electric polarization vector is derived by averaging over the intrinsic or induced electric dipole moments of these constituents; while the magnetic polarization vector is similarly derived from an averaging over their magnetic dipole moments. In particular, if it is assumed that there are no intrinsic magnetic dipole moments, the averaged magnetic dipole moment will arise entirely from the circulation of microscopic charge.

More careful treatments emphasize that these are merely the first two terms in a multipole expansion, the terms of which decrease rapidly in value; and may even evaluate the next term, the contribution to the electric polarization from the electric quadrupole moments of the atomistic constituents [9].

The divergence of the polarization tensor gives the "bound" charge-current 4-vector:

$$p^{\alpha\beta}{}_{B} = j_B{}^{\alpha} \,. \tag{6}$$

By defining a new, macroscopic field G_M:

$$G_M{}^{\alpha\beta} = G^{\alpha\beta} - p^{\alpha\beta} \tag{7}$$

and using equations (6), one can rewrite the second set of microscopic Maxwell equations (4) as

$$G_M{}^{\alpha\beta}{}_{\beta} = j_F{}^{\alpha} \,, \tag{8}$$

where the free charge-current four-vector is defined by:

$$j_F{}^\alpha = j^\alpha - j_B{}^\alpha \,. \tag{9}$$

Equations (1), the first set of Maxwell equations for the F field; (5) relating F and G; (7) defining the G_M field in terms of the G and P fields, and (8), the second set of macroscopic Maxwell equations for the G_M field with the j_F field as its source, constitute the equations defining macroscopic Maxwell theory. They must be supplemented by prescriptions for the P and j_F fields, plus boundary conditions at the interface between matter and vacuum and at infinity, in order to solve particular problems.

33.3 The Analogy: Gravitation

Let us try to treat gravitation theory in an analogous way. For the microscopic Einstein theory, we introduce two fields, the Christoffel symbols of the first kind $[\alpha\beta, \tau]$ and a symmetric connection $\Gamma_E{}^\mu{}_{\alpha\beta} = \Gamma_E{}^\mu{}_{\beta\alpha}$, which we shall call the Einstein connection. The Christoffel symbols of the first kind are derived from a set of potentials $g_{\alpha\beta}$:

$$[\alpha\beta, \tau] = \frac{1}{2}(-g_{\alpha\beta,\tau} + g_{\tau\alpha,\beta} + g_{\beta\tau,\alpha}) \,. \tag{10}$$

Equation (10) is analogous to Eq. (2), the definition of the $F_{\alpha\beta}$ field in terms of the potentials, so we take the Christoffel symbols of the first kind as the gravitational analogue of the F field. Just as in the electromagnetic case, these definitions may be interpreted as postulating the absence of a gravitational analogue of magnetic monopoles. In this sense, they are more fundamental than the set of conditions on the Christoffel symbols of the first kind that are equivalent locally to the existence of the gravitational potentials $g_{\alpha\beta}$, which are easily derived from the commutativity of the second derivatives of the metric tensor:

$$([\alpha\beta, \tau] + [\alpha\tau, \beta])_{,\delta} - ([\delta\beta, \tau] + [\delta\tau, \beta])_{,\alpha} \,. \tag{11}$$

Equation (11) is then the gravitational analogue of Eq. (1), the first set of Maxwell equations.

The analogue of Eq. (4), the second set of microscopic Maxwell equations, are the Einstein equations for the Einstein connection Γ_E:

$$G_{\alpha\beta}(\Gamma_E) = T_{\alpha\beta}, \qquad G_{\alpha\beta} = R_{\alpha\beta} - \frac{1}{2}g_{\alpha\beta}R^\tau{}_\tau, \tag{12}$$

where the Riemann tensor, Ricci tensor, Ricci scalar, and Einstein tensor are computed from the Einstein connection in the usual way (depending on the units used, there may be a coupling constant depending on G, the Newtonian gravitational constant, on the right-hand side of this equation, and in several subsequent equations). The $T_{\alpha\beta}$ field includes the gravitational analogues of both the "free" and "bound" material charges and currents. Like the $G_{\alpha\beta}$ field, the $T_{\alpha\beta}$ field is symmetric in its indices.

We see that the Einstein connection is the gravitational analogue of the G field in electrodynamics. In vacuum, the Einstein connection and the Christoffel symbols of the first kind are related by the metric tensor

$$\Gamma_E{}^{\mu}{}_{\alpha\beta} = g^{\mu\alpha}[\alpha\beta, \sigma] = \{^{\mu}{}_{\alpha\beta}\}, \tag{13}$$

where $\{^{\mu}{}_{\alpha\beta}\}$ are the Christoffel symbols of the second kind. So the Einstein connection is the symmetric, metric connection, for which $g_{\alpha\beta;\,E\mu} = 0$, where a semicolon followed by a subscript E denotes covariant differentiation with respect to the Einstein connection.

Note the dual role that, as usual, the tensor field $g_{\alpha\beta}$ plays in general relativity: it is both the metric tensor and the potentials for the Christoffel symbols of the first kind. Because of this dual role, the treatment of gauge transformations here differs somewhat from that in electromagnetism. Here, diffeomorphisms play the role of gauge transformations. A one-parameter family of such diffeomorphisms is generated by any vector field v^{μ}; if the value of the parameter is the infinitesimal ϵ, then the tensor field $g_{\alpha\beta}$ is dragged into

$$g'_{\alpha\beta} = g_{\alpha\beta} + \epsilon \pounds_v g_{\alpha\beta}, \tag{14}$$

where $\pounds_v$ is the Lie derivative with respect to the vector field v^{μ} [10]. Since the operations of Lie derivation and partial differentiation commute, Lie differentiation of Eq. (10) tells us that the Christoffel symbols of the first kind do not remain invariant, but are also dragged along by the diffeomorphism. But this is just what we should expect. Since the g field represents the metric as well as the gravitational potentials, when the metric is dragged by a diffeomorphism, any other field *must* also be dragged along by that diffeomorphism precisely in order to maintain the *same* values at each *physical* point [11].

Now we turn to the macroscopic Einstein-Cartan theory. In this theory, another connection Γ_c is introduced, which we shall call the Cartan connection, the gravitational analogue of the G_M, the macroscopic electromagnetic field. It is assumed that this connection is still metric, that is that the nonmetricity tensor, the covariant derivative of the metric tensor with respect to the Cartan connection, vanishes. That is, $g_{\alpha\beta;C\mu} = 0$, where a semicolon followed by a subscript C denotes the covariant derivative with respect to the Cartan connection. This implies that the difference between the Cartan and Einstein connections depends only on the torsion tensor S:

$$S^{\mu}{}_{\alpha\beta} = \frac{1}{2}(\Gamma_C{}^{\mu}{}_{\alpha\beta} - \Gamma_C{}^{\mu}{}_{\beta\alpha}), \tag{15}$$

which is obviously antisymmetric in its lower indices. If we define a tensor K, called the contorsion tensor [12], by

$$K^{\mu}{}_{\alpha\beta} = -S^{\mu}{}_{\alpha\beta} + S_{\alpha\beta}{}^{\mu} - S_{\beta}{}^{\mu}{}_{\alpha}, \tag{16}$$

where the indices of the torsion tensor are raised and lowered with the metric tensor, then it follows that

$$\Gamma_C{}^{\mu}{}_{\alpha\beta} = \{^{\mu}{}_{\alpha\beta}\} - K^{\mu}{}_{\alpha\beta}. \tag{17}$$

With this mathematical background, I return to the physical description of the macroscopic theory. It is assumed that the stress-energy tensor can be divided into "free" and "bound" portions. This is done formally by assuming that inside matter the torsion tensor S and hence the contorsion tensor K do not vanish. Indeed, from Eq. (17) we see that the contorsion tensor is the gravitational analogue of the polarization tensor (see Eq. (7)); like the latter, it is assumed to depend on the properties of the matter being considered as well as on the gravitational field to which the matter is subject. Just as in the electromagnetic case, its form may simply be postulated in a fully macroscopic theory; or an attempt may be made to derive it from a microscopic model of the medium. Here, I shall follow an intermediate course, postulating its form, but motivating the *Ansatz* by a microscopic argument.

I proceed again by analogy with the electromagnetism. In the case of most dielectrics, the elementary constituents (atoms and molecules) are electrically neutral. Therefore, their electric dipole moments (intrinsic or induced) are an invariant property (i.e., independent of the origin chosen for their evaluation), which can be averaged over a volume element to give a macroscopic electric dipole moment per unit volume. In the gravitational case, there is no evidence for the existence of negative mass, so that we cannot expect the elementary constituents of matter to have an invariant mass dipole moment. Indeed, by choosing the evaluation point for each such constituent at its center of mass, we can make the mass dipole moment vanish. Of course, the elementary constituents will have an invariant mass quadrupole moment (evaluated at the center of mass point), which can in principle contribute to the macroscopic contorsion tensor. We shall return to this point in the concluding section, but here we shall assume that this contribution may be neglected.

We shall here consider only the gravitational analogue of the magnetic dipole moment (we have seen that Eq. (10) implies the absence of gravitational analogues of magnetic monopoles), treated at the macroscopic level. There is a four-velocity field U^{μ} associated with each point inside matter, and I shall assume that, in the rest frame defined at each point by this velocity field, a spin vector exists, which has only spatial ("magnetic-type") components. This is equivalent to the so-called Weysenhoff *Ansatz* [13] for the form of the spin-tensor density field $s^{\mu}{}_{\alpha\beta}$:

$$s^{\mu}{}_{\alpha\beta} = U^{\mu} s_{\alpha\beta}, \quad \text{where} \quad s_{\alpha\beta} = -s_{\beta\alpha}, \quad s_{\alpha\beta}U^{\beta} = 0. \tag{18}$$

This means, by the usual association between an antisymmetric 3-dimensional tensor and a 3-vector, that in the 3-space orthogonal to the 4-velocity U^{μ} at each point, the spin tensor is equivalent to a spin vector. In the macroscopic Einstein-Cartan theory, it is assumed that inside matter the spin-tensor density field is equal to the modified torsion tensor $T^{\mu}{}_{\alpha\beta}$ [14]; but since the latter only differs from the torsion tensor by its trace, which vanishes with the Weyssenhoff *Ansatz* (Eq. (18)), it follows that $S^{\mu}{}_{\alpha\beta} = s^{\mu}{}_{\alpha\beta}$ [15]. Hence,

$$K^{\mu}{}_{\alpha\beta} = -s_{\alpha\beta}U^{\mu} + s_{\beta}{}^{\mu}U_{\alpha} - s^{\mu}{}_{\alpha}U_{\beta}, \tag{19}$$

from which it follows that

$$K^{\mu}{}_{\alpha\beta} = 0, \qquad g^{\alpha\beta}K^{\mu}{}_{\alpha\beta} = 0\,. \tag{20}$$

We can now rewrite the Einstein equations (9) inside matter in the form

$$G_{\alpha\beta}(\Gamma_C) = T_{F\alpha\beta}, \tag{21}$$

where $G_{\alpha\beta}(\Gamma_C)$ is the Einstein tensor formed from the Cartan connection, and $T_{F\alpha\beta}$ is the "free" portion of the stress-energy tensor of matter, to be defined in a moment. Note that, since the Cartan connection is not torsion-free, $G_{\alpha\beta}(\Gamma_C)$ is not symmetric.

It remains to discuss the division of the stress-energy tensor into "bound" and "free" portions. Inserting Eq. (17) into the definition of the Ricci tensor for the Einstein-Cartan connection, and utilizing Eq. (20), we get [16]

$$R_{\alpha\beta}(\Gamma_C) = R_{\alpha\beta}(\Gamma_E) - K^{\mu}{}_{\alpha\beta;E\mu} - s_{\sigma\tau}s^{\sigma\tau}U_{\alpha}U_{\beta}. \tag{22}$$

Taking the trace of this equation, and using Eq. (20), we get

$$R(\Gamma_C) = R(\Gamma_E) - s_{\sigma\tau}s^{\sigma\tau}, \tag{23}$$

so that

$$G_{\alpha\beta}(\Gamma_C) = G_{\alpha\beta}(\Gamma_E) - K^{\mu}{}_{\alpha\beta;E\mu} - s_{\sigma\tau}s^{\sigma\tau}\left(U_{\alpha}U_{\beta} - \frac{1}{2}g_{\alpha\beta}\right) \tag{24}$$

Thus, if we define the "bound" stress-energy tensor by

$$T_{B\alpha\beta} = K^{\mu}{}_{\alpha\beta;E\mu} + s_{\sigma\tau}s^{\sigma\tau}\left(U_{\alpha}U_{\beta} - \frac{1}{2}g_{\alpha\beta}\right), \tag{25}$$

and the "free" stress-energy tensor by

$$T_{F\alpha\beta} = T_{\alpha\beta} - T_{B\alpha\beta}, \tag{26}$$

then the macroscopic Einstein-Cartan field equations (21) follow from the microscopic Einstein field equations (12) and these definitions. Note that neither the "bound" nor "free" stress-energy tensors are symmetric.

We have now completed our description of the macroscopic Einstein-Cartan theory, which is based on the two sets of field equations (11)—or (10)—and (21), with the definition of the "free" stress-energy tensor given by equations (25) and (26), and the Weyssenhoff *Ansatz*, Eq. (18).

33.4 Discussion

In the linear approximation to each theory, Adamowicz has proved a correspondence theorem between solutions of the macroscopic Einstein-Cartan equations and the microscopic Einstein quations. He considers a finite source containing a static body with a spin tensor density of the form given by the Weyssenhoff *Ansatz*, Eq. (18), and shows that in this approximation it produces the same external gravitational field as does a rotating body without spin-density field but with a corresponding distribution of rotational angular momentum density in the linearized Einstein theory [18]. As a consequence, in the linear approximation at any rate, a static solution to the Einstein-Cartan equations for an axially symmetric body with a suitable Weyssenhoff spin-density field can produce the same external field as a stationary solution to the Einstein equations representing the same body without any spin-density field, but in rigid rotation about its axis. This is analogous to the situation in electrodynamics, where the external fields of a charged magnet treated by the macroscopic Maxwell equations, and of a rotating charged body, treated by the microscopic Maxwell equations, are the same [19]. The difference is inside the body. If we wanted to treat the magnet microscopically, we would have to associate an intrinsic magnetic moment with each element of the body, and these would add up to produce the same internal field as that of the rotating charged body.

Hence, the analogy here is between the solutions to the macroscopic Einstein-Cartan equations and macroscopic Maxwell equations on the one hand and the solutions to the microscopic Einstein equations and Maxwell equations on the other. We suggest this analogy between solutions is a strong argument for the analogy between the corresponding macroscopic equations on the one hand and the microscopic equations on the other.

This result, taken together with the exact formulation of the macroscopic gravitational-electromagnetic analogy developed in Sections 33.2 and 33.3, suggests that it should be possible to find exact static interior solutions of the macroscopic Einstein-Cartan equations that have the same stationary external fields as corresponding exact stationary solutions of the microscopic Einstein equations. It should even be possible to prove theorems relating entire classes of exact stationary solutions to the Einstein equations for rigidly rotating sources to classes of exact solutions to the Einstein-Cartan equations with the same exterior metric but a static interior solution representing a nonrotating source with corresponding spin tensor density distribution. The question also arises of generalizing the macroscopic approach from the Einstein-Cartan case to the case of metric-affine geometries discussed by Hehl and collaborators [20], in which the connection is no longer metric. Our discussion suggests that the Einstein-Cartan theory with the Weyssenhoff *Ansatz* is adequate to handle the gravitational analogue of magnetic polarization **M**, but cannot treat the gravitational analogue of the electric polarization **P**. We have argued above that no gravitational analogue of the electric dipole moment should exist (see Section 33.3). However, we certainly expect a gravitational analogue of the electric quadrupole moment. In the electromagnetic case, the divergence of the electric quadrupole moment tensor contributes to the electric polarization vector [21]. We suspect that, in the gravitational case, something like the covariant derivative of the quadrupole moment tensor should be related to the nonmetricity tensor, the nonvanishing covariant derivative of the metric [22]

Indeed, the nonmetricity tensor allows generalization of the right-hand side of Eq. (12) above to include a tensor symmetric in its lower indices. I plan to investigate the possibility further. Like Adamowicz I believe that the macroscopic approach to the Einstein-Cartan theory "may be used effectively for solving certain cosmological or astrophysical problems," [23] and the extension of the analogy to matter with intrinsic or induced quadrupole moments would considerably extend its range of applicability, in particular to problems involving interactions of gravitational radiation with matter.

On the formal side, this paper (at least its gravitational part) has utilized exclusively the tensor calculus. However, it is well known that the Einstein-Cartan theory can be rewritten more perspicaciously in Cartan's language of differential forms [24], and Hehl and collaborators have used this language for its generalization [25]. The results obtained here, as well as possible generalizations, should be rewritten in terms of differential forms.

Notes and References

1. Friedrich Hehl, J. Dermott McCrea, Eckehard W. Mielke, Yuval Ne'eman, "Metric-Affine Gauge Theory of Gravity: Field Equations, Noether Identities, World Spinors, and Breaking of Dilation Invariance," *Physics Reports* **258**, Nos. 1 & 2: 1–171 (1995).

2. See, e.g., Friedrich W. Hehl, Paul von der Heyde, G. David Kerlick, "General relativity with spin and torsion: Foundations and prospects," *Reviews of Modern Physics* **48**, 393–416 (1976).

3. Friedrich Hehl, Paul van der Heyde, and G. David Kerlick, General relativity with spin and torsion and its deviations from Einstein's theory," *Physical Review* **D10**, 1066–1069 (1974).

4. W. Adamowicz, "Equivalence between the Einstein-Cartan and General Relativity Theories in the Linear Approximation for a Classical Model of Spin," *Bulletin de l'Academie Polonaise des Sciences Serie des science, mathematiques astronomiques et physiques* **23**, 1203–1205 (1975).

5. In this section I follow an approach to electromagnetism well summarized in Attay Kovetz, *The Principles of Electromagnetic Theory* (Cambridge University Press, 1990), although I do not always follow his terminology and notation.

6. Strictly speaking, we should treat the *G* field as a tensor density field of weight one (see, e.g., John Stachel, "The Generally Covariant Form of Maxwell's Equations," in Melvin S. Berger, ed., *J.C. Maxwell: The Sesquicentennial Symposium* (Amsterdam/New York/Oxford: North-Holland, 1984), pp. 23–37. But since we are only introducing the electromagnic field for purposes of comparison, we avoid this complication here.

7. For a review of closed and exact forms, Poincare's lemma, and the possible topological complications, see John Stachel, "Globally stationary but locally static

space-times: A gravitational analogue of the Aharonov-Bohm effect," *Physical Review D* **26**, 1281–1290 (1982).

8. Microscopic electrodynamics in a homogeneous, isotropic medium can be treated by introducing an "optical metric," and even a nonlinear, phenomenological "vacuum polarization" can be introduced; both are discussed in the reference in note 6.

9. For such a careful treatment, see, e.g., Leon Rosenfeld, *Theory of Electrons* (Amsterdam: North-Holland, 1951), repr. with new Preface (New York: Dover, 1965).

10. For the definition of the Lie derivative, see e.g. Jan A. Schouten, *Ricci-Calculus* (Berlin/Gottingen/Heidelberg: Springer, 1954), pp. 102–111. For a review of the Lie derivative and its application to the Cauchy problem in general relativity, see John Stachel, "Covariant Formulation of the Cauchy Problem in Generalized Electrodynamics and General Relativity," *Acta Physica Polonica* **35**, 689–709 (1969).

11. This is essentially what general covariance means. See, e.g., John Stachel, "The Meaning of General Covariance: The Hole Story," in John Earman et al., eds., *Philosophical Problems of the Internal and External World/Essays on the Philosophy of Adolf Grünbaum* (Konstanz: Universitatsverlag/ Pittsburgh: University of Pittsburgh Press, 1993), pp. 129–160.

12. See, e.g., Hehl et al. (reference in footnote 2), p. 397. Note that I have placed the upper index first.

13. Jan Weyssenhoff and A. Raabe, "Relativistic dynamics of spinfluids and spin-particles," *Acta Physica Polonica* **9**, 7-xx (1947).

14. For its definition, see, e.g., the reference in note 2.

15. Depending on the units used, the proportionality factor G may be introduced here.

16. See Schouten, reference in note 10, p. 141, Eq. (4.23a), remembering that Schouten's T is defined with the opposite sign to K (see Eq. (4.20)), and that the trace of T vanishes in this case. Note that formula (4.23a) is true in an arbitrary affine-metric space, i.e., even if the nonmetricity tensor does not vanish, and we do not make the Weyssenhoff *Ansatz* for the torsion.

17. See reference in note 4.

18. Adamowicz speaks of a Weyssenhoff fluid, but the stress tensor of the body does not enter in the linear approximation, so it can be an arbitrary body.

19. We stipulate that the body be charged in order to have a better analogy with the gravitational case in which a body always has mass.

20. See the reference in note 1.

21. See, e.g., pp. 20, 22, Rosenfeld reference in note 9.

22. For the nonmetricity tensor, see the Hehl et al., reference in note 2, p. 397. Equations (2.8), (2.9) on that page show that the nonmetricity tensor would supply an additional term in the constitutive relation relating the metric-affine connection and the Christoffel symbols; and as noted in note 15, such a term could easily be included in the relation between the metric-affine and Christoffel Ricci tensors.

23. Reference in note 4, p. 1203.

24. See, e.g., Andrzej Trautman, "On the structure of the Einstein-Cartan equations," in *Differential Geometry, Symposia Mathematica*, v. **12** (Academic Press, London 1973), p. 139.

25. See the reference in note 1.

34

On Complex Structures in Physics

Andrzej Trautman

ABSTRACT Complex numbers enter fundamental physics in at least two rather distinct ways. They are needed in quantum theories to make linear differential operators into Hermitian observables. Complex structures appear also, through Hodge duality, in vector and spinor spaces associated with space-time. This paper reviews some of these notions. Charge conjugation in multidimensional geometries and the appearance of Cauchy-Riemann structures in Lorentz manifolds with a congruence of null geodesics without shear are presented in considerable detail.

34.1 Introduction

In 1960, Ivor Robinson and I studied a class of solutions of Einstein's equations on a Lorentzian manifold, foliated by a shear-free, nonrotating, and diverging congruence of null geodesics. We were surprised to find that in the coordinate system we were using, the vanishing of the Ricci tensor implied that two components of the metric tensor satisfied Cauchy-Riemann equations [16]. Since that time I have been interested in the question of how and why complex numbers and structures appear in physical problems.

Complex numbers have been used in physics so much and for so long that one is taking them, most of the time, for granted. Their origin is in "pure" mathematics: they appeared, in the 16th century, in connection with solving polynomial equations. They were used in solutions of physical problems, such as reflection and diffraction of waves, early in the 19th century. Complex roots of polynomial equations often appear in physical problems in connection with linear differential equations with constant coefficients. An early, ingenious use of complex analytic functions was made by Arnold Sommerfeld in his rigorous solution of the problem of diffraction of waves on a half-plane [20].[1] Roger Penrose, in his twistor theory, put forward convincing arguments in favor of the relevance of holomorphic structures for fundamental physics [11, 13]. His methods and ideas have been successfully used in a variety of mathematical and physical problems. Complex numbers and analytic functions now permeate all of quantum physics.

---

[1] I am grateful to Jürgen Ehlers for this reference.

Complex numbers are introduced in physical theories in several ways; it is not obvious that "all the square roots of -1 are the same." There is the $\sqrt{-1}$ of quantum mechanics that is "universal" in the sense that it appears irrespective of the details of the model under consideration. Chiral (Weyl, reduced) spinors in Minkowski space-time are also complex, but this property reflects the signature of its metric tensor. Dirac spinors can be restricted to be real, provided one uses a metric of signature (3, 1); but to write the wave equation of an electron, in any signature, one has to introduce complex numbers because electromagnetism is a gauge theory with U_1 as the structure group. The "electromagnetic" $\sqrt{-1}$ seems to have a quantum-theoretical origin: in the classical theory of a particle of charge e the potential A_μ appears in expressions such as $p_\mu + eA_\mu$; in quantum theory this becomes

$$i\partial_\mu + eA_\mu = i(\partial_\mu - ieA_\mu).$$

and the i next to eA_μ reflects the nature of the Lie algebra of U_1.

In this article, I present some thoughts on the origin of the appearance of complex numbers in physics, emphasizing the geometric, rather than the analytic, aspects of the problem. After recalling the notion of complex structures in real vector spaces (Section 34.2), I show, on a simple example, how such structures may be considered to appear in quantum mechanics (Section 34.3). For some signatures of the metric tensor, Clifford-Hodge-Kähler duality introduces complex structures in spaces of spinors and multivectors (Section 34.4). Charge conjugation is also closely related to the appearance of complex numbers in quantum theory; its generalization to higher dimensions is described in Section 34.5. In a final section, influenced by my collaboration with W. Kopczyński, P. Nurowski, and J. Tafel, I describe the geometry underlying shear-free congruences of null geodesics, its relation to Cauchy-Riemann structures in three dimensions, and the close analogy between optical geometries in Lorentzian manifolds and Hermitian geometries in proper Riemannian manifolds.

In 1961, Engelbert Schucking and I spent some time together at Syracuse University in Peter Bergmann's group, which included also Dick Arnowitt, Asim Barut, Art Komar, Ted Newman, Roger Penrose, Ivor Robinson, Ralph Schiller, and Mel Schwartz. Since that time, I have had the pleasure to see Engelbert on various occasions and to talk with him on many issues of science and life. These discussions also included the topics touched upon in this text. On one of my visits to New York, Engelbert presented me with a copy of [2], an excellent account of the history of number systems. Several times, my family and I enjoyed his very kind hospitality at Washington Square. This article is dedicated to Engelbert as a token of my friendship and respect.

34.2 Definitions and Notation

Recall that a *complex structure* in a real vector space W is a linear automorphism J of W such that $J^2 = -id_W$; if W is finite-dimensional, then its dimension is even.

A real vector space W with a complex structure J can be made into a complex vector space in two ways, by defining, for every $w \in W$, either $iw = J(w)$ or $iw = -J(w)$. The automorphism J extends, in an obvious way, to an automorphism $J_{\mathbb{C}}$ of the *complexification* of W, i.e., to the complex vector space $\mathbb{C} \otimes W$. This space can be decomposed into the direct sum,

$$\mathbb{C} \otimes W = W_+ \oplus W_-, \quad \text{where} \quad W_\pm = \{w \in \mathbb{C} \otimes W : J_{\mathbb{C}}(w) = \pm iw\}. \tag{1}$$

Considered as a complex vector space, W is isomorphic to W_+ or W_-, depending on whether the multiplication by i in W is defined as $iw = J(w)$ or $iw = -J(w)$, respectively. Note that if W is of real dimension $2n$, then its complex dimension is n. Every complex vector space of dimension n can be "realified," i.e., considered as a real vector space of real dimension $2n$. Such a realification has a natural complex structure.

Assume now that the real vector space W has a (generalized) scalar product, i.e., a map $g : W \times W \to \mathbb{R}$ which is bilinear, symmetric and nondegenerate. The scalar product g extends to a $\mathbb{C}$-bilinear scalar product $g_{\mathbb{C}}$ on $\mathbb{C} \otimes W$. If the complex structure J in W is orthogonal with respect to g, i.e., if $g(J(w_1), J(w_2)) = g(w_1, w_2)$ for every $w_1, w_2 \in W$, then the vector spaces W_+ and W_- are both totally null (isotropic) with respect to $g_{\mathbb{C}}$. The vector space W, considered as a complex vector space such that $iw = J(w)$ has a Hermitian scalar product $h : W \times W \to \mathbb{C}$ defined by

$$h(w_1, w_2) = g(w_1, w_2) + ig(J(w_1), w_2), \qquad w_1, w_2 \in W, \tag{2}$$

so that $h(w_1, iw_2) = ih(w_1, w_2)$, $\overline{h(w_1, w_2)} = h(w_2, w_1)$ and $h(w, w) = g(w, w)$.

Consider now a *complex, finite-dimensional vector space* S. Its (complex) dual S^* consists of all $\mathbb{C}$-linear maps $s' : S \to \mathbb{C}$; it is often convenient to denote here the value of s' on $s \in S$ by $\langle s', s\rangle$. If $f : S_1 \to S_2$ is a $\mathbb{C}$-linear map of complex vector spaces, then the dual (transposed) map $f^* : S_2^* \to S_1^*$ is defined by $\langle f^*(s'), s\rangle = \langle s', f(s)\rangle$ for every $s \in S_1$ and $s' \in S_2^*$. The spaces S^{**} and S can be identified. A map $h : S_1 \to S_2$ is said to be *antilinear* (semilinear) if it is $\mathbb{R}$-linear and $h(is) = -ih(s)$ for every $s \in S_1$. The *complex conjugate* $\bar{S}$ of a complex vector space S is the complex vector space of all antilinear maps of S^* into $\mathbb{C}$; there is a canonical antilinear isomorphism $S \to \bar{S}$, $s \mapsto \bar{s}$, given by $\langle \bar{s}, s'\rangle = \overline{\langle s', s\rangle}$. With every linear map $f : S_1 \to S_2$ there is associated the linear map $\bar{f} : \overline{S_1} \to \overline{S_2}$ defined by $\bar{f}(\bar{s}) = \overline{f(s)}$ for $s \in S_1$; the map $f \mapsto \bar{f}$ is antilinear. If $g : S_2 \to S_3$ is another linear map, then $(g \circ f)^* = f^* \circ g^*$ and $\overline{g \circ f} = \bar{g} \circ \bar{f}$. One often writes gf instead of $g \circ f$.

All manifolds and maps are assumed to be smooth. Einstein's summation convention over repeated indices is used. If L is a vector bundle over a manifold M, then $\Gamma(L)$ denotes the module of sections of $L \to M$. The zero bundle is denoted by 0. The tangent and cotangent bundles of M are denoted by TM and T^*M, respectively. The contraction of a vector (field) v with a p-form ω is the $(p-1)$-form $v \lrcorner \omega$ given by its value on the vectors $v_2, \ldots, v_p$, $(v \lrcorner \omega)(v_2, \ldots, v_p) = \omega(v, v_2, \ldots, v_p)$. If g is a scalar product on a vector space

V and $v \in V$, then $g(v) \in V^*$ is defined by $v' \lrcorner g(v) = g(v', v)$ for every $v' \in V$. The exterior differential of a form ω is denoted by $\mathrm{d}\omega$.

34.3 A Complex Structure Defined by Differentiation

The usual argument for complex numbers in quantum mechanics, in a simplified form, runs as follows: differential operators such as $\partial/\partial x$, because of their relation to translations, are needed to represent components of momentum; to make them (formally) self-adjoint, one has to multiply by i. One can reformulate this argument into a statement about the appearance of a complex structure in the vector space of wave functions, initially considered as a real vector (Hilbert) space. The key observation is that the Laplacian on a compact, proper Riemannian manifold is a *negative* operator.

To illustrate this argument on a simple example, and make it explicit, consider the infinite-dimensional real Hilbert space $L^2_{\mathbb{R}}(\mathbb{S}_1)$ of square-integrable functions on the circle $\mathbb{S}_1$. Let x be a coordinate on the circle, $0 \leqslant x \leqslant 2\pi$. The scalar product of two functions $\varphi, \psi : \mathbb{S}_1 \to \mathbb{R}$, is given by

$$g(\varphi, \psi) = \int_0^{2\pi} \varphi(x)\psi(x)\mathrm{d}x,$$

so that $g(\varphi, \varphi) \geqslant 0$. Let W be the vector subspace of $L^2_{\mathbb{R}}(\mathbb{S}_1)$ containing all functions orthogonal to the constants on the circle,

$$W = \left\{\varphi \in L^2_{\mathbb{R}}(\mathbb{S}_1) : \int_0^{2\pi} \varphi(x)\mathrm{d}x = 0\right\}.$$

Smooth functions in W constitute a dense subspace of that space; for every two such functions φ and ψ one has

$$g(\varphi', \psi) = -g(\varphi, \psi'),$$

where $\varphi'(x) = d\varphi(x)/dx$. The operator d^2/dx^2 is (formally) selfadjoint and negative on W: if φ is smooth and $\varphi \neq 0$, then

$$g(\varphi, \varphi'') = -g(\varphi', \varphi') < 0.$$

The set of eigenfunctions of d^2/dx^2,

$$\{\cos kx, \ \sin kx\}, \qquad k = 1, 2, \ldots,$$

is a basis in W. The operator $-d^2/dx^2$ has only positive eigenvalues; as such it has a unique positive square root X, i.e., a (formally) selfadjoint operator with positive eigenvalues such that $X^2 = -d^2/dx^2$. The operator X in W, which may be characterized by its action on the basis vectors,

$$X(\cos kx) = k\cos kx, \qquad X(\sin kx) = k\sin kx,$$

is *invertible* and *commutes* with the operator d/dx. Therefore, the linear operator

$$J = X^{-1} \circ \frac{d}{dx} \quad \text{satisfies} \quad J^2 = -id_W$$

and defines a complex structure on W. Introducing the complex vector spaces $W_\pm$, as in the previous section, one obtains

$$X = \mp i \frac{d}{dx} \quad \text{on} \quad W_\pm.$$

34.4 Complex Structures Associated with Pseudo-Euclidean Vector Spaces

Let V be a real, m-dimensional vector space with a scalar product g of signature (k, l), $k+l = m$. The Clifford algebra associated with the pair (V, g) is denoted by $\mathsf{Cl}_{k,l}$. The algebra is generated by V; by declaring the elements of V to be *odd*, one defines a $\mathbb{Z}_2$-grading of $\mathsf{Cl}_{k,l}$: one writes $\mathsf{Cl}^0_{k,l} \to \mathsf{Cl}_{k,l}$ to emphasize this grading and exhibit the even subalgebra $\mathsf{Cl}^0_{k,l}$. The *degree* $\deg a$ of an even (respectively, odd) element $a \in \mathsf{Cl}_{k,l}$ is 0 (resp., 1). Recall that if $\mathcal{A}$ and $\mathcal{B}$ are $\mathbb{Z}_2$-graded algebras, then multiplication in their *graded product* $\mathcal{A} \otimes_{\text{gr}} \mathcal{B}$ is defined, for homogeneous elements $a' \in \mathcal{A}$ and $b \in \mathcal{B}$, by $(a \otimes b)(a' \otimes b') = (-1)^{\deg b \deg a'} aa' \otimes bb'$. For every $k, l \in \mathbb{N}$ the algebras $\mathsf{Cl}_{k,l}$ and $\mathsf{Cl}^0_{k,l+1}$ are isomorphic. Denote by $\mathbb{R}(N)$ the algebra of real N by N matrices. For every algebra $\mathcal{A}$ over $\mathbb{R}$, put $2\mathcal{A} = \mathcal{A} \oplus \mathcal{A}$ and $\mathcal{A}(N) = \mathcal{A} \otimes \mathbb{R}(N)$. Every Clifford algebra $\mathsf{Cl}_{k,l}$ is isomorphic to one of the following algebras: $\mathbb{R}(2^p)$, $\mathbb{C}(2^p)$, $\mathbb{H}(2^p)$, $2\mathbb{R}(2^p)$, $2\mathbb{H}(2^p)$, $p \in \mathbb{N}$. Recall the *Chevalley theorem:* $\mathsf{Cl}_{k,l} \otimes_{\text{gr}} \mathsf{Cl}_{k',l'} = \mathsf{Cl}_{k+k',l+l'}$ and the isomorphisms: $\mathsf{Cl}_{k+4,l} = \mathsf{Cl}_{k,l+4}$, $\mathsf{Cl}_{k+1,l+1} = \mathsf{Cl}_{k,l} \otimes \mathbb{R}(2)$. Two Clifford algebras, $\mathsf{Cl}_{k,l}$ and $\mathsf{Cl}_{k',l'}$, are said to be of the same *type* if $k + l' \equiv k' + l \bmod 8$. When grading is taken into account, there are eight types of Clifford algebras; with respect to graded tensor multiplication the set of these eight types forms a group (the *Brauer-Wall group* of $\mathbb{R}$) isomorphic to $\mathbb{Z}_8$; for this reason, the algebras are conveniently represented on the *spinorial clock* [1]:

As a vector space, the algebra $\mathsf{Cl}_{k,l}$ is isomorphic, in a natural way, to the vector space $\bigwedge V$ underlying the exterior algebra of V. This isomorphism

$$\kappa : \mathsf{Cl}_{k,l} \to \bigwedge V$$

is characterized by $\kappa(1) = 1$ and

$$\kappa(va) = v \wedge \kappa(a) + g(v) \lrcorner \kappa(a) \quad \text{for every } v \in V \subset \mathsf{Cl}_{k,l} \text{ and } a \in \mathsf{Cl}_{k,l}.$$

It respects the $\mathbb{Z}_2$-grading of the vector spaces in question.

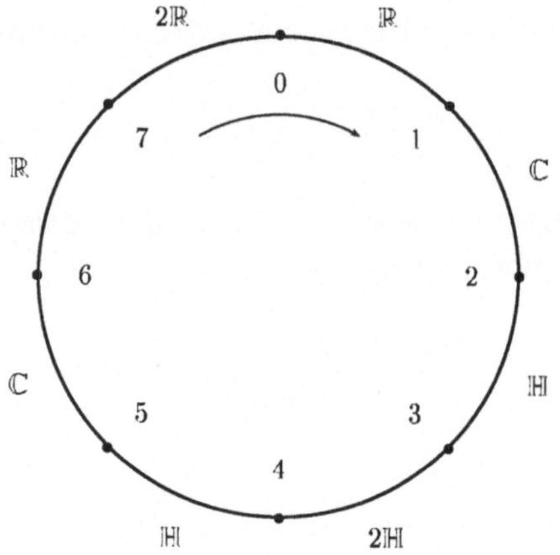

The spinorial clock can be used to find the structure of real Clifford algebras. To determine $\mathrm{Cl}^0_{k,l} \to \mathrm{Cl}_{k,l}$, compute the corresponding *hour* $h \in \{0, \ldots, 7\}$, $l - k = h + 8r$, $r \in \mathbb{Z}$. Read off the sequence $\mathcal{A}^0_h \xrightarrow{h} \mathcal{A}_h$ from the clock. If $\dim_{\mathbb{R}} \mathcal{A}_h = 2^{\nu_h}$, then $\mathrm{Cl}_{k,l} = \mathcal{A}_h(2^{1/2(k+l-\nu_h)})$, etc. The algebra $\mathcal{A}_h \otimes_{\mathrm{gr}} \mathcal{A}_{h'}$ is of the same type as $\mathcal{A}_{h+h' \bmod 8}$.

An orthonormal frame (e_μ) in $V \subset \mathrm{Cl}_{k,l}$ satisfies $e_\mu e_\nu + e_\nu e_\mu = 0$ for $\mu \neq \nu$, $e_\mu^2 = 1$ for $\mu = 1, \ldots, k$, and $= -1$ for $\mu = k+1, \ldots, m$. The square of the *volume element* $\eta = e_1 \ldots e_m$ is $\eta^2 = (-1)^{1/2(l-k)(l-k+1)}$. Hodge duality, as defined by Kähler, is given by

$$\star\kappa(a) = \kappa(\eta a), \quad \text{where} \quad a \in \mathrm{Cl}_{k,l}.$$

Whenever $\eta^2 = -1$, there is a natural complex structure in the real vector space $\bigwedge V$. Therefore,
(i) if $V = \mathbb{R}^{2n}$ and $l - k \equiv 2 \bmod 4$, then $\star$ defines a complex structure in $\bigwedge^n \mathbb{R}^{2n}$;
(ii) if $V = \mathbb{R}^{2n+1}$ and $l - k \equiv 1 \bmod 4$, then $\star$ defines a complex structure in $\bigwedge^n \mathbb{R}^{2n+1} \oplus \bigwedge^{n+1} \mathbb{R}^{2n+1}$.

If $m = k + l$ is even, $m = 2n$, then the algebra $\mathrm{Cl}_{k,l}$ is central simple and has one, up to equivalence, representation

$$\gamma : \mathrm{Cl}_{k,l} \to \mathrm{End}\, S \tag{3}$$

in a complex, 2^n-dimensional space of Dirac spinors. In particular, the contragredient representation and the complex conjugate representation are each equivalent to γ. In terms of the Dirac matrices (endomorphisms of S), $\gamma_\mu = \gamma(e_\mu)$, this equivalence may be expressed by the equations

$$\gamma_\mu^* = B\gamma_\mu B^{-1}, \qquad B : S \to S^*, \tag{4}$$

and

$$\overline{\gamma_\mu} = C\gamma_\mu C^{-1}, \qquad C : S \to \bar{S}. \tag{5}$$

The intertwining isomorphisms B and C are defined up to multiplication by non-zero complex numbers. The matrix

$$\gamma_{2n+1} = \gamma_1 \cdots \gamma_{2n}$$

anticommutes with γ_μ for $\mu = 1, \ldots, 2n$. The representation γ, restricted to $\mathrm{Cl}^0_{k,l}$, decomposes into the direct sum $\gamma_+ \oplus \gamma_-$ of representations,

$$\gamma_\pm : \mathrm{Cl}^0_{k,l} \to \mathrm{End}\, S_\pm,$$

in the spaces S_+ and S_- of Weyl spinors of opposite *chirality*,

$$S_\pm = \{\varphi \in S : \gamma_{2n+1}\varphi = \pm\iota\varphi\},$$

where

$$\iota = 1 \ \text{ for } \ \eta^2 = 1 \ \text{ and } \ \iota = i \ \text{ for } \ \eta^2 = -1. \tag{6}$$

The relevant properties of B, C, and γ_{2n+1} can be summarized in (see, for example, [1])

Proposition 34.4.1 *If $k + l = 2n$, then*

$$\begin{aligned} B^* &= (-1)^{1/2n(n-1)} B, \\ \gamma^*_{2n+1} &= (-1)^n B \gamma_{2n+1} B^{-1}, \\ \overline{\gamma_{2n+1}} &= C \gamma_{2n+1} C^{-1}. \end{aligned} \tag{7}$$

One can normalize the intertwining isomorphisms defined by (4) *and* (5) *so that*

$$\bar{C}C = (-1)^{1/8(l-k)(l-k+2)} \tag{8}$$

and

$$\bar{B}C = \bar{C}^* B^*. \tag{9}$$

The proof of (9) is based on the observation that $\bar{B}C(\bar{C}^*B^*)^{-1}$ is in the commutant of the irreducible representation γ.

From the spinorial clock one obtains, for every $p \in \mathbb{N}$, the isomorphism

$$\mathrm{Cl}_{3+p,1+p} = \mathbb{R}(2^{2+p}).$$

Since in this case $\eta^2 = -1$, there is a complex structure $J = \gamma(\eta)$ in the real space of Dirac-Majorana spinors $W = \mathbb{R}^{2^{2+p}}$. Defining, as in (1), the complex, 2^{1+p}-dimensional spaces W_+ and W_-, one sees that they can be identified with the two spaces of Weyl spinors of opposite chirality.

34.5 Charge Conjugation

Charge conjugation is intrinsically connected with the equivalence of the representations γ and $\bar{\gamma}$. The notion of charge conjugation, defined originally by physicists for spinors associated with Minkowski space, admits a generalization to higher dimensions [1]. In view of some controversy surrounding this generalization [8], I present it here, in considerable detail, for the case of an even-dimensional, flat space-time with a metric of signature $(2n-1, 1)$.

Consider first the general case of $k+l$ even, $k+l=2n$; given a representation (3) of the algebra $\mathsf{Cl}_{k,l}$ in a complex vector space S of Dirac spinors, one defines *charge conjugation* to be the antilinear map $S \to S$,

$$\varphi \mapsto \varphi_c = C^{-1}\bar{\varphi}. \tag{10}$$

If φ is a Weyl spinor, $\gamma_{2n+1}\varphi = \pm\iota\varphi$, then φ_c is also such a spinor and its chirality is the same as (respectively, opposite to) that of φ if $\eta^2=1$ (resp., if $\eta^2=-1$). If $\bar{C}C = id_S$, then the map $\varphi \mapsto \varphi_c$ is involutive, $(\varphi_c)_c = \varphi$, and there is the real vector space

$$S_{\mathbb{R}} = \{\varphi \in S : \varphi_c = \varphi\}$$

of Dirac-Majorana spinors. Charge conjugation is *not*, however, restricted to that case. If γ_μ are the Dirac matrices corresponding to a representation of $\mathsf{Cl}_{k,l}$, then the matrices $i\gamma_\mu$ correspond to a representation of $\mathsf{Cl}_{l,k}$.

Assume now that the signature is Lorentzian, $k=2n-1$ and $l=1$. In view of the previous remark, the case of signature $(1, 2n-1)$ can be easily reduced to the one under consideration. The properties of the intertwiner C, described in Proposition 34.4.1, are now expressed by the equation

$$\bar{C}C = (-1)^{1/2(n-1)(n-2)} id_S. \tag{11}$$

The Dirac equation for a particle of mass m and electric charge e can be written as

$$\gamma^\mu(\partial_\mu - ieA_\mu)\psi = m\psi, \tag{12}$$

where $\psi : \mathbb{R}^{2n} \to S$ is the wave function of the particle and A_μ, $\mu = 1, \ldots, 2n$, are the (real) components of the vector potential of the electromagnetic field. For a free particle ($A_\mu = 0$) one can consider a solution of (12) equal to a constant spinor times $\exp(ip_\mu x^\mu)$; the Dirac equation then implies that the momentum vector (p_μ) is timelike: $p_{2n}^2 = p_1^2 + \cdots + p_{2n-1}^2 + m^2$. The *charge conjugate wave function* $\psi_c : \mathbb{R}^{2n} \to S$ is defined by $\psi_c(x) = \psi(x)_c$ for every $x \in \mathbb{R}^{2n}$.

Proposition 34.5.1 *If $\psi : \mathbb{R}^{2n} \to S$ is a wave function, then*
(i) *the vector field of current defined by*

$$j^\mu(\psi) = i^{n+1}\langle B\gamma_{2n+1}\psi_c, \gamma^\mu\psi\rangle, \qquad \mu = 1, \ldots, 2n, \tag{13}$$

is real *and invariant with respect to the replacement of ψ by ψ_c,*

$$j^\mu(\psi_c) = j^\mu(\psi); \tag{14}$$

(ii) *if ψ is a solution of the Dirac equation* (12), *then the current is conserved,*

$$\partial_\mu j^\mu(\psi) = 0, \tag{15}$$

and the charge conjugate wave function satisfies the Dirac equation for a particle of charge $-e$,

$$\gamma^\mu(\partial_\mu + ieA_\mu)\psi_c = m\psi_c. \tag{16}$$

The proof of part (i) the proposition consists of simple, algebraic transformations, making use of equations (4), (5), (10), (11), and Proposition 34.4.1. Complex conjugating both sides of (12), multiplying the resulting equation by C^{-1} on the left, and using (5) and (10), one obtains that ψ_c satisfies (16); it is then easy to check that (15) holds.

These simple observations are valid irrespective of whether the algebra $\mathrm{Cl}_{2n-1,1}$ is real ($\bar{C}C = id_S$; $n \equiv 1$ or $2 \bmod 4$) or quaternionic ($\bar{C}C = -id_S$; $n \equiv 0$ or $3 \bmod 4$). Charge conjugation is not related to the existence of Majorana spinors: even if the algebra $\mathrm{Cl}_{2n-1,1}$ is *real*, one has to use *complex* spinors to write the Dirac equation for a charged particle interacting with an electromagnetic field. The invariance of the current under the replacement of ψ by ψ_c, expressed by (14), reflects the classical (or rather: first-quantized) nature of the Dirac equation under consideration here. Upon second quantization, the wave function is replaced by an *anticommuting*, spinor-valued field Ψ; anticommutativity of Ψ and Ψ_c provides a change of sign, so that (14) is replaced by $j^\mu(\Psi_c) = -j^\mu(\Psi)$.[2]

As an example, consider the case of dimension 8: one has $\mathrm{Cl}_{7,1} = \mathbb{H}(8)$ and the space of Dirac spinors is complex 16-dimensional. Let

$$\sigma_x = \begin{pmatrix} 0 & 1 \\ 1 & 0 \end{pmatrix}, \quad \sigma_y = \begin{pmatrix} 0 & -i \\ i & 0 \end{pmatrix},$$
$$\sigma_z = \begin{pmatrix} 1 & 0 \\ 0 & -1 \end{pmatrix}, \quad I = \begin{pmatrix} 1 & 0 \\ 0 & 1 \end{pmatrix}$$

be the Pauli matrices. One can take, in this case, a representation such that

$$\begin{aligned}
\gamma_1 &= \sigma_x \otimes I \otimes I \otimes I, & \gamma_2 &= \sigma_y \otimes \sigma_y \otimes I \otimes I, \\
\gamma_3 &= \sigma_y \otimes \sigma_x \otimes \sigma_y \otimes I, & \gamma_4 &= \sigma_y \otimes \sigma_x \otimes \sigma_x \otimes \sigma_y, \\
\gamma_5 &= \sigma_y \otimes \sigma_x \otimes \sigma_z \otimes \sigma_y, & \gamma_6 &= \sigma_y \otimes \sigma_z \otimes I \otimes \sigma_y, \\
\gamma_7 &= \sigma_y \otimes \sigma_z \otimes \sigma_y \otimes \sigma_x, & \gamma_8 &= \mathrm{i}\sigma_y \otimes \sigma_z \otimes \sigma_y \otimes \sigma_z,
\end{aligned}$$

and

$$C = \sigma_x \otimes \sigma_z \otimes \sigma_y \otimes \sigma_z.$$

Note that the algebra $\mathrm{Cl}^0_{7,1} \to \mathrm{Cl}_{7,1}$ is of the same type as the algebra $\mathrm{Cl}^0_{1,3} \to \mathrm{Cl}_{1,3}$.

34.6 CR Structures Associated with Integrable Optical Geometries

In this section, intended to "explain" the appearance of Cauchy-Riemann structures in the process of solving Einstein's equations for special Lorentz metrics, I restrict myself to 4-dimensional Riemannian manifolds.

---

[2] I thank Engelbert Schucking for a discussion on this aspect of charge conjugation. See also Appendix I in [22].

Consider first a Lorentz manifold, i.e., a Riemannian manifold M with a metric tensor field g of signature (3, 1). Assume that M is space and time oriented and that there is given on M a bundle $K \subset TM$ of null lines; the flow generated by $k \in \Gamma(K)$ has null curves (rays) as trajectories. Since the fibres of K are null, the bundle

$$K^{\perp} = \{u \in TM : g(k, u) = 0 \text{ for every } k \in \Gamma(K)\}$$

contains K and there is the exact sequence of homomorphisms of vector bundles,

$$0 \to K \to K^{\perp} \to K^{\perp}/K \to 0.$$

The fibres of $K^{\perp}/K$ are 2-dimensional and have a positive-definite scalar product induced by g: they are the *screen spaces* of the "optical" geometry of rays [3, 10, 13, 18]. Space and time orientation of M, together with the conformal structure of the screen spaces, induce a complex structure J in the fibres of $K^{\perp}/K$. There is a natural extension $J_{\mathbb{C}}$ of J to the complexified bundle $\mathbb{C} \otimes (K^{\perp}/K)$; the latter bundle can be identified with $(\mathbb{C} \otimes K^{\perp})/(\mathbb{C} \otimes K)$. For every $n \in \mathbb{C} \otimes K^{\perp}$, let $n + K \in (\mathbb{C} \otimes K^{\perp})/(\mathbb{C} \otimes K)$ denote the coset space containing n. The vector bundle

$$N = \{n \in \mathbb{C} \otimes K^{\perp} : J_{\mathbb{C}}(n + K) = in + K\} \tag{17}$$

is a subbundle of $\mathbb{C} \otimes TM$; its fibres are complex, totally null planes, $N^{\perp} = N$, and

$$N \cap \bar{N} = \mathbb{C} \otimes K, \qquad N + \bar{N} = \mathbb{C} \otimes K^{\perp}. \tag{18}$$

A totally null, complex plane bundle N can be also considered in other possible signatures (namely, (4, 0) and (2, 2)) of g on a 4-manifold.

If g is a *proper Riemannian* metric tensor, then

$$N \cap \bar{N} = 0 \quad \text{so that} \quad \mathbb{C} \otimes TM = N \oplus \bar{N}, \tag{19}$$

and one can define an orthogonal *almost complex* structure J on M by putting

$$J(n + \bar{n}) = i(n - \bar{n}) \quad \text{for every} \quad n \in N. \tag{20}$$

If g is *neutral* (i.e., of signature (2, 2)), then there are two possibilities: either (19) holds and there is an orthogonal, almost complex structure on M or

$$N = \bar{N} \quad \text{so that} \quad N = \mathbb{C} \otimes K, \tag{21}$$

where $K = K^{\perp}$ is now a real, totally null, *plane* subbundle of TM.

In every one of the above cases, the complex, totally null, plane bundle N can be characterized, at least locally, by (the direction of) a complex, decomposable 2-form F such that

$$n \in N \quad \text{iff} \quad n \in \mathbb{C} \otimes TM \quad \text{and} \quad n \lrcorner F = 0. \tag{22}$$

If n_1 and $n_2 \in \Gamma(N)$ are linearly independent, then one can take $F = g(n_1) \wedge g(n_2)$. If F corresponds, in the sense of (22), to N, then $\star F$ corresponds to $N^{\perp}$; since $N^{\perp} = N$, the forms F and $\star F$ are parallel. Using the notation of (6), one has

$\star F = \pm \iota F$. In signature (3, 1) one has $F \wedge \bar{F} = 0$, since F and $\bar{F}$ have $g(k)$ as a common factor; in the other two signatures, if (19) holds and $F \neq 0$, then $F \wedge \bar{F} \neq 0$.

There is also a convenient, spinorial description of the bundles N. Assume, for simplicity, that there is a spin structure Q on M; spinor and tensor fields can be then represented by equivariant maps from Q to suitable representation spaces; for example, a spinor field is given by a map $\varphi : Q \to S$ such that $\varphi(qa) = \gamma(a^{-1})\varphi(q)$ for $q \in Q$ and a in the spin structure group of the bundle $Q \to M$. Given a totally null plane bundle N on a Riemannian 4-manifold, there is a (locally defined) Weyl spinor field φ on M such that

$$N = \{n \in \mathbb{C} \otimes TM : \gamma(n)\varphi = 0\}. \tag{23}$$

The chiralities of φ and F coincide: if $\gamma_5\varphi = \iota\varphi$, then for the corresponding 2-form F one has $\star F = \iota F$. The isomorphisms κ and B of Section 34.4, together with the representation γ, induce, in dimension 4, an isomorphism of $S_+ \otimes_{\mathrm{sym}} S_+$ onto the complex space $\wedge^2_+\mathbb{C}^4$ of 2-forms F which are self-dual in the sense that $\star F = \iota F$; there is a similar isomorphism for spinors and 2-forms of the opposite chirality; these isomorphisms establish a correspondence between the descriptions of N by means of 2-forms and spinors [13]. In the Lorentzian case, the product $\varphi \otimes \varphi_c$ corresponds to $k \in \Gamma(K)$; in the proper Riemannian and neutral cases, if $\langle B\varphi_c, \varphi\rangle \neq 0$, then $\varphi \otimes \varphi_c / \langle B\varphi_c, \varphi\rangle$ corresponds to J; in the neutral case, if $\langle B\varphi_c, \varphi\rangle = 0$, then φ is (proportional to) a Weyl-Majorana spinor.

In the Lorentzian case, the real part of the 2-form F can be interpreted as representing a "null" electromagnetic field $(\mathbf{E}, \mathbf{B})$, i.e., a field such that $(\mathbf{E}+i\mathbf{B})^2 = 0$. In the 1950s, Ivor Robinson considered solutions of Maxwell's equations

$$\mathrm{d}F = 0 \tag{24}$$

for such a null field and has shown that the trajectories of the flow generated by $k \in \Gamma(K)$ constitute a congruence of null geodesics without shear [15]. He conjectured also that, given any such smooth congruence on a Lorentzian manifold, one can find a nonzero solution F of (24) such that $\star F = iF$ and $k \lrcorner F = 0$. In 1985, Jacek Tafel [21] pointed out that this need not be true, because to find such a solution one has to solve a linear partial differential equation of the first order, $\Lambda f = a$, of the type considered by Hans Lewy [7] and shown by that author not to have solutions, even locally, for some smooth, but non-analytic, functions a; see also [9]. Soon afterwards, it became clear [17] that the structure underlying shear-free congruences of null geodesics on Lorentzian manifolds is that of Cauchy-Riemann manifolds, earlier introduced into physics by Penrose, in his theory of twistors associated with Minkowski space, and its generalization to curved manifolds [11]–[13].

Proposition 34.6.1 *Let M be a Riemannian 4-manifold with a metric tensor g that is either proper Riemannian or Lorentzian or neutral. Let $N \to M$ be a totally null, complex, plane subbundle of $\mathbb{C} \otimes TM$ and let F be a 2-form such that* (22) *holds. Then*

(i) *equation* (24) *implies the* complex integrability condition:

$$[\Gamma(N), \Gamma(N)] \subset \Gamma(N); \tag{25}$$

(ii) *if* $N \cap \bar{N} = 0$, *then* (25) *is equivalent to the integrability of the almost complex structure* J *defined by* (20)*; if* g *is proper Riemannian (respectively, neutral), then* (2) *defines a proper Hermitian (respectively, Hermitian of signature* (1,1)*) tensor field* h *on* M*;*
(iii) *if* g *is Lorentzian, then* (25) *is equivalent to the statement that the trajectories of the flow generated by every* $k \in \Gamma(K)$, K *as in* (18)*, constitute a congruence of* null geodesics without shear*; moreover, if the congruence is regular in the sense that the quotient set* $M' = M/K$ *is a 3-manifold and the map* $\pi : M \to M'$ *is a submersion, then* N *projects to a complex line bundle* $H \to M'$, $H \subset \mathbb{C} \otimes TM'$, *defining a* CR-structure *on* M'*; the form* F *satisfying* (22) *and* (24) *descends to a complex 2-form* F' *on* M' *such that*

$$\mathrm{d}F' = 0, \quad Z \lrcorner F' = 0 \ \textit{for every} \ Z \in H \quad \textit{and} \quad F = \pi^* F'; \tag{26}$$

(iv) *if* g *is neutral and* (21) *holds, then* (25) *reduces to the* real integrability condition,

$$[\Gamma(K), \Gamma(K)] \subset \Gamma(K);$$

the leaves of the foliation defined by K *are 2-dimensional, totally null and totally geodesic submanifolds of* M.

The proof of Proposition 34.6.1 is straightforward; most of it can can be found in [4, 10, 13, 18, 24, 25]. There are interesting results and problems connected with the analogy between a shear-free congruence of null geodesics on a Lorentz manifold and the Hermitian geometry in the proper Riemannian case; one of them consists in the proof of the Goldberg-Sachs theorem in signatures (4, 0) and (2, 2) [14].

It is worth noting that in the Lorentzian case, the complex structure in the fibres of $K^{\perp}/K$ is determined, in a natural manner, by giving only a space and time orientation of M and the bundle of null lines K; in the proper Riemannian case, the (almost) complex structure has to be introduced explicitly, by giving either J or N. It is for this reason that the appearance of the Cauchy-Riemann equations in [16] had been somewhat unexpected.

Recall that the (abstract) Cauchy-Riemann structure on a 3-manifold M', given by the complex line bundle $H \to M'$, $H \subset \mathbb{C} \otimes TM'$, $H \cap \bar{H} = 0$, can be conveniently locally described also as follows: let Z be a nonzero section of H and let λ be a nonzero, real 1-form on M such that $Z \lrcorner \lambda = 0$. One can find a complex 1-form μ such that $\lambda \wedge \mu \wedge \bar{\mu} \neq 0$, $Z \lrcorner \mu = 0$ and $Z \lrcorner \bar{\mu} \neq 0$. These forms are defined up to transformations

$$\lambda \mapsto a\lambda, \qquad \mu \mapsto b\mu + c\lambda, \tag{27}$$

where a is a real function and b, c are complex functions on M' such that $a, b \neq 0$. The direction of the 2-form $\lambda \wedge \mu$ is invariant with respect to the changes (27) and characterizes the CR structure. In the terminology of [5], such a form is a section of the *canonical bundle* of the CR 3-manifold, defined as

$$\{\omega \in \mathbb{C} \otimes \wedge^2 T^*M' : Z \lrcorner \omega = 0 \text{ for every } Z \in H\}.$$

To simplify the language, I shall use, from now on, the expression "a CR space" instead of "a three-dimensional manifold with a CR structure."

Consider the fibration $\pi : M \to M'$. Let P and ξ be a real function and a real 1-form on M, respectively, such that $P^2\pi^*(\mu \wedge \bar{\mu} \wedge \lambda) \wedge \xi$ vanishes nowhere on M. The symmetric tensor field on M,

$$g = P^2\pi^*(\mu \otimes_{\text{sym}} \bar{\mu}) + \pi^*\lambda \otimes_{\text{sym}} \xi, \tag{28}$$

is the most general Lorentz metric admitting the fibres of π as null geodesics constituting a congruence without shear [17]. One then has $\pi^*\lambda \wedge g(k) = 0$ for $k \in \Gamma(K)$ and $K^\perp = \ker \pi^*\lambda$.

Let $f : M' \to \mathbb{C}$ be a smooth function. If the *Cauchy-Riemann equation*

$$Z \lrcorner df = 0 \tag{29}$$

has two independent (local) solutions z and w, then the 2-form $dz \wedge dw$ is a nonzero section of the canonical bundle; using the freedom implied by (27), one can choose μ to coincide with dz. The map

$$(z, w) : M' \to \mathbb{C}^2 \tag{30}$$

is a (local) embedding of M' in $\mathbb{C}^2$ and the CR structure is then said to be (locally) *embeddable*. Lewandowski, Nurowski and Tafel have shown that the CR space defined by a shear-free congruence of null geodesics on an Einstein-Lorentz manifold is so embeddable [6]. One can then introduce local coordinates (u, x, y) on M' such that $x + iy = z$ and represent the form λ and the vector field Z as

$$\lambda = du + \bar{L}dz + Ld\bar{z}, \quad Z = \frac{\partial}{\partial \bar{z}} - L\frac{\partial}{\partial u}.$$

If $L = 0$, then the CR structure is trivial in the sense that M' is foliated by a family of complex 1-manifolds of equation $u = \text{const.}$; the corresponding bundle $K^\perp$ is integrable, $\lambda \wedge d\lambda = 0$, and (29) reduces to the classical, Cauchy-Riemann equation: this is the special case of a *'hypersurface orthogonal'* congruence of shear-free null geodesics considered in [16].

According to part (iii) of Proposition 34.6.1, the general problem of finding a solution of Maxwell's equations (24) adapted to a shear-free congruence of null geodesics defined by K, i.e., such that $\star F = iF$ and $k \lrcorner F = 0$, reduces to the following: given a CR space M', find a closed section F' of its canonical bundle. If M' is embeddable, then such sections exist and are of the form

$$F' = f(z, w)dz \wedge dw$$

where z and w are as in (30) and f is an analytic function of its arguments.

It is now known that there are CR spaces that are nonembeddable, but have one solution of (29) [19]; by the results of [21], extended to higher dimensions in [5], such CR spaces do not admit closed, nonzero sections of their canonical bundle.[3] Therefore, Lorentzian manifolds constructed on the basis of these CR spaces as in (28) do not admit any nonzero solutions F of Maxwell equations such that $k \lrcorner F = 0$, $g(k) \wedge F = 0$, where $g(k) = \lambda$. There are examples of nonembeddable 7-dimensional CR manifolds that have nonzero, closed, sections of their canonical bundle, but it is not clear whether there are such examples in dimensions 3 and 5. In connection with this, I formulate the following

Conjecture 34.6.1 *A CR 3-space admits locally a closed, nonzero section of its canonical bundle if and only if it is locally embeddable.*

The conjecture can be formulated as a problem of elementary vector calculus: given a complex vector field $\mathbf{F}$ on $\mathbb{R}^3$ such that $\mathbf{F} \times \overline{\mathbf{F}} \neq 0$ and $\operatorname{div}\mathbf{F} = 0$, show that there exist two complex functions z and w such that $\mathbf{F} = \operatorname{grad} z \times \operatorname{grad} w$.

Since it is known that real analytic CR spaces are locally embeddable, the proof of the conjecture—if it is true—should concern the smooth case.

Acknowledgments

Work on this article was supported in part by the Polish Committee on Scientific Research (KBN) under grant no. 2 P03B 017 12 and by the Foundation for Polish-German cooperation with funds provided by the Federal Republic of Germany. The paper has been completed in June 1997, during the workshop on *Spaces of geodesics and complex structures in general relativity* at the Erwin Schrödinger International Institute for Mathematical Physics in Vienna. I have benefited there from discussions with C. Denson Hill, Paweł Nurowski, Roger Penrose, and Helmuth Urbantke.

References

[1] Budinich, P. and Trautman, A., *The spinorial chessboard*, Springer-Verlag, Berlin 1988.

[2] Ebbinghaus, H.-D., Hermes, H., Hirzebruch, F., Koecher, M., Mainzer, K., Prestel, A. and Remmert, R., *Zahlen*, Springer-Verlag, Berlin 1983.

[3] Ehlers, J. and Kundt, W., Exact solutions of the gravitational field equations, in *Gravitation*, edited by L. Witten, Wiley, New York 1962.

[4] Hughston, L. P. and Mason, L. J., A generalised Kerr–Robinson theorem, *Class. Quantum Grav.* **5** (1988) 275–285.

[5] Jacobowitz, H., The canonical bundle and realizable CR hypersurfaces, *Pacific J. Math.* **127** (1987) 91–101.

[6] Lewandowski, J., Nurowski, P., and Tafel, J., Einstein's equations and realizability of CR manifolds, *Class. Quantum Grav.* **7** (1990) L241–L246.

---

[3]The significance, in this context, of the examples found by Rosay and of the results of Jacobowitz has been explained at the workshop in Vienna by C. Denson Hill.

[7] Lewy, H., An example of a smooth partial differential equation without solution, *Ann. of Math.* **66** (1957) 155–158.

[8] Lounesto, P., Counter-examples in Clifford algebras, *Adv. Appl. Clifford Algebras* **6** (1996) 69–104.

[9] Nirenberg, L., On a question of Hans Lewy, *Russian Math. Surveys* **29** (1974) 251–262.

[10] Nurowski, P., Optical geometries and related structures, *J. Geom. Phys.* **18** (1996) 335–348.

[11] Penrose, R., The complex geometry of the natural world, in: *Proc. Int. Congress Math.*, pp. 189–194, Helsinki 1978.

[12] Penrose, R., Physical space-time and nonrealizable CR-structures, *Proc. Symp. Pure Math.* **39**, Part I (1983) 401–422.

[13] Penrose, R. and Rindler, W., *Spinors and space-time*, vols 1 and 2, Cambridge University Press, Cambridge 1984 and 1986.

[14] Przanowski, M. and Broda, B., Locally Kähler gravitational instantons, *Acta Phys. Polon.* **B14** (1983) 637–661.

[15] Robinson, I., Null electromagnetic fields, *J. Math. Phys.* **2** (1961) 290–291.

[16] Robinson, I. and Trautman, A., Spherical gravitational waves, *Phys. Rev. Lett.* **4** (1960) 431–432.

[17] Robinson, I. and Trautman, A., Cauchy–Riemann structures in optical geometry, in *Proc. of the Fourth Marcel Grossmann Meeting on General Relativity*, pp. 317–324, edited by R. Ruffini, Elsevier, 1986.

[18] Robinson, I. and Trautman, A., Optical geometry, in *New Theories in Physics*, Procs. of the XI Warsaw Symposium on Elementary Particle Physics, edited by Z. Ajduk et al., World Scientific, Singapore 1989.

[19] Rosay, J.-P., New examples of non-locally embeddable CR structures, *Ann. Inst. Fourier Grenoble* **39** (1989) 811–823.

[20] Sommerfeld, A., Mathematische Theorie der Diffraction, *Math. Annalen* **47** (1896) 317–374.

[21] Tafel, J., On the Robinson theorem and shear-free geodesic null congruences, *Lett. Math. Phys.* **10** (1985) 33–39.

[22] Thirring, W., *Principles of quantum electrodynamics*, Academic Press, New York 1958.

[23] Trautman, A., Optical structures in relativistic theories, in: *Élie Cartan et les mathémathiques d'aujourd'hui*, *Astérisque*, numéro hors série, (1985) 401–420.

[24] Trautman, A., Geometric aspects of spinors, in: *Clifford algebras and their applications in mathematical physics*, edited by R. Delanghe, F. Brackx, and H. Serras, Kluwer Academic Publishers, Dordrecht 1993.

[25] Trautman, A., Clifford and the 'square root' ideas, *Contemporary Mathematics*, **203** (1997) 3–24.

35

The Engelbert Experience: Pathways from the Past

C. V. Vishveshwara

It was the best of times, it was the worst of times. For me, that is. This was in the year 1969. I was working at the Institute for Space Studies in New York on a National Research Council Fellowship. I was studying the scattering of gravitational waves by black holes through computer simulation in order to find out whether the black hole left its imprint in some way on the scattered wave. This was done by bombarding the black hole with Gaussian wave packets. When the width of the Gaussian was comparable to or less than the radius of the black hole, there emerged a decaying wave pattern, later to be called the quasinormal mode of the black hole [1]. As is well known, a lot of work has been done on the black hole quasinormal modes since then. Today they are considered to be a means of detecting both black holes and gravitational waves. My original work on the quasinormal modes was terribly exciting—to me. And that is the best-of-times part of the story.

Now for the worst-of-times aspect. Unfortunately, no one at the Institute seemed to be interested in this kind of research. After all, neither black holes nor gravitational radiation had been detected, but only theoretically predicted. Why should anyone spend time investigating the interaction between two unobserved entities perhaps of doubtful existence? The consequence of this attitude, which was not at all uncommon in those days, was rather disconcerting to say the least. My contract was renewed for only three months instead of the normal one year period. As it turned out, however, it was a blessing in disguise, but most effectively disguised at the time, as Winston Churchill might have put it. Frantically I started looking for a job. Jobs were hard to come by, but not negative replies. With some hesitation I thought of writing to Professor Engelbert Schucking, who was working downtown at the New York University. I had a funny feeling that the sort of things I did were not his cup of tea or mug of beer. In reality, Engelbert's cupboard happened to hold a number of mugs filled with a variety of beers. I had listened to Engelbert a few years earlier at the historic Cornell Summer School, but had not spoken to him. Historic, because almost every well-known general relativist of the present era was there. Boyer, Carter, Chandrasekhar, Ehlers, Hawking, Kerr, Misner, Penrose, Robinson, Schild, Schücking, Taub, and so on. Most of the lectures were enormously interesting and some totally incomprehensible. Engelbert gave two or three talks on cosmology. Of all the stuff that seeped into my gray cells, it was only the opening sentence of Engelbert's lectures that was pickled permanently: "An important contribution of the general theory of relativity to cosmology has been to keep out theologians by a straightforward application of tensor analysis." Well, to return to my original story, Engelbert asked me to meet him after about a month. It was a long, dismal month indeed. For, in the meantime, my contract had run out. Unemployed, I had given up my apartment, sold my furniture, and moved in with some friends, dipping into my meager savings. At the end of the month, after a seminar at the New York University, Engelbert told me, "It would be nice if you could come and help us with our research." This was his delicate way of telling me that he was offering me a job! So, I was going to work with Engelbert and his group. The blessing had thrown off its disguise at last.

When I joined Engelbert's group, the physics department occupied a part of the Courant Institute down in the Greenwich Village. It was a modern brownish building which was quite imposing, but a bit intimidating as well. We were to acquire a brand new building of our own soon enough. Although the physicists were surrounded by a number of mathematicians, there was hardly any interaction between the two disciplines. Stranger still, it was difficult to interact with many of the students who had regular jobs during the day and surfaced at the University towards late evening. However, there were some regular graduate students.

One of them was Eli Honig, who was analyzing gyroscopic precession in curved spacetimes to identify the analogue of Thomas precession. Thomas precession arises in flat spacetime when the gyroscope moves along a nongeodesic path such as a circular orbit. However, in curved spacetimes gyroscopes following geodesic worldlines can still precess as a result of the spacetime curvature. Fokker–de Sitter precession in the Schwarzschild spacetime is an example of this. It is customary nowadays to identify automatically extra terms arising on account of nongeodesic motion as representing Thomas precession. Such a procedure was not in the least satisfactory to Engelbert with his penchant for exactitude that characterized his research. As a result, perhaps, none of the work that went in this direction was ever submitted for publication.

There was, nevertheless, an offshoot of this investigation which became a research project in its own right. As we were discussing precession of spins and magnetic moments, we started looking at the motion of charged particles in homogeneous electromagnetic fields that are, in other words, constant in both space and time [2]. In order to describe the worldlines of these charges, we utilized the Frenet-Serret formalism, which is well known to classical geometers. I had first come across it in Synge's book on general relativity, but had not seen its practical application. In this formalism, a curve is characterized by some scalar parameters that in general vary, say as functions of the arc length, along the curve. In three dimensions these parameters are κ the curvature and τ the torsion. They uniquely and invariantly define the curve. For instance, $\kappa = \tau = 0$ is a straight line,

$\tau = 0$ a planar curve with κ = constant a circle, κ = constant and τ = constant a helix and so on. In a 4-dimensional spacetime we have three such parameters, namely the curvature κ and τ_1, τ_2, the first and second torsions.

All along the curve or the worldline an orthonormal tetrad e_i $(i = 0-3)$ is defined which obeys the Frenet-Serret transport law. The vector e_0 is identical to the 4-velocity of the particle. As differentiation along the trajectory is with respect to the invariant arc-length or the proper time, curvature κ turns out to be the magnitude of the 4-acceleration acting along the unit vector e_1. How about the torsions τ_1 and τ_2? Well, they are directly related to gyroscopic precession. A gyroscope, which is by definition Fermi-Walker transported along the particle worldline, precesses with respect to the Frenet-Serret spatial triad at a rate ω_g, which has components τ_1 and τ_2 along e_3 and e_1 respectively. Thus gyroscopic precession can be described in a completely covariant and geometric manner once the torsions and the tetrad have been determined along the worldline of the particle.

In the case of charged particles moving in a constant electromagnetic field given by covariantly constant tensor F_{ab}, all the Frenet-Serret quantities can be determined in terms of F_{ab} and the 4-velocity e_0. We obtained a number of neat results. We could show, for instance, that κ, τ_1, and τ_2 were constants along the worldlines. Two equations connected these parameters to the Lorentz invariants $F_{ab}F^{ab}$ and $F^{ab}\,{}^*F_{ab}$, where ${}^*F_{ab}$ is the dual of F_{ab}:

$$\kappa^2 - \tau_1^2 - \tau_2^2 = \frac{1}{2} F^{ab} F_{ab}$$

and

$$\kappa \tau_2 = -\frac{1}{4}\,{}^*F^{ab}\,F_{ab}\,.$$

The two torsions τ_1 and τ_2, the components of gyroscopic precession, were shown to be proportional to $E \cdot B$ and $E \times B$ respectively, E and B being the electric and magnetic fields in the reference frame comoving with the charged particle.

All this was quite nice. But, then, where do you get constant electromagnetic fields in nature? John Ruskin once wrote something like "The more beautiful the object, the less useful it is." Echoing this opinion, George Ellis, who liked our work, remarked, "Pretty it is, but useful it ain't!" Nevertheless, there was one part of our work which has proved to be not only pretty, but also useful. This is the application of the formalism to particle worldlines following a Killing vector field ξ^a, or Killing trajectories for short. The analogue of the Maxwell tensor is now $F_{ab} = (\xi^c \xi_c)^{-1/2} \xi_{a;b}$. Then all the results of charged particle motion in constant electromagnetic fields follow. The first relation connecting κ, τ_1, and τ_2 as above reduces to the equation that demonstrates the existence of an ergosphere between the stationary limit and the event horizon in stationary spacetimes [3]. On the way to this result, I was able to prove that the gyroscopic precession rate was identical to the vorticity of the Killing congruence after considerable amount of manipulation. I was so excited that I rushed to Engelbert's home to show him my calculations. He was quite happy. He read through my notes, munching on a handful of Pepperidge Farm cookies dipped in lemon tea, exclaiming continually, "Beautiful, beautiful!

Beautiful geometry!" We left these considerations related to Killing trajectories at an abstract theoretical level.

After more than two decades, my colleague B. R. Iyer and I returned to this problem [4]. We developed the formalism further and applied it to a variety of examples like Thomas precession, de Sitter–Fokker precession in the Schwarzschild spacetime, Schiff correction in the Kerr spacetime, and so on. All these results hang together neatly. During the period when there was a lot of excitement about higher dimensional spacetimes, we generalized the Frenet-Serret description to such spaces as well [5].

I have worked with a lot of people. My collaborators number some thirty or so. One I never met, since he was a student of one of my colleagues abroad and we worked long distance by mail, that too before the advent of e-mail. Five of these thirty-odd relativists are senior to me, Engelbert being one of them. Usually, the way they guided their graduate students as thesis advisors was to assign a specific problem, indicate the general mode of attack and from time to time discuss the results the student had obtained. But Engelbert's methodology was considerably different from this norm. Often the problem would not be precisely formulated, but only the general area would be identified. He would then relentlessly explore this uncharted terrain until the problem for research emerged. In this process, Engelbert would continually interact with his students. I acted as a sort of multidimensional middleman: coworker, assistant advisor, and liaison officer between Engelbert and his students. Engelbert's reluctance to publish anything in haste—sometimes his disinclination to publish at all—was proverbial. Consequently, some student or the other would ask me to plead with him to have some of the results published. After all, one needed publications to get a job.

He met with each of his students at least once a week for about an hour or two. He would write down his appointments on the blackboard with boxes around them. Often they covered a major part of the blackboard, leaving very little room for writing. If the problem we were working on was interesting enough, these meetings would stretch on, sometimes into the late evening, making Engelbert forget his other engagements. Once Brenda, his companion and our friend, breezed in to whisk Engelbert away to dinner, reminding the absent-minded professor that they had already made reservation at a restaurant in his name and that his name happened to be Engelbert Schucking! Whenever we were engrossed in our discussions, Engelbert did not like to be disturbed by outside agencies. So much so that during these sessions he would not answer the telephone and would let it ring forever. On one such occasion, when he caught me and his graduate student Richard Greene exchanging glances while the telephone rang on, he explained that there was a maniac at the other end and cautioned us not to touch the instrument. How did he know, we wondered.

Richard was fortunate in completing his doctoral research very quickly, culminating in one of the shortest theses I have ever come across. His problem consisted in finding a globally hypersurface orthogonal timelike congruence in the Kerr spacetime and investigating its properties. We were not aware of Bardeen's work defining the locally nonrotating frames (LNRF) [6], which in fact implied the existence of such a congruence. Nevertheless, the global properties of this congruence, especially in a general axially symmetric stationary spacetime, had not been explored. This work was quite close to my heart, since, I felt, the starting point was contained to some extent in my earlier work on the horizon and the stationary limit in the Kerr spacetime [3]. I had shown that although the global timelike Killing vector does not become null on the horizon, the combination $\zeta = \left(\xi_t - \frac{a}{2mr_+}\xi_\phi\right)$ does, where ξ_t and ξ_ϕ are the timelike and axial Killing vectors respectively, m is the mass and a the angular momentum per unit mass, and r_+ the radius of the horizon. In order to obtain the irrotational congruence, one has to multiply ξ_ϕ not by a constant but by a function of coordinates. This combination is in fact given by

$$\chi = \xi_t - \left(\frac{\xi_t \cdot \xi_\phi}{\xi_\phi \cdot \xi_\phi}\right)\xi_\phi,$$

which is no longer a Killing vector field unlike ζ. We studied the global properties of this congruence in an arbitrary stationary spacetime with axial symmetry. We were able to prove some interesting theorems related to χ and also prove that it became null on the horizon provided conditions for orthogonal transitivity were satisfied. The congruence is orthogonal to hypersurfaces defined by t =constant, where t is the global time. These happen to be maximal hypersurfaces. Not only was our paper describing this work [7] accepted very quickly, but also we were rewarded with a glowing referee's report that considered the paper to be "impeccable." Rather a rare occurrence indeed!

The reference frame adapted to χ constitutes the general relativistic analogue of the Newtonian rest frame. It is consequently advantageous to study physical phenomena as seen by these "rest" observers. Abramowicz and coworkers have recently utilized this in defining the analogues of inertial forces [8]. In an axisymmetric stationary spacetime one has in general gravitational force G_a, centrifugal force Z_a, and Coriolis force C_a. Furthermore, they are intimately connected to gyroscopic precession. My young colleague Rajesh Nayak and I have studied this relationship between gyroscopic precession on the one hand and the inertial forces on the other [9]. These relations have striking resemblance to the results of the work on the charged particle dynamics in homogeneous electromagnetic fields. The electric field of the latter case is now replaced by a quantity proportional to the particle acceleration and the magnetic field by a combination of Z_a and C_a which reduces to only the centrifugal force Z_a in static spacetimes. This study in a sense has had its evolutionary origins in the work I did with Engelbert and his students Eli Honig and Richard Greene. Our investigations will, we expect, continue further. After all, the past, present, and future are connected by worldlines frozen in time.

All my research before joining Engelbert had been entirely in the area of black holes: geometric structure, stability, and quasi-normal modes. As I have mentioned earlier, I was not sure how much interest Engelbert had in this direction. But he was indeed concerned with some of the puzzling questions related to black hole physics.

Those were the days in which research in this field was gathering momentum and finally finding recognition. Some of the ideas, both in theoretical aspects and especially astrophysical implications, were in a fluid state and therefore to some extent confused. Engelbert gave a talk on black holes. Once again, as in the case of his Cornell lectures, one sentence of his has stuck in my memory: "Astronomers often try to make important discoveries by combining the observational techniques of ESP with the theoretical methods of science fiction."

Our work with Richard Greene belonged to black hole physics. I started working with Engelbert on another intriguing problem related to the Kerr metric. Brandon Carter [10] had shown, while studying the geodesics of the Kerr spacetime, that there existed a constant of motion quadratic in the geodesic 4-momentum p^a:

$$Q_{\text{kerr}} = K_{ab} p^a p^b,$$

where K_{ab} is a Killing tensor satisfying the equations

$$K_{ab} = K_{ba}$$

and

$$K_{(ab;c)} = 0,$$

where the parentheses denote symmetrization. We wanted to interpret this quadratic constant Q_{kerr} as something like the square of the angular momentum of the particle tracing the geodesic. The Schwarzschild spacetime, because of its inherent spherical symmetry, admits three noncommuting Killing vector fields ${}_xL$, ${}_yL$, and ${}_zL$. Here

$${}_zL \equiv \xi_\phi = \frac{\partial}{\partial\phi},$$

the axial Killing vector. The scalar products ${}_xL_a p^a$ etc. give the angular momentum components in the x, y, and z directions respectively. If we set $a = 0$ in Q_{kerr} of the Kerr spacetime, thereby going over to the Schawarzschild spacetime, we obtain

$$Q_{\text{sch}} = ({}_xL_a p^a)^2 + ({}_yL_a p^a)^2 + ({}_zL_a p^a)^2,$$

This is indeed the square of the angular momentum. Or equivalently

$$K_{ab} = {}_xL_a\, {}_xL_b + {}_yL_a\, {}_yL_b + {}_zL_a\, {}_zL_b,$$

which is a degenerate Killing tensor. In the Kerr metric, which is axially symmetric, only the Killing vector ${}_zL$ exists in addition to the timelike Killing vector but not the other two Killing vectors ${}_xL$ and ${}_yL$. The question now was whether it is still possible to construct two vector fields ${}_xL$ and ${}_yL$ in the Kerr geometry which are not Killing, but along with ${}_zL$ satisfy the angular momentum commutation relations, namely,

$$[{}_xL,\ {}_yL] = -{}_zL$$

If so, could one decompose K_{ab} of the Kerr spacetime in terms of these vector fields exactly as in the Schwarzschild spacetime? Could one further relate these fields to some sort of precession of the angular momentum vector?

We tried very hard to answer these questions, working whenever and wherever possible. We even worked in the neighborhood children's park, Engelbert keeping one eye on the calculations and the other on his little son, whom he was baby sitting. At one point another kid squirted water on us, almost washing out our precious but nonfunctional formulae. Engelbert reprimanded the kid for his attempt to sabotage science. The mothers in attendance refused to appreciate the small step we were trying to take, which would mean a giant leap for general relativity. They rebelled against us, insisting that children could do anything they liked in a children's park and suggesting that we had better take our scientific endeavour elsewhere. A shrewd strategist, Engelbert played the role of a retreating relativist with a silent minority of one, namely me, following him.

Anyway, coming back to the Killing tensor, by transforming to the Kerr-Schild coordinates K_{ab} could almost be cast into the required form, but not quite. Abbas Faridi, another graduate student, joined the quest. But no go. However, this goal can be reached not through K_{ab} but through the Killing-Yano tensor F_{ab}, which was obtained later on by Floyd [11]. This antisymmetric tensor is like the "square root" of K_{ab},

$$K_{ab} = F_a^{\ c} F_{cb}; \quad F_{ab} = -F_{ba}$$

and satisfies the equation

$$F_{a(b;c)} = 0.$$

As a result, the vector $J^a \equiv F^a_{\ b} p^b$ is parallely propogated along the geodesic.

With this new information in hand, I opened up the original problem again after returning to India for good and worked on it with Joseph Samuel [12]. The close resemblance of F_{ab} to that of the flat spacetime could be demonstrated by transforming to the Kerr-Schild coordinates (x^0, x, y, z), in which the metric takes on the familiar form,

$$g_{ab} = \eta_{ab} + 2H(x, y, z) l_a l_b.$$

In these coordinates the spatial part of F_{ab} has exactly the same form as in the flat spacetime:

$$F_{ab} = \epsilon_{abcd} x^c t^d,$$

where t^d is the timelike Killing vector and x^c stands for either the Cartesian coordinates or the Kerr-Schild coordinates (x, y, z). We can therefore define the angular momentum operators of the Kerr spacetime, in the same way as in the flat spacetime leading to the same commutation relations. The surface on which the angular momentum operators act is given by

$$\frac{1}{2} F_{ab} F^{ab} = x^2 + y^2 + z^2 = \text{constant}$$

which is the "sphere" in the Kerr spacetime. Furthermore, to first order in the Kerr parameter a, we have the equations for the precession of the angular momentum about the z-axis:

$$ {}_x\dot{L} = \omega\, {}_yL; \qquad {}_y\dot{L} = -\omega\, {}_xL; \qquad {}_z\dot{L} = 0, $$

where

$$ \dot{L} \equiv (L^a p_a)_{;b} p^b $$

and

$$ \omega = \frac{ma}{r} l^a p_a . $$

All this and some more considerations worked out quite well. In 1985 I visited Engelbert at the New York University and gave a seminar on the above findings. He was very happy. But, unfortunately, our paper ran into the usual problems with the referees. In the meantime I discovered, to my utter amazement, our old friend Abbas Faridi had not only come up with results similar to ours, but had also published them very recently [13]. This almost convinced me of the existence of telepathy. We did not pursue our efforts to publish our paper further. Questions such as the possibility of obtaining precession equations for the angular momentum without linear approximation in a as well as further investigation of the properties of the angular momentum and possible applications to astrophysical situations still remain an unexplored territory.

One of the memorable events that occurred during my tenure at the New York University was Professor S. Chandrasekhar's visit to the physics department. Chandra gave a beautiful talk on rotating fluids. I had met Chandra at the University of Maryland three years earlier. This was the time when Chandra was just getting interested in black holes. I had given him the correct equations governing perturbations of the Schwarzschild metric. Now we discussed the work I had done on scattering of gravitational radiation by the black hole, in particular the quasinormal modes. In the evening Engelbert gave a party. At that time Engelbert lived in the East Village, which was known for its Bohemian life as well as a nonnegligible crime rate. Engelbert's apartment was predominantly furnished with crates retrieved from supermarkets and topped with cushions to sit on. Normally I occupied orange crates, leaving the apple crates to senior people. Chandra, impeccably dressed in a gray suit as always, did not sit down and left early so as not to miss his customary bedtime. The party, however, continued. Somewhat late in the evening, the bell rang and Engelbert opened the door. In rushed an excited young couple. Behind them stood two burly gentlemen. Engelbert invited them in, waving his arms expansively in a gesture of welcome. To Engelbert's great disappointment, instead of accepting his invitation, the two men took to their heels. These two gentlemen happened to be muggers who had been chasing the young couple! The party went on till the wee hours in the morning.

Around five or so, Engelbert was awakened from his deep, tired slumber by the ringing of the telephone. It was Chandra calling. There was no hot water in his hotel and he could not shave. The pipes had frozen and it would be at least a couple of hours before he could have running hot water. This delay in shaving would throw Chandra's schedule into unthinkable confusion, hamper his calculations and ruin his day. So Engelbert had to arrange with Larry Spruch, another professor who lived close to Chandra's hotel, to have hot water fetched to Chandra so that he could shave in time.

It is dusk. Hemmed in on all sides by tall buildings, one never sees the sunset in New York. The sun just dissolves in the smog, leaving behind a gray twilight. I finish my light evening meal in the little German restaurant Zum Zum. Zum Zum Zum Zum .... written around the restaurant. Bratwurst with fried onion washed down with a mug of beer followed by strudel and tea. I walk down to the Avenue of the Americas and hang around the Washington Square for a while. The dying strains of a distant saxophone wafts across the square while the tired girl in the faded overalls gathers up the unsold posters. As I near the newly constructed physics building, a musty old figure suddenly materializes from the side alley. I am scared. One is always scared walking the mean streets of New York. But I have never been mugged or murdered in New York. Only happy memories. "Brother, can you spare a quarter for a glass of beer?" the man stammers. I am about to fish out two quarters out of my pocket—one for beer and one for honesty. No one admits that the handout is for a drink; it is for food, medicine or to pay the vet for fixing the cat. The man is frightened. More than I am. "Officer, please don't shoot me. I've done nothing," he pleads. I assure him that I am not a cop and that I do not carry a gun. I give him the two quarters and the alley swallows him up.

The physics building is ablaze with light. Dracula's castle on the Walpurgis Night. Evening students chattering away as they swarm in and out of class rooms. Once a week I gave informal lectures in general relativity to some of the students working with Engelbert and some who intended to do so in the near future. After the lecture, we would retire to one of the neighborhood basement bars to discuss everything under the sun except physics. Engelbert taught an extremely popular course in astronomy meant for nonscience majors. It was also televised as part of the educational program called the Sunrise Semester under the title "Astronomy and Astrology: the Heavenly Twins." In his course he taught the students how to cast horoscopes. Maybe some of them are rolling in money as professional astrologers, who knows! He set hilariously worded problems in astronomy. Something like this: "Henry Kissinger is kidnapped by a bunch of gigantic CIA agents disguised as a gaggle of Croatian dwarfs. He is left to languish in a vertical elevator shaft of an abandoned building without a roof. By consulting the ephemeris concealed in his wrist watch, which is also a two-way wireless unit, he recognizes the star above him as Sirius. What time is it in New York?" Once a week some of us met to discuss current problems in relativistic astrophysics, especially those related to the areas of gravitational collapse, neutron star models, quasars, and so on. Dressed in a lemon yellow shirt and blue jeans, one long leg crossed over the other, Engelbert would sit amidst the graduate students gathered around him and lead the discussions. Once in a while, one of us would present some classic paper. I remember discussing in detail Hermann Bondi's paper establishing the upper limit on gravitational redshift produced by a spherical mass distribution. It was a very nice paper. I presented it in all its mathematical complexity and the students left one by one, unable to stomach the calculations. Only Engelbert, his student Bill Wallace, and I remained till the bitter end.

Bill, among other things, was working on the apparent superluminal velocities displayed by some quasars like 3C 279. The model we concentrated on involved a component of the quasar moving almost directly towards the observer with relativistic speed. In such a situation, the motion of the component projected normal to the line of sight onto the celestial sphere is endowed with apparent superluminal velocity. We made detailed computations on the spectral characteristics of the quasar, magnetic fields associated with it and so on. The paper was published in *Astronomical Journal* [14]. Engelbert called our model the "three penny model." Let me quote from the introduction, which Engelbert wrote, in order to explain the reason for choosing this name:

> In our analysis one component is taken as the parent quasar while the other visible component would be a radio source ejected at a relativistic speed in our general direction. Presumably this component would be only one out of many ejects which were emitted isotropically or equatorially by the quasar but remain invisible because the relativistic intensity shifts make them inconspicuous. This model, which presumably goes back to Rees, might be described as the 'three penny model', after the Brecht and Weill opera in which the chorus sings:
>
> Therefore, some are in darkness;
> Some are in the light, and these
> You may see, but all those others
> In the darkness no one sees.

I have not seen any other technical paper that includes such a literary quotation. Engelbert carried out some very elegant calculations in the velocity space of the quasar component in motion. These were never published. They may still be languishing, buried deep in the pile of my ancient notes gathering dust. Perhaps some day I should pull them out and publish them as a joint paper.

For three years I worked as Engelbert's research associate. It was for me a period of education. I learned a lot, not just from the problems we worked on and the physics that went into them. More than that, I learned a great deal from his knowledge, insight, and thoroughness. Working in an area of physics which has become increasingly competitive, he has never wavered from his pursuit of excellence and from his fairness to other researchers. Beyond the confines of science, he is a model of warm friendship and gentleness. His wonderful sense of humour never fails to delight his friends and listeners. The experience I gained in those three years of association with Engelbert has made pathways from the past meandering into the future passing through that strange interlude called the present.

Acknowledgment

It is a pleasure to thank Ms. G. K. Rajeshwari and Mr. Rajesh Nayak for their help in preparing the manuscript.

References

[1] Vishveshwara, C. V. (1970) *Nature* **227**, 936.
[2] Honig, E, Schücking, E. L., and Vishveshwara, C. V. (1974) *J. Math. Phys.* **15**, 774.
[3] Vishveshwara, C. V. (1968) *J. Math. Phys.* **9**, 1319.
[4] Iyer, B. R., and Vishveshwara C. V. (1993) *Phys. Rev. D*, **48**, 5706.
[5] Iyer, B. R., and Vishveshwara, C. V. (1988) *Class. Quantum Grav.*, **5**, 961.
[6] Bardeen, M. (1970) *Ap. J.*, **162**, 71.
[7] Greene, R. D., Schücking, E. L., and Vishveshwara, C. V. (1975) *J. Math. Phys.*, **16**, 153.
[8] Abramowicz, M. A., Nurowski, P., and Wex, N. (1993) *Class. Quantum Grav.*. **10**, L 183.
[9] Rajesh Nayak K., and Vishveshwara, C.V. (1998) *GRG*, **30**, 593.
[10] Carter, B. (1968) *Phys. Rev.*, **174**, 1559.
[11] Floyd, R. (1973) *Dynamics of the Kerr Fields*, Ph.D. Thesis, London University.
[12] Samuel, J., and Vishveshwara, C. V. (1984) The Killing Two-form and Particle Angular Momentum in Kerr Spacetime, (unpublished preprint).
[13] Faridi, A. M. (1986) *Gen. Rel. Grav.*, **18**, 271.
[14] Behr, C., Schücking, E. L., Vishveshwara, C. V., and Wallace, W. (1976) *Astronomical Journal*, **81**, 147.

36

Curriculum Vita

Engelbert Schucking

1. The Schwarzschild Line Element and the Expansion of the Universe, *Z. Phys.*, **137**, 595 (1954).
2. Remarks on Newtonian Cosmology (with O. Heckmann), *Z. Astrophys.*, **38(2)**, 95 (1955).
3. Remarks on Newtonian Cosmology (with O. Heckmann), *Z. Astrophys.*, **40(2)**, 81, (1956).
4. A World Model of Newtonian Cosmology with Expansion and Rotation (with O. Heckmann), *Helv. Phys. Acta*, **Suppl. IV**, 114 (1956).
5. Non-Static Spherically Symmetric Solutions of the Vacuum Field Equations in (Jordan's) Extended Theory of Gravitation, *Z. Phys.*, **148** (1), 72 (1957).
6. Homogeneous World Models without Shear in Relativistic Cosmology, *Die Naturwissenschaften*, **44**, 507 (1957).
7. Remarks on the Magnitude of Extragalactic Nebulae (with J. Stock), *Astronom. J.*, **62**, 98 (1957).
8. Models of the Universe (with O. Heckmann), *Proc. Solvay Conference*, Institut International de Physique Solvay, Bruxelles (1958), pp. 1–10.
9. Newtonian and Einsteinian Cosmology (with O. Heckmann), *Encyclopedia of Physics*, **LIII**, Springer, Berlin (1959), pp. 520–537.
10. Other Cosmological Theories (with O. Heckmann), *Encyclopedia of Physics*, **LIII**, Springer, Berlin (1959), pp. 489–519.
11. Asymptotic Properties of a System with Non-Zero Total Mass (with P. G. Bergmann and I. Robinson), *Phys. Rev.*, **126**, 1227 (1962).
12. An Antimach Metric (with I. Ozsvath), *Recent Developments in Relativity*, Pergamon Press, London (1962), pp. 339–350.
13. A Finite Rotating Universe (with I. Ozsvath), *Nature*, **193**, 1163 (1962).
14. Relativistic Cosmology (with O. Heckmann), *Gravitation. An Introdution to Current Reseach*, J, Wiley and Sons, New York (1962), pp. 438–69.
15. Conference on Relativistic Theories (with I. Robinson and A. Schild), *Physics Today*, August 1963.
16. *Quasi-Stellar Sources and Gravitational Collapse* (ed. with I. Robinson and A. Schild), University of Chicago Press, Chicago (1965).

17. Relativistic Astrophysics, *Physics Today,* July 1965. p. 17.

18. Relativistic Cosmology, *Proc. of Galileo Conference*, Padova (1964).

19. *Cosmology in Relativity Theory and Astrophysics*, **I**, American Mathematical Society, (1967), pp. 3–11.

20. Relativistic Cosmology, *Proc. 5th Annual Eastern Theoretical Physics Conference*, Benjamin, New York, (1967), pp. 3–11.

21. Newtonian Cosmology, *Texas Ouarterly*, October 1967.

22. On the Hönl-Dehnen Formulation of Mach's Principle (with J. Ehlers), *Z. Physik,* 1967.

23. *Lectures on Cosmology*, *Proc. of the 1966 Les Houches Summer School*, Gordon and Breach, New York (1967).

24. Extragalactic and High Energy Astronomy, *Science Year, 1966*, Field Enterprises, Chicago (1966).

25. Science Year Report (with B. Biram), *Science Year, 1967*, Field Enterprises, (1967), pp. 60–73, 254–56.

26. Co-editor, *Proceedings of the Second Texas Symposium on Relativistic Astrophysics*, Gordon and Breach, New York (1967).

27. Relative Motion, *Proceedings of the Eighth Annual Eastern Theoretical Physics Conference*, Syracuse University, ed. F. Rohrlich

28. Thermodynamics and Cosmology (with E. Spiegel), *Comments in Astrophysics and Space Physics*, 121–125 (1970).

29. Motion of Charged Particles in Homogeneous Electromagnetic Fields (with E. Honig and C. V. Vishveshwara), *J. Math. Phys.*, **15**, 774 (1974).

30. The Rest Frame in Stationary Space-Times with Axial Symmetry (with R. D. Greene and C. V. Vishveshwara), *J. Math. Phys.*, **16**, 153 (1975).

31. Doppler Broadening at All Temperatures (with A. M. Kent), *Ann. Phys.* New York, **100**, 457 (1976).

32. Kinematics of Relativistic Ejection (with C. Behr, C. V. Vishveshwara, and W. Wallace), *Astronom. J.*, **81**, 147 (1976).

33. General Relativity and Astrophysics. *Gen. Rel. Grav.*, **7**, 113 (1976)

34. The New Boundaries of the Universe (with B. Biram), *PHYSIK*, ed. H. V. Ditfurth, Hoffmann and Campe. Hamburg (1976).

35. A Scale Transformation for Friedmann Models (with R. Spedalere), *Astronom. J.*, **85** (1979).

36. *Bull. Am. Phys. Soc.*, **26**, no. 1, 15 (1981).

37. *Symposium on the Orion Nebula*, co-Editor, *Annals N. Y. Acad. Sci,*, **395**, (1982).

38. Henry Draper, *Annals N.Y. Acad. Sci.*, **395**, 299 (1982).

39. After Einstein, *American Scientist,* **71**, 94 (1983).

40. Geometry of the SU3-Group (with J. Epstein), *Int. J. Theor. Phy.*, **23**, 197 (1984).

41. Five-Dimensional Null-cone Structure of Big Bang Singularity (with S. Lauro), *Int. J. Theor. Phys.*, **24**, 367 (1985).

42. The Homogeneous Gravitational Field, *Found. Phys.*, **15**, 571 (1985).

43. The Two-Dimensional Ivor (with J.-Z. Wang), *Gravitation and Geometry*, eds. W. Rindler and A. Trautman. (Bibliopolis, Naples, 1987).

44. A Uniform Static Magnetic Field in Kaluza-Klein Theory, *Proc. Erice 1985 International School of Physics on Cosmology*, ed. P. G. Bergmann, Plenum Press, New York.

45. Geodesics and Deformed Spheres (with Abbas M. Faridi). *Proc. Am. Math. Soc.*, **100**, 522 (1987).

46. The First Texas Symposium on Relativistic Astrophysics, *Physics Today,* August 1989. 46–52.

47. *Views from a Distant Past. General Relativity and Gravitation 1989*, ed. Neil Ashby et al., Cambridge Univ. Press, Cambridge (1990). See also Abstracts of Contributed Papers, *12th International Conference on General Relativity and Gravitation*, Boulder, CO, July 2–8, 1989, 113, 134.

48. Vortices in Tight Embrace (with E. A. Spiegel). *Proc. Seminar on Geophysical Fluid Dynamics*, Wood's Hole Oceanographic Institute, 1990.

49. Pre- Post-History of Tolman's Cosmos (with S. Lauro and J.-Z. Wang), *Gravitation and Modern Geometry*, ed. A. Zichichi, Plenum Press, (1991).

50. The introduction of the Cosmological Constant, *Gravitation and Modern Geometry*, ed. A. Zichichi, Plenum Press, New York (1991).

51. The Time before the Big Bang, *Annals N.Y. Acad. Sci.*, **655**, 340–348 (1992).

52. *Vom Urknall zum Komplexen Universum*. ed. G. Boerner et al., Piper, Munich (1993) pp. 9–32.

53. Quantum Gravity (with J. Epstein, W. Kowalski, and S. Lauro), ed. P. G. Bergmann et al. World Scientific, Singapore (1966), pp. 342–365.

54. Gravitation and Inertia, *Physics Today*, June 1996, p. 58.

55. Isometric Imbedding for Homogeneous Compact 3-Manifolds (with I. Ozsvath), *Gen. Rel. Grav.*, **28**, 999 (1996).

56. The World Viewed from Outside, *Journal of Geometry and Physics* (with I. Ozsvath). To appear.